*Nanotechnologies for the
Life Sciences
Volume 6*
**Nanomaterials for Cancer
Therapy**

*Edited by
Challa S. S. R. Kumar*

Related Titles

Kumar, C. S. S. R. (ed.)

Nanotechnologies for the Life Sciences (NtLS)

Book Series

Vol. 1

Biofunctionalization of Nanomaterials

385 pages with 153 figures
Hardcover
ISBN 3-527-31381-8

Vol. 2

Biological and Pharmaceutical Nanomaterials

400 pages with 160 figures
Hardcover
ISBN 3-527-31382-6

Vol. 3

Nanosystem Characterization Tools in the Life Sciences

413 pages with 178 figures
Hardcover
ISBN 3-527-31383-4

Vol. 4

Nanodevices for the Life Sciences

approx. 400 pages with approx. 210 figures
Hardcover
ISBN 3-527-31384-2

Vol. 5

Nanomaterials – Toxicity, Health and Environmental Issues

approx. 400 pages with approx. 100 figures
2006
Hardcover
ISBN 3-527-31385-0

zur Hausen, H.

Infections Causing Human Cancer

2006
Hardcover
ISBN 3-527-31056-8

Debatin, K.-M., Fulda, S. (eds.)

Apoptosis and Cancer Therapy

From Cutting-edge Science to Novel Therapeutic Concepts

2006
Hardcover
ISBN 3-527-31237-4

Holland, E. C. (ed.)

Mouse Models of Cancer

2004
Hardcover
ISBN 0-471-44460-X

Stuhler, G., Walden, P. (eds.)

Cancer Immune Therapy

Current and Future Strategies

Hardcover
ISBN 0-527-30441-X

Goodsell, D. S.

Bionanotechnology

Lessons from Nature

2004
Hardcover
ISBN 0-471-41719-X

Niemeyer, C. M., Mirkin, C. A. (eds.)

Nanobiotechnology

Concepts, Applications and Perspectives

2004
Hardcover
ISBN 3-527-30658-7

Nanotechnologies for the Life Sciences
Volume 6

Nanomaterials for Cancer Therapy

Edited by
Challa S. S. R. Kumar

1st Edition

WILEY-VCH Verlag GmbH & Co. KGaA

The Editor of this Book

Dr. Challa S. S. R. Kumar
The Center for Advanced Microstructures
and Devices (CAMD)
Louisiana State University
6980 Jefferson Highway
Baton Rouge, LA 70806
USA

Cover

Cover design by G. Schulz based on
micrograph courtesy of M. Heldal,
Department of Biology, University of
Bergen, Norway
(Superparamagnetic Dynabeads® by Dynal
Bead Based Separations, a part of
Invitrogen Corporation)

■ All books published by Wiley-VCH are carefully
produced. Nevertheless, authors, editors, and
publisher do not warrant the information
contained in these books, including this book,
to be free of errors. Readers are advised to keep
in mind that statements, data, illustrations,
procedural details or other items may
inadvertently be inaccurate.

Library of Congress Card No.: applied for
British Library Cataloguing-in-Publication Data:
A catalogue record for this book is available
from the British Library.

**Bibliographic information published by
Die Deutsche Bibliothek**
Die Deutsche Bibliothek lists this publication in
the Deutsche Nationalbibliografie; detailed
bibliographic data is available in the Internet at
⟨http://dnb.ddb.de⟩.

© 2006 WILEY-VCH Verlag GmbH & Co.
KGaA, Weinheim

All rights reserved (including those of
translation into other languages). No part of
this book may be reproduced in any form – by
photoprinting, microfilm, or any other means –
nor transmitted or translated into a machine
language without written permission from the
publishers. Registered names, trademarks, etc.
used in this book, even when not specifically
marked as such, are not to be considered
unprotected by law.

Printed in the Federal Republic of Germany.
Printed on acid-free paper.

Typesetting   Asco Typesetter, Hong Kong
Printing   Strauss GmbH, Mörlenbach
Binding   Litges & Dopf Buchbinderei GmbH,
Heppenheim
Cover Design   Grafik-Design Schulz,
Fußgönheim

ISBN-13: 978-3-527-31386-0
ISBN-10: 3-527-31386-9

Contents

Preface   *XIII*

List of Contributors   *XVII*

1	**Conventional Chemotherapeutic Drug Nanoparticles for Cancer Treatment** *1*	
	Loredana Serpe	
1.1	Introduction *1*	
1.2	Cancer as Drug Delivery Target *2*	
1.3	Nanoparticles as Anticancer Drug Delivery System *4*	
1.3.1	Conventional Nanoparticles *5*	
1.3.2	Sterically Stabilized Nanoparticles *6*	
1.3.3	Actively Targetable Nanoparticles *8*	
1.3.4	Routes of Drug Nanoparticles Administration *10*	
1.4	Anticancer Drug Nanoparticles *12*	
1.4.1	Anthracyclines *12*	
1.4.1.1	Reverse of P-glycoprotein Mediated Multidrug Resistance of Cancer Cells to Doxorubicin *17*	
1.4.2	Antiestrogens *19*	
1.4.3	Anti-metabolites *20*	
1.4.4	Camptothecins *20*	
1.4.5	Cisplatin *21*	
1.4.6	Paclitaxel *21*	
1.4.7	Miscellaneous Agents *26*	
1.4.7.1	Arsenic Trioxide *26*	
1.4.7.2	Butyric Acid *26*	
1.4.7.3	Cystatins *26*	
1.4.7.4	Diethylenetriaminepentaacetic Acid *26*	
1.4.7.5	Mitoxantrone *27*	
1.4.8	Gene Therapy *27*	
	References *30*	

2	**Nanoparticles for Photodynamic Therapy of Cancer** 40	
	Magali Zeisser-Labouèbe, Angelica Vargas, and Florence Delie	
2.1	Introduction 40	
2.2	Concept and Basis of Photodynamic Therapy and Photodetection 41	
2.2.1	Mechanisms of Photodynamic Therapy and Photodiagnosis 41	
2.2.2	Selective Tumor Uptake of Photosensitizers 43	
2.2.3	Photosensitizers 45	
2.2.3.1	Conventional Photosensitizers 45	
2.2.3.2	New Entities 47	
2.2.4	Photodynamic Therapy: Advantages and Limitations 51	
2.2.5	Photosensitizer Formulations 53	
2.3	Non-biodegradable Nanoparticles for Photodynamic Therapy 54	
2.3.1	Metallic Nanoparticles 54	
2.3.2	Ceramic Nanoparticles 55	
2.3.3	Nanoparticles Made of Non-biodegradable Polymers 56	
2.4	Biodegradable Polymeric Nanoparticles for Photodynamic Therapy 57	
2.4.1	Preparation of Biodegradable Polymeric Nanoparticles 58	
2.4.1.1	*In situ* Polymerization 58	
2.4.1.2	Dispersion of a Preformed Polymer 59	
2.4.1.3	"Stealth" Particles 60	
2.4.1.4	Targeted Nanoparticles 60	
2.4.2	*In Vitro* Relevance of Polymeric Nanoparticles in PDT on Cell Models 61	
2.4.2.1	Photodynamic Activity of PS-loaded Nanoparticles 61	
2.4.2.2	Uptake and Trafficking of Photosensitizers 66	
2.4.3	*In Vivo* Relevance of Polymeric Nanoparticles in PDT 70	
2.4.3.1	Biodistribution and Pharmacokinetics of Photosensitizers Coupled to Nanoparticles 70	
2.4.3.2	Vascular Effects 71	
2.4.3.3	*In Vivo* Efficacy on Tumor: Tumor Suppression Effects 73	
2.4.3.4	Adverse Effects 74	
2.5	Conclusions 75	
	Acknowledgments 75	
	Abbreviations 76	
	References 77	
3	**Nanoparticles for Neutron Capture Therapy of Cancer** 87	
	Hideki Ichikawa, Hiroyuki Tokumitsu, Masahito Miyamoto, and Yoshinobu Fukumori	
3.1	Introduction 87	
3.2	Principle of Neutron Capture Therapy of Cancer 88	
3.3	Boron Neutron Capture Therapy 89	
3.3.1	Boron Compounds 89	
3.3.2	Delivery of Boron Using Nanoparticles 90	
3.4	Approaches to GdNCT 93	

3.4.1	Typical Research on GdNCT	94
3.4.2	Delivery of Gadolinium using Lipid Emulsion (Gd-nanoLE)	95
3.4.2.1	Preparation of Gd-nanoLE	95
3.4.2.2	Biodistribution of Gadolinium after Intraperitoneal Administration of Gd-nanoLE	98
3.4.2.3	Biodistribution of Gadolinium after Intravenous Administration of Gd-nanoLE	101
3.4.3	Delivery of Gadolinium using Chitosan Nanoparticles (Gd-nanoCPs)	104
3.4.3.1	Preparation of Gd-nanoCPs	105
3.4.3.2	Gd-DTPA Release Property of Gd-nanoCPs	107
3.4.3.3	Gd-DTPA Retention in Tumor Tissue after Intratumoral Injection	107
3.4.3.4	*In vivo* Growth Suppression of Experimental Melanoma Solid Tumor	108
3.4.3.5	Bioadhesion and Uptake of Gd-nanoCP in Three Different Cell Lines	109
3.5	Conclusions	113
	References	114

4	**Nanovehicles and High Molecular Weight Delivery Agents for Boron Neutron Capture Therapy**	**122**
	Gong Wu, Rolf F. Barth, Weilian Yang, Robert Lee, Werner Tjarks, Marina V. Backer, and Joseph M. Backer	
4.1	Introduction	122
4.1.1	Overview	122
4.1.2	General Background	123
4.2	General Requirements for Boron Delivery Agents	124
4.3	Low Molecular Weight Delivery Agents	124
4.4	High Molecular Weight Boron Delivery Agents	125
4.5	Dendrimer-related Delivery Agents	125
4.5.1	Properties of Dendrimers	125
4.5.2	Boronated Dendrimers Linked to Monoclonal Antibodies	126
4.5.2.1	Boron Clusters Directly Linked to mAb	126
4.5.2.2	Attachment of Boronated Dendrimers to mAb	129
4.5.3	Boronated Dendrimers Delivered by Receptor Ligands	127
4.5.3.1	Epidermal Growth Factors (EGF)	127
4.5.3.2	Folate Receptor Targeting Agents	129
4.5.3.3	Vascular Endothelial Growth Factor (VEGF)	129
4.5.4	Other Boronated Dendrimers	130
4.6	Liposomes as Boron Delivery Agents	130
4.6.1	Overview of Liposomes	130
4.6.2	Liposomal Encapsulation of Sodium Borocaptate and Boronophenylalanine	133
4.6.2.1	Boron Delivery by Non-targeted Liposomes	133
4.6.2.2	Liposomal Encapsulation of other Boranes and Carboranes	134

4.6.3	Boron Delivery by Targeted Liposomes	137
4.6.3.1	Immunoliposomes	137
4.6.3.2	Folate Receptor-targeted Liposomes	138
4.6.3.3	EGFR Targeted Liposomes	138
4.7	Boron Delivery by Dextrans	139
4.8	Other Macromolecules used for Delivering Boron Compounds	141
4.9	Delivery of Boron-containing Macromolecules to Brain Tumors	142
4.9.1	General Considerations	142
4.9.2	Drug-transport Vectors	142
4.9.3	Direct Intracerebral Delivery	142
4.9.4	Convection-enhanced Delivery (CED)	143
4.10	Clinical Considerations and Conclusions	144
	Acknowledgments	145
	References	145

5	**Local Cancer Therapy with Magnetic Drug Targeting using Magnetic Nanoparticles**	**156**
	Christoph Alexiou and Roland Jurgons	
5.1	Introduction	156
5.2	Local Chemotherapy	156
5.3	Magnetic Drug Delivery	158
5.3.1	*In Vitro* Applications	158
5.3.2	*In Vivo* Applications	159
	References	163

6	**Nanomaterials for Controlled Release of Anticancer Agents**	**168**
	Do Kyung Kim, Yun Suk Jo, Jon Dobson, Alicia El Haj, and Mamoun Muhammed	
6.1	Introduction	168
6.2	Nanoparticles for Biomedical Applications	170
6.2.1	First Generation Nanoparticles	171
6.2.2	Second Generation Nanoparticles	171
6.2.3	Advanced Generation Nanoparticles	172
6.3	Polymer Materials for Drug Delivery Systems	174
6.4	Design of Drug Delivery Vectors and Their Prerequisites	175
6.4.1	Polymeric Nanoparticles	175
6.4.2	Inorganic Nanoparticles	180
6.4.3	Metallic Nanoparticles	181
6.5	Kinetics of the Controlled Release of Anticancer Agents	181
6.5.1	Diffusion Model	182
6.5.2	Dissolution Model	183
6.5.3	Kinetics of the Indomethacin (IMC, 1-[*p*-chlorobenzoyl]-2-methyl-5-methoxy-3-indoleacetic acid) Release	183
6.6	Controlled Release of Anticancer Agents	186

6.6.1	Alkylating Agents	*186*
6.6.1.1	Chlorambucil	*187*
6.6.1.2	Cyclophosphamide	*187*
6.6.1.3	Carmustine	*188*
6.6.2	Antimetabolic Agent	*188*
6.6.2.1	Cytarabine	*188*
6.6.2.2	Fluorouracil (FU)	*189*
6.6.2.3	Methotrexate	*189*
6.6.3	Anticancer Antibiotics	*190*
6.6.3.1	Actinomycin D	*190*
6.6.3.2	Bleomycin	*190*
6.6.3.3	Daunorubicin	*191*
6.7	Future Directions	*191*
	References	*192*

7 Critical Analysis of Cancer Therapy using Nanomaterials *199*
Lucienne Juillerat-Jeanneret

7.1	Introduction	*199*
7.2	Anticancer Therapies	*200*
7.3	Characteristics of Nanoparticles for Cancer Therapy	*202*
7.3.1	Nanovectors	*203*
7.3.2	Biological Issues	*205*
7.3.3	Nanoparticle Targeting: Passive or Active	*206*
7.4	Nanovectors in Biomedical Applications: Drug Delivery Systems (DDS) for Cancer	*207*
7.4.1	Physicochemical Drug Delivery	*208*
7.4.2	Biological Drug Delivery	*208*
7.4.3	Chemical Drug Delivery	*208*
7.4.4	Nanoparticles for Anticancer Drug Delivery	*209*
7.4.4.1	Existing Systems	*209*
7.4.4.2	Systems under Development and Challenges	*210*
7.4.5	Nanoparticles for Drug Delivery in Clinical Use or under Clinical Evaluation	*211*
7.4.5.1	Doxorubicin Family	*211*
7.4.5.2	Paclitaxel (Taxol)	*212*
7.4.5.3	5-Fluorouracil	*213*
7.4.5.4	Tamoxifen	*213*
7.4.5.5	Cisplatin	*213*
7.4.5.6	Campthotecins	*214*
7.4.5.7	Methotrexate	*214*
7.4.6	New Experimental Drugs and Therapies	*214*
7.4.6.1	Proteins, Peptides, their Inhibitors and Antagonists	*214*
7.4.6.2	New Drugs	*215*
7.4.6.3	New Therapeutic Approaches: Photodynamic Therapy (PDT)	*215*

7.4.7	Gene Therapy 215
7.4.7.1	Nanoparticle for Gene Delivery: Non-chitosan and Chitosan-type Polymers 216
7.4.8	New Approaches 216
7.4.8.1	Improvement of Biological Characteristics 217
7.4.8.2	New Technological Approaches 219
7.4.9	Superparamagnetic Iron Oxide Nanoparticles (SPIONs) as Magnetic Drug Nanovectors 219
7.5	Targeting 220
7.5.1	Passive Targeting 222
7.5.2	Active Targeting 223
7.5.2.1	Targeting Cancer-associated Cells 223
7.5.2.2	Targeting Cancer Markers 224
7.5.3	Intracellular Drug Delivery 225
7.5.4	Development of the Necessary Chemistry: Synthetic Routes and Linkers for Conjugation 226
7.6	Overcoming the Mechanisms of Resistance to Therapy of Cancers 227
7.7	Toxicity Issues 229
7.8	Conclusions 231
7.8.1	Opportunities and Challenges of Nanomedicine in Cancer 231
	References 232

8	**Nanoparticles for Thermotherapy** 242
	Andreas Jordan, Klaus Maier-Hauff, Peter Wust, and Manfred Johannsen
8.1	Introduction 242
8.2	Thermotherapy following Intratumoral Administration of Magnetic Nanoparticles 244
8.3	Ferromagnetic Embolization Hyperthermia 248
8.4	First Clinical Experiences with Thermotherapy using Magnetic Nanoparticles: MagForce Nanotherapy 249
8.4.1	Feasibility Study on Thermotherapy using Magnetic Nanoparticles in Recurrent Glioblastoma Multiforme 250
8.4.2	Feasibility Study on Thermotherapy using Magnetic Nanoparticles in Recurrent and Residual Tumors 251
8.4.3	Feasibility Study on Thermotherapy using Magnetic Nanoparticles in Recurrent Prostate Carcinoma 253
	References 254

9	**Ferromagnetic Filled Carbon Nanotubes as Novel and Potential Containers for Anticancer Treatment Strategies** 259
	Ingolf Moench, Axel Meye, and Albrecht Leonhardt
9.1	Introduction 259
9.2	Prostate Cancer 260
9.2.1	Incidence, Risk Factors and Diagnostic Criteria 260
9.2.2	Treatment Options, Outcome and Limits 261

9.2.3	MWCNT Model *263*	
9.3	Carbon Nanotubes *264*	
9.3.1	General Remarks *264*	
9.3.2	Preparation and Structure of Filled Multi-walled Carbon Nanotubes *266*	
9.3.2.1	Synthesis of Ferromagnetic Filled Multi-walled Carbon Nanotubes *266*	
9.3.2.2	Crystallographic Structure of Core Material in Filled Multi-walled Carbon Nanotubes *269*	
9.3.2.3	Growth Mechanism of Multi-walled Carbon Nanotubes *271*	
9.3.3	Post-treatment: Opening, Filling and Closing of MWCNTs *275*	
9.4	Magnetism in Nano-sized Materials *277*	
9.4.1	General Remarks *277*	
9.4.2	Magnetization in Nano-sized Materials *278*	
9.4.3	Influence of the Dimensions on the Magnetization Distribution *279*	
9.4.4	Anisotropy and Interaction *283*	
9.4.5	Magnetic Reversal *284*	
9.4.6	Magnetic Properties of Filled Multi-walled Carbon Nanotubes *285*	
9.5	Heat Generation *290*	
9.5.1	General Remarks *290*	
9.5.2	Requirements for the Development of Materials for Hyperthermia and Magnetism *296*	
9.5.3	Specific Absorption Rate (SAR) *300*	
9.6	Study Results for *In Vitro* and *In Vivo* Applications of ff-MWCNTs *309*	
9.6.1	Efficient Endocytosis *In Vitro*, Lipid-mediated Could Enhance the Internalization Rate and Efficiency *309*	
9.6.2	Production of Two Types of ff-MWCNTs for *In Vivo* Application *312*	
9.6.3	Outlook/Next Steps in Evaluation of these fff-MWCNTs *313*	
	Acknowledgments *324*	
	Abbreviations *324*	
	References *325*	

10 Liposomes, Dendrimers and other Polymeric Nanoparticles for Targeted Delivery of Anticancer Agents – A Comparative Study 338
Yong Zhang and Dev K. Chatterjee

10.1	Introduction *338*	
10.2	Cancer Chemotherapy: so Far, but not so Good *339*	
10.3	Nanoparticles and Drug Delivery in Cancer: a new Road *341*	
10.3.1	Importance of Nanoparticles in Cancer Therapy *341*	
10.3.2	An Overview of Targeting Methods *343*	
10.4	Means to the End: Methods for Targeting *343*	
10.4.1	Passive Targeting *343*	
10.4.2	Magnetic Targeting of Nanoparticles *345*	
10.4.3	Ligands for Active Targeting *346*	
10.4.3.1	Monoclonal Antibodies against Tumor-specific Antigens *347*	
10.4.3.2	Targeting the Angiogenic Process *349*	

10.4.3.3 Folic Acid and Cancer Targeting   350
10.4.3.4 Transferrin as a Targeting Ligand   354
10.4.3.5 Other Targeting Ligands   354
10.5 Targeting with Different Types of Nanoparticles   355
10.5.1 Liposomes in Cancer Targeting   355
10.5.1.1 Beyond Immunoliposomes   357
10.5.2 Dendrimers   357
10.5.3 Other Polymeric Nanoparticles   359
10.6 Conclusion   362
References   364

11 Colloidal Systems for the Delivery of Anticancer Agents in Breast Cancer and Multiple Myeloma   371

Sébastien Maillard, Elias Fattal, Véronique Marsaud, Brigitte Sola, and Jack-Michel Renoir

11.1 Introduction   371
11.2 Hormone Therapy in Breast Cancers   374
11.2.1 Molecular Mechanisms of Estrogen Action in Breast Cancers   375
11.2.1.1 Classical ER-ligand and ERE-dependent Mechanism   375
11.2.1.2 ERE-independent Pathway   377
11.2.1.3 ER-ligand-independent Pathway   377
11.2.1.4 "Non-genomic" Pathway   378
11.2.2 Differential Activity of Antiestrogens   378
11.2.3 The Need to Encapsulate Antiestrogens   379
11.3 Multiple Myeloma   380
11.3.1 Current Treatments   380
11.3.2 New Biological Therapies for MM Treatment   380
11.3.3 Incidence of Estrogens and Antiestrogens on Multiple Myeloma   381
11.4 Colloidal Systems for Antiestrogen Delivery   381
11.4.1 Nanoparticles Charged with AEs in Breast Cancer   381
11.4.2 Liposomes Charged with RU 58668 in MM   386
11.4.3 Tumor-targeted Drug-loaded Colloidal Systems   387
11.5 Conclusions and Perspectives   390
Acknowledgments   391
References   391

Index   404

Preface

Even five years into the new millennium, cancer continues to torment humanity as the second leading cause of death with 10.9 million newly diagnosed cases worldwide in the year 2005 alone. Despite new discoveries of drugs and treatment combinations as evidenced by reports of close to 200,000 experimental studies on mice, two million scientific publications and an annual spending of around 15 billion US dollars world wide, the mortality rate due to cancer did not change in the past five to six decades. Therefore, there is still a strong need for a paradigm shift in the approach to cancer diagnosis and therapy. The advent of nanotechnological revolution offers an opportunity to achieve this paradigm shift. Since the biological processes in general and those that lead to cancer in particular occur at the nanoscale, there is a great opportunity for nanotechnologists to treat cancer at an as early stage as possible. Several innovative nanoscale constructs have been demonstrated to radically change cancer therapy with capabilities to deliver large doses of chemotherapeutic agents or therapeutic genes into malignant cells while sparing healthy cells. They have also shown great promise in enabling rapid and sensitive detection of single cancer cells and cancer-related molecules. Reports of these investigations are being published in a very broad range of journals spanning several traditional disciplines. It is becoming difficult for researchers to gather all the available information on 'Cancer Nanotechnology'. I am, therefore, pleased to share with you, again on behalf of dedicated team of researchers in cancer nanotechnology, two volumes of the ten volume series on nanotechnologies for the life sciences specifically dedicated to cancer. The first of these two volumes, sixth in the series, that is being presented to you here is dedicated to cancer therapy and is aptly titled as *"Nanomaterials for Cancer Therapy."*

The book is divided into eleven chapters encompassing a number of therapeutic approaches in cancer treatment through use of a variety of nanomaterials. It begins with a chapter reviewing the progress that has been made to date in utilization of conventional chemotherapeutic drug nanoparticles for cancer treatment. The chapter *Conventional Chemotherapeutic Drug Nanoparticles for Cancer Treatment* contributed by Loredanna Serpa from the University of Turin, Italy, is an up to date review of literature on advances being made in cancer treatment through use of nanoparticle formulations containing conventional chemotherapeutic drugs such as doxorubicin, cisplatin, paclitaxel and other drugs. Moving from conventional anticancer drugs to conventional therapies that are being affected by nanotechnological tools, the second chapter starts with fundamental aspects of Photo Dynamic Therapy (PDT) that has been found to be promising in selectively treating tumors as well as metastasis without affecting the surrounding healthy tissue. The chapter *Nanoparticles for Photodynamic Therapy of Cancer*, written by Florence Delie and her team from the Laboratory of Pharmaceutical Technology and Biopharmaceutics, University of Geneva, Switzerland, provides an in-depth analysis of how nanoparticles, with special emphasis on polymeric biodegradable ones, are being developed to improve the conventional approaches to PDT. Continuing on a similar theme of improving conventional therapies, Yoshinobu Fukumori and co-workers from Kobe Gakuin University in Japan provide a general background on neutron capture therapy (NCT), a new radiotherapy that differs from the conventional radiotherapies, in the third chapter. This is followed by the authors reviewing more specifically both gadolinium neutron capture therapy (GdNCT) and boron neutron capture therapy (BNCT) with reference to use of nanomaterials. The remainder of the chapter *Nanoparticles for Neutron Capture Therapy of Cancer* provides a detailed account of the authors' experiences in developing Gd-containing lipid nanoemulsions and chitosan nanoparticles to demonstrate the usefulness of nanoparticle technology in NCT. Addressing a different facet of NCT, the fourth chapter entitled *Nanovehicles and High Molecular Weight Delivery Agents for Boron Neutron Capture Therapy* focuses on various high molecular weight (HMW) agents consisting of macromolecules and nanovehicles such as monoclonal antibodies, dendrimers, liposomes, dextrans, polylysine, avidin and folic acid, epidermal and vascular endothelial growth factors (EGF and VEGF) as delivery vehicles for introducing boron atoms. In it, Gong Wu, Rolf F. Barth, Weilian Yang, Robert Lee, Werner Tjarks, Marina V. Backer and Joseph M. Backer from the Department of Pathology at Ohio State University in Columbus, USA, have done a remarkable job in describing procedures for introducing boron atoms into HMW agents in addition to providing information on their chemical properties, bio-distribution based on in vivo studies, delivery across the blood brain barrier and various routes of their administration. Overall, the work reported in chapters three & four is very valuable and exciting not only from the scientific point of view, but also from the commercial point of view as recently clinical BNCT trials mainly for brain tumors were carried out in Japan.

Switching gears from earlier parts of the book where applications of nanotechnology to already well established treatments for cancer are presented, the rest of the chapters describe 'non-traditional' and innovative approaches completely based on nanotechnology that are being investigated for cancer therapy. Christoph Alexiou and Roland Jurgons from the Policlinic for otorhinolaryngological Illnesses of Friedrich-Alexander University, Erlangen-Nuremberg, Germany, contributed the fifth chapter, *Local Cancer Therapy with Magnetic Drug Targeting using Magnetic Nanoparticles*, in which they review current literature on the use of magnetic nanoparticles in biomedicine in general and local chemotherapies, focusing especially on regional cancer therapy, in particular. In the sixth chapter, *Nanomaterials for Controlled Release of Anticancer Agents*, the team lead by Do Kyung Kim at the

Massachusetts Institute of Technology, Cambridge, USA, has done a remarkable job in capturing nuances of nanotechnologies, particularly those based on polymeric nanomaterials, being developed for controlled release of anticancer agents. In this chapter, the authors discuss various design aspects, theoretical models and kinetics of controlled release of anticancer drugs. The seventh chapter on the other hand provides a much broader perspective to cancer therapy using nanomaterials by providing a critical analysis of various approaches. While the chapter starts with a description of the tools of nanoparticle technology that can be used to treat cancer, it goes a step further in critically reviewing and discussing the advantages and drawbacks of nanoparticles for the targeted delivery of anticancer agents to defined cells of human cancers. The chapter entitled *Critical Analysis of Cancer Therapy using Nanomaterials* contributed by Lucienne Juillerat-Jeanneret from the University Institute of Pathology in Lausanne, Switzerland, concludes the section with a final section that describes the author's own design of how an ideal nanoparticulate system should be for the targeted treatment of human cancers.

Treating cancer using heat has been known for a long time as cells are known to undergo apoptosis when exposed to temperatures around 40 °C. Thermotherapy, as it is called, as performed using conventional approaches has several drawbacks. In the eighth chapter, *Nanoparticles for Thermotherapy*, a team of cancer specialists lead by Andreas Jordan describe, based on their own experience, how some of these drawbacks are being overcome using magnetic nanoparticles. The chapter is particularly valuable as the team from the Center of Biomedical Nanotechnology (CBN), Berlin, Germany, share their experience in conducting the first ever clinical trials with thermotherapy using magnetic nanoparticles. Moving conceptually to a more elegant approach, the ninth chapter entitled *Ferromagnetic Filled Carbon Nanotubes as Novel and Potential Containers for Anticancer Treatment Strategies* explores the feasibility of different applications using magnetic multi-walled CNTs (MWCNTs) more specifically as heat mediators for hyperthermia of solid tumors. While authors Ingolf Mönch et al. from the Leibniz Institute for Solid State and Materials Research in Dresden, Germany, are upbeat about potential opportunities in the use of different types of functionalized MWCNTs, they restrict their review to only those that are novel and those that have benefits specifically in the treatment of prostate cancer.

Several types of nanomaterials are currently being investigated for targeted delivery of anticancer agents and there is a need for understanding the pros and cons of using these different types of nanomaterials. Chapter ten written by Yong Zhang and Dev K. Chatterjee from the Division of Bioengineering at the National University of Singapore provides a platform for comparing the efficacy of different types of nanomaterials for targeted delivery. The chapter, *Liposomes, Dendrimers and other Polymeric Nanoparticles for Targeted Delivery of Anticancer Agents – A Comparative Study*, provides a unique perspective on different types of nanomaterials in general and three major types in particular that are gaining importance in this field. The chapter is a must for those interested in learning about background information on mechanisms and methods of targeting cancer cells. At this point of time, liposomes seem to have an edge over the other types of nanomaterials and in the author's own words, "the most exciting news is the performance of Epeius Biotechnologies Corporation's liposomal based active targeting system that delivers genetic material to treat several cancers, including pancreatic head carcinoma, which has one of the worst prognoses among all neoplasms." In the eleventh and final chapter of the book, the team lead by Jack-Michel Renoir from University of Paris-Sud in Châtenay-Malabry, France, focuses on the benefits of encapsulating a class of anticancer drugs known as antiestrogens (AEs) to improve antitumoral activities in vivo. In addition, the authors discuss in this chapter, *Colloidal Systems for the Delivery of Anticancer Agents in Breast Cancer and Multiple Myeloma*, the use of passive targeting through long-circulating drug delivery systems in different types of xenografts. They strongly believe that the approach using colloidal systems is promising not only for the administration of antiestrogens in estrogen-dependent breast cancers and multiple myeloma (MM), but also for the delivery of much more toxic anticancer agents such as taxol, thalidomide, bortezomib, VEGF inhibitors, farnesyltransferase inhibitors, histone transferase inhibitors and hsp90 inhibitors.

I am pleased by the broad range of useful information gathered by the dedicated contributors working in the area of cancer nanotechnology, and I am hoping that the book will be a guide for all those who wish to be associated with the growing field of 'Nanomaterials for Cancer Therapy'. I am indebted to all the authors for their extraordinary efforts and as always grateful to my employer, family, friends and Wiley VCH publishers for making this book a reality. I am thankful to you, the reader, who has taken time to join the journey with fellow cancer nanotechnologists. It will be my pleasure to hear back from you and I look forward to receiving your comments, suggestions and constructive criticism in order to make further improvements in the next editions of the book.

April 2006, Baton Rouge   *Challa S. S. R. Kumar*

List of Contributors

Christoph Alexiou
Dept. for Oto-Rhino-Laryngology, Head and Neck Surgery University
Erlangen-Nuremberg
Waldstr. 1
91054 Erlangen
Germany

Joseph M. Backer
SibTech, Inc.
705 North Mountain Road
Newington, CT 06111
USA

Marina V. Backer
SibTech, Inc.
705 North Mountain Road
Newington, CT 06111
USA

Rolf F. Barth
Department of Pathology
The Ohio State University
129 Hamilton Hall, 1645 Neil Av.
Columbus, OH 43210-1218
USA

Dev K. Chatterjee
Division of Bioengineering
Faculty of Engineering
National University of Singapore
9 Engineering Drive 1
Singapore 117576
Singapore

Florence Delie
Laboratory of Pharmaceutical Technology and Biopharmaceutics
School of Pharmaceutical Sciences
University of Geneva
30, quai E. Ansermet
1211 Geneva 4
Switzerland

Jon Dobson
Institute for Science & Technology in Medicine
Keele University
Thornburrow Drive, Hartshill
Stoke-on-Trent, ST4 7QB
United Kingdom

Elias Fattal
University of Paris-Sud
UMR CNRS 8612
Faculty of Pharmacy
5, rue J. B. Clément
92296 Châtenay-Malabry
France

Yoshinobu Fukumori
Faculty of Pharmaceutical Sciences
and Cooperative Research Center
of Life Sciences
Kobe Gakuin University
Arise 518, Ikawadani-cho, Nishi-ku
Kobe 651-2180
Japan

Alicia El Haj
Institute for Science & Technology in Medicine
Keele University
Thornburrow Drive, Hartshill
Stoke-on-Trent, ST4 7QB
United Kingdom

Hideki Ichikawa
Faculty of Pharmaceutical Sciences
and Cooperative Research Center
of Life Sciences
Kobe Gakuin University
518 Arise, Ikawadani-cho, Nishi-ku
Kobe 651-2180
Japan

List of Contributors

Yun Suk Jo
Institut des Biosciences Intégratives
École Polytechnique Fédérale de Lausanne
(EPFL)
1015 Lausanne
Switzerland

Manfred Johannsen
Department of Urology
CCM, Charité – University Medicine
Schumannstr. 20/21
10117 Berlin
Germany

Andreas Jordan
Center of Biomedical Nanotechnology (CBN)
Spandauer Damm 130, Haus 30
14050 Berlin
Germany

Lucienne Juillerat-Jeanneret
University Institute of Pathology
Bugnon 25
1011 Lausanne
Switzerland

Roland Jurgons
Dept. for Oto-Rhino-Laryngology, Head
and Neck Surgery University
Erlangen-Nuremberg
Waldstr. 1
91054 Erlangen
Germany

Do Kyung Kim
Department of Electrical Engineering and
Computer Science (EECS)
Massachusetts Institute of Technology
155 Mass Ave.
Cambridge, MA 02139
USA

Robert Lee
College of Pharmacy
The Ohio State University
129 Hamilton Hall, 1645 Neil Av.
Columbus, OH 43210
USA

Albrecht Leonhardt
Leibniz Institute for Solid State and Materials
Research Dresden
IFW Dresden
Helmholtzstrasse 20
01069 Dresden
Germany

Klaus Maier-Hauff
Department of Neurosurgery
Bundeswehrkrankenhaus
Scharnhorststr. 13
10115 Berlin
Germany

Sébastien Maillard
University of Paris-Sud
UMR CNRS 8612
Faculty of Pharmacy
5, rue J. B. Clément
92296 Châtenay-Malabry
France

Véronique Marsaud
University of Paris-Sud
UMR CNRS 8612
Faculty of Pharmacy
5, rue J. B. Clément
92296 Châtenay-Malabry
France

Axel Meye
Department of Urology
Technical University Dresden
Fetscherstrasse 74
01062 Dresden
Germany

Masahito Miyamoto
Faculty of Pharmaceutical Sciences
and Cooperative Research Center
of Life Sciences
Kobe Gakuin University
518 Arise, Ikawadani-cho, Nishi-ku
Kobe 651-2180
Japan

Ingolf Moench
Leibniz Institute for Solid State and Materials
Research Dresden
IFW Dresden
Helmholtzstrasse 20
01069 Dresden
Germany

Mamoun Muhammed
Materials Chemistry Division
Royal Institute of Technology
100 44 Stockholm
Sweden

Jack-Michel Renoir
University of Paris-Sud
UMR CNRS 8612
Faculty of Pharmacy
5, rue J. B. Clément
92296 Châtenay-Malabry
France

Loredana Serpe
Department of Anatomy
University of Torino
Via Pietro Giuria 13
10125 Torino
Italy

Brigitte Sola
UFR de Medicine
CHU Côte de Nacre
14032 Caen CEDEX
France

Werner Tjarks
College of Pharmacy
The Ohio State University
129 Hamilton Hall, 1645 Neil Av.
Columbus, OH 43210
USA

Hiroyuki Tokumitsu
Faculty of Pharmaceutical Sciences
and Cooperative Research Center
of Life Sciences
Kobe Gakuin University
Arise 518, Ikawadani-cho, Nishi-ku
Kobe 651-2180
Japan

Angelica Vargas
Laboratory of Pharmaceutical Technology and
Biopharmaceutics
School of Pharmaceutical Sciences
University of Geneva
30, quai E. Ansermet
CH-1211 Geneva 4
Switzerland

Gong Wu
Department of Pathology
The Ohio State University
129 Hamilton Hall, 1645 Neil Av.
Columbus, OH 43210
USA

Peter Wust
Department of Radiology
CVK, Charité – University Medicine
Augustenburger Platz 1
13353 Berlin
Germany

Weilian Yang
Department of Pathology
The Ohio State University
129 Hamilton Hall, 1645 Neil Av.
Columbus, OH 43210
USA

Magali Zeisser-Labouèbe
Laboratory of Pharmaceutical Technology and
Biopharmaceutics
School of Pharmaceutical Sciences
University of Geneva
30, quai E. Ansermet
1211 Geneva 4
Switzerland

Yong Zhang
Division of Bioengineering
Faculty of Engineering
National University of Singapore
9 Engineering Drive 1
Singapore 117576
Singapore

1
Conventional Chemotherapeutic Drug Nanoparticles for Cancer Treatment

Loredana Serpe

1.1
Introduction

Chemotherapy is a major therapeutic approach for the treatment of localized and metastasized cancers. The selective increase in tumor tissue uptake of anticancer agents would be of great interest in cancer chemotherapy since anticancer drugs are not specific to cancer cells. Routes of administration, biodistribution and elimination of available chemotherapeutic agents can be modified by drug delivery systems to optimize drug therapy.

This chapter focuses on progress in targeted treatment of cancer through the delivery of conventional anticancer agents via microparticulate drug carriers as nanoparticles. Briefly, nanoparticles may be defined as submicronic colloidal systems that are generally made of polymers and, according to the preparation process used, nanospheres (matrix systems) or nanocapsules (reservoir systems) can be obtained. The drug can either be directly incorporated during polymerization or by adsorption onto preformed nanoparticles [1].

The first part of this chapter pays particular attention to cancer as a drug delivery target and to the development of different types of nanoparticles as drug delivery device. The interaction of drug carrier systems with the biological environment is an important basis for designing strategies; these systems should be independent in the environment and selective at the pharmacological site. If designed appropriately, nanoparticles may act as a drug vehicle able to target tumor tissues or cells, protecting the drug from inactivation during its transport. The formulation of nanoparticles and physicochemical parameters such as pH, monomer concentration, added stabilizer and ionic strength as well as surface charge, particle size and molecular weight are important for drug delivery.

For instance, poly(acrylamide) nanocapsules, due to their polymeric nature, are stable in biological fluids and during storage, and can entrap various agents in a stable and reproducible way but, since they are not lysed by lysosomal enzymes, their clinical application is restricted [2]. The development of biodegradable polymers by the polymerization of various alkyl cyanoacrylate monomers and the association of anticancer agents with these poly(alkyl cyanoacrylate) polymers has afforded further improvement. These colloidal drug carriers are biodegradable and can associate with various drugs in a non-specific manner. The binding capacity of these nanoparticles to dactinomycin (90%), vinblastin (36–85%) and methotrexate (15–40%) exceeds that of these drugs incorporated in liposomes [1].

Furthermore, certain types of nanoparticles are able to reverse multidrug resistance – a major problem in chemotherapy – and selectivity in drug targeting can be achieved by the attachment of certain forms of homing devices, such as a monoclonal antibody or lecithin [3].

Finally, the last part of the chapter reports some of the most significant and recently developed nanoparticle formulations for the delivery of specific anticancer agents.

1.2
Cancer as Drug Delivery Target

Tumor blood vessels have several abnormalities compared with physiological vessels, such as a relatively high proportion of proliferating endothelial cells, an increased tortuosity and an aberrant basement membrane formation. The rapidly expanding tumor vasculature often has a discontinuous endothelium, with gaps between the cells that may be several hundred nanometers large [4, 5].

Macromolecular transport pathways across tumor vessels occur via open gaps (interendothelial junctions and transendothelial channels), vesicular vacuolar organelles and fenestrations. However, it remains controversial which pathways are predominantly responsible for tumor hyperpermeability and macromolecular transvascular transport [6].

Tumor interstitium is also characterized by a high interstitial pressure, leading to an outward convective interstitial fluid flow, as well as the absence of an anatomically well-defined functioning lymphatic network. Hence, the transport of an anticancer drug in the interstitium will be governed by the physiological (i.e., pressure) and physicochemical (i.e., composition, structure) properties of the interstitium and by the physicochemical properties of the molecule itself (i.e., size, configuration, charge, hydrophobicity). Physiological barriers at the tumor level (i.e., poorly vascularized tumor regions, acidic environment, high interstitial pressure and low microvascular pressure) as well as at the cellular level (i.e., altered activity of specific enzyme systems, altered apoptosis regulation and transport based mechanisms) and in the body (i.e., distribution, biotransformation and clearance of anticancer agent) must be overcome to deliver anticancer agents to tumor cells *in vivo* [1].

Colloidal nanoparticles incorporating anticancer agents can overcome such resistances to drug action, increasing the selectivity of drugs towards cancer cells and reducing their toxicity towards normal cells.

The accumulation mechanism of intravenously injected nanoparticles in cancer tissues relies on a passive diffusion or convection across the hyperpermeable tumor vasculature. Additional retention of the colloidal particles in the tumor interstitium is due to the compromised clearance via lymphatics. This so-called "enhanced permeability and retention effect" results in an important intratumoral drug accumulation that is even higher than that observed in plasma and other tissues [7]. Controlled release of the drug content inside the tumoral interstitium may be achieved by controlling the nanoparticulate structure, the polymer used and the way by which the drug is associated with the carrier (adsorption or encapsulation).

However, anticancer drugs, even if they are located in the tumoral interstitium, can have limited efficacy against numerous tumor types because cancer cells are able to develop mechanisms of resistance. Simultaneous cellular resistance to multiple lipophilic drugs is one of the most important problems in chemotherapy. This drug resistance may appear clinically either as a lack of tumor size reduction or as the occurrence of clinical relapse after an initial positive response to antitumor treatment. Multidrug resistance is mainly due to overexpression of the plasma membrane P-glycoprotein, which is capable of extruding various generally positively charged xenobiotics, including some anticancer drugs, out of the cell. Multidrug resistance is always multifactorial when other mechanisms can be associated with this drug efflux pump in cancer cells, such as enzymatic function modification (topoisomerase, glutathione S-transferase) or altered intracellular drug distribution due to increased drug sequestration into cytoplasmic acidic vesicles [8]. P-glycoprotein probably recognizes the drug to be effluxed out of the cancer cell only when the drug is present in the plasma membrane, and not when it is located in the cytoplasm of lysosomes, after endocytosis. Many cancer cell types can develop resistance to doxorubicin, which is a P-glycoprotein substrate, therefore incorporation of this compound into nanoparticles to reverse multidrug resistance P-glycoprotein mediated has been extensively investigated [9]. Certain types of nanoparticles are able to overcome multidrug resistance mediated by the P-glycoprotein, such as poly(alkyl cyanoacrylate) nanoparticles [10].

The delivery of anticancer agents to a highly perfused tumoral lesion and the tumor cells response have been described through the development of a two-dimensional tumor simulator with the capability of showing tumoral lesion progression through the stages of diffusion-limited dormancy, neo-vascularization and subsequent rapid growth and tissue invasion. Two-dimensional simulations based on a self-consistent parameter estimation demonstrated fundamental convective and diffusive transport limitations in delivering anticancer drug into tumors via intravenous free drug administration or via 100 nm nanoparticles injected into the bloodstream, able to extravasate and release the drug into the tumoral tissue, or via 1–10 nm nanoparticles, able to diffuse directly and target the individual tumor cell. Even with constant drug release from the nanoparticles, homogenous drug-sensitive tumor cell type, targeted nanoparticle delivery and model parameters calibrated to ensure sufficient drug or nanoparticle blood concentration to kill all cells *in vitro*, analysis shows that fundamental transport limitations are severe and that drug levels inside the tumor are far less than *in vitro*. This leaves large parts of the tumor with inadequate drug concentration. Comparison of cell death rates predicted by simulations reveals that the *in vivo* rate of tumor reduction is several orders of magnitude less than *in vitro* for equal chemotherapeutic carrier concentrations in the blood. Small nanoparticles equipped with active transport mechanisms would overcome the predicted limitations and result in improved tumor response [11].

1.3
Nanoparticles as Anticancer Drug Delivery System

The fate of a drug after administration *in vivo* is determined by a combination of several processes, such as distribution, metabolism and elimination when given intravenously or absorption, distribution, metabolism and elimination when an extravascular route is used. The result depends mainly on the physicochemical properties of the drug and therefore on its chemical structure. In recent decades, much work has been directed towards developing delivery systems to control the fate of drugs by modifying these processes, in particular the drug distribution within the organism. Nanoparticles loaded with anticancer agents can successfully increase drug concentration in cancer tissues and also act at cellular levels, enhancing antitumor efficacy. They can be endocytosed/phagocytosed by cells, with resulting cell internalization of the encapsulated drug. Nanoparticles may consist of either a polymeric matrix (nanospheres) or of a reservoir system in which an oily or aqueous core is surrounded by a thin polymeric wall (nanocapsules). Suitable polymers for nanoparticles include poly(alkyl cyanoacrylates), poly(methylidene malonate) and polyesters such as poly(lactic acid), poly(glycolic acid), poly(ε-caprolactone) and their copolymers [3].

Nanoparticles of biodegradable polymers can provide controlled and targeted delivery of the drug with better efficacy and fewer side-effects. Lipophilic drugs, which have some solubility either in the polymer matrix or in the oily core of nanocapsules, are more readily incorporated than hydrophilic compounds, although the latter may be adsorbed onto the particle surface. Nanospheres can also be formed from natural macromolecules such as proteins and polysaccharides, from non polar lipids, and from inorganic materials such as metal oxides and silica [3].

As cancer chemotherapeutic agents are often administered systemically, numerous biological factors, associated with the tumor, influence the delivery of the drugs to the tumors. Consequently, drug delivery systems to solid tumors have been redesigned and, subsequently, injectable delivery systems (i.e., solid lipid nanoparticles) have been developed as an alternative to polymeric nanoparticles for adequate drug delivery to solid tumors [12].

The introduction of synthetic material into the body always affects different body systems, including the defense system. Synthetic polymers are usually thymus-independent antigens with only a limited ability to elicit antibody formation or to induce a cellular immune response against them. However, they influence, or can be used to influence, the immune system of the host in many other ways. Low-immunogenic water-soluble synthetic polymers sometimes exhibit significant immunomodulating activity, mainly concerning the activation/suppression of NK cells, LAK cells and macrophages. Some of them, such as poly(ethylene glycol) and poly[N-(2-hydroxypropyl)methacrylamide], can be used as effective protein carriers, as they can reduce the immunogenicity of conjugated proteins and/or reduce nonspecific uptake of nanoparticle-entrapped drugs and other therapeutic agents [13].

1.3.1
Conventional Nanoparticles

Association of a cytostatic drug to colloidal carriers modifies the drug biodistribution profile, as it is mainly delivered to the mononuclear phagocytes system (liver, spleen, lungs and bone marrow). Once in the bloodstream, surface non-modified nanoparticles, (conventional nanoparticles), are rapidly opsonized and massively cleared by the fixed macrophages of the mononuclear phagocytes system organs.

The size of the colloidal carriers as well as their surface characteristics greatly influence the drug distribution pattern in the reticuloendothelial organs, since these parameters can prevent their uptake by macrophages. The exact underlying mechanism was not fully understood, but it was rapid and compatible with endocytosis. A high curvature (resulting in size < 100 nm) and/or a hydrophilic surface (as opposed to the hydrophobic surface of conventional nanoparticles) are needed to reduce opsonization reactions and subsequent clearance by macrophages [14]. Since conventional nanoparticles are naturally concentrated within macrophages, they can be used to deliver drugs to these cells. The muramyl dipeptide, a low-molecular-weight, soluble, synthetic compound based on the structure of peptidoglycan from mucobacteria, stimulates the antitumoral activity of macrophages. Although it acts on intracellular receptors, it penetrates poorly into macrophages. Furthermore, it is eliminated rapidly after intravenous administration. These problems can be overcome by encapsulation within nanocapsules, and lipophilic derivatives such as muramyl tripeptide-cholesterol have been developed to increase encapsulation efficiency. *In vitro* studies have shown increased intracellular penetration of muramyl peptides into macrophages when they are associated with nanoparticles, increasing macrophage effector functions (i.e., nitric oxide, cytokine and prostaglandin production) and cytostatic activity against tumor cells [3].

The contribution of conventional nanoparticles to enhance anticancer drugs efficacy is limited to targeting tumors at the organs of the mononuclear phagocytes system organ. Owing to their very short circulation time (the mean half-life of conventional nanoparticles is 3–5 min after intravenous administration) addressing anticancer drug-loaded nanoparticles to other tumoral tissues is not reasonable. Moreover, penetration of such a carrier system across the tumoral endothelium would be minimum, leading to subtherapeutic concentrations of the drug near the cancer cells. This biodistribution can be of benefit for the chemotherapeutic treatment of mononuclear phagocytes system localized tumors (i.e., hepatocarcinoma, hepatic metastasis, bronchopulmonary tumors, myeloma and leukemia). Activity against hepatic metastases has been observed in mouse models, although this treatment is only therapeutic when the tumor load is low [15].

Both the polymeric composition of the nanoparticles and the associated drug greatly influence the drug distribution pattern in the mononuclear phagocytic system. Conventional nanoparticles, likely, have a better safety profile than free anticancer agents when acting on normal tissue. For example, a reduction of the cardiac accumulation of drugs [16] and of the genotoxicity of mitomycin [17] have been reported. Moreover, encapsulation of doxorubicin within nanoparticles reduces its cardiotoxicity by reducing the amount of drug that reaches the myocardium with a significant increase of drug concentrations in the liver. In one study this was not associated with any overall toxicity [18] while another study, using unloaded nanoparticles, observed a reversible decline in the phagocytic capacity of the liver after prolonged dosing, as well as a slight inflammatory response [19]. Accumulation of drug nanoparticles in the liver may also influence its elimination, since this organ is the site of metabolism and biliary excretion. Biliary clearance of indomethacin was increased three-fold by inclusion in nanocapsules [20]. Nanoparticle-associated doxorubicin accumulated in bone marrow and led to a myelosuppressive effect in one study [21]. Thus, altered distribution may generate new types of toxicity. When conventional nanoparticles are used as carriers in chemotherapy, some cytotoxicity against the Kupffer cells or other targeted macrophages can be expected, as the class of drugs being used is able to induce apoptosis in these cells. Treatments featuring frequent administrations (with intervals shorter than two weeks – the restoration period of Kupffer cells) could result in a deficiency of Kupffer cells, which in turn could lead to decreased liver uptake and a subsequent decreased therapeutic efficacy for hepatic tumors. In addition, a risk for bacteriemia can not to be excluded [22]. Moreover, conventional carriers also target the bone marrow, hence chemotherapy with such carriers may increase myelosuppressive effects. However, this tropism of carriers can also be used to deliver myelo-stimulating compounds such as granulocyte-colony-stimulating factor [23].

Certain types of nanoparticles were also found to be able to overcome multidrug resistance mediated by the P-glycoprotein efflux system localized at the cancerous cell membrane. This simultaneous cellular resistance to multiple lipophilic drugs represents a major problem in cancer chemotherapy. Such drug resistance may appear clinically either as a lack of tumor size reduction or as the occurrence of clinical relapse after an initial positive response to antitumor treatment. The resistance mechanism can have different origins, either directly linked to specific mechanisms developed by the tumor tissue or connected to the more general problem of distribution of a drug towards its targeted tissue. Poly(alkyl cyanoacrylate) nanoparticles have been developed to overcome multidrug resistance [10]. Despite encouraging results with conventional carrier systems, much research has been dedicated to designing carriers with modified distribution and new therapeutic applications.

1.3.2
Sterically Stabilized Nanoparticles

Since the usefulness of conventional nanoparticles is limited by their massive capture by the macrophages of the mononuclear phagocytes system after intravenous administration, systems with modified surface properties to reduce the disposition of plasma proteins and the recognition by phagocytes have been developed [14]. These are known as sterically stabilized carriers or "stealth carriers" and may remain in the blood compartment for a considerable time. The hydrophilic polymers poly(ethylene glycol), poloxamines, poloxamers, polysaccharides have been used to coat efficiently conventional nanoparticles' surface [24]. These coating provide a dynamic cloud of hydrophilic and neutral chains at the particle surface that keep away plasma proteins [25]. Poly(ethylene glycol) (PEG) has been introduced at the surface either by adsorption of surfactants [24] or by using block or branched copolymers, usually with poly(lactide) [26, 27]. These Stealth™ nanoparticles are characterized by a prolonged half-life in the blood compartment and by extravasation into sites where the endothelium is more permeable, such as solid tumors, regions of inflammation and infection. Consequently, such long-circulating nanoparticles are supposed to be able to target directly most tumors located outside the mononuclear phagocytes system [22]. A higher tumor uptake, thanks to the small size and the hydrophilicity of the carrier device, as well as a sustained release of the drug could improve the efficacy of anticancer chemotherapy. The complement-rejecting properties of nanocapsules are superior to those of nanospheres at equivalent surface area [28].

Coating conventional nanoparticles with surfactants to obtain a long-circulating carrier, the first strategy used to direct tumor targeting *in vivo*, has been demonstrated using photosensitizers like phthalocyanines in PEG-coated nanoparticles [29] and *meta*-tetra(hydroxyphenyl)chlorin in poly(lactide)-PEG nanocapsules [30].

PEG chains have been attached covalently to poly(alkyl cyanoacrylate) polymers by two chemical strategies; both types of particle have shown long-circulating properties *in vivo* [31]. Such particles have been loaded with tamoxifen with a view to their use in treating hormone-dependent tumors [32]. Poly(alkyl cyanoacrylate) nanoparticles allow the delivery of several drugs, including doxorubicin, across the blood–brain barrier after coating with surfactants. However, only the surfactants polysorbate (Tween 20, 40, 60 and 80) and some poloxamers (Pluronic F68) can induce this uptake. A therapeutic effect with this sterically stabilized carrier system has been noted in rats bearing intracranial glioblastoma [33]. The delivery mechanism across the blood–brain barrier is most likely endocytosis via the LDL receptor by the endothelial cells lining the brain blood capillaries after injection of the nanoparticles into the blood stream. This endocytotic uptake seems to be mediated by the adsorption of apolipoprotein B and/or E adsorption from the blood. The drug, then, may be released either within these cells followed by passive diffusion into the brain or be transported into the brain by transcytosis [34]. Long-circulating nanoparticles have also been used in tumor targeting followed by irradiation of tumor site in the case of photodynamic therapy. To improve anticancer efficacy and avoid systemic phototoxicity this combination was evaluated in a EMT-6 tumor-bearing mice model. Unfortunately, poly(lactic acid) nanospheres covered by PEG did not allow a higher intratumoral accumulation of its incorporated photosensitizer (hexadecafluoro zinc phthalocyanine). The fragility of the adsorbed coating and the large size (>900 nm) could be responsible for this failure. However, formulation of the photosensitizer in the biodegradable nanospheres improved photodynamic therapy response, probably by influencing the intratumoral distribution pattern of the photosensitizer [35]. The biodistribution of non-biodegradable poly(methyl methacrylate) nanospheres coated or not with different surfactants (Polysorbate 80, Poloxamer 407 and Polaxamine 908) have been investigated in several mice bearing tumor models, including a murine B16-melanoma (inoculated intramuscularly), a human breast cancer MaTu (engrafted subcutaneously) and an U-373 glioblastoma (implanted intracerebrally). The coated poly-(methyl methacrylate) nanospheres in circulation showed a prolonged half-life, especially with the Poloxamer 407 and Polaxamine 908 coatings. Moreover, an accumulation and retention of the coated nanospheres in the B-16 and MaTu tumors were observed, depending on the particle surface hydrophilicity and the specific growth difference of the tumor [36].

Although interesting results have been obtained with adsorbed surfactant, a covalent linkage of amphiphilic copolymers is generally preferred to obtain a protective hydrophilic cloud on nanoparticles that avoids the possibility of rapid coating desorption upon either dilution or after contact with blood components. This approach has been employed with poly(lactic acid), poly(caprolactone) and poly(cyanoacrylate) polymers, which were chemically coupled to PEG [27, 31, 37].

The surface characteristics of nanospheres prepared from poly(lactide-*co*-glycolide) copolymers have been optimized to reduce their interactions with plasma proteins and to increase their circulating half-life [26, 27, 38]. Biodegradable nanospheres prepared from poly(lactide-*co*-glycolide) coated with poly(lactide)- PEG diblock copolymers showed a significant increase in blood circulation time and reduced liver uptake in a rat model compared with non-coated poly(lactide-*co*-glycolide) nanospheres [39]. Nanocapsules containing poly(lactide)-PEG showed reduced association with a macrophage-like cell line (J774 A1) irrespective of dilution and incubation time up to 24 h [40]. This formulation, which showed a lower capture by macrophages *in vitro*, also gave good results *in vivo*, yielding a plasma area under the curve 15-fold higher than that obtained with nanocapsules stabilized by adsorbed Poloxamer F68. Persistence in the blood compartment was accompanied by delayed liver and spleen uptake [41].

1.3.3
Actively Targetable Nanoparticles

Nanoparticles can be targeted to the particulate region of capillary endothelium, to concentrate the drug within a particular organ and allow it to diffuse from the carrier to the target tissue. The folate receptor is overexpressed on many tumor cells. Folinic acid has some advantages over transferrin or antibodies as a ligand for long-circulating carriers because it is a much smaller molecule that is unlikely to interact with opsonines and can be coupled easily to a poly(ethylene glycol) (PEG) chain without loss of receptor-binding activity. This targeting strategy has also been applied to long-circulating nanoparticles prepared from a cyanoacrylate-based polymer. Folate grafted to PEG cyanoacrylate nanoparticles has a ten-fold higher apparent affinity for the folate-binding protein than the free folate. Indeed, the particles represent a multivalent form of the ligand folic acid and folate receptors are often disposed in clusters. Thus conjugated nanoparticles could display a stronger interaction with the surface of malignant cells [42]. Bovine serum albumin nanoparticles have been reacted with the activated folic acid to conjugate folate via amino groups of the nanoparticle to improve their intracellular uptake to target cells. The nanoparticles were taken up to SKOV3 cells (human ovarian cancer cell line) and levels of binding and uptake were increased with the time of incubation until 4 h. The levels of folate-conjugated bovine serum albumin nanoparticles were higher than those of non-conjugated nanoparticles and saturable. Association of folate conjugated bovine serum albumin nanoparticles to SKOV3 cells was inhibited by an excess amount of folic acid, suggestive of binding and/or uptake mediated by the folate receptor [43].

Folate-linked microemulsions of the antitumor antibiotic aclacinomycin have been developed and investigated both *in vitro* and *in vivo*. Three kinds of folate-linked microemulsions with different PEG chain lengths loading aclacinomycin were formulated. *In vitro* studies were performed in a human nasopharyngeal cell line, KB, which overexpresses the folate receptor and in a human hepatoblastoma cell line, folate receptor (−), HepG2. *In vivo* experiments were carried out in a KB xenograft by systemic administration of folate-linked microemulsions loading aclacinomycin. This afforded selective folate receptor mediated cytotoxicity in KB but not in HepG2 cells. Association of the folate-PEG 5000 linked and folate-PEG 2000 linked microemulsions with the cells was 200- and 4-fold higher, respectively, whereas their cytotoxicity was 90- and 3.5-fold higher than those of non-folate microemulsion. The folate-PEG 5000 linked microemulsions showed a 2.6-fold higher accumulation in solid tumors 24 h after intravenous injection and greater tumor growth inhibition than free aclacinomycin. This study showed how folate modification with a sufficiently long PEG-chain on emulsions can be an effective way of targeting emulsion to tumor cells [44].

Transferrin is another well-studied ligand for tumor targeting due to upregulation of transferrin receptors in numerous cancer cell types. A transferrin-modified, cyclodextrin polymer-based gene delivery system has been developed. The nanoparticle is surface-modified to display PEG for increasing stability in biological fluids and transferrin for targeting of cancer cells that express transferrin receptor. At low transferrin modification, the particles remain stable in physiologic salt concentrations and transfect K562 leukemia cells with increased efficiency over untargeted particles. The transferrin-modified nanoparticles are appropriate for use in the systemic delivery of nucleic acid therapeutics for metastatic cancer applications [45].

Nanoparticles conjugated with an antibody against a specific tumor antigen have been developed to obtain selective drug delivery systems for the treatment of tumors expressing a specific tumor antigen. For instance, biodegradable nanoparticles based on gelatin and human serum albumin have been developed. The surface of the nanoparticles was modified by covalent attachment of the biotin-binding protein NeutrAvidin, enabling the binding of biotinylated drug targeting ligands by avidin–biotin complex formation. Using the HER2 receptor specific antibody trastuzumab (Herceptin) conjugated to the surface of these nanoparticles, a specific targeting to HER2-overexpressing cells could be shown. Confocal laser scanning microscopy demonstrated an effective internalization of the nanoparticles by HER2-overexpressing cells via receptor-mediated endocytosis [46].

There is a tremendous effort to develop and test gene delivery vectors that are efficient, non-immunogenic, and applicable for cancer gene systemic therapy. Since internalization of colloidal carriers usually leads to the lysosomal compartment, in which hydrolytic enzymes will degrade both the carrier and its content, the intracellular distribution of the carrier must be modified when the encapsulated drug is a nucleic acid. Thus, systems have been developed that either fuse with the plasma membrane or have a pH-sensitive configuration that changes conformation in the lysosomes and allows the encapsulated material to escape into the cytoplasm [3]. Rapid endo-lysosomal escape of biodegradable nanoparticles formulated from the copolymers of poly(D,L-lactide-co-glycolide) has been reported. This occurred by selective reversal of the surface charge of nanoparticles (from anionic to cationic) in the acidic endo-lysosomal compartment, which causes the nanoparticles to interact with the endo-lysosomal membrane and escape into the cytosol. These nanoparticles can deliver various therapeutic agents, including macromolecules such as DNA at a slow rate, resulting in a sustained therapeutic effect. Thermo-responsive, pH-responsive and biodegradable nanoparticles have been developed by grafting biodegradable poly(D,L-lactide) onto N-isopropyl acrylamide and methacrylic acid. It may be sufficient for a carrier system to concentrate the drug in the tissue of interest. However, for hydrophilic molecules such as nucleic acid, which cross the plasma membrane with difficulty, intracellular delivery is required. Clinical trials for deadly pancreatic cancer have recently begun on two continents. The aim is to evaluate the safety and efficacy of engineered nanoparticles guided by a targeted delivery system to overcome dilution, filtration and inactivation encountered in the human circulatory system to deliver a killing designer gene to metastatic tumors that are refractory to conventional chemotherapy. The first patients receiving multiple intravenous infusions of the targeted delivery system-encapsulated genetic bullets have all responded favorably [47].

1.3.4
Routes of Drug Nanoparticles Administration

The most convenient route of drug administration is the oral one but this presents several barriers to the use of colloidal carrier owing to conditions within the gastrointestinal tract. Duodenal enzymes and bile salts destroy the lipid bilayers of most types of liposome, releasing the drug. Polymeric nanoparticles are more stable, although there is some evidence that polyesters can be degraded by pancreatic lipases [48]. They may be able to improve bioavailability, particularly for highly insoluble drugs, by increasing the surface area for dissolution and as a result of bioadhesion. However, nanoparticles can be used to protect a labile drug from degradation in the gastrointestinal tract or to protect it from toxicity due to the drug.

Polymeric nanoparticles, due to their bioadhesive properties, may be immobilized within the mucus or, when in contact with the epithelial cells, show a slower clearance from the gastrointestinal tract [49]. Nanoparticles of biodegradable polymers containing alpha-tocopheryl PEG 1000 succinate (vitamin E TPGS) have been proposed to replace the current method of clinical administration and to provide an innovative solution for oral chemotherapy. Vitamin E TPGS could be a novel surfactant as well as a matrix material when blended with other biodegradable polymers – it has great advantages for the manufacture of polymeric nanoparticles for controlled release of paclitaxel and other anticancer drugs [50].

Potential applications of colloidal drug carriers by the intravenous route can be summarized as concentrating drugs in accessible sites, rerouting drugs away from sites of toxicity and increasing the circulation time of labile or rapidly eliminated drugs.

After subcutaneous or intraperitoneal administration, nanoparticles are taken up by regional lymph nodes [51].

Subcutaneously or locally injected (in the peri-tumoral region) nanoparticles can be used for lymphatic targeting as a tool for chemotherapy against lymphatic tumors or metastases since they penetrate the interstitial space around the injection site and are gradually absorbed by the lymphatic capillaries into the lymphatic system. Aclarubicin adsorbed onto activated carbon particles has been tested after subcutaneous injection in mice, against a murine model (P288 leukemia cells) of lymph node metastases [52]. The same system has also been used as a locoregional chemotherapy adjuvant for breast cancer in patients after intratumoral and peritumoral injections [53]. In both cases, this carrier system distributed selectively high levels of free aclarubicin to the regional lymphatic system and low levels to the rest of the body. However, this carrier system is not biodegradable and it is rather big (>100 nm), impeding drainage from the injection site. In addition, the drug is associated to the particles by adsorption, leading to a rapid release with possible systemic absorption. Biodegradable systems coated by adsorption of the surfactant Poloxamine 904 or poly(isobutyl cyanoacrylate) nanocapsules [51, 54] have been developed to improve lymphatic targeting. Poloxamine 904 caused an increased sequestration of the particles in lymph nodes, reducing the systemic absorption of any encapsulated drug [54]. Poly(isobutyl cyanoacrylate) nanocapsules were also able to retain the lipophilic indicator 12-(9-anthroxy)stearic acid in the regional lymph nodes for 168 h after intramuscular administration [51]. A study combining intratumoral administration of gadolinium-loaded chitosan nanoparticles and neutron-capture therapy has been performed on the B16F10 melanoma model subcutaneously implanted in mice. Gadolinium retention in tumor tissue increased when it was encapsulated with respect to the free drug. Tumor irradiation 8 h after the last intratumoral injection of gadolinium nanoparticles prevented further tumor growth in the animals treated, thereby increasing their life expectancy [55].

Retention of carriers instilled into the eye also occurs, leading to important therapeutic potential in this area. As well as a bioadhesive effect, some evidence has been presented to show that nanoparticles can penetrate through the corneal epithelium [56].

1.4
Anticancer Drug Nanoparticles

1.4.1
Anthracyclines

One of the most powerful and widely used anticancer drugs is doxorubicin, an anthracyclinic antibiotic that inhibits the synthesis of nucleic acids. This drug has a very narrow therapeutic index as its clinical use is hampered by several undesirable side-effects like cardiotoxicity and myelosuppression [57]. Thus, much effort has been made to target doxorubicin to cancer tissues, improving its efficacy and safety.

Poly(isohexyl cyanoacrylate) nanospheres incorporating doxorubicin have been developed. On an hepatic metastases model in mice (M5076 reticulum cell sarcoma) such doxorubicin nanospheres afforded a greater reduction in the number of metastases than when free doxorubicin was used and it appeared to increase the life span of the metastasis-bearing mice [58]. Another study found higher concentrations of doxorubicin in mice liver, spleen and lungs with doxorubicin incorporated into poly(isohexyl cyanoacrylate) than in mice treated with free doxorubicin. At the same time, the concentration of doxorubicin in heart and kidneys of the mice were lower than when free doxorubicin was used [18]. Tissue pharmacokinetic studies showed that the underlying mechanism responsible for the increased therapeutic efficacy of the nanoparticle formulation was a transfer of doxorubicin from the healthy hepatic tissue, acting as a drug reservoir, to the malignant cells. Kupffer cells, after a massive uptake of nanoparticles by phagocytosis, were able to induce the release of doxorubicin, leading to a gradient concentration, favorable for a prolonged diffusion of the free and still active drug towards the neighboring metastatic cells [59]. Thus, this biodistribution profile can be of benefit for chemotherapeutic treatment of tumors localized in the mononuclear phagocytic system.

Clinical pharmacokinetics after a single intravenous administration of doxorubicin adsorbed onto poly(methacrylate) nanospheres has been investigated in hepatoma patients. This type of conventional carrier, although of limited use *in vivo* because not biodegradable, allowed a reduction in both the volume of distribution and the elimination half-life of doxorubicin [60]. Another phase I clinical investigation has been carried out on 21 cancer patients with doxorubicin associated to biodegradable poly(isohexyl cyanoacrylate) nanospheres. Pharmacokinetic studies conducted in 3/21 patients revealed important interindividual variation for doxorubicin (encapsulated or not) plasma levels. Clinical toxicity of encapsulated doxorubicin consisted of dose-dependent myelosuppression of different grades in all patients, in pseudo-allergic reactions in 3/21 patients and in diffuse bone pain in 3/21 patients. However, neither cardiac toxicity nor hepatotoxicity were encountered among 18 patients treated with the nanoparticles formulation. According to WHO criteria, there were only 2/21 stable diseases lasting 4–6 months. All the other patients had progressive disease after the first course of doxorubicin-loaded poly(isohexyl cyanoacrylate) nanospheres. This could be due to tumor localization, as they were rarely located at sites of the mononuclear phagocytic system, resulting in subtherapeutic anticancer drug concentration exposure [61].

As doxorubicin-loaded poly(isobutyl cyanoacrylate) nanoparticles are more efficient than free drug in mice bearing hepatic metastasis of the M5076 tumor, and Kupffer cells could have acted as a drug reservoir after nanoparticle phagocytosis, the role of macrophages in mediating the cytotoxicity of doxorubicin-loaded nanoparticles on M5076 cells was studied. After direct contact, free doxorubicin and doxorubicin-loaded nanoparticles had the same efficacy against M5076 cell growth. Co-culture experiments with macrophages J774.A1 led to a five-fold increase in the IC_{50} for both doxorubicin and doxorubicin-loaded nanoparticles. The activation of macrophages by IFN-γ in co-culture significantly decreased the IC_{50}s. After phagocytosis of doxorubicin-loaded nanoparticles, J774.A1 cells were able to release active drug, allowing it to exert its cytotoxicity against M5076 cells. Drug efficacy was potentiated by the activation of macrophages releasing cytotoxic factors such as NO, which resulted in increased tumor cell death [62]. Similarly, the use in a murine tumor model (implanted subcutaneously J774A.1 macrophages) of a dextran–doxorubicin conjugate incorporated into small chitosan nanospheres was reported to outperform the free conjugate, especially in relation to life expectancy [63].

The body distribution of nanoparticle incorporating doxorubicin injected into the hepatic artery of hepatoma bearing rats was investigated. The nanoparticle formulation and free doxorubicin were injected into the hepatic artery of Walker-256 hepatoma bearing rats on the seventh day after tumor implantation. Survival time, tumor enlargement ratio and tumor necrosis degree were compared. Nanoparticles containing doxorubicin substantially increased the drug concentrations in liver, spleen, and tumor of rats compared to free drug, whereas the concentrations in plasma, heart and lungs were significantly decreased. The nanoparticulate formulation brought on a more significant tumor inhibition and more extensive tumor necrosis. The prolonged life span ratio was 109.22% as compared with rats that accepted normal saline [64, 65].

The nanoparticle modified biodistribution may generate new types of toxicity, as has been observed with doxorubicin incorporated in poly(isobutyl cyanoacrylate) and poly(isohexyl cyanoacrylate) nanospheres, whose hematopoietic toxicity was generally more pronounced and long-lasting than that of free doxorubicin [21]. Acute renal toxicity was another murine-reported doxorubicin toxicity, which was amplified by the association of the drug to poly(isobutyl cyanoacrylate) nanospheres. Proteinuria was probably the result of a modified biodistribution of the associated drug, which resulted in glomerular damage [66]. In another instance, doxorubicin was incorporated into gelatin nanospheres by a covalent bond on its amino group via glutaraldehyde [67]. Unfortunately, no or only marginal antitumor activity against a C26 tumor (mouse colon adenocarcinoma) *in vivo* was showed, and in certain cases there was even an increased doxorubicin cardiotoxicity. Once again this lack of antitumor activity may due to slow dissociation of the complex due to the covalent linkage or to the slow diffusion of the complex across the cellular membrane.

Polysorbate 80-coated poly(isobutyl cyanoacrylate) nanospheres provided no major changes in doxorubicin biodistribution and pharmacokinetic parameters in rats when compared to those of conventional poly(isobutyl cyanoacrylate) nanospheres. However, the coated nanospheres did transport a significant amount of incorporated doxorubicin to the brain of healthy rats, reaching the highest drug levels in this organ 2–4 h after intravenous administration. Consequently, an active transport from the blood to the brain was suspected since cerebral accumulation occurred against a concentration gradient. A mechanism involving an endocytosis by brain endothelial cells was hypothesized, as the brain uptake was inhibited at 4 °C and after a pretreatment with cytochalasin B [68]. Then, polysorbate 80-coated poly(isobutyl cyanoacrylate) nanoparticles were shown to enable the transport of doxorubicin across the blood–brain barrier to the brain after intravenous administration and to considerably reduce the growth of brain tumors in rats. The acute toxicity of doxorubicin associated with polysorbate 80-coated nanoparticles in healthy rats was studied and a therapeutic dose range for this formulation in rats with intracranially implanted 101/8 glioblastoma was established. Single intravenous administration of empty poly(isobutyl cyanoacrylate) nanoparticles in the dose range 100–400 mg kg^{-1} did not cause mortality within the period of observation. Association of doxorubicin with poly(isobutyl cyanoacrylate) nanoparticles did not produce significant changes of quantitative parameters of acute toxicity of the antitumor agent. Likewise, the presence of polysorbate 80 in the formulations was not associated with changes in toxicity compared with free or nanoparticulate drug. The results in tumor-bearing rats were similar to those in healthy rats. The toxicity of doxorubicin bound to nanoparticles was similar or even lower than that of free doxorubicin [69]. As doxorubicin bound to polysorbate-coated nanoparticles crossed the intact blood–brain barrier, reaching therapeutic concentrations in the brain, the therapeutic potential of this formulation *in vivo* was studied using an animal model created by implantation of 101/8 glioblastoma tumor in rat brains. Rats treated with doxorubicin bound to polysorbate-coated nanoparticles had significantly higher survival times than with free doxorubicin; over 20% of the animals showed a long-term remission (Fig. 1.1). All animals treated with polysorbate-containing formulation also had a slight inflammatory reaction to the tumor. There was no indication of neurotoxicity [70].

Doxorubicin has also been conjugated chemically to a terminal end group of poly(D,L-lactic-*co*-glycolic acid) by an ester linkage and then formulated into nanoparticles. The conjugated doxorubicin nanoparticles showed increased uptake within a HepG2 cell line. They exhibited a slightly lower IC$_{50}$ against the HepG2 cell line than did free doxorubicin. An *in vivo* antitumor activity assay also showed that a single injection of the nanoparticles had comparable activity to that of free doxorubicin administered by daily injection [71]. Doxorubicin-loaded poly(butyl cyanoacrylate) nanoparticles have been used to enhance the delivery of the drug to Dalton's lymphoma solid tumor. ^{99m}Tc-labeled complexes of doxorubicin and doxorubicin-loaded poly(butyl cyanoacrylate) nanoparticles were administered subcutaneously below the Dalton's lymphoma tumor. The distribution of doxorubicin-loaded poly(butyl cyanoacrylate) nanoparticles to the blood, heart and organs of the

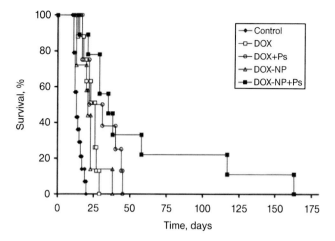

Fig. 1.1. Kaplan–Meier survival curves. Percentage of survival of the rats with intracranially transplanted glioblastoma after intravenous injection of 2.5 mg kg^{-1} doxorubicin each on day 2, 5 and 8 using one of the following formulations: DOX, doxorubicin in saline; DOX+Ps, doxorubicin in saline plus polysorbate 80; DOX-NP, doxorubicin bound to poly(butyl cyanoacrylate) nanoparticles; DOX-NP+Ps, doxorubicin bound to poly(butyl cyanoacrylate) nanoparticles coated with polysorbate 80. (Reproduced with permission from Ref. [70].)

reticuloendothelial system was biphasic with a rapid initial distribution, followed by a significant decrease later at 6 h post-injection. The distribution of doxorubicin to tissues was very low initially and increased significantly at 6 h post-injection, indicating its accumulation at the injection site for a longer time. The concentration of doxorubicin-loaded poly(butyl cyanoacrylate) nanoparticles was also high in tissues at 6 h post-injection, indicating their accumulation at the subcutaneous site and consequent disposition to tissues with time. A significantly high tumor uptake of doxorubicin-loaded poly(butyl cyanoacrylate) nanoparticles (approx. 13-fold higher at 48 h post-injection) was found compared to free doxorubicin. Tumor concentrations of both doxorubicin and doxorubicin-loaded poly(butyl cyanoacrylate) nanoparticles increased with time, indicating their slow penetration from the injection site into tumor [72].

Poly(ethylene glycol)-coated poly(hexadecyl cyanoacrylate) nanospheres also displayed a significant accumulation within an orthotopic 9L gliosarcoma model, after intravenous administration to rats. Hence, in the same model, the pre-clinical efficacy of this carrier when loaded with doxorubicin was evaluated. The cumulative maximum tolerated dose of nanoparticulate doxorubicin was 1.5-fold higher than that of free doxorubicin. Nevertheless, encapsulated doxorubicin was unable to elicit a better therapeutic response in the 9L gliosarcoma. A biodistribution study revealed that the doxorubicin-loaded nanospheres accumulated to a 2.5-fold lesser extent in the 9L tumor than did the unloaded nanospheres and that they were mainly localized in the lungs and the spleen. Such a typical profile indicated aggregation with plasma proteins as a consequence of the positive surface charge of these loaded particles; this ionic interaction resulting from drug encapsulation was mainly responsible for 9L treatment failure [73].

The ability of doxorubicin-loaded solid lipid nanoparticles to achieve prolonged drug plasma levels was also investigated. One study observed a low uptake of solid lipid nanoparticles by the liver and spleen macrophages. This might be explained by a low surface hydrophobicity of the nanoparticles avoiding the adsorption of any blood proteins mediating the uptake by liver and spleen macrophages [74]. Moreover, uptake of the solid lipid nanoparticles (SLNs) by the brain might be explained by adsorption of a blood protein mediating adherence to the endothelial cells of the blood–brain barrier [75]. Pharmacokinetic studies of doxorubicin incorporated into SLNs showed higher blood levels in comparison to a commercial drug solution after intravenous injection in rats. Concerning the body distribution, doxorubicin-loaded SLNs caused higher drug concentrations in lung, spleen and brain, while the solution led to a distribution more into liver and kidneys [76]. Furthermore, incorporation of doxorubicin into SLNs strongly enhanced its cytotoxicity in several cell lines [77, 78]. In particular, in the human colorectal cancer cell line HT-29, the intracellular doxorubicin content was double after 24 h exposure to loaded SLNs versus the conventional drug formulation [78]. Pharmacokinetics and tissue distribution of doxorubicin incorporated in non-sterically stabilized SLNs and in sterically stabilized SLNs at increasing concentration of stearic acid-PEG 2000 as stabilizing agent after intravenous administration to rabbits have been studied. The doxorubicin area under the concentration–time curve increased as a function of the amount of sterically stabilizing agent present in the SLNs. Doxorubicin was present in the brain only after doxorubicin-loaded SLNs administration and the increase in the stabilizing agent affected the doxorubicin transported into the brain. There was always less doxorubicin in the liver, lungs, spleen, heart and kidneys after injection of any of the types of SLNs than after the doxorubicin solution. In particular, all SLNs formulations significantly decreased heart and liver concentrations of doxorubicin [79]. Interestingly, both non-coated and coated SLNs showed a similar low uptake by liver and spleen macrophages [80]. The bioavailability of idarubicin can be also improved by administering idarubicin-loaded SLNs duodenally to rats. The pharmacokinetic parameters of idarubicin found after duodenal administration of idarubicin solution and idarubicin-loaded SLNs were different. The area under the concentration–time curve and the elimination half-life were 2- and 30-fold higher after administration of idarubicin-loaded SLNs than after solution administration, respectively. Tissue distribution also differed: idarubicin and its main metabolite idarubicinol concentrations were lower in heart, lung, spleen and kidneys after idarubicin-loaded SLNs administration than after solution administration. Furthermore, the drug and its metabolite were detected in the brain only after idarubicin-loaded SLNs administration, indicating that SLNs were able to cross the blood–brain barrier. After intravenous idarubicin-loaded SLNs administration, the area under the time–concentration curve of idarubicin was lower than after duodenal administration of the same formulation [81].

1.4.1.1 Reverse of P-glycoprotein Mediated Multidrug Resistance of Cancer Cells to Doxorubicin

Different types of nanoparticles have been developed to reverse the P-glycoprotein mediated multidrug resistance of cancer cells to doxorubicin, an important problem in its clinical use.

Conventional poly(cyanoacrylate) nanoparticles allow doxorubicin P-glycoprotein mediated multidrug resistance to be overcome *in vitro* only when the nanoparticles and the resistant cancer cell line are in close contact [82]. The nanoparticle-associated drug accumulated within the cells and appeared to avoid P-glycoprotein dependent efflux. This reversal was only observed with poly(alkyl cyanoacrylate) nanoparticles and was not due to particle endocytosis. The formation of a complex between positively charged doxorubicin and negatively charged polymer degradation products seemed to favor diffusion across the plasma membrane [82].

When doxorubicin was coupled via an ionic interaction to non-biodegradable poly(methacrylate) nanospheres, the P-glycoprotein mediated multidrug resistance reversal differed from that of doxorubicin incorporated into biodegradable poly(cyanoacrylate) nanoparticles.

When adsorbed onto the surface of poly(methacrylate) nanospheres, doxorubicin was demonstrated to be cell internalized by an endocytic process in cultured rat hepatocytes and in a human monocyte-like cancer cell line expressing P-glycoprotein (U-937). Once internalized, poly(methacrylate) nanospheres generated an intracellular sustained release of doxorubicin in U-937 cells. A higher intracellular accumulation, related to a more important cytotoxicity on U-937 cells, was noted for encapsulated doxorubicin than for free doxorubicin. However, such a carrier, despite its ability to mask the positive charge of doxorubicin, is of limited use *in vivo* since it is not biodegradable [83].

When doxorubicin was incorporated into rapidly biodegraded poly(isobutyl cyanoacrylate) nanospheres and tested in a resistant murine leukemia sub-line overexpressing P-glycoprotein (P388/ADR), a higher cellular uptake was observed. Furthermore, the cell uptake kinetics of doxorubicin nanoparticles were unchanged in the presence of cytochalasin B, an endocytosis inhibitor. Efflux studies showed a similar profile for doxorubicin in nanoparticulate or free form. This suggests that poly(isobutyl cyanoacrylate) nanospheres did not enter the cells. In an *in vitro* model of doxorubicin-resistant rat glioblastoma (C6 cells sub-lines), doxorubicin incorporated into poly(isobutyl cyanoacrylate) nanospheres was always more cytotoxic, and also had a lower intracellular concentration, than the free drug. The polymer constituting the nanospheres did not inhibit the P-glycoprotein by direct interaction with the protein. It was also observed that on C6 cell sub-lines with different expression of P-glycoprotein that doxorubicin nanospheres were only efficient on pure P-glycoprotein mediated multidrug resistance phenotype cells and not on the additional mechanisms of resistance to doxorubicin [84]. Thus, the mechanism of P-gp reversion by nanoparticles could only be explained by a local delivery of the drug in high concentration close to the cell membrane, after degradation of the polymeric carrier. Such local microconcentration of doxorubicin was supposed to be able to saturate P-glycoprotein [85]. P-glycoprotein mediated multidrug resistance is overcome with poly(isobutyl cyanoacrylate) and poly(isobutyl cyanoacrylate) doxorubicin-loaded nanospheres not only due to the adsorption of nanoparticles to the cell surface but also due to an increased diffusion of doxorubicin across the plasma membrane, thanks to the formation of an ion pair between negatively charged cyanoacrylic acid (a nanoparticle degradation product) and the positively charged doxorubicin. Such ion-pair formation has been evidenced by Raman spectroscopy and by ion-pair reversed-phase HPLC [86]. Masking the positive charge of the amino sugar of doxorubicin appears to be key to overcoming P-gp mediated multidrug resistance. At the same time, the cytotoxic activity of doxorubicin was only slowly compromised after chemical modifications of the amino sugar. Consequently, some studies focused on developing systems featuring covalent linkage between the polymers and the amino sugar of doxorubicin [87]. Such a complex mechanism for overcoming P-glycoprotein mediated multidrug resistance was only observed with cyanoacrylate nanoparticles.

The failure to overcome P-glycoprotein mediated multidrug resistance with nanoparticles designed with other polymers could be explained by an appropriate release mechanism of drug (diffusion could lead to the release of the active compound without the polymeric counter ion), by degradation kinetics of the polymer that are too slow, or by the size of the polymeric counter ion, which could be the limiting factor for diffusion across the cell membrane.

Interestingly, the association of doxorubicin with poly(alkyl cyanoacrylate) nanoparticles also reversed the resistance to doxorubicin in numerous multidrug-resistant cell lines [88].

Furthermore, doxorubicin encapsulated in poly(isohexyl cyanoacrylate) nanospheres can circumvent P-glycoprotein mediated multidrug resistance. K562 and MCF7 cell lines were more resistant to free doxorubicin than to doxorubicin poly(isohexyl cyanoacrylate) nanospheres. The MCF7 sub-lines selected with doxorubicin poly(isohexyl cyanoacrylate) nanospheres exhibited a higher level of resistance to both doxorubicin formulations than those selected with free doxorubicin. Different levels of overexpression of several genes involved in drug resistance occurred in the resistant variants. MDR1 gene overexpression was consistently higher in free doxorubicin selected cells than in doxorubicin poly(isohexyl cyanoacrylate) nanosphere selected cells, while this was the reverse for the BCRP gene. Overexpression of the MRP1 and TOP2 alpha genes was also observed in the selected variants. Thus, drug encapsulation markedly alters or delays the several mechanisms involved in the acquisition of drug resistance [89].

Other strategies to bypass doxorubicin multidrug resistance, such as the use of Stealth™ poly(cyanoacrylate) nanoparticles, could be considered, as well as the co-administration of doxorubicin with chemo-sensitizing agents, generally acting as P-glycoprotein inhibitors. Co-encapsulation of the reversing agent cyclosporin and doxorubicin into poly(isobutyl cyanoacrylate) nanospheres was investigated to reduce the side-effects of both drugs while enhancing their efficacy. This formulation, compared to incubation of cyclosporin and doxorubicin, or doxorubicin nanoparticles and cyclosporin, elicited the most effective growth rate inhibition on

P388/ADR cells. Such a high efficacy was supposed to result from the synergistic effect of the rapid release of both doxorubicin and cyclosporin at the surface of cancer cells, allowing a better internalization of doxorubicin, while inhibiting its efflux by blocking the P-gp with the cyclosporin [90].

1.4.2
Antiestrogens

Nanospheres and nanocapsules made of biodegradable copolymers and coated with poly(ethylene glycol) (PEG) chains have been developed as parenteral delivery system for the administration of the anti-estrogen 4-hydroxytamoxifen RU 58668 (RU). Coating with PEG chains lengthened the anti-estrogen activity of RU, with prolonged antiuterotrophic activity of the encapsulated drug into PEG-poly(D,L-lactic acid) nanospheres as compared with non-coated nanospheres. In mice bearing MCF-7 estrogen-dependent tumors, free RU injected by the intravenous route slightly decreased estradiol-promoted tumor growth while RU-loaded PEG-poly-(D,L-lactic acid) nanospheres injected at the same dose strongly reduced it. The antitumoral activity of RU encapsulated within PEGylated nanocapsules was stronger than that of RU entrapped with PEGylated nanospheres loaded at an equivalent dose. The PEGylated nanocapsules decreased the tumor size in nude mice transplanted with estrogen receptor-positive but estrogen-independent MCF-7/Ras breast cancer cells at a concentration 2.5-fold lower than that of the PEGylated nanospheres [91, 92].

To increase the local concentration of tamoxifen in estrogen receptor-positive breast cancer cells poly(ε-caprolactone) nanoparticle formulation has been developed. Poly(ε-caprolactone) nanoparticles labeled with rhodamine123 were incubated with MCF-7 estrogen receptor-positive breast cancer cells. A significant fraction of the administered rhodamine123-loaded poly(ε-caprolactone) nanoparticles was found in the perinuclear region of the MCF-7 cells, where estrogen receptors are also localized, after 1 h of incubation. These nanoparticles were rapidly internalized in MCF-7 cells and intracellular tamoxifen concentrations followed a saturable process [93]. In another study, the biodistribution profile of tamoxifen encapsulated in polymeric nanoparticulate formulations after intravenous administration was evaluated, with or without surface-stabilizing agents. *In vivo* biodistribution studies of tamoxifen-loaded poly(ethylene oxide)-modified poly(ε-caprolactone) nanoparticles were carried out in Nu/Nu athymic mice bearing a human breast carcinoma xenograft, MDA-MB-231, using tritiated [^{3}H]tamoxifen as radio-marker for quantification. After intravenous administration the drug-loaded nanoparticles accumulated primarily in the liver, though up to 26% of the total activity could be recovered in tumor at 6 h post-injection for poly(ethylene oxide)-modified poly(ε-caprolactone) nanoparticles. In comparison with free drug and uncoated nanoparticles, the modified nanoparticles exhibited a significantly increased level of accumulation of the drug within the tumor with time as well as prolonged drug presence in the systemic circulation [94].

1.4.3
Anti-metabolites

Poly(amidoamine) dendritic polymers coated with poly(ethylene glycol) have been developed to deliver 5-fluorouracil. In rats after intravenous administration this nanoparticle formulation showed a lower drug clearance than after the free drug administration [95]. In addition, poly(D,L-lactide)-g-poly(N-isopropyl acrylamide-co-methacrylic acid) nanoparticles have been studied as drug carrier for intracellular delivery of 5-fluorouracil [96].

Methotrexate has been incorporated in modified poly(amidoamine) dendritic polymers conjugated to folic acid as a targeting agent. These conjugates were injected intravenously into immunodeficient mice bearing human KB tumors that overexpress the folic acid receptor. Targeting methotrexate increased its antitumor activity and markedly decreased its toxicity, allowing therapeutic responses not possible with the free drug [97].

1.4.4
Camptothecins

Solid lipid nanoparticles (SLNs) are a promising sustained release system for camptothecin after oral administration. The pharmacokinetics and body distribution of camptothecin after intravenous injection in mice have been studied. Two plasma peaks were observed after administration of camptothecin-loaded SLNs. The first was attributed to the presence of free drug, the second peak can be attributed to controlled release or potential gut uptake of the SLNs. In comparison to the drug solution, SLNs lead to a much higher area under the concentration–time curve/dose and mean residence times, especially in brain, heart and organs containing reticuloendothelial cells. The highest area under the concentration time curve ratio of SLNs to drug solution among the tested organs was found in the brain. Incorporation of camptothecin into SLNs also prevented its hydrolysis [98].

The *in vitro* and *in vivo* antitumor characteristics of methoxy poly(ethylene glycol)-poly(D,L-lactic acid) nanoparticles containing camptothecin have been examined. After intravenous administration in rats, camptothecin-loaded nanoparticles showed a longer plasma retention than camptothecin solution and high and long tumor localization. In both single and double administration to mice bearing sarcoma 180 solid tumor, camptothecin-loaded nanoparticles were much more effective than camptothecin solution, in particular the tumor disappeared completely in three of the four mice after double administration of camptothecin-loaded nanoparticles [99].

Irinotecan-containing nanoparticles have been prepared by co-precipitation with addition of water to an acetone solution of poly(D,L-lactic acid), PEG-*block*-poly(propylene glycol)-*block*-PEG and irinotecan. When the antitumor effect was examined using mice bearing sarcoma 180 subcutaneously, only nanoparticles suppressed tumor growth significantly. After intravenous injection in rats, nanoparticles maintained irinotecan plasma concentration longer than irinotecan aqueous solution [100].

Lipid based nanoparticles incorporating the irinotecan analog SN-38 have been developed and studied *in vitro* and *in vivo*. Interestingly, incorporation of SN-38 into nanoparticles improved the stability of the active drug, the lactone form, in serum-containing medium. Furthermore, studies in nude mice showed a prolonged half-life of the active drug in whole blood and increased efficacy compared to irinotecan in a mouse xenograft tumor model [101].

1.4.5
Cisplatin

Nanoparticles prepared from poly(lactide-*co*-glycolide) copolymers increase the circulating half-life of cisplatin [102]. A system for the local delivery of chemotherapy to malignant solid tumors has been developed based on calcium phosphate nanoparticles containing cisplatin. Cytotoxicity was investigated in a K8 clonal murine osteosarcoma cell line. Drug activity was retained after adsorption onto the apatite crystals and the apatite/cisplatin formulation exhibited cytotoxic effects with a dose-dependent decrease of cell viability [103].

1.4.6
Paclitaxel

Paclitaxel, a microtubule-stabilizing agent that promotes polymerization of tubulin causing cell death by disrupting the dynamics necessary for cell division, is effective against a wide spectrum of cancers, including ovarian cancer, breast cancer, small and non-small cell lung cancer, colon cancer, head and neck cancer, multiple myeloma, melanoma and Kaposi's sarcoma. In clinical practice high incidences of adverse reactions of the drug such as neurotoxicity, myelosuppression and allergic reactions have been reported. Since its clinical administration is hampered by its poor solubility in water, excipients such as Cremophor EL (polyethoxylated castor oil) and ethanol are used in the pharmaceutical drug formulation of the current clinical administration [104].

Cremophor EL and polysorbate 80 (Tween 80) are widely used as drug formulation vehicles for the taxane anticancer agents paclitaxel and docetaxel. Both solubilizers are biologically and pharmacologically active compounds, and their use as drug formulation vehicles has been implicated in clinically important adverse effects, including acute hypersensitivity reactions and peripheral neuropathy. Cremophor EL and Tween 80 have also been demonstrated to influence the disposition of solubilized drugs that are administered intravenously. The overall resulting effect is a highly increased systemic drug exposure and a simultaneously decreased clearance, leading to alteration in the pharmacodynamic characteristics of the solubilized drug. Kinetic experiments revealed that this effect is caused primarily by reduced cellular uptake of the drug from large spherical micellar-like structures with a highly hydrophobic interior, which act as the principal carrier of circulating drug.

The existence of Cremophor EL and Tween 80 in blood as large polar micelles has also raised additional complexities in the case of combination chemotherapy regimens with taxanes, such that the disposition of several co-administered drugs, including anthracyclines and epipodophyllotoxins, is significantly altered. In contrast to the enhancing effects of Tween 80, addition of Cremophor EL to the formulation of oral drug preparations seems to result in significantly diminished drug uptake and reduced circulating concentrations [105]. Nanoparticles of biodegradable polymers can provide an ideal solution to such an adjuvant problem and realize a controlled and targeted delivery of paclitaxel with better efficacy and less side-effects. With further development, such as particle size optimization and surface coating, nanoparticle formulation of paclitaxel can promote full paclitaxel efficacy, thereby improving the quality of life of the patients.

Paclitaxel incorporated into poly(vinylpyrrolidone) nanospheres has been assayed on a B16F10 murine melanoma transplanted subcutaneously in C57B1/6 mice. Mice treated with repeated intravenous injections of paclitaxel-loaded nanospheres showed a significant tumor regression and higher survival rates than mice treated with free paclitaxel [106]. Sterically stabilized SLNs were also prepared to prolong the blood circulation time following intravenous administration [107]. Incorporation of paclitaxel into SLNs enhanced paclitaxel cytotoxicity on the human breast adenocarcinoma cell line MCF-7, but not on the human promyelocytic leukemia cell line HL-60 [77].

In a phase I study, ABI-007 – a novel formulation prepared by high-pressure homogenization of paclitaxel in the presence of human serum albumin, which results in a nanoparticle colloidal suspension – was found to offer several clinical advantages, including a rapid infusion rate and a high maximum tolerated dose. Furthermore, the absence of Cremophor EL allowed ABI-007 to be administered to patients without a need for premedication that is routinely used to prevent the hypersensitivity reactions associated with the conventional formulation of paclitaxel. Hematologic toxicities were mild throughout treatment, while most non-hematologic toxicities were grade 1 or 2 [108].

The feasibility, maximum tolerated dose, and toxicities of intraarterial administration of ABI-007 were studied in patients with advanced head and neck and recurrent anal canal squamous cell carcinoma in 43 patients (31 with advanced head and neck and 12 with recurrent anal canal squamous cell carcinoma). Patients were treated intraarterially with paclitaxel-albumin nanoparticles every four weeks for three cycles. Paclitaxel albumin nanoparticles were compared preliminarily with paclitaxel for *in vitro* cytostatic activity. Significantly, pharmacokinetic profiles after intraarterial administration were obtained (Fig. 1.2). The dose-limiting toxicity of the nanoparticles formulation was myelosuppression consisting of neutropenia in three patients. Non-hematologic toxicities included total alopecia, gastrointestinal toxicity, skin toxicity, neurologic toxicity, ocular toxicity, flu-like. The maximum tolerated dose in a single administration was 270 mg m^{-2}. Most dose levels showed considerable antitumor activity (80.9% complete response and partial response) [109].

The effectiveness of intraarterial infusion of paclitaxel incorporated into human

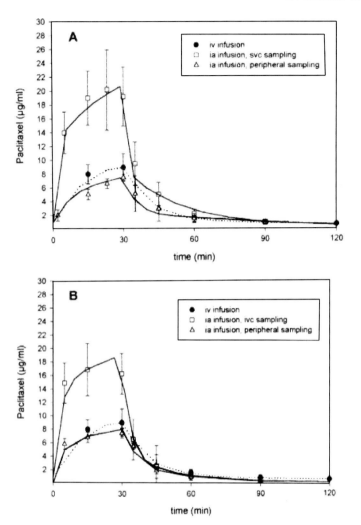

Fig. 1.2. Concentration–time curves. Mean paclitaxel concentration versus time profiles in patients with head and neck (A) or anal canal (B) carcinomas during and after 30 min constant infusion of ABI-007 (250 mg m^{-2} of paclitaxel); iv, intravenous; ia, intraarterial; svc, superior vena cava; ivc, inferior vena cava. (Reproduced with permission from Ref. [109].)

albumin nanoparticles for use as induction chemotherapy before definitive treatment of advanced squamous cell carcinoma of the tongue was also evaluated on 23 untreated patients who had carcinoma of the tongue. Each patient received two to four infusions, with a three-week interval between infusions. Sixteen patients underwent surgery, and of the remaining seven patients one received chemotherapy alone, four received radiotherapy alone, one received chemotherapy plus radiotherapy and one refused any further treatment. Eighteen patients (78%) had a clinical and radiologic objective response (complete, 26%; partial, 52%). Three patients (13%) showed stable disease and two (9%) showed disease progression. The toxicities encountered were hematologic in two patients (8.6%) and neurologic in two patients (8.6%) [110].

Furthermore, a recent phase III trial in patients with metastatic breast cancer compared ABI-007 with paclitaxel. ABI-007 resulted in significantly higher response rates and time to tumor progression than paclitaxel. Toxicity data showed that ABI-007 resulted in less grade 4 neutropenia than paclitaxel, and although the incidence of grade 3 sensory neuropathy was higher with ABI-007, the time until the neuropathy decreased to grade 2 was significantly less with ABI-007 compared with paclitaxel [111].

A cholesterol-rich microemulsion or nanoparticle termed LDE has been developed and its cytotoxicity, pharmacokinetics, toxicity to animals and therapeutic action has been compared with those of the commercial paclitaxel. The cytostatic activity of the drug in the complex was diminished compared with commercial paclitaxel due to the cytotoxicity of the vehicle Cremophor EL used in the commercial formulation. Competition experiments in neoplastic cultured cells showed that paclitaxel oleate and LDE were internalized together by the LDL receptor pathway. LDE-paclitaxel oleate arrested the G_2/M phase of cell cycle, similarly to commercial paclitaxel. Tolerability to mice was remarkable, such that the lethal dose (LD_{50}) was nine-fold greater than that of the commercial formulation. Furthermore, LDE concentrated paclitaxel oleate in the tumor four-fold relative to the normal adjacent tissues. At equimolar doses, the association of paclitaxel oleate with LDE resulted in remarkable changes in the drug pharmacokinetic parameters when compared to commercial paclitaxel, with increasing half-life, area under the time–concentration curve and diminishing clearance [112].

The *in vitro* antitumoral activity of a developed poly(lactic-*co*-glycolic acid) nanoparticle formulation incorporating paclitaxel-loaded has been assessed on a human small cell lung cancer cell line (NCI-H69 SCLC) and compared to the *in vitro* antitumoral activity of the commercial formulation. The release behavior of paclitaxel from the nanoparticles exhibited a biphasic pattern characterized by an initial fast release during the first 24 h, followed by a slower and continuous release. Incorporation of paclitaxel in these nanoparticles strongly enhanced the cytotoxic effect of the drug as compared to free drug, this effect being more relevant for prolonged incubation times [113]. In addition, nanoparticles of poly(lactic-*co*-glycolic acid) have been developed by a modified solvent extraction/evaporation technique, in which natural emulsifiers, such as phospholipids, cholesterol and vitamin E TPGS were applied. These natural emulsifiers showed great advantages for nanoparticle formulation of paclitaxel over the traditional macromolecular emulsifiers such as poly(vinyl alcohol). In the human adenocarcinoma cell line HT-29 the cytotoxicity caused by the drug administered after 24 h incubation was 13-fold higher than that caused by the free drug [114].

To improve paclitaxel use in intravesical therapy of superficial bladder cancer, paclitaxel-loaded gelatine nanoparticles have been developed. The excipient Cremophor EL contained in the commercial formulation of paclitaxel forms micelles that entrap the drug, reducing its partition across the urothelium. The paclitaxel-loaded nanoparticles were active against human RT4 bladder transitional cancer cells with IC_{50}s nearly identical to those of the commercial solution of paclitaxel. In dogs given an intravesical dose of paclitaxel-loaded particles, drug concentrations in the urothelium and lamina propria tissue layers, where Ta and T1 tumors would be located, were 2.6-fold greater than those reported for dogs treated with the Cremophor EL formulation [115].

The *in vitro* cytotoxicity, *in vivo* antitumor activity, pharmacokinetics, pharmacodynamics, and neurotoxicity of a micellar nanoparticle formulation of paclitaxel, NK105, has been compared with those of free paclitaxel. NK105 showed significantly potent antitumor activity on a human colorectal cancer cell line (HT-29) xenograft as compared with paclitaxel. The area under the time–concentration curve was approximately 90-fold higher for NK105 than for free paclitaxel. Leakage of paclitaxel from normal blood vessels was minimal and its capture by the reticuloendothelial system minimized. Thus, the tumor area under the time–concentration curve was 25-fold higher for NK105 than for free paclitaxel. Neurotoxicity was significantly weaker with NK105 than with free paclitaxel [116].

Paclitaxel-loaded biodegradable nanoparticles following conjugation to transferrin ligand have been developed to enhance the therapeutic efficacy of the encapsulated drug in the treatment of prostate cancer. The antiproliferative activity of nanoparticles was determined in a human prostate cancer cell line (PC3) and the effect on tumor inhibition in a murine model of prostate cancer. The IC_{50} of the drug with transferrin ligand conjugated nanoparticles was about five-fold lower than that with unconjugated nanoparticles or drug in solution. Animals that received an intratumoral injection of paclitaxel loaded transferrin ligand conjugated nanoparticles demonstrated complete tumor regression and greater survival rate than those that received either unconjugated nanoparticles or paclitaxel Cremophor EL formulation [117].

Poly(ethylene glycol)-coated biodegradable poly(cyanoacrylate) nanoparticles were also conjugated to transferrin for paclitaxel delivery. This nanoparticle formulation exhibited a markedly delayed blood clearance in mice and the paclitaxel level remained much higher at 24 h compared with that of free drug after paclitaxel intravenous injection. In S-180 solid tumor-bearing mice, tumor regression was significant with the actively targetable nanoparticles, and complete tumor regression occurred for five out of nine mice. In addition, the life span of tumor-bearing mice was significantly increased when they were treated with the nanoparticle formulation [118].

Paclitaxel is active against gliomas and various brain metastases, though its use in treating brain tumors is limited due to low blood–brain barrier permeability. The lack of paclitaxel brain uptake is thought to be associated with the P-glycoprotein efflux transporter. To improve paclitaxel brain uptake, paclitaxel was entrapped in ethyl alcohol/polysorbate nanoparticles. The paclitaxel nanoparticles cytotoxicity profile was monitored in two different cell lines, U-118 and HCT-15. Brain uptake of paclitaxel nanoparticles was evaluated using an *in situ* rat brain perfusion model. The results suggest that entrapment of paclitaxel in nanoparticles significantly increases the drug brain uptake and its toxicity toward P-glycoprotein-expressing tumor cells. It was hypothesized that paclitaxel nanoparticles could mask paclitaxel characteristics and thus limit its binding to P-glycoprotein, which consequently would lead to higher brain and tumor cell uptake of the otherwise effluxed drug [119].

1.4.7
Miscellaneous Agents

1.4.7.1 Arsenic Trioxide
Arsenic trioxide was considered as a novel antitumor agent. However, it also showed a severe toxicity effect on normal tissue. To improve its therapeutic efficacy and decrease its toxicity, arsenic trioxide-loaded albuminutes immuno-nanospheres targeted with monoclonal antibody (McAb) BDI-1 have been developed and its specific killing effect against bladder cancer cells (BIU-87) investigated. The albuminutes immuno-nanospheres were tightly junctioned with the BIU-87 cells and specific killing activity of bladder tumor cells was observed [120].

1.4.7.2 Butyric Acid
Butyric acid, a short-chain fatty acid naturally present in the human colon, regulates cell proliferation. It specifically modulates the expression of oncogenes such as c-myc, c-fos and H-ras, and various genes involved in the activation of apoptosis like p53 and bcl-2. The clinical applicability of the sodium salt of butyric acid is limited because of its short half-life of approximately 5 min. To improve its efficacy the pro-drug cholesterylbutyrate has been used as a lipid matrix to develop SLNs. The *in vitro* antiproliferative effect of cholesterylbutyrate SLNs was stronger than that of sodium butyrate in several human cell lines [121–123].

1.4.7.3 Cystatins
Cystatins can inhibit the tumor-associated activity of intracellular cysteine proteases cathepsins B and L and have been suggested as potential anticancer drugs. Chicken cystatin, a model protein inhibitor of cysteine proteases, in poly(lactide-*co*-glycolide) nanoparticles has been developed to improve its bioavailability and delivery into tumor cells. Poly(lactide-*co*-glycolide) nanoparticles and cystatin-loaded poly(lactic-*co*-glycolic acid) (PLGA) nanoparticles were cytotoxic towards mammary MCF-10A neoT cells, but free cystatin at the same concentrations was not. Poly(lactide-*co*-glycolide) nanoparticles were rapidly internalized into MCF-10A neoT cells, whereas the uptake of free cystatin was very slow. These nanoparticles are a useful carrier system. Delivery of the protein inhibitor into tumor cells was rapid and useful to inhibit intracellular proteolysis [124].

1.4.7.4 Diethylenetriaminepentaacetic Acid
Since chelating agents exhibit anticancer effects, the cytotoxicity of the extracellular chelator diethylenetriaminepentaacetic acid (DTPA) has been evaluated in breast

cancer cells, MCF-7, and neuroblastoma cells, UKF-NB-3. DTPA inhibited cancer cell growth in three-fold lower concentrations compared to human foreskin fibroblasts. The anticancer activity of chelating agents is caused by intracellular complexation of metal ions. DTPA was covalently coupled to human serum albumin nanoparticles and gelatin type B (GelB) nanoparticles to increase its cellular uptake. Coupling of DTPA to this drug carrier system increased its cytotoxic activity by five-fold [125].

1.4.7.5 Mitoxantrone

Mitoxantrone-loaded poly(butyl cyanoacrylate) nanoparticles have been tested in leukemia- or melanoma-bearing mice after intravenous injection. Efficacy and toxicity of mitoxantrone nanoparticles were compared with a drug solution and with a mitoxantrone-liposome formulation. The poly(butyl cyanoacrylate) nanoparticles and liposomes influenced the efficacy of mitoxantrone in cancer therapy in different ways. Liposomes prolonged survival time in P388 leukemia, whereas nanoparticles led to a significant tumor volume reduction in B16 melanoma. Neither nanoparticles nor liposomes were able to reduce the toxic side-effects caused by mitoxantrone, specifically leucocytopenia [126].

In a different study, mitoxantrone adsorbed onto poly(isobutyl cyanoacrylate) nanospheres, coated or not with poloxamine 1508, was administered by intravenous injection in B16 melanoma-bearing mice. In both cases, the observed tumor concentrations of mitoxantrone were high. However, the influence of the hydrophilic coating of the nanoparticles on the biodistribution and pharmacokinetics was minor. Moreover, the non-adsorbed drug was not removed from the nanoparticle preparation and the hydrophilic coating was rapidly desorbed *in vivo* [127].

1.4.8
Gene Therapy

Delivery systems are necessary for molecules such as antisense oligonucleotides since they are susceptible to nuclease-mediated degradation in the circulation and penetrate poorly through the membranes. They are also susceptible to nuclease attack within the lysosomes and their site of action is either in the cytoplasm in the case of an antisense strategy or in the nucleus for gene replacement or antigene therapy.

Antisense oligonucleotides are molecules that can inhibit gene expression being potentially active for the treatment of cancer. Short nucleic acid sequences specific to oncogene targets such as bcl-2, bcr-abl and c-myc have been shown to exhibit specific anticancer activity *in vitro* through antigene or antisense activity. However, their negative charge seriously hinders the intracellular penetration of these short fragments of nucleic acid. Efficient *in vivo* delivery of oligonucleotides remains a major limitation for their therapeutic application.

To prevent the degradation of oligonucleotides and improve their intracellular capture, it was proposed to associate them with nanoparticles. Oligonucleotides associated to nanoparticles were shown to be protected against degradation and to penetrate more easily into different types of cells. Poly(alkyl cyanoacrylate) nanoparticles mainly release their drug content by biodegradation, thus rendering the release profile of an entrapped compound independent of its physicochemical characteristics. After intratumoral injection, the bioavailability of oligonucleotides is seriously reduced due to their fast degradation by ubiquitous exo- and endonucleases [128].

For instance, a carrier system consisting of a cationic hydrophobic detergent (cetyltrimethylammonium bromide, CTAB), which interacted with the oligonucleotides by ion-pairing, was developed. *In vivo*, HBL100ras1 cells implanted in nude mice were treated intratumorally with several formulations of oligonucleotides targeted against Has-ras oncogene. Tumor growth inhibition was achieved at concentrations 100-fold lower than those needed with free oligonucleotides, when the oligonucleotides-CTAB complex was adsorbed onto the surface of the poly(isohexyl cyanoacrylate) nanospheres. Interestingly, the oligonucleotides-CTAB complex alone exerted no effect on HBL1000ras1 cell proliferation. Analysis of the amount of intact intracellular oligonucleotides, in cell culture experiments, revealed concentrations 100-fold higher in cells treated with oligonucleotides-CTAB adsorbed onto nanospheres [129].

The nanospheres were able to enhance oligonucleotides cell internalization and to protect oligonucleotides from both rapid cell internalization and rapid intracellular breakdown, which led to a considerably higher intracellular concentration of intact oligonucleotides and to a more efficient antisense activity. However, the oligonucleotides release was followed by the release of the detergent CTAB, which, at high intracellular concentrations, could induce cell toxicity.

Cholesterol-modified oligonucleotides, capable of direct adsorption onto poly(alkyl cyanoacrylate) nanospheres without the need for potentially toxic intermediates, have also been tested, but they proved less able to inhibit T24 human bladder carcinoma cells proliferation in culture than the system previously described [130].

To circumvent CTAB toxicity while maintaining an efficient biological activity, another approach was developed. Functional nanospheres were obtained by free radical emulsion polymerization of methyl methacrylate using the quaternary ammonium salt of 2-(dimethylamino)ethyl methacrylate as the reactive emulsifier [131].

The cationic dextran derivative diethylaminoethyl-dextran, by formation of an ion pair at the oligonucleotides surface, was developed to replace CTAB and was associated with poly(isohexyl cyanoacrylate) nanospheres. Poly(isohexyl cyanoacrylate) nanoparticles, with an aqueous core containing oligonucleotides, were prepared by interfacial polymerization of isobutyl cyanoacrylate in water/oil emulsion. Further studies also demonstrated that nanoencapsulation was able to protect oligonucleotides against degradation by serum nucleases contained in the cell culture medium [132].

For *in vivo* studies, phosphorothioate oligonucleotides against EWS Fli-1 chimeric RNA were encapsulated into cyanoacrylate nanocapsules and their efficacy was tested on experimental Ewing sarcoma, after intratumoral administration. Oligonucleotides nanocapsules led to more efficient tumor growth inhibition. The encapsulation of oligonucleotides provided enhanced protection against *in vivo* degradation, resulting in a higher number of oligonucleotides available. Cyanoacrylate nanocapsules yielded a higher ratio of oligonucleotides targeting the tumor cells [133].

In vivo, poly(alkyl cyanoacrylate) nanoparticles were able to efficiently distribute the oligonucleotides to the liver and to improve, in mice, the treatment of RAS cells expressing the point mutated Ha-ras gene. As all the oligonucleotides studied had a specific therapeutic efficacy, this means that the oligonucleotides delivered by nanoparticles escape from the lysosome compartment before degradation [134].

Preclinical studies have shown that DOTAP:cholesterol nanoparticles are effective systemic gene delivery vectors that efficiently deliver tumor suppressor genes to disseminated lung tumors. A phase-I trial for systemic treatment of lung cancer using a novel tumor suppressor gene, FUS1, has been initiated. Although DOTAP:cholesterol nanoparticles complexed to DNA are efficient vectors for systemic therapy, induction of an inflammatory response in a dose-dependent manner has also been observed, thus limiting its use. Systemic administration of DNA nanoparticles induced multiple signaling molecules both *in vitro* and *in vivo* that are associated with inflammation. Use of small molecule inhibitors against the signaling molecules resulted in their suppression and thereby reduced inflammation without affecting transgene expression [135]. Nanoparticles coated with ligands such as transferrin and epidermal growth factor (EGF) have also been developed in gene delivery to target selectively the tumor cells [136]. Complexes for DNA delivery composed of polyethylenimine (polyplexes) linked to poly(ethylene glycol) and coated with transferrin or EGF were prepared with fixed nanoparticle diameters. Intravenous injection of the transferrin-coated polyplexes resulted in gene transfer to subcutaneous neuroblastoma tumors in syngenic mice, and intravenous injection of the EGF-coated polyplexes targeted human hepatocellular carcinoma xenografts in SCID mice. In these models, luciferase marker gene expression levels in tumor tissues were 10- to 100-fold higher than in other organ tissues. Repeated systemic application of the transferrin polyplexes encoding tumor necrosis factor alpha (TNF-α) into tumor-bearing mice induced tumor necrosis and inhibition of tumor growth in different murine tumor models [137].

Transferrin-modified nanoparticles containing DNAzymes (short catalytic single-stranded DNA molecules) for tumor targeting have been developed. Linear, β-cyclodextrin-based polymers were complexed with DNAyzme molecules to form sub-50 nm particles termed "polyplexes". Adamantane forms inclusion complexes with the surface cyclodextrins of the polyplexes, allowing a sterically stabilizing layer of poly(ethylene glycol) to be added. The stabilized polyplexes were also modified with transferrin to increase the targeting to tumor cells expressing transferrin receptors. Administration by intraperitoneal bolus, infusion, intravenous bolus and subcutaneous injection were studied in tumor-bearing nude mice. DNAzymes packaged in polyplex formulations were concentrated and retained in tumor tissue and other organs, whereas unformulated DNAzyme was eliminated from the body within 24 h post-injection. Tumor cell uptake was observed with intravenous bolus injection only and intracellular delivery requires transferrin targeting [138].

Potent sequence selective gene inhibition by siRNA is also hindered by poor intracellular uptake, limited blood stability and non-specific immune stimulation. Thus, ligand-targeted, sterically stabilized nanoparticles have been developed for siRNA. Self-assembling nanoparticles with siRNA were constructed with poly-(ethyleneimine) that is PEGylated with an Arg-Gly-Asp peptide ligand, as a means to target tumor neovasculature expressing integrins, and used to deliver siRNA inhibiting vascular endothelial growth factor receptor-2 (VEGF R2) expression and thereby tumor angiogenesis. Intravenous administration into tumor-bearing mice gave selective tumor uptake, siRNA sequence-specific inhibition of protein expression within the tumor and inhibition of both tumor angiogenesis and growth rate [139].

References

1 I. Brigger, C. Dubernet, P. Couvreur, Nanoparticles in cancer therapy and diagnosis, *Adv. Drug Deliv. Rev.* **2002**, 54, 631–651.
2 R.K.Y. Zee-Cheng, C.C. Cheng, Delivery of anticancer agents, *Meth. Find. Exp. Clin. Pharmacol.* **1989**, 11, 439–529.
3 G. Barratt, Colloidal drug carriers: Achievements and perspectives, *Cell. Mol. Life Sci.* **2003**, 60, 21–37.
4 R.K. Jain, Transport of molecules in the tumor interstitium: A review, *Cancer Res.* **1987**, 47, 3039–3051.
5 D. Baban, L.W. Seymour, Control of tumor vascular permeability, *Adv. Drug Deliv. Rev.* **1998**, 34, 109–119.
6 S.K. Hobbs, W.L. Monsky, F. Yuan, W.G. Roberts, L. Griffith, V.P. Torchilin, R.K. Jain, Regulation of transport pathway in tumor vessels: Role of tumor type and microenvironment, *Proc. Natl. Acad. Sci. U.S.A.* **1998**, 95, 4607–4612.
7 H. Maeda, The enhanced permeability and retention (EPR) effect in tumor vasculature: The key role of tumor-selective macromolecular drug targeting, *Adv. Enzyme Regul.* **2001**, 41, 189–207.
8 R. Krishna, L.D. Mayer, Multidrug resistance (MDR) in cancer- mechanisms, reversal using modulators of MDR and the role of MDR modulators in influencing the pharmacokinetics of anticancer drugs, *Eur. J. Cancer Sci.* **2000**, 11, 265–283.
9 A.K. Larsen, A.E. Escargueil, A. Skladanowski, Resistance mechanisms associated with altered intracellular distribution of anticancer agents, *Pharmacol. Ther.* **2000**, 88, 217–229.
10 C. Vauthier, C. Dubernet, C. Chauvierre, I. Brigger, P. Couvreur, Drug delivery to resistant tumors: The potential of poly(alkyl cyanoacrylate) nanoparticles, *J. Control. Release* **2003**, 93, 151–160.
11 J. Sinek, H. Frieboes, X. Zheng, V. Cristini, Two-dimensional chemotherapy simulations demonstrate fundamental transport and tumor response limitations involving nanoparticles, *Biomed. Microdevices* **2004**, 6, 297–309.
12 V.S. Shenoy, I.K. Vijay, R.S. Murthy, Tumour targeting: Biological factors and formulation advances in injectable lipid nanoparticles, *J. Pharm. Pharmacol.* **2005**, 57, 411–422.
13 B. Rihova, Immunomodulating activities of soluble synthetic polymer-

bound drugs, *Adv. Drug Deliv. Rev.* **2002**, 54, 653–674.

14 G. Storm, S.O. Belliot, T. Daemen, D.D. Lasic, Surface modification of nanoparticles to oppose uptake by the mononuclear phagocyte system, *Adv. Drug Deliv. Rev.* **1995**, 17, 31–48.

15 G.M. Barrat, F. Puisieux, W.P. Yu, C. Foucher, H. Fessi, J.P. Devissaguet, Anti-metastatic activity of MDP-L-alanyl-cholesterol incorporated into various types of nanocapsules, *Int. J. Immunopharm.* **1994**, 457–461.

16 P. Couvreur, B. Kante, L. Grislain, M. Roland, P. Speiser, Toxicity of polyalkykyanoacrylate nanoparticles II: Doxorubicin-loaded nanoparticles, *J. Pharm. Sci.* **1982**, 71, 790–792.

17 P.M. Blagoeva, R.M. Balansky, T.J. Mircheva, M.I. Simeonova, Diminished genotoxicity of mitomycin C and farmorubicin included in polybutylcyanoacrylate nanoparticles, *Mut. Res.* **1992**, 268, 77–82.

18 C. Verdun, F. Brasseur, H. Vranckx, P. Couvreur, M. Roland, Tissue distribution of doxorubicin associated with polyisohexylcyanoacrylate nanoparticles, *Cancer Chemother. Pharmacol.* **1990**, 26, 13–18.

19 R. Fernandez-Urrusuno, E. Fattal, J.M. Rodrigues, J. Féger, P. Bedossa, P. Couvreur, Effect of polymeric nanoparticle administration on the clearance activity of the mononuclear phagocyte system in mice, *J. Biomed. Mater. Res.* **1996**, 31, 401–408.

20 F. Fawaz, F. Bonini, M. Guyot, A.M. Lagueny, H. Fessi, J.P. Devissaguet, Influence of poly(DL-lactide) nanocapsules on the biliary clearance and enterohepatic circulation of indomethacin in the rabbit, *Pharm. Res.* **1993**, 10, 750–766.

21 S. Gibaud, J.P. Andreux, C. Weingarten, M. Renard, P. Couvreur, Increased bone marrow toxicity of doxorubicin bound to nanoparticles, *Eur. J. Cancer* **1994**, 30A, 820–826.

22 S.M. Moghimi, A.C. Hunter, J.C. Murray, Long-circulating and target-specific nanoparticles: Theory to practice, *Pharmacol. Rev.* **2001**, 53, 283–318.

23 S. Gibaud, C. Rousseau, C. Weingarten, R. Favier, L. Douay, J.P. Andreux, P. Couvreur, Polyalkylcyanoacrylate nanoparticles as carriers for granulocyte colony stimulating factor (G-CSF), *J. Control. Release* **1998**, 52, 131–139.

24 L. Illum, L.O. Jacobsen, R.H. Müller, R. Mak, S.S. Davis, Surface characteristics and the interaction of colloidal particles with mouse peritoneal macrophages, *Biomaterials* **1987**, 8, 113–117.

25 S.I. Jeon, J.H. Lee, J.D. Andrade, P.G. de Gennes, Protein-surface interactions in the presence of polyethylene oxide: Simplified theory, *J. Colloid. Interface Sci.* **1991**, 142, 149–158.

26 R. Gref, A. Domb, P. Quellec, T. Blunk, R.H. Müller, J.M. Verbavatz, R. Langer, The controlled intravenous delivery of drugs using PEG-coated sterically stabilized nanospheres, *Adv. Drug Deliv. Rev.* **1995**, 16, 215–233.

27 D. Bazile, C. Prud'Homme, M.T. Bassoulet, M. Marlard, G. Spenlehauer, M. Veillard, Stealth Me.PEG-PLA nanoparticles avoid uptake by the mononuclear phagocyte system, *J. Pharm. Sci.* **1995**, 84, 493–498.

28 V.C.F. Mosqueira, P. Legrand, A. Gulik, O. Bourdon, R. Gref, D. Labarre, G. Barratt, Relationship between complement activation, cellular uptake and surface physicochemical aspects of novel PEG-modified nanocapsules, *Biomaterials* **2001**, 22, 2967–2979.

29 V. Lenaerts, A. Labib, F. Choinard, J. Rousseau, H. Ali, J. van Lier, Nanocapsules with a reduced liver uptake: Targeting of phthalocyanines to EMT-6 mouse mammary tumor *in vivo*, *Eur. J. Pharm. Biopharm.* **1995**, 41, 38–43.

30 O. Bourdon, V. Mosqueira, P. Legrand, J. Blais, A comparative study of the cellular uptake,

localization and phototoxicity of meta-tetra(hydroxyphenyl) chlorin encapsulated in surface-modified submicronic oil/water carriers in HT29 tumor cells, *J. Photochem. Photobiol. B* **2000**, 55, 164–171.

31 M.T. PERACCHIA, C. VAUTHIER, F. PUISIEUX, P. COUVREUR, Development of sterically stabilized poly(isobutyl 2-cyanoacrylate) nanoparticles by chemical coupling of poly(ethylene glycol), *J. Biomed. Mater. Res.* **1997**, 34, 317–326.

32 I. BRIGGER, P. CHAMINADE, V. MARSAUD, M. APPEL, M. BESNARD, R. GURNY, M. RENOIR, P. COUVREUR, Tamoxifen encapsulation within polyethylen glycol-coated nanospheres: A new antiestrogen formulation, *Int. J. Pharm.* **2001**, 214, 37–42.

33 S.E. GHELPERINA, Z.S. SMIRNOVA, A.S. KHALANSKIY, I.N. SKIDAN, A.I. BOBRUSKIN, J. KREUTER, Chemotherapy of brain tumours using doxorubicin bound to polysorbate 80-coated nanoparticles, *Proceedings of 3rd World Meeting APV/APGI*, Berlin, Germany **2000**, 441–442.

34 J. KREUTER, Influence of the surface properties on nanoparticle-mediated transport of drugs to the brain, *J. Nanosci. Nanotechnol.* **2004**, 4, 484–488.

35 E. ALLÉMANN, J. ROUSSEAU, N. BRASSEUR, S.V. KUDREVICH, K. LEWIS, J.E. VAN LIER, Photodynamic therapy of tumours with hexadecafluoro zinc phthalocyanine formulated in PEG-coated poly(lactic acid) nanoparticles, *Int. J. Cancer* **1996**, 66, 821–824.

36 J. LODE, I. FICHTNER, J. KREUTER, A. BERNDT, J.E. DIEDERICHS, R. RESZKA, Influence of surface-modifying surfactants on the pharmacokinetic behaviour of 14C-poly(methylmethacrylate) nanoparticles in experimental tumor models, *Pharm. Res.* **2001**, 18, 1613–1619.

37 M.T. PERACCHIA, C. VAUTHIER, D. DESMAËLE, A. GULIK, J.C. DEDIEU, M. DEMOY, J. D'ANGELO, P. COUVREUR, Pegylated nanoparticles from a novel methoxypolyethylene glycol cyanoacrylate-hexadecyl cyanoacrylate amphiphilic copolymer, *Pharm. Res.* **1998**, 15, 550–556.

38 S.E. DUNN, A.G.A. COOMBES, M.C. GARNETT, S.S. DAVIS, M.C. DAVIES, L. ILLUM, In vitro interaction and in vivo biodistribution of poly(lactide-glycolide) nanospheres surface modified by poloxamer and poloxamine copolymers, *J. Control. Release* **1997**, 44, 65–76.

39 S. STOLNIK, S.E. DUNN, M.C. GARNETT, M.C. DAVIES, A.G.A. COOMBES, D.C. TAYLOR, M.P. IRVING, S.C. PURKISS, T.F. TADROS, S.S. DAVIS, Surface modification of poly(lactide-co-glycolide) nanospheres by biodegradable poly(lactide)-poly(ethylene glycol) copolymers, *Pharm. Res.* **1994**, 11, 1800–1808.

40 V.C.F. MOSQUEIRA, P. LEGRAND, R. GREF, B. HEURTAULT, M. APPEL, G. BARRATT, Interactions between a macrophage cell line (J774A1) and surface-modified poly(D,L-lactide) nanocapsules bearing poly(ethylene glycol), *J. Drug Target.* **1999**, 7, 65–78.

41 V.C.F. MOSQUEIRA, P. LEGRAND, A. GULIK, O. BOURDON, R. GREF, D. LABARRE, G. BARRATT, Biodistribution of novel long circulating PEG-grafted nanocapsules in mice: Effects of PEG chain length and density, *Pharm. Res.* **2001**, 18, 1411–1419.

42 B. STELLA, S. ARPICCO, M.T. PERACCHIA, D. DESMAËLE, J. HOEBENE, M. RENOIR, J. D'ANGELO, L. CATTEL, P. COUVREUR, Design of folic acid-conjugated nanoparticles for drug targeting, *J. Pharm. Sci.* **2000**, 89, 1452–1464.

43 L. ZHANG, S. HOU, S. MAO, D. WEI, X. SONG, Y. LU, L. ZHANG, S. HOU, S. MAO, D. WEI, X. SONG, Y. LU, Uptake of folate-conjugated albumin nanoparticles to the SKOV3 cells, *Int. J. Pharm.* **2004**, 287, 155–162.

44 T. SHIOKAWA, Y. HATTORI, K. KAWANO, Y. OHGUCHI, H. KAWAKAMI, K. TOMA, Y. MAITANI, Effect of polyethylene glycol linker chain length of folate-linked microemulsions loading aclacinomycin A on targeting ability and antitumor effect *in vitro*

and *in vivo*, *Clin. Cancer Res.* **2005**, 11, 2018–2025.
45 N.C. BELLOCQ, S.H. PUN, G.S. JENSEN, M.E. DAVIS, Transferrin-containing, cyclodextrin polymer-based particles for tumor-targeted gene delivery, *Bioconj. Chem.* **2003**, 14, 1122–1132.
46 H. WARTLICK, K. MICHAELIS, S. BALTHASAR, K. STREBHARDT, J. KREUTER, K. LANGER, Highly specific HER2-mediated cellular uptake of antibody-modified nanoparticles in tumour cells, *J. Drug Target.* **2004**, 12, 461–471.
47 E.M. GORDON, F.L. HALL, Nanotechnology blooms, at last, *Oncol. Rep.* **2005**, 13, 1003–1007.
48 F.B. LANDRY, D.V. BAZILE, G. SPENLEHAUER, M. VEILLARD, J. KREUTER, Peroral administration of 14C-poly-(D,L-lactic acid) nanoparticles coated with human serum albumin or polyvinyl alcohol to guinea pigs, *J. Drug Target* **1998**, 6, 293–307.
49 G. PONCHEL, J.M. IRACHE, Specific and non-specific bioadesive particulate systems for oral delivery to the gastrointestinal tract, *Adv. Drug Deliv. Rev.* **1998**, 34, 191–219.
50 L. MU, S.S. FENG, A novel controlled release formulation for the anticancer drug paclitaxel (Taxol): PLGA nanoparticles containing vitamin E TPGS, *J. Control. Release* **2003**, 86, 33–48.
51 Y. NISHIOKA, H. YOSHINO, Lymphatic targeting with nanoparticulate system, *Adv. Drug Deliv. Rev.* **2001**, 47, 55–64.
52 C. SAKAKURA, T. TAKAHASHI, K. SAWAI, A. HAGIWARA, M. ITO, S. SHOBAYASHI, S. SASAKI, K. OZAKI, M. SHIRASU, Enhancement of therapeutic efficacy of aclarubicin against lymph node metastases using a new dosage form: Aclarubicin adsorbed on activated carbon particles, *Anti-Cancer Drugs* **1992**, 3, 233–236.
53 A. HAGIWARA, T. TAKAHASHI, K. SAWAI, C. SAKAKURA, M. SHIRASU, M. OHGAKI, T. IMANASHI, J. YAMASAKI, Y. TAKEMOTO, N. KAGEYAMA, Selective drug delivery to peri-tumoral region and regional lymphatics by local injection of aclarubicin adsorbed on activated carbon particles in patients with breast cancer – a pilot study. *Anti-Cancer Drugs* **1997**, 8, 666–670.
54 A.E. HAWLEY, S.S. DAVIS, L. ILLUM, Targeting of colloids to lymph nodes: Influence of lymphatic physiology and colloidal characteristics, *Adv. Drug Deliv. Rev.* **1995**, 17, 129–148.
55 H. TOKUMITSU, J. HIRATSUKA, Y. SKURAI, T. KOBAYASHI, H. ICHIKAWA, Y. FUKUMORI, Gadolinium neutron-capture therapy using novel gadopentetic acid-chitosan complex nanoparticles: *In vivo* growth suppression of experimental melanoma solid tumor, *Cancer Lett.* **2000**, 150, 177–182.
56 P. CALVO, J.L. VILA-JATO, M.J. ALONSO, Evaluation of cationic polymer-coated nanocapsules as ocular drug carriers, *Int. J. Pharm.* **1997**, 153, 41–50.
57 G. MINOTTI, P. MENNA, E. SALVATORELLI, G. CAIRO, L. GIANNI, Anthracyclines: Molecular advances and pharmacological developments in antitumor activity and cardiotoxicity. *Pharmacol. Rev.* **2004**, 56, 185–229.
58 N. CHIANNILKULCHAI, Z. DRIOUICH, J.P. BENOIT, A.L. PARODI, P. COUVREUR, Doxorubicin-loaded nanoparticles: Increased efficiency in murine hepatic metastasis, *Sel. Cancer Ther.* **1989**, 5, 1–11.
59 N. CHIANNILKULCHAI, N. AMMOURY, B. CAILLOU, J.Ph. DEVISSAGUET, P. COUVREUR, Hepatic tissue distribution of doxorubicin-loaded particles after i.v. administration in reticulosarcoma M 5076 metastasis-bearing mice, *Cancer Chemother. Pharmacol.* **1990**, 26, 122–126.
60 A. ROLLAND, Clinical pharmaco-kinetics of doxorubicin in hepatoma patients after a single intravenous injection of free or nanoparticle-bound anthracycline, *Int. J. Pharm.* **1989**, 54, 113–121.
61 J. KATTAN, J.P. DROZ, P. COUVREUR, J.P. MARINO, A. BOUTAN-LAROZE, P. ROUGIER, P. BRAULT, H. VRANCKX, J.M. GROGNET, X. MORGE, H. SANCHO-GARNIER, Phase I clinical trial and pharmacokinetics evaluation of doxorubicin carried by poly-

isohexylcyanoacrylate nanoparticles, *Invest. New Drugs* **1992**, 10, 191–199.
62 C.E. SOMA, C. DUBERNET, G. BARRATT, S. BENITA, P. COUVREUR, Investigation of the role of macrophages on the cytotoxicity of doxorubicin and doxorubicin-loaded nanoparticles on M5076 cells *in vitro*, *J. Control. Release* **2000**, 68, 283–289.
63 S. MITRA, U. GAUR, P.C. GOSH, A.N. MAITRA, Tumor targeted delivery of encapsulated dextran-doxorubicin conjugate using chitosan nanoparticles as carrier, *J. Control. Release* **2001**, 74, 317–323.
64 J.H. CHEN, R. LING, Q. YAO, L. WANG, Z. MA, Y. LI, Z. WANG, H. XU, Enhanced antitumor efficacy on hepatoma-bearing rats with adriamycin-loaded nanoparticles administered into hepatic artery. *World J. Gastroenterol.* **2004**, 10(13), 1989–1991.
65 J.H. CHEN, L. WANG, R. LING, Y. LI, Z. WANG, Q. YAO, Z. MA, Body distribution of nanoparticle-containing adriamycin injected into the hepatic artery of hepatoma-bearing rats, *Dig. Dis. Sci.* **2004**, 49, 1170–1173.
66 L. MANIL, P. COUVREUR, P. MAHIEU, Acute renal toxicity of doxorubicin (adryamycin)-loaded cyanoacrylate nanoparticles, *Pharm. Res.* **1995**, 12, 85–87.
67 E. LEO, R. ARLETTI, F. FORNI, R. CAMERONI, General and cardiac toxicity of doxorubicin-loaded gelatin nanoparticles, *Il Farmaco* **1997**, 52, 385–388.
68 J. KREUTER, Nanoparticulate systems for brain delivery of drugs, *Adv. Drug Deliv. Res.* **2000**, 47, 65–81.
69 S.E. GELPERINA, A.S. KHALANSKY, I.N. SKIDAN, Z.S. SMIRNOVA, A.I. BOBRUSKIN, S.E. SEVERIN, B. TUROWSKI, F.E. ZANELLA, J. KREUTER, Toxicological studies of doxorubicin bound to polysorbate 80-coated poly(butyl cyanoacrylate) nanoparticles in healthy rats and rats with intracranial glioblastoma, *Toxicol. Lett.* **2002**, 126, 131–141.
70 S.C. STEINIGER, J. KREUTER, A.S. KHALANSKY, I.N. SKIDAN, A.I. BOBRUSKIN, Z.S. SMIRNOVA, S.E. SEVERIN, R. UHL, M. KOCK, K.D. GEIGER, S.E. GELPERINA, Chemotherapy of glioblastoma in rats using doxorubicin-loaded nanoparticles, *Int. J. Cancer* **2004**, 109, 759–767.
71 H.S. YOO, K.H. LEE, J.E. OH, T.G. PARK, *In vitro* and *in vivo* anti-tumor activities of nanoparticles based on doxorubicin-PLGA conjugates, *J. Control. Release* **2000**, 68, 419–431.
72 L.H. REDDY, R.K. SHARMA, R.S. MURTHY, Enhanced tumour uptake of doxorubicin loaded poly(butyl cyanoacrylate) nanoparticles in mice bearing Dalton's lymphoma tumour, *J. Drug Target* **2004**, 12, 443–451.
73 I. BRIGGER, J. MORIZET, L. LAUDANI, G. AUBERT, M. APPEL, V. VELASCO, M.J. TERRIER-LACOMBE, D. DESMAELE, J. D'ANGELO, P. COUVREUR, G. VASSAL, Negative preclinical results with stealth nanospheres-encapsulated doxorubicin in an orthotopic murine brain tumor model, *J. Control. Release* **2004**, 100, 29–40.
74 R.H. MÜLLER, K. MÄDER, S. GOHLA, Solid lipid nanoparticles (SLN) for controlled drug delivery – a review of the state of the art, *Eur. J. Pharm. Biopharm.* **2000**, 50, 161–177.
75 R.N. ALYAUTDIN, V.E. PETROV, K. LANGER, A. BERTHOLD, D.A. KHARKEVICH, J. KREUTER, Delivery of loperamide across the blood-brain barrier with Polysorbate 80-coated polybutylcyanoacrylate nanoparticles, *Pharm. Res.* **1997**, 14, 325–328.
76 G.P. ZARA, R. CAVALLI, A. FUNDARÒ, A. BARGONI, O. CAPUTO, M.R. GASCO, Pharmacokinetics of doxorubicin incorporated in solid lipid nanospheres (SLN), *Pharm. Res.* **1999**, 44, 281–286.
77 A. MIGLIETTA, R. CAVALLI, C. BOCCA, L. GABRIEL, M.R. GASCO, Cellular uptake and cytotoxicity of solid lipid nanospheres (SLN) incorporating doxorubicin or paclitaxel, *Int. J. Pharm.* **2000**, 210, 61–67.
78 L. SERPE, M.G. CATALANO, R. CAVALLI, E. UGAZIO, O. BOSCO, R. CANAPARO, E. MUNTONI, R. FRAIRIA, M.R. GASCO, M. EANDI, G.P. ZARA, Cytotoxicity of

79. G.P. Zara, R. Cavalli, A. Bargoni, A. Fundarò, D. Vighetto, M.R. Gasco, Intravenous administration to rabbits of non stealth and stealth doxorubicin-loaded solid lipid nanoparticles at increasing concentrations of stealth agent: Pharmacokinetics and distribution of doxorubicin in brain and other tissues, *J. Drug Target* **2002**, 10, 327–335.

anticancer drugs incorporated in solid lipid nanoparticles on HT-29 colorectal cancer cell line, *Eur. J. Pharm. Biopharm.* **2004**, 58, 673–680.

80. A. Fundarò, R. Cavalli, A. Bargoni, D. Vighetto, G.P. Zara, M.R. Gasco, Non-stealth and stealth solid lipid nanoparticles (SLN) carrying doxorubicin: Pharmacokinetics and tissue distribution after i.v. administration to rats, *Pharm. Res.* **2000**, 42, 337–343.

81. G.P. Zara, A. Bargoni, R. Cavalli, A. Fundarò, D. Vighetto, M.R. Gasco, Pharmacokinetics and tissue distribution of idarubicin-loaded solid lipid nanoparticles after duodenal administration to rats, *J. Pharm. Sci.* **2002**, 91, 1324–1333.

82. A.C. de Verdière, C. Dubernet, F. Némati, E. Soma, M. Appel, J. Ferté, S. Bernard, F. Puisieux, P. Couvreur, Reversion of multidrug resistance with polyalkylcyanoacrylate nanoparticles: Towards a mechanism of action, *Br. J. Cancer* **1997**, 76, 198–205.

83. A. Astier, B. Doat, M.J. Ferrer, G. Benoit, J. Fleury, A. Rolland, R. Leverge, Enhancement of adriamycin antitumor activity by its binding with an intracellular sustained-release form, polymethacrylate nanospheres, in U-937 cells, *Cancer Res.* **1988**, 48, 1835–1841.

84. S. Bennis, C. Chapey, P. Couvreur, J. Robert, Enhanced cytotoxicity of doxorubicin encapsulated in polyisohexylcyanoacrylate nanospheres against multi-drug-resistant tumour cells in culture, *Eur. J. Cancer* **1994**, 30A, 889–893.

85. A.C. de Verdière, C. Dubernet, F. Némati, M.F. Poupon, F. Puisieux, P. Couvreur, Uptake of doxorubicin from loaded nanoparticles in multidrug-resistant leukemic murine cells, *Cancer Chemother. Pharmacol.* **1994**, 33, 504–508.

86. X. Pépin, L. Attali, C. Domrault, S. Gallet, J.M. Metreau, Y. Reault, P.J. Cardot, M. Imalalen, C. Dubernet, E. Soma, P. Couvreur, On the use of ion-pair chromatography to elucidate doxorubicin release mechanism from polyalkylcyanoacrylate nanoparticles at the cellular level, *J. Chromatogr. B* **1997**, 702, 181–187.

87. J. Nafziger, G. Averland, E. Bertounesque, G. Gaudel, C. Monneret, Synthesis and antiproliferative effects of a 4′-morpholino-9-methyl anthracycline, *J. Antibiot.* **1995**, 48, 1185–1187.

88. C.E. Soma, C. Dubernet, G. Barratt, F. Nemati, M. Appel, S. Benita, P. Couvreur, Ability of doxorubicin-loaded nanoparticles to overcome multidrug resistance of tumor cells after their capture by macrophages, *Pharm. Res.* **1999**, 16, 1710–1716.

89. A. Laurand, A. Laroche-Clary, A. Larrue, S. Huet, E. Soma, J. Bonnet, J. Robert, Quantification of the expression of multidrug resistance-related genes in human tumour cell lines grown with free doxorubicin or doxorubicin encapsulated in poly-isohexylcyanoacrylate nanospheres, *Anticancer Res.* **2004**, 24, 3781–3788.

90. C.E. Soma, C. Dubernet, D. Bentolila, S. Benita, P. Couvreur, Reversion of multidrug resistance by co-encapsulation of doxorubicin and cyclosporin A in polyalkylcyanoacrylate nanoparticles, *Biomaterials* **2000**, 21, 1–7.

91. T. Ameller, V. Marsaud, P. Legrand, R. Gref, J.M. Renoir, In vitro and in vivo biologic evaluation of long-circulating biodegradable drug carriers loaded with the pure antiestrogen RU 58668, *Int. J. Cancer* **2003**, 106, 446–454.

92. S. Maillard, T. Ameller, J. Gauduchon, A. Gougelet, F. Gouilleux, P. Legrand, V. Marsaud, E. Fattal, B. Sola, J.M. Renoir,

Innovative drug delivery nanosystems improve the anti-tumor activity in vitro and in vivo of anti-estrogens in human breast cancer and multiple myeloma, *J. Steroid Biochem. Mol. Biol.* **2005**, 94, 111–121.

93 J.S. CHAWLA, M.M. AMIJI, Cellular uptake and concentrations of tamoxifen upon administration in poly(epsilon-caprolactone) nanoparticles, *AAPS Pharm. Sci.* **2003**, 5, E3.

94 D.B. SHENOY, M.M. AMIJI, Poly(ethylene oxide)-modified poly(epsilon-caprolactone) nanoparticles for targeted delivery of tamoxifen in breast cancer, *Int. J. Pharm.* **2005**, 293, 261–270.

95 D. BHADRA, S. BHADRA, S. JAIN, N.K. JAIN, A PEGylated dendritic nanoparticulate carrier of fluorouracil, *Int. J. Pharm.* **2003**, 257, 111–124.

96 C.L. LO, K.M. LIN, G.H. HSIUE, Preparation and characterization of intelligent core-shell nanoparticles based on poly(D,L-lactide)-g-poly(N-isopropyl acrylamide-co-methacrylic acid), *J. Control. Release* **2005**, 104, 477–488.

97 J.F. KUKOWSKA-LATALLO, K.A. CANDIDO, Z. CAO, S.S. NIGAVEKAR, I.J. MAJOROS, T.P. THOMAS, L.P. BALOGH, M.K. KHAN, J.R. JR BAKER, Nanoparticle targeting of anticancer drug improves therapeutic response in animal model of human epithelial cancer, *Cancer Res.* **2005**, 65, 5317–5324.

98 S. YANG, J. ZHU, B. LU, B. LIANG, C. YANG, Body distribution of camptothecin solid lipid nanoparticles after oral administration, *Pharm. Res.* **1999**, 16, 751–757.

99 H. MIURA, H. ONISHI, M. SASATSU, Y. MACHIDA, Antitumor characteristics of methoxypolyethylene glycol-poly(DL-lactic acid) nanoparticles containing camptothecin, *J. Control. Release* **2004**, 97, 101–113.

100 H. ONISHI, Y. MACHIDA, Y. MACHIDA, Antitumor properties of irinotecan-containing nanoparticles prepared using poly(DL-lactic acid) and poly(ethylene glycol)-block-poly(propylene glycol)-block-poly(ethylene glycol), *Biol. Pharm. Bull.* **2003**, 26, 116–119.

101 J. WILLIAMS, R. LANSDOWN, R. SWEITZER, M. ROMANOWSKI, R. LABELL, R. RAMASWAMI, E. UNGER, Nanoparticle drug delivery system for intravenous delivery of topoisomerase inhibitors, *J. Control. Release* **2003**, 91, 167–172.

102 K. AVGOUSTAKIS, A. BELETSI, Z. PANAGI, P. KLEPETSANIS, A.G. KARYDAS, D.S. ITHAKISSIOS, PLGA-mPEG nanoparticles of cisplatin: In vitro nanoparticle degradation, in vitro drug release and in vivo drug residence in blood properties, *J. Control. Release* **2002**, 79, 123–135.

103 A. BARROUG, L.T. KUHN, L.C. GERSTENFELD, M.J. GLIMCHER, Interactions of cisplatin with calcium phosphate nanoparticles: In vitro controlled adsorption and release, *J. Orthop. Res.* **2004**, 22, 703–708.

104 A.K. SINGLA, A. GARG, D. AGGARWAL, Paclitaxel and its formulations, *Int. J. Pharm.* **2002**, 235, 179–192.

105 A.J. TEN TIJE, J. VERWEIJ, W.J. LOOS, A. SPARREBOOM, Pharmacological effects of formulation vehicles: Implications for cancer chemotherapy, *Clin. Pharmacokinet.* **2003**, 42, 665–685.

106 D. SHARMA, T.P. CHELVI, J. KAUR, K. CHAKRAVORTY, T.K. DE, A. MAITRA, R. RALHAN, Novel Taxol formulation: Polyvinylpyrrolidone nanoparticle-encapsulated Taxol for drug delivery in cancer therapy, *Oncol. Res.* **1996**, 8, 281–286.

107 R. CAVALLI, O. CAPUTO, M.R. GASCO, Preparation and characterization of solid lipid nanoparticles incorporating paclitaxel, *Eur. J. Pharm. Sci.* **2000**, 10, 305–309.

108 N.K. IBRAHIM, N. DESAI, S. LEGHA, P. SOON-SHIONG, R.L. THERIAULT, E. RIVERA, B. ESMAELI, S.E. RING, A. BEDIKIAN, G.N. HORTOBAGYI, J.A. ELLERHORST, Phase I and pharmacokinetic study of ABI-007, a cremophor-free, protein-stabilized, nanoparticle formulation of paclitaxel, *Clin. Cancer Res.* **2002**, 8, 1038–1044.

109 B. Damascelli, G. Cantu, F. Mattavelli, P. Tamplenizza, P. Bidoli, E. Leo, F. Dosio, A.M. Cerrotta, G. Di Tolla, L.F. Frigerio, F. Garbagnati, R. Lanocita, A. Marchiano, G. Patelli, C. Spreafico, V. Ticha, V. Vespro, F. Zunino, Intraarterial chemotherapy with polyoxyethylated castor oil free paclitaxel, incorporated in albumin nanoparticles (ABI-007): Phase II study of patients with squamous cell carcinoma of the head and neck and anal canal: Preliminary evidence of clinical activity, *Cancer* **2001**, 92, 2592–2602.

110 B. Damascelli, G.L. Patelli, R. Lanocita, G. Di Tolla, L.F. Frigerio, A. Marchiano, F. Garbagnati, C. Spreafico, V. Ticha, C.R. Gladin, M. Palazzi, F. Crippa, C. Oldini, S. Calo, A. Bonaccorsi, F. Mattavelli, L. Costa, L. Mariani, G. Cantu, A novel intraarterial chemotherapy using paclitaxel in albumin nanoparticles to treat advanced squamous cell carcinoma of the tongue: Preliminary findings, *Am. J. Roentgenol.* **2003**, 181, 253–260.

111 W.J. Gradishar, The future of breast cancer: The role of prognostic factors, *Breast Cancer Res. Treat.* **2005**, 89, 17–26.

112 D.G. Rodrigues, D.A. Maria, D.C. Fernandes, C.J. Valduga, R.D. Couto, O.C. Ibanez, R.C. Maranhao, Improvement of paclitaxel therapeutic index by derivatization and association to a cholesterol-rich microemulsion: *in vitro* and *in vivo* studies, *Cancer Chemother. Pharmacol.* **2005**, 55, 565–576.

113 C. Fonseca, S. Simoes, R. Gaspar, Paclitaxel-loaded PLGA nanoparticles: Preparation, physicochemical characterization and *in vitro* antitumoral activity, *J. Control. Release* **2002**, 83, 273–286.

114 S.S. Feng, L. Mu, K.Y. Win, G. Huang, Nanoparticles of biodegradable polymers for clinical administration of paclitaxel, *Curr. Med. Chem.* **2004**, 11, 413–424.

115 Z. Lu, T.K. Yeh, M. Tsai, J.L. Au, M.G. Wientjes, Paclitaxel-loaded gelatin nanoparticles for intravesical bladder cancer therapy, *Clin. Cancer Res.* **2004**, 10, 7677–7684.

116 T. Hamaguchi, Y. Matsumura, M. Suzuki, K. Shimizu, R. Goda, I. Nakamura, I. Nakatomi, M. Yokoyama, K. Kataoka, T. Kakizoe, NK105, a paclitaxel-incorporating micellar nanoparticle formulation, can extend *in vivo* antitumour activity and reduce the neurotoxicity of paclitaxel, *Br. J. Cancer* **2005**, 92, 1240–1246.

117 S.K. Sahoo, W. Ma, V. Labhasetwar, Efficacy of transferrin-conjugated paclitaxel-loaded nanoparticles in a murine model of prostate cancer, *Int. J. Cancer* **2004**, 112, 335–340.

118 Z. Xu, W. Gu, J. Huang, H. Sui, Z. Zhou, Y. Yang, Z. Yan, Y. Li, *In vitro* and *in vivo* evaluation of actively targetable nanoparticles for paclitaxel delivery, *Int. J. Pharm.* **2005**, 288, 361–368.

119 J.M. Koziara, P.R. Lockman, D.D. Allen, R.J. Mumper, Paclitaxel nanoparticles for the potential treatment of brain tumors, *J. Control. Release* **2004**, 99, 259–269.

120 J. Zhou, F.Q. Zeng, C. Li, Q.S. Tong, X. Gao, S.S. Xie, L.Z. Yu, Preparation of arsenic trioxide-loaded albuminutes immuno-nanospheres and its specific killing effect on bladder cancer cell *in vitro*, *Chin. Med. J.* **2005**, 118, 50–55.

121 C. Pellizzaro, D. Coradini, S. Morel, E. Ugazio, M.R. Gasco, M.G. Daidone, Cholesteryl butyrate in solid lipid nanoparticles as an alternative approach for butyric acid delivery, *Anticancer Res.* **1999**, 15, 3921–3926.

122 B. Salomone, R. Ponti, M.R. Gasco, E. Ugazio, P. Quaglino, S. Osella-Abate, M.G. Bernengo, *In vitro* effects of cholesteryl butyrate solid lipid nanoparticles as a butyric acid pro-drug on melanoma cells: Evaluation of antiproliferative activity and apoptosis induction, *Clin. Exp. Metastasis* **2001**, 18, 663–673.

123 L. Serpe, S. Laurora, S. Pizzimenti, E. Ugazio, R. Ponti, R. Canaparo, F. Briatore, G. Barrera, M.R.

Gasco, M.G. Bernengo, M. Eandi, G.P. Zara, Cholesteryl butyrate solid lipid nanoparticles as a butyric acid pro-drug: Effects on cell proliferation, cell-cycle distribution and c-myc expression in human leukemic cells, *Anticancer Drugs* **2004**, 15, 525–536.

124 M. Cegnar, A. Premzl, V. Zavasnik-Bergant, J. Kristl, J. Kos, Poly(lactide-co-glycolide) nanoparticles as a carrier system for delivering cysteine protease inhibitor cystatin into tumor cells, *Exp. Cell Res.* **2004**, 301, 223–231.

125 M. Michaelis, K. Langer, S. Arnold, H.W. Doerr, J. Kreuter, J. Jr. Cinatl, Pharmacological activity of DTPA linked to protein-based drug carrier systems, *Biochem. Biophys. Res. Commun.* **2004**, 323, 1236–1240.

126 P. Beck, J. Kreuter, R. Reszka, I. Fichtner, Influence of polybutylcyanoacrylate nanoparticles and liposomes on the efficacy and toxicity of the anticancer drug mitoxantrone in murine tumour models, *J. Microencapsul.* **1993**, 10, 101–114.

127 R. Reszka, P. Beck, I. Fichtner, M. Hentschel, L. Richter, J. Kreuter, Body distribution of free, liposomal and nanoparticle-associated mitoxantrone in B16-melanoma-bearing mice, *J. Pharm. Exp. Ther.* **1997**, 280, 232–237.

128 R.L. Juliano, S. Alahari, H. Yoo, R. Kole, M. Cho, Antisense pharmacodynamics: Critical issues in the transport and delivery of antisense oligonucleotides, *Pharm. Res.* **1999**, 16, 494–502.

129 G. Schwab, C. Chavany, I. Duroux, G. Goubin, J. Lebeau, C. Hélène, T. Saison-Behmoaras, Antisense oligonucleotides adsorbed to polyalkylcyanoacrylate nanoparticles specifically inhibit mutated Ha-*ras*-mediated cell proliferation and tumorigenicity in nude mice, *Proc. Natl. Acad. Sci. U.S.A.* **1994**, 91, 10 460–10 464.

130 G. Godard, A.S. Boutorine, E. Saison-Behmoaras, C. Hélèn, Antisense effect of cholesterol-oligodeoxynucleotide conjugates associated with poly(alkylcyanoacrylate) nanoparticles, *Eur. J. Biochem.* **1995**, 232, 404–410.

131 L. Tondelli, A. Ricca, M. Laus, M. Lelli, G. Citro, Highly efficient cellular uptake of c-myb antisense oligonucleotides through specifically designed polymeric nanospheres, *Nucleic Acids Res.* **1998**, 26, 5425–5431.

132 G. Lambert, E. Fattal, H. Pinto-Alphandary, A. Gulik, P. Couvreur, Polyisobutylcyanoacrylate nanocapsules containing an aqueous core as a novel colloidal carrier for the delivery of oligonucleotides, *Pharm. Res.* **2000**, 17, 707–714.

133 G. Lambert, J.R. Bertrand, E. Fattal, F. Subra, H. Pinto-Alphandary, C. Malvy, C. Auclair, P. Couvreur, EWS Fli-1 antisense nanocapsules inhibits Ewing sarcoma-related tumor in mice, *Biochem. Biophys. Res. Commun.* **2000**, 279, 401–406.

134 G. Lambert, E. Fattal, P. Couvreur, Nanoparticulate systems for the delivery of antisense oligonucleotides, *Adv. Drug Deliv. Rev.* **2001**, 47, 99–112.

135 B. Gopalan, I. Ito, C.D. Branch, C. Stephens, J.A. Roth, R. Ramesh, Nanoparticle based systemic gene therapy for lung cancer: Molecular mechanisms and strategies to suppress nanoparticle-mediated inflammatory response, *Technol. Cancer Res. Treat.* **2004**, 3, 647–657.

136 L. Brannon-Peppas, J.O. Blanchette, Nanoparticle and targeted systems for cancer therapy, *Adv. Drug Deliv. Rev.* **2004**, 56, 1649–1659.

137 M. Ogris, G. Walker, T. Blessing, R. Kircheis, M. Wolshek, E. Wagner, Tumor-targeted gene-therapy: Strategies for the preparation of ligand-polyethylene glycol-polyethyleneimine/DNA complexes, *J. Control. Release* **2003**, 91, 173–181.

138 S.H. Pun, F. Tack, N.C. Bellocq, J. Cheng, B.H. Grubbs, G.S. Jensen, M.E. Davis, M. Brewster, M. Janicot, B. Janssens, W. Floren, A. Bakker, Targeted delivery of RNA-

cleaving DNA enzyme (DNAzyme) to tumor tissue by transferrin-modified, cyclodextrin-based particles, *Cancer Biol. Ther.* **2004**, 3, 641–650.
139 R.M. SCHIFFELERS, A. ANSARI, J. XU, Q. ZHOU, Q. TANG, G. STORM, G. MOLEMA, P.Y. LU, P.V. SCARIA, M.C. WOODLE, Cancer siRNA therapy by tumor selective delivery with ligand-targeted sterically stabilized nanoparticle, *Nucleic Acids Res.* **2004**, 32, 149.

2
Nanoparticles for Photodynamic Therapy of Cancer

Magali Zeisser-Labouèbe, Angelica Vargas, and Florence Delie

2.1
Introduction

Currently, the limiting factor in cancer chemotherapy is still the lack of selectivity of anticancer drugs towards neoplastic cells. Photosensitizers (PS), the active compounds used for photodynamic therapy (PDT), have the intrinsic advantage of distributing primarily in highly regenerative tissues after intravenous or topical administration. Therefore, they will accumulate preferentially in tumor tissue when present. In addition, these molecules are inactive as such; indeed, the anticancer effect is only attained after irradiation with light at the right wavelength. Compared to other current cancer therapies such as surgery, radiotherapy or chemotherapy, PDT is an effective and selective means of suppressing diseased tissues without altering the surrounding healthy tissue. It also offers a unique opportunity to reach unseen metastasis. As fluorescent molecules, PS may also be used as a tool, namely photodetection (PD), to reveal tumor tissues that remain unseen by other conventional methods.

Interestingly, PD and PDT are mutually beneficial. Ideally, a PS could be used to detect tumors and then to treat them. Quantification of PS fluorescence allows us to follow PS uptake and pharmacokinetics [1–3]. Finally, after treatment, the tissue can be examined by PD to evaluate disease control or possible recurrence [4, 5].

The first compound on the market was Photofrin®, a synthetic haematoporphyrin derivative characterized by a pronounced skin photosensitivity. Second-generation PS have since been designed with less pronounced adverse effects. There are, currently, about ten molecules marketed worldwide. The most potent PS currently under development are hydrophobic molecules with a high tendency to localize in cancer tissue. However, water-insoluble products are difficult to administer to patients, especially when the intravenous route is considered. Therefore several strategies have developed, among which polymeric nanoparticles offer multiple advantages.

This chapter briefly presents the basis of PDT and the most used PS and then reviews the interest in developing nanoparticles (NPs) to improve current PDT, with special interest on polymeric biodegradable NPs. First, the main preparation methods used to load nanoparticles with PS will be introduced. The major achievements obtained both *in vitro* and *in vivo* with encapsulated PS are then critically appraised, especially with regards to methodologies. The primary outcomes are discussed, highlighting the interest in polymeric nanoparticles as delivery systems for PS. To our knowledge, this is the first published review on the use of polymeric nanoparticles for the delivery of PS in the framework of cancer therapy or detection.

2.2
Concept and Basis of Photodynamic Therapy and Photodetection

Photodynamic therapy, an innovative alternative to conventional therapies, is based on the systemic or topical administration of a photosensitizing drug, also known as a photosensitizer. After biodistribution of the drug, the diseased tissue is illuminated with light at an appropriate wavelength. Light will activate the PS and, in the presence of molecular oxygen, will generate cytotoxic species. In turn, those highly oxidizing species will damage cellular constituents, leading to tumor destruction. Fluorescence PD of cancer is also based on the administration of a PS, and takes advantage of the fluorescence emission of these substances. Cancer diagnosis is of major importance, as early detection of malignancies and metastasis can improve the chance of success for anticancer therapies. In addition, PD is a valuable tool to help guide biopsy and resection to minimize the removal of non-cancerous tissue. Although PS are usually administered systemically, the therapeutic effect of PDT is local rather than systemic. The selectivity of the treatment is not only due to the preferential biodistribution of the PS in cancer tissues but also to the precise activation of the drug by a light beam, usually from a laser source directed to the target tissue.

2.2.1
Mechanisms of Photodynamic Therapy and Photodiagnosis

The phototoxic effects on which PDT is based and the fluorescence used for PD are both initiated by the absorption of light by a PS, leading to its excitation from the ground state to the singlet state (Fig. 2.1). The singlet state lifetime, which is on the nanosecond time scale, is too short to allow significant interaction with surrounding molecules [6]. The singlet state can be deactivated via three pathways. The first, fluorescence, results in the emission of photons of a wavelength longer than the excitation light (Fig. 2.1A), and thus allows fluorescence photodetection. The use of highly sensitive imaging devices in PD permits visualization of diseased tissues over normal tissues due to the preferential accumulation of PS in the former [7]. The singlet state can also be deactivated by the release of heat (Fig. 2.1B) or by undergoing intersystem crossing to generate the triplet state (Fig. 2.1C). Since the lifetime of the triplet state is in the micro- to millisecond range, the transfer energy to surrounding molecules is possible [6]. The triplet state is the

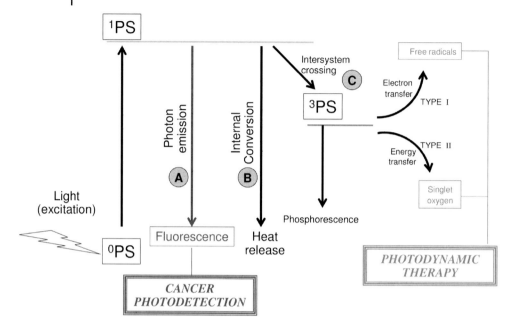

Fig. 2.1. Schematic representation of the photophysical and photochemical mechanisms associated with photodiagnosis and photodynamic therapy. After light irradiation, photosensitizer in its ground state (^{0}PS) is excited to the singlet state (^{1}PS). This state can be deactivated by (A) emitting fluorescence, (B) releasing heat or (C) undergoing intersystem crossing, which results in the generation of PS in its triplet state (^{3}PS). The PS triplet state induces phototoxicity via type I and type II reactions, generating free radical species and singlet oxygen, respectively. These entities are toxic in biological systems, inducing cellular death and vascular occlusion.

key to initiating the photochemical process that induces phototoxicity, and it occurs via two main mechanisms [8, 9]. The type I mechanism involves hydrogen-atom abstraction or electron-transfer between the triplet state and neighboring molecules, generating free radicals. These radicals react with oxygen and generate a mixture of highly reactive oxygen intermediates, such as $\cdot O_2^-$, H_2O_2 and $\cdot OH$, which are highly oxidizing [10, 11]. In the type II mechanism, energy is transferred from the PS in its triplet state to molecular oxygen to form highly reactive singlet oxygen ($^1O_2$), which is presumed to be the most reactive species in PDT. Reactive intermediate oxygen species, including both radicals and non-radicals, are called reactive oxygen species (ROS). Type I and type II pathways are not mutually exclusive and both ultimately lead to the formation of oxidized products and radical chain reactions, which can trigger cascades of biochemical, immunological, and physiological reactions, finally resulting in the destruction of the irradiated tissue [7, 12–14].

At a cellular level, PDT induces cytotoxic effects through photodamage to subcellular organelles and biomolecules. Various cellular components can react with singlet oxygen, such as amino acid residues (especially cysteine, methionine, tyrosine, histidine and tryptophan), nucleosides (mainly guanine) and unsaturated lipids [12]. PS can localize in lysosomes, mitochondria, plasma membrane, Golgi apparatus and endoplasmic reticulum of tumor cells, as well as in tumor vasculature. Interestingly, most PS do not accumulate in cell nuclei; thus PDT has a low potential for causing DNA damage, mutation and carcinogenesis [15]. Two distinct types of cell death may be induced by PDT. First, PDT can trigger apoptosis, a form of programmed cell death that involves the activity of proteolytic caspases, whose action dismantles the cell and results in cell death. Apoptosis begins internally with condensation and subsequent fragmentation of the cell nucleus while the plasma membrane remains intact. Afterwards, apoptotic cells are ultimately fragmented into multiple membrane-enclosed spherical vesicles, which are scavenged by phagocytes [16, 17]. The second mode of cell death induced by PDT is necrosis, characterized by cytoplasm swelling, destruction of organelles and disruption of plasma membrane, leading to the release of intracellular content and inflammatory factors. The cell type, PS subcellular localization and applied light dose determine the type of cell death. PS with tropism for mitochondria are more likely to induce apoptosis, whereas PS localized in the plasma membrane are expected to cause necrosis [15].

There are three main mechanisms for suppressing malignant tissue when using PDT:

1. Direct cellular damage by necrotic and/or apoptotic mechanisms [16, 17].
2. Alteration of tumor vascularization such as occlusion, stasis and/or increase in vascular permeability, thereby depriving cancer tissue of oxygen and nutrients [18–20].
3. Stimulation of inflammatory and immune responses against the tumor [21–23].

The impact of these pathways in the therapeutic effects of PDT depends on the PS and its formulation, the route of administration, and the time interval between administration and light irradiation.

2.2.2
Selective Tumor Uptake of Photosensitizers

One of the key aspects of PDT is the preferential accumulation of PS into tissues with a high rate of regeneration such as cancer tissue and neovasculature [15, 24, 25]. Although the exact underlying mechanisms that drive this tropism have not been completely elucidated, the abnormal physiology of tumors is the main contributor to the selectivity of PS. The affinity of PS for tumors and their surrounding stroma has been related to some of the unique characteristics of hyperproliferative tissues listed in Table 2.1. Of great importance is the increased vascular permeability of tumors, which facilitates the crossing of PS, or their carriers to the interstitial space. This effect, called the enhanced permeability and retention (EPR) effect, is potentiated by an impaired lymphatic drainage reducing macromolecule clearance

Tab. 2.1. Characteristics of tumor tissues.

- Decreased pH of interstitial fluid, due to an increased glycolysis and a decreased supply of oxygen.
- Increased number of low-density lipoprotein receptors at the cell surface.
- Abundance of macrophages.
- Abnormal stromal composition, due to the newly formed collagen.
- Leaky vasculature.
- High rate of angiogenesis (neovasculature development).
- Poor lymphatic drainage.

---

from the tumors [26]. Indeed, macromolecules can be entrapped and retained in solid tumors at high concentration for an extended period (>100 h) whereas low-molecular weight substances return to circulating blood by diffusion. Consequently, entrapping PS in macromolecular drug carriers, such as nanoparticles, is thought to increase PS concentration within tumors.

The physicochemical characteristics of PS molecules, such as molecular weight, hydrophile–lipophile balance value [1, 27–29], ionic charge [30], and protein binding characteristics, also influence their biodistribution. In a literature review, Boyle and Dolphin have attempted to correlate PS structure with its biodistribution and pharmacokinetics [31]. Evaluation of this relationship was, however, made impossible by the disparity between delivery vehicles, animals and tumor types used in the different studies. The PS hydrophile–lipophile balance value appears to be a key factor that is able to regulate pharmacokinetic profiles. Indeed, hydrophobic PS induce preferential damage to tumor cells, whereas hydrophilic PS mainly cause damage to the tumor vasculature. This selectivity will be determined by the nature of the binding between the PS and plasma proteins. Hydrophobic PS are bound to the lipid moiety of lipoproteins [32–36] whereas hydrophilic PS are bound to albumin and globulins [37]. Hydrophobic PS bound to low-density lipoproteins (LDL) can be endocytosed via LDL receptors [34, 38], which are overexpressed in tumor cells. Thus, they accumulate in the lipophilic compartment of tumor cells, including plasma, mitochondrial, endoplasmic reticulum, nuclear and lysosomal membranes. In this case, photodamage of tumor cells occurs preferentially. Conversely, hydrophilic PS will be preferentially localized within the interstitial space and the vascular stroma of the tumor tissue. Owing to their hydrophilic character, diffusion across the plasma membrane into the cytoplasm is limited. Subsequently, these PS cause extensive impairment of the vascular system, promoting tumor ischemia and hypoxia [15, 39].

Since ROS have a short life-time and act close to their site of generation, the sites of initial cell and tissue damage induced by PDT are closely related to both the localization of the PS within cells and the site of illumination. As a strategy to improve photosensitizer efficiency regardless of the properties of the PS itself, several pharmaceutical formulations have emerged. Indeed, PS delivery systems influence the pharmacokinetic profiles and tissue distribution of PS [40–45].

2.2.3
Photosensitizers

A photosensitizer is defined as a chemical entity that, upon absorption of light, induces a chemical or physical alteration of another chemical entity [14]. Photosensitizers used against cancer must be able to absorb light and then to transfer the absorbed energy to molecular oxygen in order to cause biological effects. Several publications have described the characteristics of the ideal PS [7, 12, 14, 46, 47]. Since the triplet state of the PS initiates the photochemical process, a good PS requires a long-lived triplet state with a high quantum yield to allow enough time to interact with neighboring target molecules [8]. Ideally, a PS should absorb photons efficiently in the red part of the spectrum because light with long wavelengths have an increased penetration depth in tissues. High chemical purity, good solubility in pharmaceutical acceptable formulations and low aggregation tendency are also mandatory. PS used for photodetection should be photostable, have a high fluorescence quantum yield with low interference with tissue autofluorescence and induce minimal photodamage. Allison et al. have reviewed the clinically relevant properties of PS intended for PDT of cancer [48]. PS should not induce dark toxicity, which is defined as toxicity in the absence of light. Furthermore, PS should have a preferential biodistribution in the tumor as selectively as possible and be eliminated quickly enough to avoid generalized skin photosensitization and systemic toxicity.

2.2.3.1 Conventional Photosensitizers

Originally, PDT was based on the topical application of dyes such as eosin, methylene blue (MB) and rose bengal with remarkable success [46]. Most compounds able to reach triplet states and to produce ROS have either tricyclic, heterocyclic or porphyrin-like ring structures with conjugated double bonds (π-electron system) [14]. Figure 2.2 shows examples of these structures. PS used in clinics have been classified as first, second and third-generation PS. Haematoporphyrin derivative (HpD) or Photofrin®, a first-generation PS, is a very complex mixture of monomers and oligomers. It was the first PS approved by the U.S. Food and Drug Administration (FDA), in 1995. Photofrin® has been used widely in clinics to treat esophageal, papillary bladder and endobronchial cancers. It is also indicated for ablation of high-grade dysplasia associated with Barrett's esophagus, which is a precancerous condition, thus reducing the risk of progression to esophageal cancer [49]. However, HpD and its analogues not only have poor tumor selectivity, resulting in long-lasting skin photosensitivity [50, 51], but also lack strong absorption bands in the red region of the spectrum [52]. Furthermore, they are complex mixtures and their synthesis and biological activity are difficult to reproduce [12, 48,

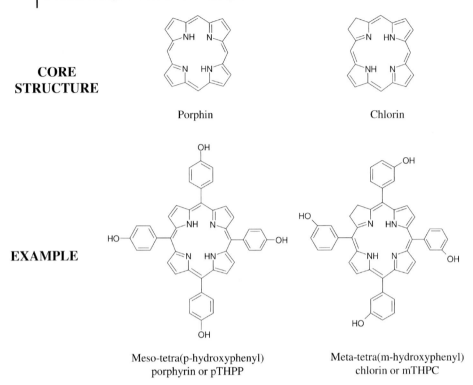

Fig. 2.2. Structure of some photosensitizers. Porphins consist of four pyrrole subunits linked together by four methane bridges. Derivatives of porphin are named porphyrins. Reduction of one of the pyrrole units on the porphin ring leads to a class of porphyrin derivatives called chlorins.

53]. Therefore, considerable effort has been devoted to developing new second-generation photosensitizers, characterized by greater selectivity for tumor tissue, rapid excretion from the body, and improved purity, stability, solubility and photophysical properties. Owing to the high oxygen quantum yield of the porphyrin skeleton, most second-generation PS belong to this family. The porphyrin structure provides 12 positions that can potentially be substituted. Furthermore, the porphyrin cycle can be oxidized, extended and/or a central ion may be introduced to modulate the pharmacological, as well as the photophysical, properties of the molecule [7, 52]. Porphyrins, chlorins, texaphyrins, purpurins and phthalocyanines have been most actively investigated [24, 47].

Recently, considerable interest in the use of 5-aminolevulinic acid (5-ALA) in PDT has risen. 5-ALA itself is not a photosensitizer, but it can induce the formation of protoporphyrin IX (PpIX), a potent PS. 5-ALA is a substrate in the biosynthetic pathway of heme, the iron(II) complex of PpIX. In contrast to heme, PpIX is a fluorescent molecule and a potent PS. When 5-ALA is administered, its biosynthesis leads to the selective transient accumulation of PpIX in neoplastic cells [7, 54]. If tissues containing the 5-ALA-induced PpIX are irradiated, photochemical processes are triggered, resulting in tissue destruction. Although the underlying mechanisms of the preferential accumulation of 5-ALA-induced PpIX in tumors and other proliferating tissues are not fully understood, the 5-ALA approach is one of the most selective modalities currently available in anticancer therapy. Collaud et al. have recently reviewed the mechanisms of the selective formation of PpIX in neoplastic tissue after 5-ALA administration [55]. Topical administration (e.g., skin and bladder) of 5-ALA and 5-ALA ester derivatives induced no skin photosensitivity and had short half-life in the treated area [56–59]. 5-ALA and its derivatives can also be used for photodiagnosis [60–62], because PpIX emits red fluorescence upon irradiation with blue light.

Recently, the third generation of PS has emerged. These are second-generation photosensitizers with fine tuned properties that allow selective delivery to tumor tissue. Different strategies have been proposed, such as conjugation to biomolecules like monoclonal antibodies [63].

Although the clinical potential of PDT has been recognized for over 25 years, it is only now starting to be used clinically [24]. Table 2.2 summarizes clinically relevant PS for PDT and the PD of cancer. There are few PS intended for PD, only 5-ALA-hexyl ester has been recently approved for detection of bladder cancer in Europe and has been submitted to the FDA (www.photocure.com). Local administration of hypericin has also shown promising results for the PD of bladder carcinoma [64]. Other PS evaluated for PD such as haematoporphyrin derivative and meta-tetra(hydroxyphenyl)chlorin induced prolonged skin and eye photosensitivity [50, 51, 65, 66] and are not yet approved for photodiagnosis.

2.2.3.2 New Entities

New systems such as semiconductors, fullerenes and transition metal complexes are promising photosensitizers, as shown by several studies measuring singlet oxygen production and *in vitro* phototoxicity against tumor cells. However, since little is known about their biocompatibility, their use in clinical PDT of cancer is still unexplored.

Semiconductors, such as titanium dioxide (TiO_2), can sensitize the photogeneration of ROS, thus inducing damage similar to that found in traditional PDT [67]. Particles of TiO_2 have shown phototoxicity against HeLa cells *in vitro* and were not toxic when administered orally or parenterally to nude mice. After UV irradiation, TiO_2 particles significantly suppressed the growth of HeLa [68] and T-24 cells [69] implanted in nude mice. However, this approach is limited because UV light, used to irradiate TiO_2 particles, does not penetrate deep in the skin [70].

When the size of a semiconductor particle is decreased to the nanometric scale, these materials are called quantum dots (QDs). QDs are single crystals of semiconductor material, a few nanometers in diameter. Their size and shape can be controlled precisely [71]. A quantum dot can have anything from a single electron to a collection of several thousands. Therefore, light absorption of QDs can be precisely

Tab. 2.2. Relevant photosensitizers under clinical trials or approved for PDT and PD of cancer [5, 15, 24, 48, 163, 164].

Photosensitizer/ Generic name	Commercial name	Formulation/ Administration	Approved oncological indications	Clinical trials for different types of cancer	Drug–light interval[a]	Skin photosensitivity[b]
Haematoporphyrin derivative (HpD)/ Porfimer sodium	Photofrin®	Powder for solution/ i.v. or topical	Esophageal adenocarcinoma and high-grade dysplasia in Barrett's esophagus. Cervical, superficial gastric, bronchial, bladder, and advanced and early lung cancer	Intraperitoneal, hilar bile-duct, head and neck, intestinal, lung, larynx, skin, bladder and metastatic breast	40–50 h	1–3 months
Meta-tetra(hydroxyphenyl) chlorin (mTHPC)/ Temoporfin	Foscan®	Solution in ethanol and propylene glycol/i.v.	Palliative advanced head and neck cancer. Squamous cell carcinoma	Brain, gastric, prostate and oral cavity. Mesothelioma	24–96 h	Up to 6 weeks
5-Aminolevulinic acid (5-ALA)[c]	Levulan®	Powder for solution/ Topical. Oral and i.v. have been also evaluated	Actinic keratosis	Basal-cell carcinoma, esophageal, gastrointestinal and non-melanoma skin	2 h	1–2 days
5-ALA-methyl ester[c]	Metvix®	Cream/Topical	Actinic keratosis, superficial basal-cell carcinoma and basal cell-carcinoma. Bowen's disease		3 h	Uncommon

5-ALA-hexyl ester[c]	Hexvix®	Powder for solution/ Intravesical injection	Bladder cancer diagnosis. Surgeons can easily define bladder lesions and ablate them		1–2 h	Uncommon
Benzoporphyrin-derivative monoacid ring A (BPD-MA)/ Verteporfin	Visudyne®	Liposomes/i.v.	None. It is approved for choroidal neovascularization associated to age-related macular degeneration (CNV-AMD), a non-malignant disease	Skin and Barrett's esophagus	30–150 min	3–5 days
Tin-ethyl etiopurpurin (SnET2)	Purlytin™	Lipid emulsion		Skin, prostate, metastasic breast, Kaposi's sarcoma and CNV-AMD	24 h	2–3 weeks
Lutetium texaphyrin/Lutex	Lutrin™	Powder for solution		Skin and breast		1–2 days

[a] Interval between drug administration and light irradiation.
[b] Skin photosensitivity reactions include erythema, edema, blistering, hyperpigmentation and sunburn. Ultraviolet sunscreens provide no protection. Patients must avoid exposure of eyes and skin to direct sunlight and wear protective clothing and dark sunglasses when outdoors.
[c] Prodrug of protoporphyrin IX.

tuned from the UV to the infrared (IR) region of the spectrum by changing their size and composition. At the same time, QDs have narrow emission spectra that can be tuned to emit in the near-IR region, in contrast to the visible emission of most conventional PS [72–74]. QDs are highly resistant to degradation and their fluorescence is remarkably stable. Most work on semiconductor QDs has focused on fluorescence imaging and diagnosis applications [71, 74–76]. However, they are seen as suitable candidates for PDT because of their capacity to generate ROS after light irradiation [72, 73, 77]. Interestingly, QDs can enhance the effect of conventional photosensitizers if light-mediated energy transfer between both molecules is possible. Consequently, in addition to their intrinsic efficacy as photosensitizers, QDs have been used to potentiate conventional PS. Semiconductor QDs of cadmium selenide (CdSe) conjugated to anti-CD90 antibodies potentiate the activity of the PS trifluoperazine against leukemia cells [77]. Moreover, CdSe QDs linked to silicon phthalocyanine have enabled the use of an excitation wavelength where the PS alone does not absorb [72]. Unwanted potential toxicity of QDs is a key issue inhibiting their development as a therapeutic tool. Since the primary site of acute injury within the body after exposure to Cd is the liver, the cytotoxicity of CdSe QDs was investigated *in vitro* on primary hepatocytes isolated from rats [78]. QDs were cytotoxic due to the slow release of Cd. Surface oxidation of QDs after air and UV light exposure leads to the formation of reduced Cd on the QDs surfaces. Surface coating of QDs with either zinc sulfide or bovine serum albumin decreased the oxidation and consequently the cytotoxicity of the nanoparticles [78]. However, even when QDs are coated, there is still the risk of Cd release into the body after *in vivo* oxidation of these particles.

Before considering the use of QDs in clinics, some concerns regarding the propensity of QDs to aggregate, their toxicity profile and potential to release heavy metals should be addressed.

Fullerenes are a class of spherically-shaped molecules made exclusively of carbon atoms. Fullerene C_{60} and C_{70} efficiently generate singlet oxygen when irradiated with light. However, fullerenes are practically insoluble in both aqueous and most polar media – not to mention their poor absorption in the red region of the visible spectrum [79, 80]. These properties hamper the development of fullerene-based agents for PDT. Furthermore, the mechanism of action of photoexcited fullerenes in biological systems is not at all understood [81]. Nevertheless, intense research has been devoted to evaluate the potential applications of fullerenes in medicine and biology [79, 80].

Some inorganic complexes are also efficient photosensitizers, among them transition metal complexes of ruthenium(II), osmium(II), iridium(III), chromium(III), platinum(II) and palladium(II) have been investigated [46].

As described above, the selectivity of PDT can be increased using a PS that preferentially accumulates in cancer tissues. The affinity of PS for neoplastic cells is in part governed by the lipoprotein transport of PS and subsequent cellular uptake of these protein-PS complexes [34, 38]. However, little is known about the interaction of semiconductors, fullerenes and transition metal complexes with serum proteins. Besides the ability of such PS to photogenerate singlet oxygen, their toxicity, biodistribution and selectivity towards cancer tissues should be addressed before clinical application.

2.2.4
Photodynamic Therapy: Advantages and Limitations

PDT is effective, minimally invasive and offers several advantages over other cancer therapies, such as surgery, radiation therapy and chemotherapy. Generation of cytotoxic species after PDT is only due to the combination of PS, light and oxygen; therefore, great selectivity towards diseased tissues is achieved. Indeed, singlet oxygen has a short lifetime in the biological environment (<40 ns) [82] and cannot diffuse far from its point of origin (radius of action ≤ 70 nm) [83]. Differences in PS clearance between tissues enable optimization of the interval between PS administration and light irradiation, which should be performed when the drug has reached the maximum concentration in the tumor. Additionally, the ability of cells to recover from photodynamic damage also contributes to the selectivity of PDT. Indeed, healthy tissues are able to recover better than tumor tissues after PDT. For example, in the treatment of skin cancers, even if healthy tissue is damaged during PDT, the cosmetic results are usually excellent with little or no scarring, as has been demonstrated with topically applied 5-ALA [59] and intravenously administered verteporfin (benzoporphyrin derivative monoacid ring A) [84]. Additionally, PDT is a photochemical process without tissue heating, thus connective tissues such as collagen and elastin stay largely unaffected [53]. PDT can be repeated if necessary, and performed after surgery, chemotherapy or radiotherapy. Last but not least, PDT can be used to treat different types of cancers, including tumor resistant to other treatments [85].

Even though PS are expected to be retained preferentially by neoplastic tissues and their activity is triggered by light activation at a specific wavelength, the drug is still distributed throughout the whole body [53]. Therefore, PS administration can induce side effects in tissues exposed to daylight such as skin and eyes. Skin photosensitivity reactions are characterized by erythema, edema, blistering, hyperpigmentation and sunburn. Some precautions are strongly recommended during the period in which the PDT patient remains photosensitive. Physical barriers, particularly protective clothing and sun glasses, provide some protection against UV and visible light, but by far the optimum safety for these patients is complete sun avoidance. The period during which these safety measures have to be applied depends specifically on the nature of the photosensitizer [86]. Depending on the molecule, it ranges from a few days to up to three months. PDT can induce also occasional systemic and metabolic disturbances, and excessive tissue destruction at the treated site [12]. Adverse effects of PDT depend mainly on the nature of the PS, the route of administration and the type and localization of the malignancy. Table 2.3 summarizes the main adverse effects induced by PS used in PDT and PD of cancer. Intravenously administered PS, such as temoporfin (Foscan®) and HpD (Photofrin®), induce prolonged skin and eye photosensitivity [50, 51, 65, 66]. Conversely, 5-ALA (Levulan®) and 5-ALA-hexyl ester (Hexvix®), which are topically ap-

Tab. 2.3. Adverse effects induced by commonly used PS in clinical oncology [5, 48, 49].

Generic and commercial name/route of administration	Manufacturer and website/PS approval date	Adverse effects[a] General	Specific to the pathology to be treated
Porfimer sodium Photofrin®/i.v.	Axcan Pharma, Inc www.axcan.com/FDA (1995)	Skin and eye photosensitivity. Local swelling and inflammation in and around the treated area. Pain in the chest, back, or abdomen. Breathing difficulties. Nausea and constipation	(S) *Papillary bladder cancer.* Urination alterations (frequency, haematuria, dysuria and nocturia), genital edema, suprapubic pain and urinary tract infection. (S) *Partially-obstructing esophageal cancer.* Pleural effusion, respiratory insufficiency, fever and anemia. (S) *Endobronchial cancer.* Respiratory disorders (dyspnoea, coughing, pneumonia, bronchitis, increased sputum, chest pain and respiratory insufficiency), and fever [50, 51]. (S) *High-grade dysplasia associated with Barrett's esophagus.* Abdominal problems (esophageal narrowing, vomiting, upper or lower abdominal pain, dysphagia, diarrhea), pain chest, dyspnoea, fever and headache
Temoporfin Foscan®/i.v	Biolitec Pharma, Ltd www.biolitecpharma.com/ European Union, Norway and Iceland (2001)	Skin and eye photosensitivity, injection site pain, constipation, and vomiting	(S) *Advanced head and neck cancer.* Pain, hemorrhage, pain in face, scar, mouth necrosis, dysphagia, and face edema [65, 66]
5-ALA Levulan®/Topical (skin)	DUSA Pharmaceuticals, Inc. www.dusapharma.com/ FDA (1999)	Pain during treatment. Photosensitivity is only reported in skin under treatment. No adverse effects are reported in the body system [59].	(M) *Actinic keratosis.* Only in the treated region of the skin: scaling, crusting, itching and hypo- or hyperpigmentation [57, 59, 165]
5-ALA-hexyl ester Hexvix®/Intravesical	PhotoCure ASA www.photocure.com/ European Union (2005)		(M) *Bladder cancer photodiagnosis.* Bladder spasm and pain, dysuria, headache, nausea and vomiting [56, 58]

[a] (S) = severe and (M) = mild to moderate adverse effects.

plied on either skin or bladder, induce no skin photosensitivity and have a short residence time at the application site [56–59].

The therapeutic outcome of PDT is limited by the penetration of light in tissues. Light is either scattered or absorbed when it enters tissues and the extent of both processes depends on the tissue type and the light wavelength. Between 600 and 1000 nm, however, light is scattered to a relatively small extent and is poorly absorbed by important endogenous chromophores such as melanin and hemoglobin [87]. As a consequence, red light possesses a high penetration power into human tissues. Ochsner has compared the penetration of light as a function of the wavelength. At an equal incident light intensity, the penetration depth in human skin is of 6.8 mm at 800 nm, whereas it is only 0.4 mm at 400 nm [88]. The deeper penetration of longer wavelengths is a major incentive for the development of PS absorbing at these wavelengths. Additionally, to improve the outcome of PDT, new light delivery devices have been developed for this particular application. The traditional argon-dye and copper-dye lasers can be replaced by diode lasers, which are cheaper, very stable, reliable and easily transportable [89]. However, they are not tunable and may only be used at fixed wavelengths [90]. Optical fibers can be used to deliver light to the target tissue. They facilitate the illumination in various directions using either cylindrical or spherical diffusers. Furthermore, the versatility of optical fibers enables illumination of the skin or inside a body cavity [91]. PDT is usually performed with external illumination of the target site; however, deeply localized tumors should be treated with special light delivery devices that are inserted percutaneously. This technique, namely interstitial PDT, uses multiple laser fibers that are inserted directly into tumors through needles positioned under image guidance. Therefore, it is possible to use PDT for the treatment of internal tumors [92, 93].

The dependence of PDT on the oxygenation of the irradiated tissue represents a limitation of this treatment. Indeed, the efficacy of PDT depends on the amount of singlet oxygen produced within the tumor, which in turn depends on the concentration of oxygen in the tissue [94, 95]. Consequently, hypoxic tumor cells are generally more resistant to PDT, and may contribute to treatment failures.

Finally, most PS are hydrophobic, which is a key factor contributing to their selectivity for cancer tissues. However, PS lipophilicity makes formulation difficult due to the lack of physiologically acceptable solvents, especially when intravenous administration is considered. Furthermore, hydrophobic PS can aggregate in aqueous media, leading to quenching; thus in their aggregated form PS are less efficient than in their monomeric form [96–98]. Additionally, some PS lack selectivity for accumulating in cancerous tissues. Therefore, the design of adequate PS delivery systems is critical to improving the outcome and acceptability of PDT and PD in a clinical context.

2.2.5
Photosensitizer Formulations

Different formulation approaches have been proposed, such as the incorporation of PS into liposomes, micelles, polymeric particles, and LDL, as well as the development of hydrophilic polymer–drug complexes, as reviewed by Konan et al. [99]. The delivery carrier can influence the PS biodistribution and hence the mechanisms and kinetics of PS transport to tissues, as well as PS subcellular distribution [15]. Among the different approaches, nanoparticles (NPs) offer numerous advantages, including high drug loading, controlled release, and a large variety of carrier materials and manufacturing processes. Nanoparticles are defined as particles in the nanometer scale, typically <1 µm. NPs appear to be suitable delivery systems for PS because encapsulation of PS into NPs would make it possible to disperse hydrophobic PS in aqueous media. Moreover, NPs have large surface areas, and their surface can be modified with functional groups to modulate their biochemical and physicochemical properties. Owing to their size, direct targeting of tumor tissues is also possible by taking advantage of the tumor vasculature enhanced permeability [100, 101]. Biodegradable and non-biodegradable materials can be used to produce NPs. The use of biodegradable materials enables the controlled release of the encapsulated drug. Conversely, non-biodegradable materials offer the advantage of enhancing the direct interaction of PS with molecular oxygen, either within the nanoparticles or at their surface. The use of non-biodegradable NPs made of metals, ceramics and non-biodegradable polymers are discussed in the next section. The final section is devoted to biodegradable polymeric NPs used for photodynamic activity. The biodegradability and biocompatibility of polymeric NPs bring them closer to clinical application than non-biodegradable carriers.

2.3
Non-biodegradable Nanoparticles for Photodynamic Therapy

2.3.1
Metallic Nanoparticles

This approach involves the coating of metallic nanoparticles, mainly made of gold or magnetic materials, with photosensitizers. The design of PS-coated metallic NPs for PDT has been primarily developed in two directions. First, the adhesion of hydrophobic PS to gold NPs enables an aqueous PS suspension, where the PS photophysical properties are enhanced. This concept has been demonstrated with phthalocyanine derivative-coated NPs that were able to generate singlet oxygen with higher quantum yield than the free PS [102]. In the second approach, the development of magnetic nanoparticles coated with PS allowed either the targeting of the pathological tissue by directing the NPs by an external magnetic field or cancer diagnosis by using the NPs as magnetic resonance (MR) contrast agents. For instance, magnetic NPs made of Fe_3O_4 have been coated with haematoporphyrin [103]. Likewise, pheophorbide-a has been complexed to Fe_3O_4 NPs and the spectroscopic and photophysical properties of this complex characterized [104]. The authors hypothesized that these NPs might be used to combine the action of hyperthermia therapy (HT) and PDT synergistically. Similarly, Gu et al. have conjugated porphyrin to Fe_3O_4 NPs for the same combination of anticancer therapies [105]. The conjugation of porphyrin to the NPs induced a blue-shift in the fluorescence emission spectrum of the PS. No dark toxicity on HeLa cells was seen 5 h after incubation with NPs at 37 °C. Fluorescence microscopy observations showed that NPs were taken up and localized intracellularly. Irradiation of HeLa cells, incubated with the PS-conjugated NPs for 10 min, induced changes in cell morphology. Although the authors interpreted this data as a qualitative sign of phototoxicity, further experiments should be performed to assess the potential of such NPs.

So far, the efficacy of metallic nanoparticles in photodynamic experiments with cancer animal models has not been yet evaluated, although the biocompatibility of metallic nanoparticles has been tested. Neither gold nor magnetic nanoparticles made of iron oxides were toxic *in vivo*. Indeed, 2-nm-gold NPs were administered intravenously to Balb/C mice (2.7 g-Au kg^{-1}) and no lethality was observed [106]. It was also shown that gold NPs were largely cleared from the body through the kidneys. Furthermore, blood analysis two weeks after injection from mice treated with 0.8 g-Au kg^{-1} demonstrated no signs of toxicity as far as haematocrits and plasmatic enzymes are concerned. Similarly, iron oxides particles seem to be generally well tolerated [107, 108] and are intended for several medical applications, as recently reviewed by Ito et al. [109]. The safety of colloidal dispersions of magnetic nanoparticles made of iron oxides has been also demonstrated [110, 111]. In fact, magnetic resonance agents, such as Feridex I.V.™ (Advanced Magnetic, Cambridge, MA, USA) and Resovist® (Schering, AG, Germany), have already been approved for the detection of focal hepatic lesions by MR imaging.

2.3.2
Ceramic Nanoparticles

Ceramic NPs made of silica have been developed as an alternative to polymeric NPs. Ceramic NPs are resistant to microbial attack [112] and their particle size, shape, and porosity can be easily controlled during the preparation process [113]. Ceramic NPs do not release encapsulated compounds even at extreme pH conditions and temperature [114]. This feature represents a limitation for the delivery of common drugs, but can be suitably used in PDT. Since ceramic matrices are generally porous, molecular oxygen can diffuse through the pores and interact with the PS entrapped within the NPs [115]. The photogenerated singlet oxygen can diffuse out of the particle to generate the cytotoxic effect. This approach has been evaluated by entrapping 2-devinyl-2-(1-hexyloxyethyl) pyropheophorbide (HPPH) into 30 nm silica NPs; unfortunately, the percentage of PS loaded into the NPs was not reported. *In vitro* studies with HeLa and UCI-107 cells demonstrated that HPPH-loaded silica NPs were taken up by tumor cells and induced significant cell death, similarly to Tween-80 micelles, which were used as a control [113]. *m*-tetra(Hydroxyphenyl)chlorin (mTHPC) has been embedded in amine-functionalized silica NPs of 180 nm to deliver singlet oxygen instead of releasing PS molecules [116]. The results showed that singlet oxygen production from mTHPC embedded in silica NPs exceeds that of free mTHPC. However, the tests were run in a mixture of water and ethanol in which mTHPC is soluble. Thus, it is possible that mTHPC could be released from the core of the NPs during oxygen sensitization. As a result, more molecules of mTHPC would be soluble and the amount of singlet oxygen would be higher than if mTHPC would have been a solid dispersion inside the NP core. Ideally, singlet oxygen production should be evaluated in aqueous media simulating biological environments. These experiments are far from physiological conditions and may not reflect what would be observed in the cellular environment. Methylene blue, a water-soluble PS of low molecular weight, was encapsulated into three types of sub-200 nm NPs, achieving different MB loadings: polyacrylamide (20–30 nm; loading 0.1%), sol–gel silica (190 nm; loading 3.0%) and organically modified silicate (160 nm; loading 0.4%) [117]. Polyacrylamide NPs were, *in vitro*, the most efficient delivery of singlet oxygen per MB molecule. Moreover, these particles gave the most stable aqueous suspension and therefore were used for *in vitro* photodynamic experiments on rat C6 glioma tumor cells. MB-loaded polyacrylamide NPs were more active than free MB. Notably, concerning silica NPs, the potential toxicity of ceramic nanoparticles is controversial. Toxicological data from studies investigating silica NPs as DNA delivery systems suggest that these carriers have little toxicity [118]. Likewise, organically modified silica NPs have been used for *in vivo* gene delivery in mice and no toxicity was reported up to four weeks after transfection [119]. Conversely, Chen and von Mikecz showed that the uptake of silica NPs (40–70 nm) by the cell nucleus of HEp-2 and RPMI 2650 cells induced nuclear damages close to those seen in neurodegenerative disorders [120]. The nuclear architecture was altered, probably as a result of the formation of nucleoplasmic protein aggregates. Furthermore, silica NPs (4–40 nm) induced inflammatory reactions in cultured human endothelial cells, as shown by the overexpression of interleukin-8 [121].

2.3.3
Nanoparticles Made of Non-biodegradable Polymers

Polyacrylamide (PAA) and amine-functionalized PAA have been used to encapsulate the disulfonated PS 4,7-diphenyl-1,10-phenanthroline ruthenium into nanoparticles of 40–50 nm [122]. Incorporation of the PS into the polyacrylamide matrix did not affect the singlet oxygen production, allowing it to be released into the aqueous media in which the NPs were suspended. PS delivery from amine-functionalized PAA was slower than from PAA NPs. Furthermore, less singlet oxygen was produced than with both free PS and PAA nanoparticles [122]. Recently, the same group developed polyacrylamide NPs to perform simultaneously magnetic resonance imaging and PDT of a rat brain cancer model, providing a real-time tumor death measurement [123]. In this approach, Photofrin® and a magnetic resonance contrast agent were encapsulated together within a polyacrylamide core, resulting in 30–60 nm NPs. The NPs were surface-coated with poly(ethylene glycol) (PEG) to increase the plasma half-life of the carrier. Singlet oxygen production and *in vitro* photoactivity against 9L rat gliosarcoma cells were evaluated. Although the PS was not released from the NPs, as demonstrated by *in vitro* degradation studies in phosphate buffer at 37 °C, the photoactivity of the encapsulated PS was retained. The photoactivity and magnetic resonance contrast ability of this formulation have been evaluated after intravenous administration in intracerebral 9L tumor bearing rats, but the dose of Photofrin® is not mentioned. The activity of the free compound was not tested; however, rats receiving no treatment and treated only with laser were used as controls. The evolution of tumor volume was followed by magnetic resonance imaging. Photofrin®-loaded NPs induced tumor shrinkage, whereas tumors not treated or treated only with laser continued to grow. However, tumor re-growth was observed 12 days post-PDT treatment with Photofrin®-loaded NPs. Polyacrylamide NPs showed no toxicity, in terms of alterations in histopathology or clinical chemistry values, after administration of doses up to 500 mg of NPs per kg over four weeks. The authors suggest that these NPs, namely a multifunctional nanoparticle platform, might enable simultaneous cancer detection, therapy and monitoring. Additionally, particles coated with an integrin ligand for the recognition of the tumor neovasculature were prepared. The specific binding of these NPs was demonstrated *in vitro* with MDA 435 cells expressing integrins. The authors hypothesized that these multifunctional nanoparticles stay in the extracellular compartment and do not release the PS; only singlet oxygen would be delivered after light irradiation [123]. However, the intracellular localization of these NPs after *in vivo* administration should be further studied to confirm the advantages of such a system.

Most research on non-biodegradable materials for the administration of PS has focused on the development of carriers delivering singlet oxygen without releasing the PS. Additionally, non-biodegradable NPs are thought to protect the entrapped PS from the biological environment. The internalization of such NPs into target cells is thought to be unnecessary. In this context, only the external contact of NPs with the cell membrane is required. However, none of these systems have undoubtedly demonstrated the absence of internalization of such NPs. Despite the promising results encountered with these materials, their use in PS delivery has not yet been fully explored, probably due to toxicity concerns in the administration of non-biodegradable materials, particularly if chronic or repeated administrations are needed. Degradation is desired to prevent accumulation of extraneous material and possible subsequent toxicity. Indeed, recent histopathological studies of human biopsies indicate that the development of kidney and liver pathologies, such as chronic inflammation and granulomas, was associated with the presence of non-biodegradable micro- and nanoparticles in these organs [124]. These particles probably originated from debris of implants and prostheses. Certainly, studies of the long-term toxicity non-biodegradable nanoparticles should be undertaken before clinical investigations can be launched.

2.4
Biodegradable Polymeric Nanoparticles for Photodynamic Therapy

Polymeric nanoparticles, as drug delivery systems, have been investigated for over three decades. Several polymers and preparation methods have been developed and the choice of both depends on the physicochemical nature of the drug, as well as

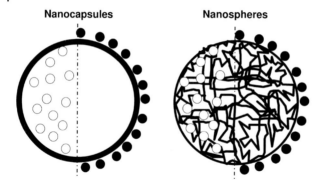

Fig. 2.3. Different types of drug-loaded nanoparticles: drugs may be either adsorbed at the surface of the polymer (●) or encapsulated within the particle (○).

on the type of controlled release kinetic being sought, and the desired target site. Polymeric nanoparticles used for drug delivery are defined as submicron (<1000 nm) colloidal systems made of solid polymers that may be classified according to their preparation processes and internal structure (Fig. 2.3). Nanocapsules (NCs) are composed of a polymeric wall containing a liquid inner core, while nanospheres (NS) are made of a polymeric matrix in which the drug can be dispersed. Active substances may be either adsorbed at the surface of the polymer or encapsulated in the particle. After administration, the drug is released by diffusion from the particles to the surrounding medium or after particle erosion resulting from polymer degradation. Ester or amide functions can be hydrolyzed, and the rate of this process depends on the nature of the polymer (chemical composition and molecular weight), and is triggered by water or the presence of enzymes.

2.4.1
Preparation of Biodegradable Polymeric Nanoparticles

Particles may be produced by polymerization of synthetic monomers, or dispersion of synthetic polymers or natural macromolecules. The preparation methods have been extensively reviewed in the literature [125–127] and will be described only briefly here.

Nanoencapsulation of PS has been considered primarily for hydrophobic molecules that are difficult to formulate in aqueous media, which are mostly used for parenteral administration. The development of a solid suspension offers an interesting alternative. Owing to their hydrophobicity, organosoluble polymers have been mainly used as encapsulating material.

2.4.1.1 *In situ* Polymerization
In situ polymerization of monomers has been used mainly with poly(alkyl cyanoacrylates) (PACA) to prepare either nanoparticles or nanocapsules. The different

methods of preparation as well as medical application of these polymers have been reviewed recently [128].

The preparation of nanoparticles is based on an emulsion-polymerization process in which the cyanoacrylic monomer is dissolved in an organic solvent and dispersed in an aqueous phase containing a surfactant. Anionic polymerization is then induced by hydroxide ions present in water. The polymerization rate is mainly determined by the surrounding pH or the presence of inhibitors. PS can be encapsulated directly during the reaction [129, 130] or adsorbed on the particle surface by incubation with the nanoparticles after neutralization of the aqueous medium [131, 132]. For PACA, the major degradation pathway is based on enzymatic hydrolysis, the rate of degradation being governed by the length of the side chain.

Nanocapsules are usually prepared by interfacial polymerization, where an organic phase containing the monomer and the PS is emulsified in an aqueous phase [129, 130]. Concomitantly, solvent diffusion and anionic polymerization will occur, creating a polymeric wall around the oil core. These particles are especially well adapted for the encapsulation of lipophilic material.

One of the critical concerns with these techniques is the purification step to remove all the residual monomers and the surfactant which may induce undesirable effects.

2.4.1.2 Dispersion of a Preformed Polymer

To reduce toxicity related to the presence of monomer residues or traces of polymerization initiators, preparation methods based on the use of preformed polymers have been developed. These are based on the formation of an emulsion in which the polymer is solubilized in an organic solvent immiscible with the external phase. Polymer precipitation is initiated by the removal of the organic solvent. Another approach, called nanoprecipitation, is based on direct precipitation of the solubilized polymer when in contact with a non-solvent. However, to our knowledge, this method has not been yet reported in the literature for PS.

Polymers used for nanoparticle preparation may be of natural origin. For instance, Zhao et al. have used gelatine to encapsulate hypocrellin B [133]. Particles were made by a modified salting-out coagulation process. An organic solution of the photosensitizer is added to an aqueous solution of the polymer and a surfactant. After nanoparticle formation, glutaraldehyde, a crosslinking agent, is added to the NP suspension. The suspension is then dialyzed to eliminate the glutaraldehyde. However, for hydrophobic photosensitizers, using water-soluble polymers might not be the best choice for their encapsulation. Indeed, no encapsulation rate was reported; therefore, it is not possible to assess how much drug was actually entrapped in the particles. The fluorescence quantum yields decreased as compared to the free drug, but this may also be the result of quenching due to the high number of molecules in the particles. Furthermore, the use of gelatine is warranted due to the potential allergenic reactions induced by this protein; thus, biocompatibility may be a concern at least with certain patients prone to allergies.

To promote entrapment of hydrophobic compounds into polymeric particles, synthetic hydrophobic polymers are frequently used. Polyesters such as poly(lactic acid) (PLA) and copolymers such as poly(lactic-*co*-glycolic acid) (PLGA) are widely used due to their good biocompatibility and because they are accepted by the authorities (FDA) as suture threads. Several methods have been described for PLA and PLGA particle preparation. They are based on the formation of an emulsion of an organic solvent containing the polymer (and the drug) solubilized in an aqueous phase containing a surfactant. The solvent is then removed to induce polymer precipitation and particle formation. The size of the particles is governed by the size of the emulsion and the rate of solvent removal. Several methods have been developed to eliminate the solvent, including evaporation, diffusion [134–138] and dilution after salting-out [138–141]. These methods allow high encapsulation rates since usually more than 80% of the compound is entrapped in the polymeric matrix.

Encapsulation of sensitizers has also been reported in PLA nanocapsules [142, 143]; they were obtained by a solvent-displacement process. The polymer was dissolved in acetone while the hydrophobic photosensitizer was dissolved in Miglyol® (caprilic/capric diglyceryl myristate) and added to the polymer solution. The organic solution was then poured into a water solution containing surfactants. Solvent removal leads to precipitation of the polymer around the oil core containing the active compound.

The emulsion-diffusion method has also been used to prepare particles from a complex made of poly(sebacic anhydride) and phthalocyanine, where the drug was conjugated with the polymer before particle formation [144]. Different complexes were made with various amounts of phthalocyanine, and were characterized by UV/visible spectra. Depending on the degree of aggregation of the photosensitizer in the copolymers, different spectra were found.

Complexation of a photosensitizer and a polymer has been further studied by associating poly(ε-caprolactone) and silicon phthalocyanine [145]. Particles 30–90 nm in diameter were prepared by an emulsion-diffusion process.

2.4.1.3 "Stealth" Particles

When administered *in vivo*, polymeric nanoparticles are rapidly taken up by the reticuloendothelial system (RES) due to the adsorption of proteins at their surface [146–148]. Thus, the biodistribution of particles is mainly directed towards liver and spleen where they are sequestrated and made unavailable to other target tissues. This propensity to localize in RES has been related to the hydrophobicity of the particle surface. Therefore, "stealth" particles have been designed to limit this drawback. The principle is based on "hydrophilization" of the surface. The first approach reported was the coating of the particles with polymers such as poloxamer [130, 142] or PEG [139, 142]. Another approach is to use a directly modified polymer such PLA-PEG [142, 143] or PLGA-PEG.

2.4.1.4 Targeted Nanoparticles

Even though biodistribution of PS is characterized by a preferential accumulation into target tissues such as cancer cells and neovasculature, their distribution in normal tissues induces adverse side effects. Therefore, a more specific distribution may be sought by using active targeting. One way to increase the biodistribution of colloidal carriers to the target site is to covalently bind a recognition molecule to their surface that will drive the carrier to the target site. This approach was used by Kopelman et al. with non-biodegradable particles [123], but to our knowledge it has not yet been developed with PS-loaded biodegradable nanoparticles. Several methods are available to covalently bind ligands to the surface of colloidal systems [149] and numerous recognition molecules are available to target either cancer cells or neovasculature surrounding tumor sites. This aspect of active targeting strategies will be further developed in chapter 10.

2.4.2
In Vitro Relevance of Polymeric Nanoparticles in PDT on Cell Models

In vitro studies with cultured cells are usually an easy way to evaluate the efficacy of new drug delivery systems. Two main issues are evaluated with *in vitro* studies: either the activity of the drug-loaded carriers on cancer cells or the cellular uptake and trafficking of the photosensitizers encapsulated into nanoparticles.

2.4.2.1 Photodynamic Activity of PS-loaded Nanoparticles

First, the potential activity of encapsulated photosensitizers has to be verified on the targeted cancer cells. For this purpose, several *in vitro* models are used, corresponding to different cancer cell lines, but the principle is generally similar. The activity of PS formulations is evaluated by measuring the inhibition of cell growth. Cells are incubated with either the free drug or the drug-loaded carriers, PDT is applied to the cells by illumination at the right wavelength, and the damage, namely cell death, is evaluated by a simple viability test, such as the colorimetric 3-(4,5-dimethylthiazol-2-yl)-2,5-diphenyltetrazolium bromide (MTT) assay (Fig. 2.4).

Nanoencapsulation does not affect the activity of photosensitizers. Indeed, the phototoxicity of loaded nanoparticles, when compared to free drug under the same experimental conditions, was similar or even better.

The effect of mTHPC in different formulations: in solution, PLA NCs, poloxamer-coated NCs, PEG-grafted PLA (PLA-PEG) NCs, and oil/water nanoemulsion (NE) has been compared by Bourdon et al. [142]. HT29 human adenocarcinoma cells were treated with increasing concentrations of PS (0.125 to 1.25 µg mL^{-1}). A long incubation time (300 min) and high light dose (25 J cm^{-2}) were used, as fluences of 5 or 10 J cm^{-2} seemed to be inefficient. Phototoxicity increased when PS concentrations increased, regardless of formulation. In their experimental setting, all formulations showed similar photoactivity, except PLA-PEG NCs, which yielded a lower phototoxicity.

The activity of loaded nanocarriers was demonstrated with mTHPC on HT29 tumor cells, but the same concentration-dependent profile was obtained for other nanocarriers or photosensitizers such as meso-tetra(*p*-hydroxyphenyl)porphyrin (pTHPP) used by Konan et al. [136]. In this case, the influence of drug concentration on cellular toxicity with pTHPP-loaded nanoparticles (PLA or PLGA) was compared to free pTHPP. For this purpose, EMT-6 mammary tumor cells were used

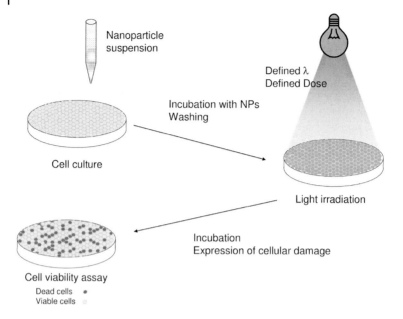

Fig. 2.4. Schematic representation of *in vitro* efficacy assay on cell culture. Cells are incubated with NPs; after washing, the plate is irradiated with light at the right wavelength. The light dose depends on the power of the light and time of exposure. Photodamage is assessed by cell viability assay after incubation in fresh medium.

under the following conditions: 1 h of incubation with increasing concentrations of PS (3 to 10 µg mL^{-1}) and a light exposure of 6 J cm^{-2}. For all delivery systems tested, the phototoxicity increased with PS concentration. The most important difference between formulations was observed at 3 µg mL^{-1}, where loaded NPs exhibited a two-fold higher activity than free pTHPP. In contrast, at 6 µg mL^{-1} all formulations reached a plateau of cell death (90% dead cells). Thus, encapsulation of a photosensitizer is an advantage, as the same photodynamic effect could be obtained with a lower concentration, thereby minimizing possible side effects.

Before analyzing the phototoxicity of photosensitizers, the following controls have to be performed: dark toxicity of photosensitizers, effect of irradiation on untreated cells and finally toxicity of unloaded nanoparticles. With this intention, Bourdon et al. [142] related no dark toxicity with mTHPC-loaded NCs in human colorectal adenocarcinoma (HT29) cells after incubation times of up to 18 h. Light or photosensitizer alone did not lead to a significant decrease in survival fraction according to Konan et al. [136] on EMT-6 mammary tumor cells. Moreover, unloaded nanoparticles, after a 24 h-incubation and an irradiation at 9 J cm^{-2}, induced no phototoxicity in concentrations up to 20 µg mL^{-1}. Thus, the observed phototoxic effects were only triggered by the combination of a specific photosensitizer and its activation by light.

The research by Bourdon et al. and Konan et al. have validated the proof of concept of using polymeric NPs for PDT [136, 142]. It is, however, difficult to compare the ability of two different photosensitizers when the carriers and the experimental conditions are different. Indeed, several parameters can influence the photoactivity of a drug: experimental conditions (incubation time, light dose), nature of drugs and delivery systems (free or encapsulated), size of carriers, mechanisms of uptake (diffusion, endocytosis), subcellular localization and mechanism of cells destruction.

Influence of Experimental Settings The survival rate is determined by a viability assay, which is generally done one day after irradiation. Konan et al. have studied the influence of the time delay between irradiation and viability assay [136]. For this purpose, the cell viability was determined either immediately or 18 h after irradiation. The effect of the photosensitizer (pTHPP) was undervalued when the MTT assay was carried out immediately after irradiation: in this case the drug dose needed to kill 50% of cells (IC_{50}) was doubled. Cell damage resulting from photochemical reactions may not be immediately lethal. Indeed, cell death implicates several cascades of reactions such as activation of enzymatic processes. Therefore, cell death can not be observed immediately after illumination. This study on the time delay is the only one in the literature on PDT with loaded-nanoparticles. However, a study carried out on 5-ALA in solution by Betz et al. [150], evaluating cell viability 18 or 24 h after irradiation, seemed to be reasonable to compare different carriers or experimental conditions. It is, however, difficult to deduce a general trend from these results as the needed post-irradiation incubation will depend at least on the PS and the cell line.

The incubation time of the PS with cells and irradiation parameters also have an effect on the phototoxicity of photosensitizers on cancer cells. Indeed, phototoxicity generally increases with incubation time and this correlation is intensified at higher drug concentrations. For example, for pTHPP-nanoparticles, cell viability began to show a decrease after 30 min incubation at 3 µg mL^{-1}; however, this was reduced to 15 min when using 6 and 8 µg mL^{-1} [136]. At higher concentrations, more NPs are available, so cellular uptake may be improved and then the time required for a good efficiency is reduced.

If the incubation time is long enough, an increase in light dose can also play a considerable role on phototoxicity. PDT is performed at a fixed wavelength, where the photosensitizer absorbs photons, from a laser source [142] or from white light passing through an aqueous filter (e.g., rhodamine) [136]. The fluence rate is determined using a photometer and the irradiation times are adjusted accordingly to the desired light doses. The homogeneity of the light delivered to cells is very important to allow comparison between different treatments. Moreover, the temperature during irradiation has to be controlled to avoid thermal effects on cell viability. Therefore, irradiation times should not be too long. For PLA-nanoparticles, an increase from 6 to 9 J cm^{-2} triggered a decrease in cell viability from 85 to 18% after 15 min incubation [136]. Such a fall in viability was observed for all polymeric systems, whereas the influence of light dose with free drug was quite low, 100 to 73%, respectively, for the same dose. Perhaps the free drug needs more time to penetrate the cells, and thus 15 min is not long enough to obtain a reasonable phototoxicity. Moreover, encapsulation may change the intracellular distribution into the different compartments of the cell, bringing the PS closer to the targets.

Influence of Incubation Medium Photoactivity of loaded NPs is generally affected by the presence of serum proteins. When cells are incubated with a photosensitizer in the presence of proteins from fetal calf serum (FCS) or fetal bovine serum (FBS), phototoxicity is improved [135, 141]. This trend was observed for all nanoparticle formulations, whereas free drug did not seem to be influenced by serum. For example, pTHPP-loaded nanoparticles [135] were two-fold more efficient in the presence of 10% FBS regardless of the polymer (PLA, 50:50 and 75:25 PLGA). The serum may enhance the intracellular accumulation of photosensitizer but also favor the monomeric form of the photosensitizer. Indeed, as hydrophobic entities, photosensitizer molecules tend to aggregate in aqueous solution or within the nanoparticles. In this aggregated state, the photosensitizer exhibits a low photodynamic activity and has to be first dispersed to become efficient [151]. Proteins could enhance this dispersion process mainly by lipoprotein association with PS. Improvements in phototoxicity were observed according to the formulations by Konan et al. with pTHPP and verteporfin [135, 141] and Bourdon et al. with mTHPC [142]. Indeed, when compared to nanoparticle formulations, free drug and drug formulated in oil/water NE were not affected by the presence of proteins. The association of PS molecules and proteins can be studied by fluorescence. When aggregated, PS emits very little fluorescence, so the state of dispersion can be observed by following the increase in fluorescence after adding serum. A rapid transfer of verteporfin seems to occur from the nanoparticles to the serum proteins, as Konan-Kouakou et al. observed a rapid increase in fluorescence immediately after the injection of 5% FBS in the suspension medium [141]. In contrast, verteporfin was slightly transferred from DMSO/PBS formulation, which correlates well with the results observed for *in vitro* phototoxicity.

Influence of Carrier Characteristics Different types of polymers are available for NP manufacture. The characteristics of the surface can also be changed either by using block copolymers or by coating particles with different excipients. Finally, managing the preparation processes parameters allows control of NP size. It is therefore possible to design customized particles that exhibit very different characteristics in terms of physicochemical properties as well as delivery features. Several studies have evaluated the influence of these parameters.

Polymer and Surface Modifications The carrier nature can modulate the photoactivity. The nature of the polymer influences the phototoxicity of drug-loaded nanoparticles. Nanoparticles having the same characteristics (size, drug loading and polymer molecular weight) can exhibit different *in vitro* toxicities that depend on the polymer they are made of. Konan et al. have shown that PLGA-nanoparticles are more efficient than PLA-nanoparticles [136]. Even the lactic/glycolic ratio played a role, since 50:50 PLGA induced a drop in viability at lower doses than 75:25

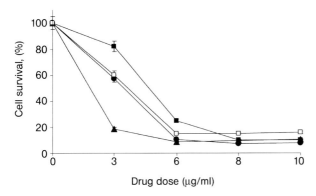

Fig. 2.5. Influence of drug concentration on photocytotoxicity of free pTHPP (■) or pTHPP-loaded nanoparticles (▲, 50:50 PLGA; □, 75:25 PLGA; ●, PLA). The EMT-6 tumor cells were incubated for 1 h, at equivalent drug concentrations, ranging from 3 to 10 µg mL^{-1}, for 1 h and irradiated at a light dose of 6 J cm^{-2} (655 nm). The MTT assay was performed 18 h after light exposure. Each data point represents the mean ($\pm$S.D.) of six values. (Redrawn from [136].)

PLGA (Fig. 2.5). Thus, with the same photosensitizer, *in vitro* activity can be affected by the polymer hydrophile–lipophile balance value, degradation rate and drug release profile. The hydrolysis-labile ester linkages are more accessible to water in PLGA than PLA. In the same way, the higher the content of glycolide in the polymer, the more accessible it would be to water, resulting in a faster hydrolysis. Thus, a faster intracellular release of drug is expected from 50:50 PLGA nanoparticles, which explains their higher efficiency.

By coating nanoparticles with poloxamer or using PEG-grafted polymers, different efficacies can also be obtained with the same polymer (PLA). Naked PLA NCs or poloxamer NCs exhibited quite similar toxicities as determined by Bourdon et al. [142]. However, PLA-PEG NCs induced a slight decrease in phototoxicity.

Particle Size The influence of particle size is a key factor regarding cellular internalization [141, 152, 153]. Usually, smaller particles tend to be more readily internalized, and thus higher activity is expected. This parameter has not been yet extensively studied for PDT. The efficiency of a photosensitizer seems to be improved by encapsulating it into small nanoparticles rather than larger nanoparticles. Konan-Kouakou et al. have compared three formulations of verteporfin on EMT-6 mammary tumor cells [141]. Free verteporfin and verteporfin-loaded nanoparticles with different mean sizes (167 and 370 nm) resulted in different survival rates after 1 h-incubation when irradiated at 6 J cm^{-2}. Treatment with 70 ng mL^{-1} of free or entrapped verteporfin into large nanoparticles yielded only 11 and 29% cell death, respectively. At the same concentration, almost 69% of cells were killed by small nanoparticles. The small size can increase the particles capacity to be taken up by cells and allow a faster drug release into cells. Furthermore, the distri-

on whether it is encapsulated or free. This was further demonstrated in a recent publication where the degree of internalization of pTHPP-loaded nanoparticles sharply dropped when incubation was performed at +4 °C whereas no affect on the uptake of the free drug was observed [135]. Thus, an energy-dependent process is involved in the uptake of loaded-nanoparticles. Endocytosis of NPs leads to higher intracellular concentration whereas free pTHPP, owing to its hydrophobic nature, tends to diffuse passively into the cell membrane, where it is less active. Therefore, encapsulation favors the PS internalization; however, PLA nanoparticles were less likely to be taken up than PLGA NP. In this case, pTHPP is a PS able to cross the membrane by diffusion; however, other PS, which are less hydrophobic or too polar to diffuse through the plasma membrane, are taken up, similar to the NPs, by endocytosis.

ROS have a short life-time, and their activity is limited to sites close to ROS generation; thus, PS uptake by cancer cells is crucial for effective PDT. To a certain degree, the type of damage that occurs in cells depends on the subcellular localization of the PS. The localization of PS in cancer cells, studied by fluorescence microscopy, seems to be formulation-dependent. The intracellular localization of PS might differ between free PS and loaded-carriers, permitting induction of different photochemical lesions in irradiated cells. A specific subcellular localization could determine the mechanism of cell death. For example, localization in mitochondria has been associated with the tendency of PS to produce apoptosis [12].

Precise PS localization can be determined by confocal laser scanning microscopy and by using co-staining of cellular organelles using fluorescent markers or indirect immunofluorescence. Bachor et al. first showed that microspheres loaded with Ce_6 can be visualized in the cytoplasm and more precisely in phagolysosomes [155]. In contrast, unconjugated PS, due to their lipophilicity, seemed to stay in cellular membranes, including plasma, nuclear and mitochondrial membranes. The cellular distribution within HT29 cells reported by Bourdon et al. was also affected by encapsulation [142]. Cells treated with free drug showed a diffused distribution throughout the cytoplasm, with a non-fluorescent nuclear area. As for NCs, even if the nuclear area is still non-fluorescent, PS was localized in the Golgi system.

In contrast to these studies reporting different localization for free and encapsulated PS, Konan et al. observed, in EMT-6 cells, similar localization in early and late lysosomes for free and encapsulated pTHPP [135]. However, this discrepancy might be related to the use of different cell lines and experimental parameters. Indeed, as far as uptake is concerned, a specific behavior of EMT-6 cells has been reported compared to other cell lines [156].

The photochemical reactions induced by irradiation may damage membrane integrity and therefore cause the release of PS from their primary site of localization. The efflux of pTHPP from EMT-6 cells was studied in a comparative manner after treatment with PLGA loaded-nanoparticles and free drug at 6 µg mL^{-1} for 15 min incubation [135]. After irradiation at 6 J cm^{-2}, the rate of the pTHPP escape was evaluated by loss of fluorescence as a function of time. Indeed, free or loaded drug gradually escaped from cells as a function of time. This trend was faster for a treatment with NPs. This may be due to a higher uptake and thus to more severe photochemical damage, including important membrane disruption. Before total efflux from cells, redistribution inside the cells could be observed, e.g., from lysosomes to cytoplasm or nuclei [157]. This phenomenon has been demonstrated with PS, such as porphyrins or phthalocyanines derivatives [158, 159], but not yet with PS-loaded nanoparticles.

One of the limitations of nanocarriers is their rapid uptake by the reticuloendothelial system (RES), which results from the adsorption of opsonins (plasma proteins) on these carriers. Then, they are taken up by the cells of the immune system located mainly in the liver and the spleen. Thus, a strategy has been developed to avoid the adsorption of opsonins. Bourdon et al. have studied this phenomenon with macrophage-like cells (J774) [142]. For this purpose they compared different mTHPC formulations with the aim of evaluating the capability of surface-modified NCs to reduce phagocytosis. Compared with naked PLA NCs, drug uptake by macrophages is indeed decreased by poloxamer coating of particles, or by using PEGylated polymers (Fig. 2.7). This reduction in uptake is better achieved with PLA-PEG NCs. These results are a positive indication that RES clearance *in vivo* could be limited with such PEGylated carriers (as discussed below). Interaction between carriers and opsonins is of van der Waals type, and coating nanoparticles with a hydrophilic chain of poloxamer or PEG could increase the circulating time in the body.

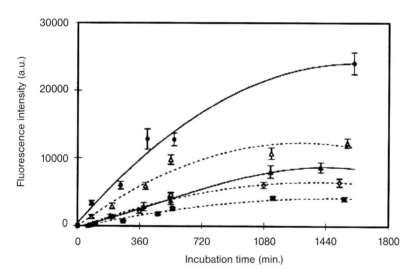

Fig. 2.7. Uptake of mTHPC (0.25 µg mL^{-1}) by macrophage-like J774 cells as determined by microspectrofluorimetry. mTHPP formulations: PLA NCs (●), nanoemulsion (△), solution (▲), poloxamer-coated PLA NCs (○) and PLA–PEG NCs (●). Cellular fluorescence intensities were measured at 654 nm. For each experiment, data have been averaged from intensity values determined on 30 individual living cells. Experiments were carried out in triplicate (bars represent S.E.). (Reprinted from Ref. [142] with permission.)

2.4.3
In Vivo Relevance of Polymeric Nanoparticles in PDT

Encapsulation may lead to a different intracellular localization of the hydrophobic photosensitizers, favoring light activated phototoxicity. One must, however, be very careful in extrapolating previous data to an *in vivo* situation. Indeed, *in vitro* experiments are obtained on cell monolayers directly in contact with light and PS, conditions rarely encountered *in vivo*. These studies are a proof of concept as they document the ability of the NP approach to improve PDT efficacy. Encapsulation would allow one not only to reduce the PS dose administered to the patients but also to use a lower light dose, thus reducing potential collateral damage to neighboring tissues. To demonstrate these advantages of NPs, *in vivo* studies have to be carried out with several goals: first, to compare the body biodistribution of PS according to their formulations, to assess the efficacy of PS-loaded nanocarriers as far as vascular or tumor suppression effects are concerned, and finally to investigate possible side effects, such as skin photosensitization.

2.4.3.1 Biodistribution and Pharmacokinetics of Photosensitizers Coupled to Nanoparticles

As with therapies against cancer, one of the main challenges in photodynamic therapy is enhancement of the PS concentration ratio between tumor and other organs [7]. However, the localization of NP in specific tissues may depend on the intrinsic characteristics of the carrier (e.g., nature of the polymer, size, and surface properties). The tissue distribution and pharmacokinetics of a PS can be influenced by its incorporation into nanoparticles. Generally, following i.v. administration, nanoparticles are rapidly and extensively taken up by the RES [125]. Accordingly, as soon as a few minutes after intravenous injection of nanoparticles, the PS mainly accumulates in the liver and spleen. Thereafter, the RES drug level gradually decreases over several days, depending on the biodegradability of the polymer and on the drug release kinetics.

Although few *in vivo* studies have been carried out, the advantages of surface modification to decrease the accumulation of the photosensitizer in the liver have been demonstrated. The body distribution of poloxamer-coated-PIHCA NCs containing radiolabeled tetraiodinated zinc-phthalocyanine ($ZnPcI_4$) were studied in healthy Balb-C mice and mice bearing the EMT-6 mammary tumor [130]. The accumulation of PS in the liver was significantly lower with poloxamer-coated NCs than with free $ZnPcI_4$ in solution. As a consequence, the main fraction of PS was present in blood when NCs are used as a carrier. The experiments in tumor-bearing mice confirmed not only the reduced liver uptake with NCs but also showed a higher uptake by the tumor. Photodynamic therapy should be performed when PS concentrations have a maximum value in the tumor, thus, in this study, tumor-to-blood ratios > 200 were obtained as early as 12 h post-injection. This high tumor selectivity achieved by surface modification of nanocarriers confirmed that NPs could be an efficient drug delivery system for cancer treatment.

PLA biodegradable nanoparticles coated with PEG-20 000 have also been suggested to enhance tumor uptake of the encapsulated compound [139]. Coating PLA NPs with PEG-20 000 substantially enhanced the blood circulation time of the photosensitizer hexadecafluoro zinc phthalocyanine (ZnPcF$_{16}$), as compared to plain particles. After 24 and 168 h, the cumulated uptake of the compound in the liver and spleen represented 61% and 44% for plain NPs versus 50% and 29% for PEG-coated NPs, respectively. The reduction of the uptake of the NPs by the RES and the resulting longer blood circulation time were associated with a threefold increase of the compound concentration in the tumor after 24 h. Such coated NPs yielded advantageous tumor-to-skin and tumor-to-muscle ratios, which is important in predicting the risk of damage to adjacent tissues during PDT.

Similar results were also observed by fluorescence with poloxamer-coated nanocapsules and PEG-grafted PLA nanocapsules loaded with mTHPC [143]. A decrease in liver distribution was observed with coated particles and was more pronounced for PLA-PEG NCs. Tumor distribution was also affected, and PLA-PEG particles seemed to better accumulate in the tumor tissue. Nonetheless, the influence of surface modification on particle biodistribution is difficult to compare between studies because results are frequently presented in arbitrary units of fluorescence, and the relative fluorescence of each formulation is not given.

2.4.3.2 Vascular Effects

Vascular damage and blood flow stasis are consequences of PDT on solid tumors. The irreversible destruction of the tumor vasculature, with the subsequent ischemia, is primarily responsible for an effective PDT of solid tumors and contributes to the long-term tumor control [18, 160]. Vascular events observed after PDT include release of vasoactive molecules, enhanced leakage and platelet aggregation, followed by occlusion of the blood vessels. Apart from oncological applications, the vascular occlusion induced by PDT has been used to treat the wet form of age-related macular degeneration, characterized by choroidal neovascularization. The mechanisms underlying the vascular effects of PDT differ greatly according to the nature of the PS and have been studied in different animal models [25]. However, the effect of incorporating photosensitizers in nanoparticles on the PDT-induced vascular occlusion is still unexplored and has only been studied *in vivo* using the chick chorioallantoic (CAM) model. The chorioallantoic membrane is a highly vascularized organ of the chick embryo, which allows the evaluation of vascular occlusion induced by PDT [161]. The pharmacokinetics of intravenously injected PS in the CAM is followed by measurement of the fluorescence of the vascularized and non-vascularized tissues. The photodynamic activity of the PS can be assessed by evaluation of the vascular occlusion achieved after irradiation of the CAM. Vargas et al. have compared the vascular effects of pTHPP, either in solution or encapsulated in 120 nm PLGA NPs, on the CAM vessels [137]. Vascular occlusion was greatly enhanced by the incorporation of pTHPP into NPs (Fig. 2.8), probably because of the reduced diffusion of NPs out of vessels during irradiation. Indeed, non-encapsulated PS quickly leaked out of the vasculature, whereas PS-

a. Evaluation of PDT induced damage on CAM vessels

Damage Scale	Criterion
0	No damage
1	Partial closure of capillaries of diameter < 10 μm
2	Closure of capillary system, partial closure of blood vessels of diameter < 30 μm and size reduction of larger blood vessels
3	Closure of vessels of diameter < 30 μm and partial closure of higher order vessels
4	Total closure of vessels of diameter < 70 μm and partial closure of larger vessels
5	Total occlusion of vessels in the irradiated area

b. pTHPP-loaded nanoparticles

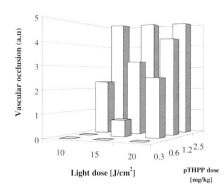

c. pTHPP non-encapsulated

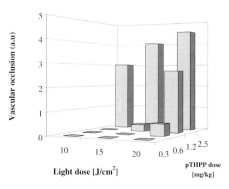

Fig. 2.8. Top. (a) Evaluation criteria of vascular occlusion induced in CAM vessels. Bottom: Comparison of the vascular damage induced by (b) pTHPP-loaded nanoparticles and (c) pTHPP dissolved in a mixture of ethanol, poly(ethylene glycol) and water. CAM was irradiated with various light doses. Mean ($n = 3$). (Adapted from [137].)

loaded NPs remained longer intravascularly during the 25 min observation time. Using the CAM model, Pegaz et al. have evaluated the influence of the encapsulation of PS with different degrees of lipophilicity on the vascular effects of PDT [138]. Porphyrin derivatives, such as meso-tetraphenylporphyrin (TPP) and meso-tetra(4-carboxyphenyl)porphyrin (TCPP), and chlorin derivatives, such as pheophorbide-a (pheo-a) and Ce_6, were encapsulated in PLA nanoparticles of around 200 nm. The PS loading increased with the lipophilicity of the encapsulated PS, ranging from 0.5% for Ce_6 to 4.6% for TPP. The more hydrophobic PS, TPP for porphyrins and pheo-a for chlorins, extravasated to a lesser extent than the more hydrophilic derivatives. At 1 mg per kg of chick embryo body weight, the extent of vascular occlusion induced by the NPs decreased with decreasing lipophilicity of the PS: TPP > TCPP > pheo-a > Ce_6. The authors suggest that the hydrophile–

lipophile balance value of the PS plays an important role in the release of porphyrins and chlorins from these NPs. However, NPs with different drug loadings were compared in this study, and the loading rate may also influence the phototoxicity of NPs. Furthermore, controls using the PS in solution were not performed. Recently, the effect of the NP size on the extent of vascular occlusion has been investigated in the same model. pTHPP was incorporated into PLGA NPs of about 100, 300 and 600 nm [154]. Although the nanoparticles had similar porphyrin loading, the phototoxic effects and the pharmacokinetic profile of the drug were influenced by the size of the nanocarrier. Vascular occlusion decreased as the NPs size rose. Although *in vivo* cancer models should be used to further evaluate the vascular effects of photosensitizers loaded in NPs, observations on the CAM model suggest that the pharmacokinetic profile and the vascular effects induced by PS can be enhanced and modulated by their incorporation into NPs.

2.4.3.3 *In Vivo* Efficacy on Tumor: Tumor Suppression Effects

Encapsulation of $ZnPcF_{16}$ into PEG-coated PLA NPs greatly improved its photodynamic activity against EMT-6 mouse mammary tumor implanted in Balb/c mice [140]. Light irradiation was performed 24, 48 and 72 h after intravenous injection of $ZnPcF_{16}$ in PEG-coated NPs or Cremophor® [polyoxethylated castor oil (CRM)] based emulsion. At a concentration of 1 µmol kg^{-1}, the best tumor response was obtained when irradiation was carried out 24 h post-injection. Indeed, 63% of Balb/c mice showed no macroscopic sign of tumor progression one week after PDT, and they were completely cured three-weeks post-treatment. In contrast, treatment with CRM formulation led to only 14% tumor regression. Thus, uptake is not the only important parameter as the two formulations were taken up by tumor at a similar rate at this time point (24 h), but the distribution inside the tumor might be different, depending on the formulation. To improve early tumor response, Allémann et al. [140] increased the PS doses (2 and 5 µmol kg^{-1}). For both doses, early tumor response, in terms of edema, was observed for all treated mice. At 2 µmol kg^{-1}, no effect of the time delay between NP injection and irradiation was observed. In this case, three-weeks post-PDT, 40% of the mice showed complete healing when PDT was performed either at 24, 48 or 72 h post-injection. At 5 µmol kg^{-1}, the treatment with NPs is even more efficient as all mice were cured as compared to only 60% when treated with CRM. Unfortunately, no comparison with uncoated NPs was performed. Nonetheless, uncoated NPs appear to be efficient as well, since Konan-Kouakou et al. have shown the ability of verteporfin-loaded PLGA NPs to control rhabdomyosarcoma (M1) tumor growth implanted in DBA/2 mice [141]. At 1.4 µmol kg^{-1}, tumor suppression was obtained with irradiation as early as 15 and 30 min after injection, with 66 and 75% tumor-free animals, respectively, on day 20. However, irradiation performed 60 min post-injection led to only 33% tumor-free animals. This suggests rapid *in vivo* clearance of the PS when administered in NPs. Moreover, the size of NPs might influence their efficacy *in vivo*, as demonstrated *in vitro* [141]. Indeed, small NPs could reach the target sites and thus be active more rapidly than large NPs. But in this part of the study, Konan-Kouakou et al. showed no control formulations, such as free PS in solution or emulsion in order to compare the possible differences in efficacy *in vivo* [141].

2.4.3.4 Adverse Effects

Despite promising results in the treatment of cancer, the use of photosensitizers is associated with undesirable side effects, such as prolonged cutaneous photosensitivity that persists for a long time (up to several months). New sensitizers have thus been developed to combine acceptably low rates of skin phototoxicity and clinically useful tumor tissue specificity.

Konan-Kouakou et al. have evaluated the skin photosensitivity caused by verteporfin when administered in NPs to SKH1 hairless mice [141]. Mice were exposed to 60 J cm^{-2} of light 15 or 60 min after PS injection. Photosensitivity was assessed 1 and 3 days later using a scoring system for erythema/eschar and edema formation (Table 2.4). The total skin photosensitivity score was calculated as the sum of scores from injury observations. The highest skin photosensitivity scores were observed one day after light exposure when mice were exposed to light 15 min after injection. The average score was then 1.7 ± 0.6, and 3 days after exposure the value came back to the normal. No skin photosensitivity was observed when mice were irradiated 60 min after injection. Thus, the animals are photosensitive for only a short period, which is very important in limiting the risk of adverse side effects. However, this might be limited to verteporfin, for which a rapid clearance has been described [162]. Although NPs appear to prevent adverse effects, this aspect has only been evaluated in one report – confirmation requires more experimental data.

Tab. 2.4. Skin photosensitivity scoring system[a].

Erythema plus eschar
0 No observable reaction
1 Minimally detectable erythema
2 Slightly visible pale pink erythema, no vessels broken, no red spots
3 Blanching, few broken vessels, no eschar formation
4 Definite erythema, more broken vessels leading to yellow eschar formation
5 Severe reaction, many broken vessels, eschar formation – but less than 50% of site
6 Very severe rosette, eschar formation on more than 50% of site

Edema
0 No observable reaction
1 Slight edema within exposure site
2 Mild edema within exposure site (skin fold measurement < 1 mm)
3 Moderate edema (skin fold measurement 1–2 mm thick)
4 Severe edema extending beyond exposure site (skin fold measurement > 2 mm thick)

---

[a] Total skin photosensitivity score is the sum of scores from erythema plus eschar and edema observations (minimum = 0, maximum = 10).

2.5 Conclusions

Photodynamic therapy and photodetection are innovative and newly developed approaches for the treatment and detection of cancer. It implies the photoactivation of photosensitizers to induce cellular damage in the target tissue. This approach to cancer chemotherapy, due to the nature of the drug itself, is characterized by better targeting of the treatment, thus limiting side effects usually encountered with conventional therapies. The development of this therapy is somehow limited by the fact that the more potent molecules are hydrophobic, thus requiring potentially harmful solvents to have injectable formulations. The use of biodegradable or non-biodegradable particles as a photosensitizer formulation allows injectable suspensions to be obtained. Encapsulation of different types of photosensitizers does not reduce their photoactivity and in most cases leads to a better activity than shown by the free compounds.

Promising results have been found with non-biodegradable materials. However, non-biodegradable nanoparticles do not offer the possibility of drug release patterns achieved with biodegradable polymers. Further, toxicological aspects have not been yet fully addressed.

In vivo and *in vitro* literature data have reported interest in the encapsulation of PS into polymeric nanoparticles, a system offering long-term stable shelf life. The main strength of polymeric nanoparticles for PS delivery is that they are well tolerated both *in vivo* and *in vitro*. Furthermore, encapsulation of PS preserves their pharmacological activity and may decrease the occurrence of adverse effects. Currently, the mechanistic approach behind pharmacological efficiency has not yet been explored and it would be interesting to better understand how photosensitizers can be still active when incorporated into a polymeric shell. Indeed, is the release of the PS necessary for the PDT effect to happen? Since PDT is a two-step process with, first, drug administration and, second, light activation, release kinetics from the polymeric matrix may have a determining influence on treatment dose and schedule. Another unevaluated aspect is the possibility of using particles to actively target cancer or neovascularized tissues. Active targeting with nanoparticles is still at an early stage and its development in the field of PDT will increase tremendously the interest in this type of carrier.

Photodetection with PS loaded NPs has not yet been evaluated. However, there are many opportunities to develop delivery systems explicitly for PD. For instance, PD is mostly used to detect superficial cancers, but the detection of deeper localized tumor, using optic fibers, could take advantage of the tumor localizing properties of NPs.

Acknowledgments

The authors are very grateful to Dr. Marino Campo for interesting discussions and for critical reading of the manuscript.

Abbreviations

5-ALA	5-Aminolevulinic acid
BPD-MA	Benzoporphyrin derivative monoacid ring A
CAM	Chick chorioallantoic membrane
Ce_6	Chlorine e_6
CNV-AMD	Choroidal neovascularization associated with age-related macular degeneration
CRM	Cremophor®
DMSO	Dimethyl sulfoxide
EPR	Enhanced permeability and retention
FBS	Fetal bovine serum
FCS	Fetal calf serum
FDA	U.S. food and drug administration
HpD	Haematoporphyrin derivative
HPPH	2-Devinyl-2-(1-hexyloxyethyl)pyropheophorbide
HT	Hyperthermia therapy
IC_x	Inhibitory concentration (drug dose needed to kill x% of cells)
i.v.	Intravenous
IR	Infrared
LDL	Low-density lipoproteins
MB	Methylene blue
MR	Magnetic resonance
mTHPC	meta-tetra(Hydroxyphenyl)chlorin
MTT	3-(4,5-Dimethylthiazol-2-yl)-2,5-diphenyltetrazolium bromide
NC(s)	Nanocapsule(s)
NE	Nanoemulsion
NP(s)	Nanoparticle(s)
NS	Nanosphere
PAA	Polyacrylamide
PACA	Poly(alkyl cyanoacrylate)
PD	Photodetection
PDT	Photodynamic therapy
PEG	Poly(ethylene glycol)
pheo-a	Pheophorbide-a
PIHCA	Poly(isohexyl cyanoacrylate)
PLA	Poly(lactic acid)
PLGA	Poly(lactic-*co*-glycolic acid)
PpIX	Protoporphyrin IX
PS	Photosensitizer
pTHPP	meso-tetra(*p*-Hydroxyphenyl)porphyrin
QD(s)	Quantum dot(s)
RES	Reticuloendothelial system
ROS	Reactive oxygen species
SnET2	Tin-ethyl etiopurpurin

TCPP	meso-tetra(4-Carboxyphenyl)porphyrin
TPP	meso-Tetraphenylporphyrin
UV	Ultraviolet
$ZnPcF_{16}$	Hexadecafluoro zinc phthalocyanine
$ZnPcI_4$	Tetraiodinated zinc phthalocyanine
$ZnPcS_2$	Disulfonated zinc phthalocyanine
$ZnPcS_4$	Tetrasulfonated zinc phthalocyanine

References

1 A. Ruck, G. Beck, R. Bachor, N. Akgun, M. H. Gschwend, R. Steiner, Dynamic fluorescence changes during photodynamic therapy in vivo and in vitro of hydrophilic Al(III) phthalocyanine tetrasulphonate and lipophilic Zn(II) phthalocyanine administered in liposomes, J. Photochem. Photobiol. B, 1996, 36, 127–133.

2 L. Morlet, V. Vonarx-Coinsmann, P. Lenz, M. T. Foultier, L. X. de Brito, C. Stewart, T. Patrice, Correlation between meta(tetra-hydroxyphenyl)chlorin (m-THPC) biodistribution and photodynamic effects in mice, J. Photochem. Photobiol. B, 1995, 28, 25–22.

3 T. Glanzmann, M. Forrer, S. A. Blant, A. Woodtli, P. Grosjean, D. Braichotte, H. van den Bergh, P. Monnier, G. Wagnieres, Pharmacokinetics and pharmacodynamics of tetra(m-hydroxyphenyl)chlorin in the hamster cheek pouch tumor model: Comparison with clinical measurements, J. Photochem. Photobiol. B, 2000, 57, 22–32.

4 G. A. Wagnieres, W. M. Star, B. C. Wilson, In vivo fluorescence spectroscopy and imaging for oncological applications, Photochem. Photobiol., 1998, 68, 603–632.

5 R. R. Allison, H. C. Mota, C. H. Sibata, Clinical PD/PDT in North America: An historical review, Photodiagn. Photodyn. Therap., 2004, 1, 263–277.

6 J. E. van Lier, Photosensitization: Reaction pathways, in D. P. Valenzeno, R. H. Pottier, P. Mathis, R. H. Douglas (eds.), Photobiological Techniques, Plenum Press, New York, 1991, ch. 7.

7 N. Lange, Controlled drug delivery in photodynamic therapy and fluorescence-based diagnosis of cancer, in M.-A. Mycek, B. W. Pogue (eds.), Handbook of Biomedical Fluorescence, Marcel Dekcer, Inc., New York, 2003, ch. 16.

8 C. S. Foote, Mechanisms of photosensitized oxidation, Science, 1968, 162, 963–970.

9 C. S. Foote, Definition of type I and type II photosensitized oxidation, Photochem. Photobiol., 1991, 54, 659.

10 C. Hadjur, G. Wagnieres, F. Ihringer, P. Monnier, H. van den Bergh, Production of the free radicals O_2^- and OH by irradiation of the photosensitizer zinc(II) phthalocyanine, J. Photochem. Photobiol. B, 1997, 38, 196–202.

11 T. Shutova, T. Kriska, A. Németh, V. Agabekov, D. Gàl, Physico-chemical modeling of the role of free radicals in photodynamic therapy, Biochem. Biophys. Res. Commun., 2000, 270, 125–130.

12 A. P. Castano, T. N. Demidova, M. R. Hamblin, Mechanisms in photodynamic therapy: Part one – photosensitizers, photochemistry and cellular localization, Photodiagn. Photodyn. Ther., 2004, 1, 279–293.

13 H. M. Chen, C. T. Chen, H. Yang, M. Y. Kuo, Y. S. Kuo, W. H. Lan, Y. P. Wang, T. Tsai, C. P. Chiang,

Successful treatment of oral verrucous hyperplasia with topical 5-aminolevulinic acid-mediated photodynamic therapy, *Oral Oncol.*, **2004**, 40, 630–637.

14 K. Berg, P. K. Selbo, A. Weyergang, A. Dietze, L. Prasmickaite, A. Bonsted, B. O. Engesaeter, E. Angell-Petersen, T. Warloe, N. Frandsen, A. Hogset, Porphyrin-related photosensitizers for cancer imaging and therapeutic applications, *J. Microsc.*, **2005**, 218, 133–147.

15 T. J. Dougherty, C. J. Gomer, B. W. Henderson, G. Jori, D. Kessel, M. Korbelik, J. Moan, Q. Peng, Photodynamic therapy, *J. Natl. Cancer Inst.*, **1998**, 90, 889–905.

16 N. Oleinick, R. Morris, I. Belichenko, The role of apoptosis in response to photodynamic therapy: What, where, why, and how, *Photochem. Photobiol. Sci.*, **2002**, 1, 1–21.

17 A. P. Castano, T. N. Demidova, M. R. Hamblin, Mechanisms in photodynamic therapy: Part two – Cellular signaling, cell metabolism and modes of cell death, *Photodiagn. Photodyn. Ther.*, **2005**, 2, 1–23.

18 V. H. Fingar, Vascular effects of photodynamic therapy, *J. Clin. Laser Med. Surg.*, **1996**, 14, 323–328.

19 J. V. Moore, C. M. L. West, C. Whitehurst, The biology of photodynamic therapy, *Phys. Med. Biol.*, **1997**, 42, 913–935.

20 B. Krammer, Vascular effects of photodynamic therapy, *Anticancer Res.*, **2001**, 21, 4271–4277.

21 J. Krutmann, Therapeutic photoimmunology: Photoimmunological mechanisms in photo(chemo) therapy, *J. Photochem. Photobiol. B*, **1998**, 44, 159–164.

22 M. Korbelik, G. J. Dougherty, Photodynamic therapy-mediated immune response against subcutaneous mouse tumors, *Cancer Res.*, **1999**, 59, 1941–1946.

23 G. Canti, A. De Simone, M. Korbelik, Photodynamic therapy and the immune system in experimental oncology, *Photochem. Photobiol. Sci.*, **2002**, 1, 79–80.

24 S. B. Brown, E. A. Brown, I. Walker, The present and future role of photodynamic therapy in cancer treatment, *Lancet Oncol.*, **2004**, 5, 497–508.

25 A. P. Castano, T. N. Demidova, M. R. Hamblin, Mechanisms in photodynamic therapy: Part three – Photosensitizer pharmacokinetics, biodistribution, tumor localization and modes of tumor destruction, *Photodiagn. Photodyn. Ther.*, **2005**, 2, 91–106.

26 H. Maeda, J. Wu, Y. Sawa, Y. Matsumura, K. Hori, Tumor vascular permeability and the EPR effect in macromolecular therapeutics: A review, *J. Controlled Release*, **2000**, 65, 271–284.

27 J. P. Rovers, A. E. Saarnak, M. de Jode, H. J. C. M. Sterenborg, O. T. Terpstra, M. F. Grahn, Biodistribution and bioactivity of tetra-pegylated meta-tetra(hydroxyphenyl)chlorin compared to native meta-tetra(hydroxyphenyl)chlorin in a rat liver tumor model, *Photochem. Photobiol.*, **2000**, 71, 210–217.

28 T. Reuther, A. C. Kubler, U. Zillmann, C. Flechtenmacher, H. Sinn, Comparison of the in vivo efficiency of photofrin II-, mTHPC-, mTHPC- PEG- and mTHPCnPEG-mediated PDT in a human xenografted head and neck carcinoma, *Lasers Surg. Med.*, **2001**, 29, 314–322.

29 F. Rancan, A. Wiehe, M. Nobel, M. O. Senge, S. A. Omari, F. Bohm, M. John, B. Roder, Influence of substitutions on asymmetric dihydroxychlorins with regard to intracellular uptake, subcellular localization and photosensitization of Jurkat cells, *J. Photochem. Photobiol. B*, **2005**, 78, 17–28.

30 D. J. Ball, S. R. Wood, D. I. Vernon, J. Griffiths, T. M. Dubbelman, S. B. Brown, The characterisation of three substituted zinc phthalocyanines of differing charge for use in photodynamic therapy. A comparative study of their aggregation and photosensitising ability in relation to mTHPC and polyhaematoporphyrin,

J. Photochem. Photobiol. B, **1998**, 45, 28–35.

31 R. W. BOYLE, D. DOLPHIN, Structure and biodistribution relationships of photodynamic sensitzers, *Photochem. Photobiol.*, **1996**, 64, 469–485.

32 M. KONGSHAUG, J. MOAN, S. B. BROWN, The distribution of porphyrins with different tumour localising ability among human plasma proteins, *Br. J. Cancer*, **1989**, 59, 184–188.

33 J. C. MAZIERE, P. MORLIERE, R. SANTUS, New trends in photobiology: The role of the low density lipoprotein receptor pathway in the delivery of lipophilic photosensitizers in the photodynamic therapy of tumours, *J. Photochem. Photobiol. B*, **1991**, 8, 351–360.

34 B. A. ALLISON, P. H. PRITCHARD, J. G. LEVY, Evidence for low-density lipoprotein receptor-mediated uptake of benzoporphyrin derivative, *Br. J. Cancer*, **1994**, 69, 833–839.

35 A. S. SOBOLEV, D. A. JANS, A. A. ROSENKRANZ, Targeted intracellular delivery of photosensitizers, *Prog. Biophys. Mol. Biol.*, **2000**, 73, 51–90.

36 A. K. HAYLETT, J. V. MOORE, Comparative analysis of foetal calf and human low density lipoprotein: Relevance for pharmacodynamics of photosensitizers, *J. Photochem. Photobiol. B*, **2002**, 66, 171–178.

37 G. JORI, In vivo transport and pharmacokinetic behavior of tumour photosensitizers, *Ciba Found. Symp.*, **1989**, 146, 78–86.

38 L. POLO, G. VALDUGA, G. JORI, E. REDDI, Low-density lipoprotein receptors in the uptake of tumour photosensitizers by human and rat transformed fibroblasts, *Int. J. Biochem. Cell Biol.*, **2002**, 34, 10–23.

39 M. OCHSNER, Photophysical and photobiological processes in the photodynamic therapy of tumours, *J. Photochem. Photobiol. B*, **1997**, 39, 1–18.

40 K. WOODBURN, C. K. CHANG, S. LEE, B. HENDERSON, D. KESSEL, Biodistribution and PDT efficacy of a ketochlorin photosensitizer as a function of the delivery vehicle, *Photochem. Photobiol.*, **1994**, 60, 154–159.

41 M. M. ZUK, B. D. RIHTER, M. E. KENNEY, M. A. J. RODGERS, Effect of delivery system on the pharmacokinetics and tissue distribution of bis(Di-isobutyl octadecylsiloxy)silicon 2,3-naphthalocyanine (*iso*BOSINC), a photosensitizer for tumor therapy, *Photochem. Photobiol.*, **1996**, 63, 132–140.

42 E. REDDI, Role of delivery vehicles for photosensitizers in the photodynamic therapy of tumours, *J. Photochem. Photobiol. B*, **1997**, 37, 189–195.

43 Z.-J. WANG, Y.-Y. HE, C.-G. HUANG, J.-S. HUANG, Y.-C. HUANG, J.-Y. AN, Y. GU, L.-J. JIANG, Pharmacokinetics, tissue distribution and photodynamic therapy efficacy of liposomal-delivered hypocrellin A, a potential photosensitizer for tumor therapy, *Photochem. Photobiol.*, **1999**, 70, 773–780.

44 D. WÖRLE, S. MULLER, M. SHOPOVA, V. MANTAREVA, G. SPASSOVA, F. VIETRI, F. RICCHELLI, G. JORI, Effect of delivery system on the pharmacokinetic and phototherapeutic properties of bis(methyloxyethyleneoxy)silicon-phthalocyanine in tumor-bearing mice, *J. Photochem. Photobiol. B*, **1999**, 50, 124–128.

45 A. CASAS, H. FUKUDA, G. DI VENOSA, A. M. BATLLE, The influence of the vehicle on the synthesis of porphyrins after topical application of 5-aminolaevulinic acid. Implications in cutaneous photodynamic sensitization, *Br. J. Dermatol.*, **2000**, 143, 564–572.

46 M. C. DEROSA, R. J. CRUTCHLEY, Photosensitized singlet oxygen and its applications, *Coord. Chem. Rev.*, **2002**, 233–234, 351–371.

47 E. S. NYMAN, P. H. HYNNINEN, Research advances in the use of tetrapyrrolic photosensitizers for photodynamic therapy, *J. Photochem. Photobiol. B*, **2004**, 73, 1–28.

48 R. R. ALLISON, G. H. DOWNIE, R. CUENCA, X. H. HU, C. J. H. CHILDS, C. H. SIBATA, Photosensitizers in

clinical PDT, *Photodiagn. Photodyn. Ther.*, **2004**, 1, 27–42.
49 S.-J. TANG, N. E. MARCON, Photodynamic therapy in the esophagus, *Photodiagn. Photodyn. Ther.*, **2004**, 1, 65–74.
50 S. LAM, E. C. KOSTASHUK, E. P. COY, E. LAUKKANEN, J. C. LERICHE, H. A. MUELLER, I. J. SZASZ, A randomized comparative study of the safety and efficacy of photodynamic therapy using Photofrin II combined with palliative radiotherapy versus palliative radiotherapy alone in patients with inoperable obstructive non-small cell bronchogenic carcinoma, *Photochem. Photobiol.*, **1987**, 46, 893–897.
51 K. MOGHISSI, K. DIXON, J. A. C. THORPE, C. OXTOBY, M. R. STRINGER, Photodynamic therapy (PDT) for lung cancer: the Yorkshire Laser Centre experience, *Photodiagn. Photodyn. Ther.*, **2004**, 1, 253–262.
52 G. JORI, Tumor photosensitizers: Approaches to enhance the selectivity and efficiency of photodynamic therapy, *J. Photochem. Photobiol. B*, **1996**, 36, 87–93.
53 C. HOPPER, Photodynamic therapy: A clinical reality in the treatment of cancer, *Lancet Oncol.*, **2000**, 1, 212–219.
54 P. UEHLINGER, M. ZELLWEGER, G. WAGNIERES, L. JUILLERAT-JEANNERET, H. VAN DEN BERGH, N. LANGE, 5-Aminolevulinic acid and its derivatives: Physical chemical properties and protoporphyrin IX formation in cultured cells, *J. Photochem. Photobiol. B*, **2000**, 54, 72–80.
55 S. COLLAUD, A. JUZENIENE, J. MOAN, N. LANGE, On the selectivity of 5-aminolevulinic acid-induced protoporphyrin IX formation, *Curr. Med. Chem. Anti-Canc. Agents*, **2004**, 4, 301–316.
56 P. JICHLINSKI, Hexyl aminolevulinate fluorescence cystoscopy: A new diagnostic tool for photodiagnosis of superficial bladder cancer – a multicenter study, *J. Urol.*, **2003**, 170, 226–229.
57 V. MATZI, A. MAIER, O. SANKIN, J. LINDENMANN, P. REHAK, F. M. SMOLLE-JUTTNER, 5-Aminolaevulinic acid compared to polyhematoporphyrin photosensitization for photodynamic therapy of malignant bronchial and esophageal stenosis: Clinical experience, *Photodiagn. Photodyn. Ther.*, **2004**, 1, 137–143.
58 J. SCHMIDBAUER, Improved detection of urothelial carcinoma in situ with hexaminolevulinate fluorescence cystoscopy, *J. Urol.*, **2004**, 171, 135–138.
59 A. SIERON, A. KAWCZYK-KRUPKA, M. A. W. CEBULA, M. SZYGULA, W. ZIELEZNIK, M. GRUK, B. SUWALA-JURCZYK, Photodynamic therapy (PDT) using topically applied [delta]-aminolevulinic acid (ALA) for the treatment of malignant skin tumors, *Photodiagn. Photodyn. Ther.*, **2004**, 1, 311–317.
60 C. FRITSCH, K. LANG, W. NEUSE, T. RUZICKA, P. LEHMANN, Photodynamic diagnosis and therapy in dermatology, *Skin Pharmacol. Appl. Skin Physiol.*, **1998**, 11, 358–373.
61 C. J. KELTY, N. J. BROWN, M. W. REED, R. ACKROYD, The use of 5-aminolaevulinic acid as a photosensitiser in photodynamic therapy and photodiagnosis, *Photochem. Photobiol. Sci.*, **2002**, 1, 158–168.
62 S. JAIN, R. C. KOCKELBERGH, The role of photodynamic diagnosis in the contemporary management of superficial bladder cancer, *BJU Int.*, **2005**, 96, 17–21.
63 G. A. M. S. VAN DONGEN, G. W. M. VISSER, M. B. VROUENRAETS, Photosensitizer-antibody conjugates for detection and therapy of cancer, *Adv. Drug Deliv. Rev.*, **2004**, 56, 31–52.
64 M. A. D'HALLEWIN, A. R. KAMUHABWA, T. ROSKAMS, P. A. M. DE WITTE, L. BAERT, Hypericin-based fluorescence diagnosis of bladder carcinoma, *BJU Int.*, **2002**, 89, 760–763.
65 A. K. D'CRUZ, M. H. ROBINSON, M. A. BIEL, mTHPC-mediated photodynamic therapy in patients with advanced, incurable head and neck cancer: A multicenter study of 128

patients, *Head Neck*, **2004**, 26, 232–240.
66. C. HOPPER, A. KUBLER, H. LEWIS, I. B. TAN, G. PUTNAM, mTHPC-mediated photodynamic therapy for early oral squamous cell carcinoma, *Int. J. Cancer*, **2004**, 111, 138–146.
67. A. MILLS, S. LE HUNTE, An overview of semiconductor photocatalysis, *J. Photochem. Photobiol. A*, **1997**, 108, 35.
68. R. CAI, Y. KUBOTA, T. SHUIN, H. SAKAI, K. HASHIMOTO, A. FUJISHIMA, Induction of cytotoxicity by photo-excited TiO_2 particles, *Cancer Res.*, **1992**, 52, 2346–2348.
69. Y. KUBOTA, T. SHUIN, C. KAWASAKI, M. HOSAKA, H. KITAMURA, R. CAI, H. SAKAI, K. HASHIMOTO, A. FUJISHIMA, Photokilling of T-24 human bladder cancer cells with titanium dioxide, *Br. J. Cancer*, **1994**, 70, 1107–1111.
70. K. HOFFMANN, K. KASPAR, P. ALTMEYER, T. GAMBICHLER, UV transmission measurements of small skin specimens with special quartz cuvettes, *Dermatology*, **2000**, 201, 307–311.
71. X. MICHALET, F. F. PINAUD, L. A. BENTOLILA, J. M. TSAY, S. DOOSE, J. J. LI, G. SUNDARESAN, A. M. WU, S. S. GAMBHIR, S. WEISS, Quantum dots for live cells, in vivo imaging, and diagnostics, *Science*, **2005**, 307, 538–544.
72. A. C. S. SAMIA, X. CHEN, C. BURDA, Semiconductor quantum dots for photodynamic therapy, *J. Am. Chem. Soc.*, **2003**, 125, 15 736–15 737.
73. R. BAKALOVA, H. OHBA, Z. ZHELEV, M. ISHIKAWA, Y. BABA, Quantum dots as photosensitizers?, *Nat. Biotechnol.*, **2004**, 22, 1360–1361.
74. M. OZKAN, Quantum dots and other nanoparticles: What can they offer to drug discovery?, *Drug Discov. Today*, **2004**, 9, 1065–1071.
75. X. GAO, Y. CUI, R. M. LEVENSON, L. W. K. CHUNG, S. NIE, In vivo cancer targeting and imaging with semiconductor quantum dots, *Nat. Biotechnol.*, **2004**, 22, 969–976.
76. K. K. JAIN, Nanotechnology in clinical laboratory diagnostics, *Clin. Chim. Acta*, **2005**, 358, 37–54.
77. R. BAKALOVA, H. OHBA, Z. ZHELEV, M. ISHIKAWA, Y. BABA, Quantum dot anti-CD conjugates: Are they potential photosensitizers or potentiators of classical photosensitizing agents in photodynamic therapy of cancer?, *Nano Lett.*, **2004**, 4, 1567–1573.
78. A. M. DERFUS, W. C. W. CHAN, S. N. BHATIA, Probing the cytotoxicity of semiconductor quantum dots, *Nano Lett.*, **2004**, 4, 11–18.
79. A. W. JENSEN, S. R. WILSON, D. I. SCHUSTER, Biological applications of fullerenes, *Bioorg. Med. Chem.*, **1996**, 4, 767–779.
80. T. DA ROS, M. PRATO, Medicinal chemistry with fullerenes and fullerene derivatives, *Chem. Commun.*, **1999**, 663–669.
81. Y. YAMAKOSHI, N. UMEZAWA, A. RYU, K. ARAKANE, N. MIYATA, Y. GODA, T. MASUMIZU, T. NAGANO, Active oxygen species generated from photoexcited fullerene (C60) as potential medicines: O2-* versus 1O2, *J. Am. Chem. Soc.*, **2003**, 125, 12 803–12 809.
82. J. MOAN, K. BERG, The photodegradation of porphyrins in cells can be used to estimate the lifetime of singlet oxygen, *Photochem. Photobiol.*, **1991**, 53, 549–553.
83. J. MOAN, On the diffusion length of singlet oxygen in cells and tissues, *J. Photochem. Photobiol. B*, **1990**, 6, 343–344.
84. H. LUI, L. HOBBS, W. D. TOPE, P. K. LEE, C. ELMETS, N. PROVOST, A. CHAN, H. NEYNDORFF, X. Y. SU, H. JAIN, I. HAMZAVI, D. MCLEAN, R. BISSONNETTE, Photodynamic therapy of multiple nonmelanoma skin cancers with verteporfin and red light-emitting diodes: Two-year results evaluating tumor response and cosmetic outcomes, *Arch. Dermatol.*, **2004**, 140, 26–32.
85. M. A. CAPELLA, L. S. CAPELLA, A light in multidrug resistance: Photo-dynamic treatment of multidrug-resistant tumors, *J. Biomed. Sci.*, **2003**, 10, 361–366.
86. M. TREHAN, C. R. TAYLOR, Cutaneous photosensitivity and photoprotection for photodynamic therapy patients, in

Reddy, A. Rehemtulla, B. D. Ross, M. Philbert, R. J. Schneider, R. Kopelman, Production of singlet oxygen by Ru(dpp(SO$_3$)$_2$)$_3$ incorporated in polyacrylamide PEBBLES, *Sens. Actuators, B*, **2003**, 90, 82–89.

123 R. Kopelman, Y. E. L. Koo, M. Philbert, B. A. Moffat, G. R. Reddy, P. McConville, D. E. Hall, T. L. Chenevert, M. S. Bhojani, S. M. Buck, A. Rehemtulla, B. D. Ross, Multifunctional nanoparticle platforms for in vivo MRI enhancement and photodynamic therapy of a rat brain cancer, *J. Magn. Magn. Mater.*, **2005**, 293, 404–410.

124 A. M. Gatti, F. Rivasi, Biocompatibility of micro- and nanoparticles. Part I: In liver and kidney, *Biomaterials*, **2002**, 23, 2381–2387.

125 E. Allémann, R. Gurny, E. Doelker, Drug-loaded nanoparticles – Preparation methods and drug targeting issues, *Eur. J. Pharm. Biopharm.*, **1993**, 39, 173–191.

126 P. Couvreur, C. Dubernet, F. Puisieux, Controlled drug-delivery with nanoparticles – Current possibilities and future-trends, *Eur. J. Pharm. Biopharm.*, **1995**, 41, 2–13.

127 M. J. Alonso, Nanoparticulate drug carrier technology, *Drugs Pharm. Sci.*, **1996**, 77, 203–242.

128 C. Vauthier, C. Dubernet, E. Fattal, H. Pinto-Alphandary, P. Couvreur, Poly(alkylcyanoacrylates) as biodegradable materials for biomedical applications, *Adv. Drug Deliv. Rev.*, **2003**, 55, 519–548.

129 A. Labib, V. Lenaerts, F. Chouinard, J. C. Leroux, R. Ouellet, J. E. van Lier, Biodegradable nanospheres containing phthalocyanines and naphthalocyanines for targeted photodynamic tumor therapy, *Pharm. Res.*, **1991**, 8, 1027–1031.

130 V. Lenaerts, A. Labib, F. Chouinard, J. Rousseau, H. Ali, J. Van Lier, J. E. van Lier, Nanocapsules with a reduced liver uptake: Targeting of phthalocyanines to EMT-6 mouse mammary tumour in vivo, *Eur. J. Pharm. Biopharm.*, **1995**, 41, 38–43.

131 N. Brasseur, D. Brault, P. Couvreur, Adsorption of hematoporphyrin onto polyalkylcyanoacrylate nanoparticles: Carrier capacity and drug release, *Int. J. Pharm.*, **1991**, 70, 129–135.

132 S. Steiniger, S. E. Gelperina, I. N. Skidan, A. I. Bobruskin, S. E. Severin, J. Kreuter, Optimization of photosensitizer pharmacokinetics with biodegradable nanoparticles, *Proceedings – 28th International Symposium on Controlled Release of Bioactive Materials and 4th Consumer & Diversified Products, San Diego, CA, United States*, Controlled Release Society, Minneapolis, **2001**.

133 B. Zhao, J. Xie, J. Zhao, A novel water-soluble nanoparticles of hypocrellin B and their interaction with a model protein: C-phycocyanin, *Biochim. Biophys. Acta*, **2004**, 1670, 113–120.

134 Y. N. Konan, R. Cerny, J. Favet, M. Berton, R. Gurny, E. Allémann, Preparation and characterization of sterile sub-200 nm meso-tetra(4-hydroxylphenyl)porphyrin-loaded nanoparticles for photodynamic therapy, *Eur. J. Pharm. Biopharm.*, **2003**, 55, 115–124.

135 Y. N. Konan, J. Chevallier, R. Gurny, E. Allémann, Encapsulation of p-THPP into nanoparticles: Cellular uptake, subcellular localization and effect of serum on photodynamic activity, *Photochem. Photobiol.*, **2003**, 77, 638–644.

136 Y. N. Konan, M. Berton, R. Gurny, E. Allémann, Enhanced photodynamic activity of meso-tetra(4-hydroxyphenyl)porphyrin by incorporation into sub-200 nm nanoparticles, *Eur. J. Pharm. Sci.*, **2003**, 18, 241–249.

137 A. Vargas, B. Pegaz, E. Debefve, Y. N. Konan-Kouakou, N. Lange, J. P. Ballini, H. van den Bergh, R. Gurny, F. Delie, Improved photodynamic activity of porphyrin loaded into nanoparticles: An in vivo evaluation using chick embryos, *Int. J. Pharm.*, **2004**, 286, 131–145.

138 B. Pegaz, E. Debefve, F. Borle, J. P.

Ballini, H. van den Bergh, Y. N. Kouakou-Konan, Encapsulation of porphyrins and chlorins in biodegradable nanoparticles: The effect of dye lipophilicity on the extravasation and the photothrombic activity. A comparative study, *J. Photochem. Photobiol. B*, **2005**, 80, 19–27.

139 E. Allémann, N. Brasseur, O. Benrezzak, J. Rousseau, S. V. Kudrevich, R. W. Boyle, J. C. Leroux, R. Gurny, J. E. van Lier, PEG-coated poly(lactic acid) nanoparticles for the delivery of hexadecafluoro zinc phthalocyanine to EMT-6 mouse mammary tumours, *J. Pharm. Pharmacol.*, **1995**, 47, 382–387.

140 E. Allémann, J. Rousseau, N. Brasseur, S. V. Kudrevich, K. Lewis, J. E. van Lier, Photodynamic therapy of tumours with hexadecafluoro zinc phthalocynine formulated in PEG-coated poly(lactic acid) nanoparticles, *Int. J. Cancer*, **1996**, 66, 821–824.

141 Y. N. Konan-Kouakou, R. Boch, R. Gurny, E. Allémann, In vitro and in vivo activities of verteporfin-loaded nanoparticles, *J. Controlled Release*, **2005**, 103, 83–91.

142 O. Bourdon, V. Mosqueira, P. Legrand, J. Blais, A comparative study of the cellular uptake, localization and phototoxicity of meta-tetra(hydroxyphenyl) chlorin encapsulated in surface-modified submicronic oil/water carriers in HT29 tumor cells, *J. Photochem. Photobiol. B*, **2000**, 55, 164–171.

143 O. Bourdon, I. Laville, D. Carrez, A. Croisy, Ph. Fedel, A. Kasselouri, P. Prognon, P. Legrand, J. Blais, Biodistribution of meta-tetra(hydroxyphenyl)chlorin incorporated into surface-modified nanocapsules in tumor-bearing mice, *Photochem. Photobiol. Sci.*, **2002**, 1, 709–714.

144 J. Fu, X. Y. Li, D. K. P. Ng, C. Wu, Encapsulation of phthalocyanines in biodegradable poly(sebacic anhydride) nanoparticles, *Langmuir*, **2002**, 18, 3843–3847.

145 P. P. S. Lee, T. Ngai, J.-D. Huang, C. Wu, W.-P. Fong, D. K. P. Ng, Synthesis, characterization, biodegradation, and in vitro photodynamic activities of silicon(IV) phthalocyanines conjugated axially with poly(e-caprolactone), *Macromolecules*, **2003**, 36, 7527–7533.

146 R. H. Müller, K. H. Wallis, S. D. Tröster, J. Kreuter, In vitro characterization of poly(methyl-methacrylate) nanoparticles and correlation to their in vivo fate, *J. Controlled Release*, **1992**, 20, 237–246.

147 G. Borchard, J. Kreuter, Interaction of serum components with poly(methyl methacrylate) nanoparticles and the resulting body distribution after intravenous injection in rats, *J. Drug Target.*, **1993**, 1, 15–19.

148 J. C. Leroux, P. Gravel, L. Balant, B. Volet, B. M. Anner, E. Allémann, E. Doelker, R. Gurny, Internalization of poly(D,L-lactic acid) nanoparticles by isolated human leukocytes and analysis of plasma proteins adsorbed onto the particles, *J. Biomed. Mater. Res.*, **1994**, 28, 471–481.

149 L. Nobs, F. Bucheger, R. Gurny, E. Allémann, Current methods for attaching targeting ligands to liposomes and nanoparticles, *J. Pharm. Sci.*, **2004**, 93, 1980–1992.

150 C. S. Betz, J. P. Lai, W. Xiang, P. Janda, P. Heinrich, H. Stepp, R. Baumgartner, A. Leunig, In vitro photodynamic therapy of nasopharyngeal carcinoma using 5-aminolevulinic acid, *Photochem. Photobiol. Sci.*, **2002**, 1, 315–319.

151 B. M. Aveline, T. Hasan, R. W. Redmond, The effects of aggregation, protein binding and cellular incorporation on the photophysical properties of benzoporphyrin derivative monoacid ring A (BPDMA), *J. Photochem. Photobiol. B*, **1995**, 30, 161–169.

152 R. Raghuvanshi, Y. Katare, K. Lalwani, M. Ali, O. Singh, A. Panda, Improved immune response from biodegradable polymer particles entrapping tetanus toxoid by use of different immunization protocol and

adjuvants 585, *Int J. Pharm.*, **2002**, 245, 109.

153 P. JANI, G. W. HALBERT, J. LANGRIDGE, A. T. FLORENCE, Nanoparticle uptake by the rat gastrointestinal mucosa: Quantitation and particle size dependency, *J. Pharm. Pharmacol.*, **1990**, 42, 821–826.

154 A. VARGAS, R. GURNY, F. DELIE, Photosensitizer-loaded nanoparticles: A strategy to enhance the vascular occlusion in response to photodynamic therapy, *Proceedings – 32nd Annual Meeting & Exposition of the Controlled Release Society, Miami Beach, FL, United States*, **2005**.

155 R. BACHOR, C. R. SHEA, R. GILLES, T. HASAN, Photosensitized destruction of human bladder carcinoma cells treated with chlorin e_6-conjugated microspheres, *Proc. Natl. Acad. Sci. U.S.A.*, **1991**, 88, 1580–1584.

156 K. BERG, J. C. MAZIERE, M. GEZE, R. SANTUS, Verapamil enhances the uptake and the photocytotoxic effect of PII, but not that of tetra(4-sulfonatophenyl)porphine, *Biochim. Biophys. Acta*, **1998**, 1370, 317–324.

157 A. A. ROSENKRANZ, D. A. JANS, A. S. SOBOLEV, Targeted intracellular delivery of photosensitizers to enhance photodynamic efficiency, *Immunol. Cell Biol.*, **2000**, 78, 452–464.

158 K. BERG, K. MADSLIEN, J. C. BOMMER, R. OFTEBRO, J. W. WINKELMAN, J. MOAN, Light induced relocalization of sulfonated meso-tetraphenylporphines in NHIK 3025 cells and effects of dose fractionation, *Photochem. Photobiol.*, **1991**, 53, 203–210.

159 S. R. WOOD, J. A. HOLROYD, S. B. BROWN, The subcellular localization of Zn(II) phthalocyanines and their redistribution on exposure to light, *Photochem. Photobiol.*, **1997**, 65, 397–402.

160 C. ABELS, Targeting of the vascular system of solid tumours by photodynamic therapy (PDT), *Photochem. Photobiol. Sci.*, **2004**, 3, 765–771.

161 N. LANGE, J. P. BALLINI, G. WAGNIERES, H. VAN DEN BERGH, A new drug-screening procedure for photosensitizing agents used in photodynamic therapy for CNV, *Invest. Ophthalmol. Vis. Sci.*, **2001**, 42, 38–46.

162 A. M. RICHTER, S. YIP, E. WATERFIELD, P. M. LOGAN, Mouse skin photosensitization with benzoporphyrin derivative and photofrin: Macroscopic and microscopic evaluation, *Photochem. Photobiol.*, **1991**, 53, 281–286.

163 W. M. SHARMAN, C. M. ALLEN, J. E. VAN LIER, Photodynamic therapeutics: Basic principles and clinical applications, *Drug Discov. Today*, **1999**, 4, 507–517.

164 T. J. DOUGHERTY, An update on photodynamic therapy applications, *J. Clin. Laser Med. Surg.*, **2002**, 20, 3–7.

165 C. A. MORTON, S. B. BROWN, S. COLLINS, S. IBBOTSON, H. JENKINSON, H. KURWA, K. LANGMACK, K. MCKENNA, H. MOSELEY, A. D. PEARSE, M. STRINGER, D. K. TAYLOR, G. WONG, L. E. RHODES, Guidelines for topical photodynamic therapy: Report of a workshop of the British Photodermatology Group, *Br. J. Dermatol.*, **2002**, 146, 552–567.

3
Nanoparticles for Neutron Capture Therapy of Cancer

Hideki Ichikawa, Hiroyuki Tokumitsu, Masahito Miyamoto, and Yoshinobu Fukumori

3.1
Introduction

Conventional radiotherapy using X-rays or γ-rays has been applied to cancer treatment due to its non-invasive methodology. However, some types of tumor are radioresistive; even if radiosensitive, the radiation dose that can be delivered to the tumor is limited by the tolerance of surrounding normal tissues within the treatment volume. In addition, some special techniques are required to treat tumors in deeper parts of the body or in organs because the radiation dose is in general highest on the surface of the body. In particle radiotherapy using beams of accelerated protons or carbon nuclei, the dose has the Brag peak, i.e., becomes maximal at the depth depending on the energy of the particles. Despite this benefit of particle radiotherapy, the dose is also limited by the tolerance of surrounding normal tissues within the treatment volume. Targeting radiotherapy uses radioactive isotopes that can be selectively targeted to tumor after injection, but special attention has to be paid to handling of the radioactive isotopes.

This chapter describes neutron capture therapy (NCT), a new radiotherapy that differs from the conventional radiotherapies described above. Coderre and Morris have reviewed the radiation biology of boron neutron capture therapy (BNCT), including the biodistribution of currently used boron compounds, in detail, but they limited the coverage to BNCT [1]. Consequently, the present chapter also covers gadolinium neutron capture therapy (GdNCT), and drug delivery issues in BNCT and GdNCT are discussed with special focus on the roles of nanoparticle technology. Section 3.2 explains the principle of NCT, and Section 3.3 reviews the present status of BNCT, including two boron compounds clinically used at present and past investigations on their nanoparticulate delivery systems. Section 3.4 deals with the approaches to GdNCT carried out so far, including a detailed account of the present authors' experiences in developing Gd-containing lipid nanoemulsions and chitosan nanoparticles to demonstrate the usefulness of nanoparticle technology in NCT.

Tab. 3.1. Neutron capture cross section of typical atoms.

Atom	Cross section (barn)	Reaction
^{16}O	0.00019	
^{12}C	0.0035	
^{1}H	0.333	^{1}H(n, γ)^{2}H
^{14}N	1.83	^{14}N(n, p)^{14}C
^{10}B	3840	^{10}B(n, α)^{7}Li
^{157}Gd	254000	^{157}Gd(n, γ)^{158}Gd

3.2
Principle of Neutron Capture Therapy of Cancer

NCT is a cancer therapy that utilizes the radiations emitted *in vivo* as a result of the nuclear neutron capture reaction (NCR) between radiation-producing agents administered in the body and thermal or epithermal neutrons irradiated from outside of the body. NCT is a binary treatment system, consisting of dosing the radiation-producing agents and neutrons. Neutrons have in general been classified according to their energies, E, as thermal neutrons ($E < 0.4$ eV), epithermal neutrons (0.4 eV $< E <$ 10 keV), and fast neutrons ($E > 10$ keV) [1]. Table 3.1 shows the NCRs and neutron capture cross sections of typical elements [1]. Commonly existing O, C, H and N in the body or tissues have small cross sections, but boron (^{10}B) and gadolinium (^{157}Gd) have extremely large cross sections.

The most common radiation-producing element for NCT is ^{10}B at present [2]. In clinical experiences, NCT with ^{10}B (BNCT) has achieved encouraging results using the mercaptoundecahydro-*closo*-dodecaborate dianion ([B(12)H(11)SH]$^{2-}$, BSH) in patients with grades III–IV glioma and using the boronophenylalanine (BPA) in patients with malignant melanoma (Fig. 3.1). In BNCT, the ^{10}B compounds administered have to be delivered to tumor intracellularly to obtain an antitumor effect, because ^{10}B emits α-particles whose range is nearly equal (9 μm) to a cell diameter or shorter [3]: success in clinical BNCT trials depends on the selective accumulation of ^{10}B compounds into individual tumor cells.

^{157}Gd causes the following neutron capture reaction by thermal neutron irradiation [4]:

^{157}Gd + thermal neutron → ^{158}Gd + γ-rays + internal conversion electrons → ^{158}Gd + γ-rays + Auger electrons + characteristic X-rays

^{157}Gd NCR (Gd-NCR) results in emission of long-range prompt γ-rays, internal conversion electrons, X-rays and Auger electrons with a large total kinetic energy (7.94 MeV) [4]. The γ-rays and the electrons thus generated provide a tumor-killing

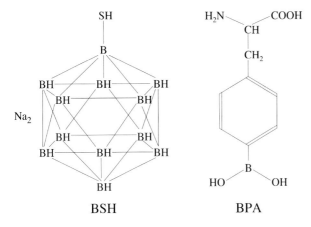

Fig. 3.1. Compounds in clinical use for BNCT.

effect [5, 6]. GdNCT has the following theoretical and possible advantages over typical BNCT: (a) ^{157}Gd has the highest thermal neutron capture cross section (254 000 barns) among naturally occurring isotopes, 66× larger than that of ^{10}B [7]; (b) the γ-rays released by Gd-NCR have a long range (>100 µm) in contrast with the α-particles, so that they may extensively affect tumors even if Gd exists extracellularly in tumor tissue [8]; (c) Auger electrons with a short-range and high linear energy transfer may lead to a local and intensive execution of DNA in neoplastic cells [9]; (d) since Gd has been used as a magnetic resonance imaging (MRI) diagnostic agent [10, 11], it will be possible in future to integrate GdNCT with MRI diagnosis by using Gd-loaded dosage forms.

NCT has certain advantages over traditional cancer chemotherapy. Unlike chemotherapy that uses antitumor drugs, NCT does not need to use pharmacologically active substances in a traditional sense, since it is the neutron capture element itself that contributes to tumor inactivation. Therefore, a large amount of the radiosensitizer can be administered, provided the elements are modified so as to be non-toxic compounds. Thus, severe side effects, which are often experienced in cancer chemotherapy, are not a major concern in NCT.

3.3
Boron Neutron Capture Therapy

3.3.1
Boron Compounds

NCT was first postulated by Locher in 1936 [12]. The earliest clinical treatments of malignant glioma were carried out at the Brookhaven National Laboratory in 1951–1961, and the Massachusetts Institute of Technology in 1959–1961, but no successful result could be obtained, possibly because of the poor tumor-selectivity of the compounds used. Thereafter, many boron compounds have been synthesized for this purpose and their potential to accumulate boron in tumors has been evaluated. However, most compounds could not be used clinically due to chemical instability and toxicity even if they would exhibit a high tumor accumulation. Nonetheless, two compounds, i.e., BSH and BPA (Fig. 3.1), were found to be applicable to clinical treatment.

Hatanaka et al. at Teikyo University, Japan, started clinical trials on brain tumors by using BSH in 1968 [13–15]. They reported one case where a brain tumor seemed to be completely treated. Thereafter, they have employed BNCT for over 200 patients to date. In 1987 Mishima et al. at Kobe University, Japan, also began BNCT of malignant melanoma, subsequently reporting 18 excellent, almost completely treated, results in the treatment of 22 patients [16, 17].

BSH and BPA are the only boron compounds that can be applied to clinical trials at present. They are intravenously (i.v.) administered only as solutions; however, the selectivity in tumor accumulation has to be increased to gain a more efficient treatment outcome. Consequently, new compounds having a high potential for selective tumor-accumulation are still actively sought [18].

BNCT of melanoma has been carried out clinically as follows. Fukuda et al. in Mishima's group have analyzed the neutron dose in 22 melanoma patients with primary or metastatic melanomas who received BPA and subsequently underwent BNCT [19]. The blood concentration in nine patients receiving 179.7 ± 14.9 mg-BPA per kg body weight (BW) increased with time during i.v. infusion, peaked at the end of administration and decreased thereafter. The peak values at the end of administration were 9.4 ± 2.6 μg-^{10}B per gram of blood, and half-lives for the initial and second components of the blood clearance were 2.8 and 9.2 h, respectively. Skin concentrations in ten patients varied from case to case; however, skin-to-blood (Sk/B) ratios were relatively constant at 1.31 ± 0.22 during the 6 h after the end of administration. Boron concentrations in the tumors resected from the seven patients who were operated on decreased in parallel with the blood values, the tumor-to-blood (T/B) ratio being relatively constant at 3.40 ± 0.83. Based on these analytical data of BPA pharmacokinetics, they optimized the timing of irradiation and the setting of the neutron flux large enough for tumor eradication but still tolerable for normal skin.

3.3.2
Delivery of Boron Using Nanoparticles

Drug delivery system (DDS) is most often associated with fine particulate carriers, such as emulsion, liposomes and nanoparticles, which are designed to localize drugs in the target site. They have actively been studied as injectable devices that may enhance therapeutic potency and reduce side effects. From a clinical viewpoint, they might have to be biodegradable and/or highly biocompatible. In addition, high drug content is desirable, because in many cases actual drug-loading efficiency is often too low to secure an effective dose at the target site. Biodegradable liposomes have received considerable attention as potent vehicles for targeting a site and controlled release of a drug.

The successful treatment of cancer by BNCT requires the selective, very high concentration of ^{10}B within malignant tumors. To achieve this, many approaches relating to the liposomal delivery of boron have been carried out [20–51]. The results from typical studies are described in the following.

Shelly et al. [40] have carried out model studies on ^{10}B delivery to murine tumors with small unilamellar liposomes of 70 nm or less in tumor-bearing mice. The liposomes were composed of a pure synthetic phospholipid (distearoyl phosphatidylcholine) and cholesterol, encapsulating high concentrations of water-soluble ionic boron-rich compounds, with hydrolytically stable borane anions such as $[B(20)H(18)]^{2-}$. Unlike the boron compounds themselves, which exhibited no affinity for tumors and are normally rapidly cleared, liposomes selectively delivered the borane anions to tumors. The highest tumor concentrations after i.v. injection of the two isomers of $[B(20)H(18)]^{2-}$ reached the therapeutic range (>15 μg-B per g tumor) while maintaining high T/B ratios (>3). The authors suggested that these boron compounds might have the capability to react with intracellular components after they had been deposited within tumor cells by the liposome, thereby preventing the borane ion from being released into blood.

Feakes et al. [41] in the same group have investigated the newly synthesized $[B(20)H(17)NH(3)]^{3-}$ as a water-soluble boron-delivery agent. The [ae-B(20)H(17)NH(3)]$^{3-}$ anion was encapsulated in liposomes prepared with 5% poly(ethylene glycol) (PEG)-2000–distearoyl phosphatidylethanolamine in the liposome membrane. As expected, these liposomes exhibited a longer circulation lifetime in the biodistribution experiment, resulting in the continued accumulation of boron in the tumor over the entire 48 h experiment, reaching a maximum of 47 μg-B per g tumor.

Subsequently, these authors have synthesized the acylated nido-carborane species K[$nido$-7-CH$_3$(CH$_2$)$_{15}$-7,8-C2B(9)H(11)] for use as an addend for the bilayer membrane of liposomes [42]. Low injected doses of approximately 5–10 mg-B per kg BW afforded a peak tumor boron concentration of approximately 35 μg-B per g tumor and a T/B boron ratio of approximately 8. These values are sufficiently high for the successful application of BNCT. Further, the incorporation of both hydrophilic and hydrophobic species within the same liposomes demonstrated significantly enhanced biodistribution characteristics, as exemplified by the maximum tumor boron concentration of approximately 50 μg-B per g tumor and a T/B ratio of approximately 6. The same authors have also synthesized the thiol derivative [43]. At low i.v. injected doses, the tumor boron concentration increased throughout the time-course experiment, resulting in a maximum observed boron concentration of 46.7 μg-B per g tumor at 48 h and a T/B boron ratio of 7.7. They also reported that the most favorable results were obtained with the polyhedral borane Na$_3$[a2-B(20)H(17)NH$_2$CH$_2$CH$_2$NH$_2$] [44]. Liposomes encapsulating this species produced a tumor boron concentration of 45 μg-B per g tumor at 30 h post-injection, at which time the T/B ratio was 9.3.

Watson-Clark et al. have reported model studies directed toward the application of BNCT to rheumatoid arthritis using the above liposomes incorporating K[nido-7-$CH_3(CH_2)_{15}$-7,8-$C_2B(9)H(11)$] as an addend in the lipid bilayer and encapsulated $Na_3[a2\text{-}B(20)H(17)NH_2CH_2CH_2NH_2]$ in the aqueous core [45]. With low i.v. injected doses of 13–18 mg-B per kg BW, the peak boron concentration observed in arthritic synovium was 29 µg-B per g tumor. The highest synovium/B ratio observed was 3.0, when the synovial boron concentration was 22 µg-B per g tumor.

Yanagie et al. have reported BNCT using ^{10}B entrapped anti-CEA (carcinoembryonic antigen) immunoliposome. A new murine monoclonal antibody (2C-8) was prepared by immunizing mice i.p. with a CEA-producing human pancreatic cancer cell line, AsPC-1 [46]. This anti-CEA monoclonal antibody was conjugated with large multilamellar liposomes incorporating BSH. AsPC-1 cells were incubated with the ^{10}B-Lip-MoAb (CEA) for 8 h. After irradiation with thermal neutrons (10^{11}–10^{13} neutrons cm^{-2}), AsPC-1 cells showed decreasing uptake of ^{3}H-TdR compared with control group, indicating that the immunoliposomes could exert cytotoxic effect by thermal neutrons. Further, the boronated anti-CEA immunoliposome was applied to tumor cell growth inhibition in an *in vitro* BNCT model [47]. The liposomes were shown to bind selectively to cells bearing CEA on their surface. The immunoliposomes attached to tumor cells suppressed growth *in vitro* upon thermal neutron irradiation, and suppression was dependent upon the concentration of the ^{10}B compound in the liposomes and on the density of antibody conjugated to the liposomes. The cytotoxic effects of locally injected ^{10}B compound, multilamellar liposomes containing ^{10}B compound or ^{10}B immunoliposomes (anti-CEA) on human pancreatic carcinoma xenografts in nude mice have been evaluated with thermal neutron irradiation [48]. Injection of ^{10}B immunoliposomes caused the greatest tumor suppression with thermal neutron irradiation *in vivo*. Histopathologically, hyalinization and necrosis were found in boron-treated tumors, while tumor tissue injected with saline or saline-containing immunoliposomes showed neither destruction nor necrosis, suggesting that BNCT with intratumoral (i.t.) injection of immunoliposomes was able to destroy malignant cells in the marginal portion between normal tissues and cancer tissues.

Yanagie et al. have also extended the use of BNCT assisted by liposomal boron delivery to breast cancer [49]. In addition, they have employed neutron capture autoradiography (NCAR) of the sliced whole-body samples of tumor-bearing mice [50]. They obtained NCAR images for mice i.v. injected by ^{10}B-PEG liposome, ^{10}B-transferrin (TF)-PEG liposome, or ^{10}B-bare liposome. This study demonstrated the increased accumulation of ^{10}B atoms in the tumor tissues by binding PEG-chains to the surface of liposome, which increased retention in the blood flow and escaped phagocytosis by the reticuloendothelial system (RES).

Maruyama et al. have prepared unilamellar TF-PEG liposomes less than 200 nm in diameter for intracellular targeting of BSH to solid tumors [51]. When TF-PEG liposomes were injected at 35 mg-^{10}B per kg BW, a prolonged residence time in the circulation and a low uptake by RES in Colon 26 tumor-bearing mice were observed. TF-PEG liposomes maintained a high ^{10}B level in the tumor with concentrations above 30 µg g^{-1} for at least 72 h after injection, indicating that binding and concomitant cellular uptake of the extravasated TF-PEG liposomes occurred by TF receptor and receptor-mediated endocytosis, respectively. On the other hand, the plasma level of ^{10}B decreased, resulting in a tumor/plasma ratio of 6.0 at 72 h after injection. Administration of BSH encapsulated in TF-PEG liposomes at a dose of 5 or 20 mg-^{10}B per kg BW and irradiation with 2×10^{12} neutrons cm^{-2} produced tumor growth suppression and improved long-term survival compared with controls. Thus, intravenous injection of TF-PEG liposomes could increase the tumor retention of ^{10}B atoms, which were introduced by receptor-mediated endocytosis of liposomes after binding, causing tumor growth suppression *in vivo* upon thermal neutron irradiation.

Koning et al. have tried to target ^{10}B to the tumor vasculature for NCT [24]. Alpha (v)-integrin specific RGD-peptides were coupled to liposomes that encapsulated BSH. These RGD-liposomes strongly associated with human umbilical vein endothelial cells (HUVEC) expressing this integrin and were internalized. Irradiation of RGD-^{10}B-liposome-incubated HUVEC with neutrons strongly inhibited endothelial cell viability.

Low-density lipoproteins (LDLs) are internalized by the cell through receptor-mediated mechanisms. A boronated analogue of LDL has been synthesized for possible application in BNCT by Laster et al. [52]. The analogue was tested in cell culture for uptake and biological efficacy in the thermal neutron beam. The boron concentration found was 10× higher than that required in tumors for BNCT, 240 µg-^{10}B per gram of cells.

3.4
Approaches to GdNCT

Therapeutic potential in GdNCT has been explored theoretically and experimentally [53–69]. Magnevist® (gadopentetate dimeglumine aqueous solution), an MRI contrast agent, has been most used as gadolinium source in GdNCT studies. Because of the lack of its targeting ability, however, an adequate amount of gadolinium required for an efficient therapeutic index could not be delivered through the i.v. route. In addition, even i.t. injection did not give rise to significantly prolonged retention of gadolinium in tumor tissues. Thus, a key for success in current GdNCT trails is the use of a device by which gadolinium can be delivered efficiently and retained in a high level inside tumor tissues and/or cells during thermal neutron irradiation. In addition, this may extend NCT to wider types of tumors. From these perspectives, the present authors have prepared delayed-release type ethyl cellulose-coated microcapsules containing gadopentetate dimeglumine for preliminary GdNCT trials [61]. GdNCT using the microcapsules demonstrated a significant effectiveness in survival time of the murine Ehrlich ascites tumor model [62]. This result, which first demonstrated the potential of GdNCT *in vivo*, led us develop a more elaborate gadolinium-loaded particulate system and to establish its potential application to GdNCT, as described below.

3.4.1
Typical Research on GdNCT

Hofmann et al. have reported GdNCT of melanoma cells and solid tumors with a neutral macrocyclic gadolinium complex (Gadobutrol), a magnetic resonance imaging contrast agent [53]. In mice, i.t. administration of 1.2 mmol-Gd per kg BW of the Gd complex, corresponding to about 23 000 µg-Gd per mL of tumor, before neutron irradiation (3.6×10^{12} neutrons cm^{-2}) resulted in a significant delay in tumor growth with respect to control groups.

Kobayashi et al. have developed avidin-dendrimer-(1B4M-Gd)(254) (Av-G6Gd) as a tumor-targeting therapeutic agent for GdNCT of intraperitoneal disseminated tumor that can be monitored by MRI in order to deliver large quantities of Gd atoms into tumor cells [63]. An *in vitro* internalization study showed that Av-G6Gd accumulated and was internalized into SHIN3 cells, a human ovarian cancer, 50- and 3.5-fold greater than gadolinium diethylenetriaminepentaacetic acid (Gd-DTPA, Magnevist®) and G6Gd. The accumulation of Gd in the cells was also detected by the increased signal on T1-weighted MRI. Av-G6Gd showed specific accumulation in the SHIN3 tumor 366- and 3.4-fold greater than Gd-DTPA and G6Gd one day after intraperitoneal (i.p.) injection. Thus, a sufficient amount of Av-G6Gd (162 ppm of Gd) was accumulated and internalized into the SHIN3 cells *in vivo*.

De Stasio et al. have observed directly the microdistribution of Gd in cultured human glioblastoma cells exposed at 1–25 mg of Gd-DTPA per mL, corresponding to 300–7100 µg-Gd mL^{-1} [64]. They demonstrated that Gd-DTPA penetrated the plasma membrane, and observed no deleterious effect on cell survival and a higher Gd accumulation in cell nuclei compared with cytoplasm. They also exposed Gd-containing cells to thermal neutrons (3.6×10^{12} neutrons cm^{-2}) and demonstrated the effectiveness of Gd-NCR in inducing cell death. However, the efficacy of Gd-DTPA and Gd-DOTA as GdNCT agents *in vivo* was predicted to be low due to the insufficient number of tumor cell nuclei incorporating Gd [65]. The authors then suggested that although multiple administration schedules *in vivo* might induce Gd penetration into more tumor cell nuclei, a search for new Gd compounds with higher nuclear affinity would be warranted before planning GdNCT in animal models or clinical trials.

Oyewumi and Mumper have used microemulsions (oil-in-water) as templates to engineer stable emulsifying wax or Brij 72 (polyoxyl 2 stearyl ether) nanoparticles [66]. The technique was simple, reproducible, and amenable to large-scale production of stable nanoparticles having diameters below 100 nm. The emulsifying wax and Brij 72 nanoparticles (2 mg mL^{-1}) made together with polyoxyl-20 stearyl ether and polysorbate 80, respectively, were the most stable based on retention of nanoparticle size over time. Gadolinium acetylacetonate (GdAcAc), a potential anticancer agent for NCT, was entrapped in the nanoparticles. Challenges of these cured nanoparticles in biologically relevant media at 37 °C for 60 min demonstrated that these nanoparticles were stable. The results showed the ease of preparation of these very small, stable nanoparticles and the ability to entrap lipophilic drugs such as GdAcAc with high efficiency.

Oyewumi's group further synthesized gadolinium hexanedione (GdH) by complexation of Gd^{3+} with hexane-2,4-dione as NCT agent, and a folate ligand by chemically linking folic acid to distearoylphosphatidylethanolamine (DSPE) through a PEG (MW 3350) spacer [67]. To obtain folate-coated nanoparticles, the folate ligand (0.75 to 15% w/w) was added either to the microemulsion templates at 60 °C or to nanoparticle suspensions at 25 °C. Cell uptake studies were carried out in KB cells (human nasopharyngeal epidermal carcinoma cell line), which are known to overexpress folate receptors. The uptake of folate-coated nanoparticles was about ten-fold higher than uncoated nanoparticles after 30 min at 37 °C. The uptake of folate-coated nanoparticles at 4 °C was 20-fold lower than the uptake at 37 °C and comparable to that of uncoated nanoparticles at 37 °C. Folate-mediated endocytosis was further verified by the inhibition of the uptake of folate-coated nanoparticles by free folic acid. Folate-coated nanoparticle uptake decreased to approximately 2% of its initial value with the co-incubation of 0.001 mm of free folic acid. The authors suggested that these tumor-targeted nanoparticles containing high concentrations of Gd may have potential for NCT. Thiamine was also effective as a tumor-specific ligand for gadolinium nanoparticles in a methotrexate-resistant breast cancer cell line, MTX(R)ZR75, transfected with thiamine transporter genes (THTR1 and THTR2) [68].

Using the folate-coated and PEG-coated gadolinium (Gd) nanoparticles, Oyewumi et al. carried out *in vivo* studies in KB tumor-bearing athymic mice [69]. Gd nanoparticles did not aggregate platelets or activate neutrophils. The retention of nanoparticles in the blood 8, 16 and 24 h post-injection of nanoparticles (1.6 mg-Gd per kg BW) was 60, 13 and 11% of the injected dose (ID), respectively. The maximum Gd tumor localization was 33 ± 7 µg-Gd g^{-1}. Both folate-coated and PEG-coated nanoparticles had comparable tumor accumulation. However, the cell uptake and tumor retention of folate-coated nanoparticles was significantly enhanced over PEG-coated nanoparticles. Thus, the folate-ligand coating shows beneficial facilitation of tumor cell internalization and retention of Gd-nanoparticles in the tumor tissue.

3.4.2
Delivery of Gadolinium using Lipid Emulsion (Gd-nanoLE)

3.4.2.1 Preparation of Gd-nanoLE
Oil-in-water (o/w) emulsions stabilized with emulsifiers such as phospholipids have attracted much attention as drug carriers because they are biodegradable and biocompatible and, unlike liposomes, they can be prepared on an industrial scale and are relatively stable below 25 °C for long periods [70]. One problem with using emulsion particles as drug carriers is how they can leave the vascular space and reach their site of action. The extravascular transfer of particulate carriers largely depends on their size. A diameter of approximately 100 nm is the cut-off value for drug carriers able to pass through the discontinuous capillary endothelium of tumors [71]. In addition, drug carriers <100 nm are expected to more easily avoid uptake by RES and to circulate for longer periods in blood, as estimated from the

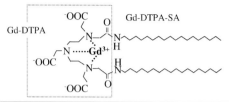

	Gd-DTPA	Gd-DTPA-SA
	hydrophilic	amphiphilic
Solubility (water)	highly soluble	insoluble
Solubility (soybean oil)	insoluble	insoluble
Clearance from body	immediate	very slow

Fig. 3.2. Chemical structure and properties of Gd-DTPA-SA.

finding that small unilamellar liposomes of about 70–100 nm are cleared more slowly from the circulation than larger ones of the same composition [72].

The authors' first study on developing nanoparticulate systems for GdNCT was to prepare Gd-containing emulsions with reduced particle size, surface properties exhibiting prolonged blood retention, and a high gadolinium content [73]. As a Gd-source, a water-insoluble and oil-insoluble Gd-DTPA derivative, distearylamide (Gd-DTPA-SA) (Fig. 3.2), was synthesized. Gd-DTPA-SA has two hydrophobic tails (side-chains) consisting of stearylamines that are connected to the Gd-DTPA moiety through amide linkages. It can be incorporated into the membrane of liposomal vesicles as a component of the liposomal lamella [74].

The particle structure is schematically shown in Fig. 3.3. Emulsions containing soybean oil, water, Gd-DTPA-SA, as an amphiphilic drug, and hydrogenated egg yolk phosphatidylcholine (HEPC), as an emulsifier, in a weight ratio of 7.36:92:1:2 were prepared by the thin-layer hydration method using a bath-type sonicator (Table 3.2). The mean particle size of the emulsions was 250 nm.

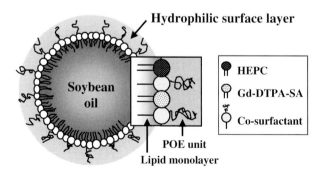

Fig. 3.3. Structure of Gd-nanoLP.

Tab. 3.2. Formulation, Gd content and particle size of the standard- and high-Gd-nanoLE.

	Formulation		
	Standard-Gd-nanoLE		High-Gd-nanoLE
	Plain	With co-surfactant	
HEPC[a] (mg)	500	500	250
Gd-DTPA-SA (mg)	250	250	500
Soybean oil (mL)	2	2	2
Co-surfactant[b] (mg)	–	750	750
Water (mL)	23	23	23
Gd-content[c] (mg mL^{-1})	–	1.5	3.0
Particle size (nm)	250	78	84

[a] L-Phosphatidylcholine hydrogenated from egg yolk.
[b] HCO-60.
[c] Theoretical Gd content.

To prepare o/w emulsions, oils other than the commonly used soybean oil have been employed. The particle size of triolein-emulsion (240 nm) was almost the same as that of soybean oil-emulsion. Lipiodol Ultra-Fluide® (Guerbert Laboratories, France) produced a mean droplet size of 195 nm. When caster oil (448 mPa s/25 °C; 262 mPa s/37 °C [75a]) with a higher viscosity was selected as an oil component, no emulsion could be prepared by the procedure used here, while stable emulsions could be prepared using oils with a lower viscosity such as soybean oil (49 mPa s/25 °C, 29 mPa s/37 °C [75a]), Lipiodol (37 mPa s/25 °C, 25 mPa s/37 °C [75a]) and ethyl oleate (4 mPa s/37 °C [75b]). The mean particle size of the emulsions seemed to be closely correlated to the viscosity of the oils: the particle size of the emulsions fell on reducing the viscosity of the oil. This indicated that oil viscosity is an important factor, influencing the particle size of the Gd-DTPA-SA emulsions. Despite this, soybean oil was used in this study because of its widespread application. To make the droplet size of the emulsions smaller than 100 nm, as well as to modify the emulsion surfaces, a co-surfactant, Tween® 80, HCO®-60, Pluronic® F68, polyoxyethylene alkyl ether (Brij®) or polyoxyethylene alkyl ester (Myrj®), was introduced into the standard system. Figure 3.4 shows the chemical structures of typical co-surfactants. Tween 80, HCO-60, Brij 76, 78 and 700 were effective in reducing the particle size to below 100 nm when the co-surfactant weight ratio (CWR), defined as co-surfactant/(HEPC+Gd-DTPA-SA) (w/w), was larger than 0.67. The particle size with Tween 80 and HCO-60 was reduced to 53 and 78 nm, respectively, at a CWR of 1.0 (w/w) (Table 3.2). To increase the gadoli-

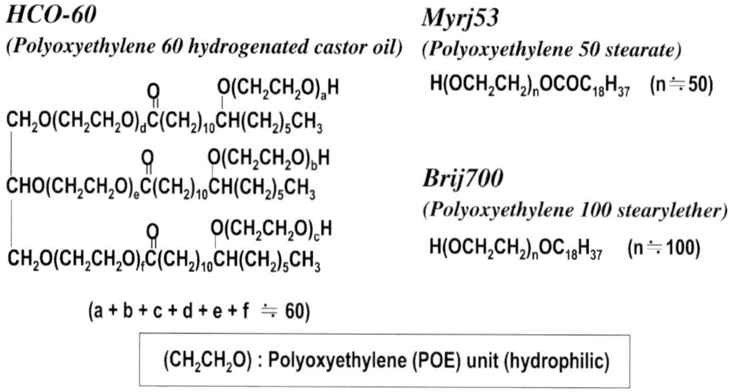

HCO-60
(Polyoxyethylene 60 hydrogenated castor oil)

$$CH_2O(CH_2CH_2O)_d\overset{O}{\overset{\|}{C}}(CH_2)_{10}\overset{O(CH_2CH_2O)_aH}{\overset{|}{CH}}(CH_2)_5CH_3$$

$$CHO(CH_2CH_2O)_e\overset{O}{\overset{\|}{C}}(CH_2)_{10}\overset{O(CH_2CH_2O)_bH}{\overset{|}{CH}}(CH_2)_5CH_3$$

$$CH_2O(CH_2CH_2O)_f\overset{O}{\overset{\|}{C}}(CH_2)_{10}\overset{O(CH_2CH_2O)_cH}{\overset{|}{CH}}(CH_2)_5CH_3$$

$(a + b + c + d + e + f \risingdotseq 60)$

Myrj53
(Polyoxyethylene 50 stearate)

$H(OCH_2CH_2)_nOCOC_{18}H_{37}$ $(n \risingdotseq 50)$

Brij700
(Polyoxyethylene 100 stearylether)

$H(OCH_2CH_2)_nOC_{18}H_{37}$ $(n \risingdotseq 100)$

(CH_2CH_2O) : Polyoxyethylene (POE) unit (hydrophilic)

Fig. 3.4. Chemical structure of co-surfactants.

nium content, the weight ratio of Gd-DTPA-SA to HEPC was increased from 1:2 of the standard-Gd formulation to 2:1 of the high-Gd formulation (Table 3.2). The measured particle size of the HCO-60 high-Gd emulsions was 84 nm when the CWR was 1.0 (w/w). In this case, the calculated gadolinium content reached 3.0 mg-Gd mL^{-1}. These results indicate that HCO-60 is an effective co-surfactant not only in terms of particle size reduction but also with respect to gadolinium enrichment.

3.4.2.2 Biodistribution of Gadolinium after Intraperitoneal Administration of Gd-nanoLE

Tokuuye et al. have reported the effect of ^{157}Gd concentration on tumor inactivation in GdNCT [76]. The neutron fluence required for 10% survival of the Chinese hamster cells (V79) decreased rapidly between 0 and 100 µg-^{157}Gd mL^{-1}, but leveled off above that. This result indicates that the optimal tumor inactivation effect would be achieved around 100 µg-^{157}Gd mL^{-1}. To achieve this level *in vivo*, Gd-nanoLEs were i.p. administrated in Greene's melanotic melanoma (D$_1$-179)-bearing hamsters as a model system and the biodistribution of gadolinium was investigated [77].

The inferior surface of the diaphragm is very rich in lymphatic capillaries. The lymphatic lumen is separated from the abdominal cavity by its own endothelium, a fenestrated basement membrane and the peritoneal mesothelium. Hirano and Hunt have reported that, at least in rats, most compounds of molecular weight smaller than 20 000 were exclusively absorbed via splenic or intestinal blood capillaries into the portal vein [78]. In contrast, the lymphatic system was a major absorption route for compounds with a molecular weight larger than 70 000 that were impermeable to blood capillary membranes. Klein et al. have reported that even particles administered through the abdominal cavity were transported through pores in the diaphragm directly to lymphatic and then to the venous system [79]. Thus, lymphatic capillaries could accommodate and transport large materials such as erythrocytes of about 10 μm in diameter [80] and particles with a diameter of up to 22 μm [81]. Hirano and Hunt also reported that the use of liposomes 50–720 nm in diameter gave no size effect on absorption from the abdominal cavity [78]. These results suggested that almost all the emulsion particles so-prepared might have the potential to be absorbed through the lymphatic system.

For our *in vivo* experiments, a Greene's melanotic melanoma (D_1-179) [82] fragment ($2 \times 2 \times 2$ mm) was subcutaneously inoculated on the left thigh of Syrian hamster. The experiments were performed at 10 d after inoculation, when the diameter of the tumor mass became about 10 mm (0.81 ± 0.60 g) and the body weight was 99.7 ± 10.6 g. Gd-DTPA derivative-containing lipid emulsion (Table 3.2) was i.p. injected at a dose of 2.0 mL per hamster (30 mg-Gd/kg BW for the standard-Gd formulation and 60 mg-Gd/kg BW for the high-Gd formulation). Tissue concentration was measured by inductively coupled plasma atomic emission spectroscopy (ICP-AES) at 355.047 nm.

Effect of Co-surfactant In addition to HCO-60, polyoxyethylene (POE) stearyl ether (Brij, C18POEm) and POE stearyl ester (Myrj, C18POEm′) were also selected as co-surfactant to build up the steric hindrance on the emulsion particle surface. Figure 3.5 shows time-courses of gadolinium levels in blood and tumor after i.p. injection of Gd-DTPA-SA-containing plain, HCO-60, Myrj 53 (C18POE50) and Brij 700 (C18POE100) emulsions prepared in the standard-Gd formulation (Table 3.2). The Gd levels in blood (Fig. 3.5A) with the co-surfactant-containing emulsions were significantly higher than those with the plain emulsion. The gadolinium level in blood with the Brij 700 emulsion was prolonged during 12–48 h at the highest level. The high blood gadolinium level with Brij 700 emulsion is related to the stable nature of its ether linkage, in contrast to the ester linkage of the Myrj family (Fig. 3.4).

The Gd concentration in tumor with all emulsions leveled off 24 h after i.p. injection (Fig. 3.5B). Although the gadolinium level in tumor with the Myrj 53 emul-

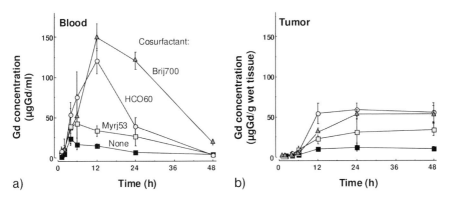

Fig. 3.5. Effect of co-surfactant on Gd distribution after i.p. injection of the standard Gd-nanoLE at 3 mg/2 mL. Data are represented as the mean $\pm$ SD ($n = 3$–9).

sion was 31 µg-Gd per g tumor (wet) at 24 h, those with the HCO-60 and Brij 700 emulsions were 59 µg-Gd per g tumor (wet) and 53 µg-Gd per g tumor (wet) at 24 h, respectively. Although gadolinium retention in blood with the Brij 700 emulsion was prolonged at a higher level (Fig. 3.5A), the gadolinium level in tumor increased more slowly, and the final level was not higher than that with the HCO-60 emulsion (Fig. 3.5B). HCO-60 was the most effective co-surfactant for gadolinium accumulation in tumor.

The T/B ratio at 48 h was >2.8 in every emulsion, reaching 21.4 with the HCO-60 emulsion. T/B with the Brij 700 emulsion was as low as that with the plain emulsion. This was related to the prolonged and higher blood gadolinium level. The T/Sk values with the Brij 700 and HCO-60 emulsions were larger than 6.0. Conversely, the Sk/B ratios at 24 h were low (0.05–0.5).

In NCT, thermal neutrons are irradiated from outside the body. As a result, the T/Sk ratio, T/B ratio and/or Sk/B ratio of radiation sensitizer (gadolinium or boron) concentration are important factors influencing in the therapeutic effects. For BNCT, Mishima [83a] and Honda et al. [83b] have reported that when ^{10}B-BPA fructose complex was administered by the drip infusion method, the average T/B ratio of ^{10}B concentration was 3–4, T/Sk was 3–6 and Sk/B was around 1.2 [19]. The T/B ratio at 48 h in the present study using the high-Gd-DTPA-SA was far higher (13.2), with the sensitizer concentration in tumor being kept very high (Fig. 3.5), whereas the T/Sk and Sk/B ratios were comparable with the corresponding values in BNCT. These results indicate that the methodology developed here can be used efficiently when a high and long retention of sensitizer in tumor and a high T/B ratio are required.

Effect of Gadolinium Content in Emulsion and Derivative Type The Gd content in the HCO-60 emulsion could be increased while the particle size was kept small (Table 3.2). Gadolinium levels in tumor, blood, liver and spleen with the high-Gd emulsion were almost twice as high as those with the standard-Gd emulsion. At 48 h after i.p. injection, the Gd level in tumor with the high-Gd-DTPA-SA HCO-60 emulsion reached 107 µg-Gd per g tumor (wet).

The amide linkages of Gd-DTPA-SA were not expected to degrade *in vivo*. Thus, the stearylester derivative of Gd-DTPA (Gd-DTPA-SE) was synthesized. The Gd levels in tumor and blood with the high-Gd-DTPA-SE HCO-60 emulsion were far lower than those with the high-Gd-DTPA-SA HCO-60 emulsion. Conversely, gadolinium levels in liver and spleen in the early period (2–6 h) with the high-Gd-DTPA-SE HCO-60 emulsion were higher than those with the high-Gd-DTPA-SA HCO-60 emulsion. However, both liver and spleen gadolinium levels decreased to levels lower than those with the high-Gd-DTPA-SA HCO-60 emulsion thereafter. Gd-DTPA-SA was very stable in serum: no transfer to serum proteins and no metabolism to small-molecular-weight compounds occurred up to 24 h at 37 °C [74b, 84]. However, the Gd-DTPA-SE complex was not stable, showing both the transfer and metabolism in a time-dependent manner [74b, 84]. The more rapid elimination of gadolinium from liver and spleen with the Gd-DTPA-SE emulsion could be well explained by the faster degradation of Gd-DTPA-SE.

The pharmacokinetic parameters of MRT, Vdss and Cltot with the standard- and the high-Gd-DTPA-SA HCO-60 emulsion were similar. In addition, the AUC with the high-Gd-DTPA-SA HCO-60 emulsion was twice that with the standard-Gd-DTPA-SA HCO-60 emulsion. Since the gadolinium content in the high-Gd-DTPA-SA HCO-60 emulsion was just twice that with the standard-Gd-DTPA-SA HCO-60 emulsion (Table 3.2), these results indicated that particles of both types of emulsion, with different gadolinium content, behaved similarly *in vivo*. This led to the high level in tumor reaching 107 µg-Gd per g tumor (wet) at 48 h, which was the level reported by Tokuuye et al. [76] as an optimal level for tumor inactivation when ^{157}Gd would be used.

3.4.2.3 Biodistribution of Gadolinium after Intravenous Administration of Gd-nanoLE

The administration route is one of the most important factors in the biodistribution and pharmacokinetics of drug carriers such as liposomes and lipid-emulsions. Previously, the i.p. route was adopted as an administration route of Gd-nanoLE because it allows the injection of a relatively large amount and, consequently, delivers a large amount of Gd to the tumor via the systemic circulation [85]; as a result, the Gd concentration in the tumor reached 107 µg-Gd per g wet tumor at a dose of 60 mg-Gd per kg BW (2 mL as an administration volume of the Gd-nanoLE). However, even with i.p. injection, many factors affect the biodistribution and pharmacokinetics of the drug carriers, including the absorption of these carriers from the abdominal cavity and their localization in the lymph nodes, making estimation of the *in vivo* fate of the drug carriers difficult. In contrast, i.v. injection delivers drugs more simply and, therefore, has been widely employed instead of i.p. injection in clinical treatments. Consequently, our next study aimed to evaluate i.v. as an alternative to i.p. injection with respect to tumor accumulation of Gd incorporated in Gd-nanoLE.

Comparison of I.V. with I.P. Injection of Gd-nanoLE The biodistribution of Gd after i.v. or i.p. injection of the standard-Gd-nanoLE with HCO-60 (particle size, 78 nm) was determined at a dose of 15 mg-Gd/kg BW, half that of the previous case [85]. Table 3.3 summarizes the calculated pharmacokinetic parameters. The Gd concentration in the blood after i.v. injection of the Gd-nanoLE consistently decreased from the initial high concentration of 136 µg-Gd per mL blood at the first sampling time, whereas i.p. injection showed a peak concentration of 50 µg-Gd per mL blood at 4 h after administration (Table 3.3).

The AUC of Gd after i.p. injection was only 57% of that after i.v. injection (Table 3.3). It was reported earlier that the blood concentration of a marker incorporated into liposomes containing sphingomyelin and cholesterol after i.p. injection peaked at several hours after administration and subsequently declined [86]. In addition, some reports have shown that the absorption route of the conventional liposomes from the peritoneal cavity to the bloodstream is mediated by the lymphatic system [87, 88], and part of the i.p. injected liposomes are localized in the lymphatic system over 24 h [88]. These findings can probably be extended to interpret the behavior of the lipid-nanoemulsions in the peritoneal cavity. Accordingly, the

Tab. 3.3. Pharmacokinetic parameters after i.p. or i.v. administration of the standard-Gd-nanoLE with HCO-60 at a dose of 1.5 mg Gd in 1 mL.

Route	C_{max}[a] (µg mL^{-1})	T_{max}[a] (h)	AUC[b] (µg h mL^{-1})	MRI[b] (h)
i.p.	50.0	4.0	612	9.2
i.v.	(135.9)	(0.5)	1071	6.3

[a] The maximum blood concentration, C_{max}, and the time of maximum blood concentration, T_{max}, were derived directly from the mean blood concentration–time curve.
[b] Area under the blood concentration time curve. Each value was calculated from the mean values of blood concentration up to infinite time ($n = 3$–5).

delayed increase of Gd concentration in the blood after i.p. injection might be ascribed to the transport time of the Gd-nanoLE from the peritoneal cavity to the bloodstream through the lymphatic system. In addition, the lower AUC of Gd after i.p. injection, compared with i.v. injection might be explained by the partial localization of Gd-nanoLE in the lymphatic system, since the Gd-nanoLE remaining in the peritoneal cavity, as determined by visual observation, was negligible at 12 h after administration.

In terms of the tumor accumulation of Gd, i.v. injection of the Gd-nanoLE had an advantage over i.p. injection, namely, faster accumulation. The Gd concentration in the tumor after i.v. injection rapidly increased for 6 h after administration, and thereafter it almost leveled off. In contrast, the Gd concentration after i.p. injection remained at a low level for the first 6 h. The maximum Gd tumor concentrations after i.v. and i.p. injections of the Gd-nanoLE were 30 µg-Gd per g wet tumor at 24 h and 22 µg-Gd per g wet tumor at 12 h, respectively.

Effect of Repeated Dosing and Gd Content of Gd-nanoLE To achieve a higher Gd accumulation in the tumor, two i.v. injections at a 24 h interval were made at a dose of 15 and 30 mg-Gd/kg BW per injection by using the standard- and high-Gd-nanoLE, respectively. Biodistribution was determined at 12 h after administration, because the tumor accumulation almost leveled off thereafter with Gd-nanoLEs prepared with HCO-60 (Fig. 3.5) [77]. Two i.v. injections of the standard-Gd-nanoLE made the Gd concentration in the tumor higher, reaching a level of 50 µg-Gd per g wet tumor at 12 h after the second injection. Unfortunately, Gd concentrations in the liver and spleen concurrently increased to almost twice those observed after a single i.v. injection.

In certain cases of the repeated administration of liposomes, an accelerated blood clearance and altered biodistribution of the liposomes were observed at the second or later administration for a certain period after the administration, resulting from the induction of an immunoreaction [89]. Oussoren and Storm, though,

Tab. 3.4. Effect of dosing frequency and Gd content of the Gd-nanoLE with HCO-60 on blood and tumor concentrations of Gd (μg-Gd wet tissue) at 12 h after the final i.v. administration at a dose of 1.5 mg Gd per injection for the standard- or 3.0 mg Gd per injection for the high-Gd formulation.

Tissue	Standard-Gd-nanoLE		High-Gd-nanoLE
	Single (1.5)[a]	Double (3.0)[a]	Double (6.0)[a]
Blood	29.7 ± 2.8	25.3 ± 5.7	45.1 ± 11.4*
Tumor	26.9 ± 1.4	49.7 ± 30.9	100.7 ± 35.9**

[a] Values in parentheses are total dose of Gd (mg). Each value represents the mean ± S.D. ($n = 3$–5). *$p < 0.05$ and **$p < 0.01$, significantly different from the Gd concentration of the standard-Gd-nanoLE injected twice.

demonstrated that the kinetic profiles of the first and second injections of the PEG-liposomes with a 24 h interval were virtually identical [90]. In our case, the Gd concentrations in tissues after two administrations of the standard-Gd-nanoLE seemed to be almost twice as high as those after a single administration (Table 3.4). This implied that, by employing the same schedule as Oussoren and Storm, the repeated administration had no significant effect on the biodistribution mechanisms of the Gd-nanoLE.

With the administration of two i.v. injections of the high-Gd-nanoLE, the Gd concentration in the tumor reached 101 μg per g wet tissue (Table 3.4). This was comparable to the level achieved by a single i.p. injection of 2 mL of the high-Gd-nanoLE at a dose of 60 mg-Gd per kg body weight per injection [85]. These results indicated that even i.v. injection of the Gd-nanoLE whose tolerable volume was only 1 mL could result in the accumulation of Gd in the tumor at a high concentration when an appropriate dosing schedule was employed.

Many researchers have demonstrated that the charge and fluidity of the liposomal membrane affect the biodistribution even in conventional liposomes [91, 92]. For instance, Nagayasu et al. [92] reported that the tumor-to-bone marrow accumulation ratio of the HEPC-containing liposomes increased remarkably with a decrease in cholesterol content. In the present study, it was anticipated that the membrane property of the high-Gd-nanoLE would differ from that of the standard-Gd-nanoLE. However, the biodistribution of Gd-nanoLE after i.v. injection of each formulation was likely to be almost identical, as it had been after i.p. injection [85]. Thus, the biodistribution of the Gd-nanoLE was hardly affected by the membrane property. In concurrence, it was also observed that the Gd tumor concentration was proportional to the Gd content of the lipid particles in the Gd-nanoLE. According to this finding, a higher Gd tumor concentration could be achieved by loading a highly lipophilic Gd compound into the core component, though the further introduction of the Gd-DTPA-SA into the membrane of the Gd-nanoLE led to instability of the emulsion.

3.4.3
Delivery of Gadolinium using Chitosan Nanoparticles (Gd-nanoCPs)

A major problem with previous GdNCT trials using an MRI contrast agent such as Magnevist® was that a sufficient quantity of gadolinium could not be retained in the tumor tissue during neutron irradiation [53, 54]. The commercially available gadolinium agent does not exhibit such a selective accumulation in the tumor after i.v. injection as BPA in BNCT [17] and is eliminated rapidly from the tumor tissue after i.t. injection. Therefore, gadolinium compounds that can be efficiently accumulated in the tumor have been sought [93]. We have developed novel Gd-nanoCPs in order to retain gadolinium in the tumor tissue during a GdNCT trial.

As is well-known, chitosan (poly[β-(1 → 4)-2-amino-2-deoxy-D-glucopyranose]) is a hydrophilic, cationic polysaccharide derived by the deacetylation of chitin (Fig. 3.6), which is the second most abundant polysaccharide, next to cellulose, in the world and a promising resource, originating from crustaceans shells and insects [94]. Chitosan has some interesting properties such as bioadhesive (cationic), biocompatible (nontoxic) and biodegradable (bioerodible) capabilities. Therefore, it has been investigated in depth and used in various industrial and medical applications [95]. Furthermore, these interesting properties make it one of the most promising biopolymers for drug delivery [96–107]. Indeed, chitosan has been studied as a drug carrier in various forms, such as tablets, beads, granules, microparticles and nanoparticles. In particular, micro- and nano-particles are being most widely studied as a drug carrier for the purpose of protein, peptide, vaccine and DNA delivery. A wide variety of preparation methods of chitosan particles have been investigated, such as solvent evaporation techniques, multiple emulsion methods, spray drying methods, electrostatic complex-formation with anionic materials (ionotropic gelation) and block copolymerization. These methods often require a crosslinking agent, such as glutaraldehyde. While such crosslinking agents afford hardened par-

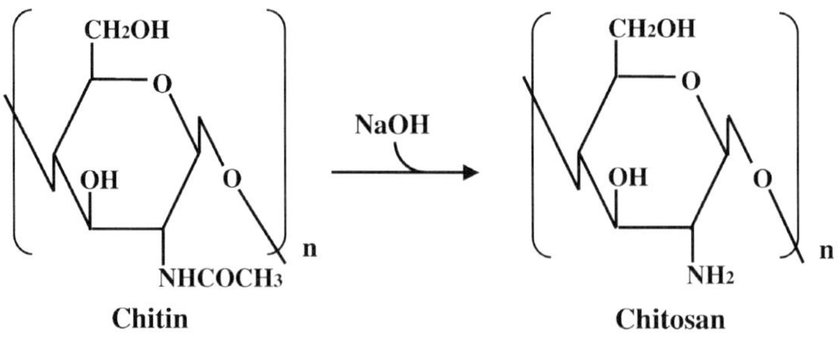

Fig. 3.6. Chemical structures of chitin and chitosan (100% deacetylated).

ticles as well as the possibility of controlling the drug-release rate through the degree of crosslinking, the toxicity of the crosslinking agent would become a major concern. In addition, it leads to low loading of anionic drugs because amino groups in chitosan responsible for electrostatic interaction with anionic drugs become unavailable due to the crosslinking reaction between the amino groups in chitosan and aldehyde groups in crosslinking agents. As an alternative approach, a novel emulsion-droplet coalescence technique has been developed in our laboratory to prepare non-crosslinked chitosan nanoparticles. In fact, this technique offered a useful methodology for the preparation of Gd-nanoCPs specially designed for GdNCT as an i.t. injectable device.

3.4.3.1 Preparation of Gd-nanoCPs

Figure 3.7 shows the preparation of Gd-nanoCPs [108, 109]. Chitosan (2.5% w/v) was dissolved in a Gd-DTPA aqueous solution (5–15% w/v) and an aliquot (1 mL) was added to 10 mL of liquid paraffin containing 5% v/v sorbitan sesquioleate (Arlacel C). The mixture was then stirred to form a water-in-oil (w/o) emulsion A using a high-speed homogenizer. Similarly, a w/o emulsion B was prepared by adding 3 M sodium hydroxide solution (1.5 mL) to liquid paraffin (10 mL) containing 5% v/v Arlacel C. As emulsion B was added to emulsion A, they were mixed and stirred vigorously. As a result of coalescence of droplets, chitosan was deposited as nanoparticles. Gd-nanoCPs in the mixed emulsion were washed and separated by centrifugation at 3000 rpm for 60 min using toluene, ethanol and water

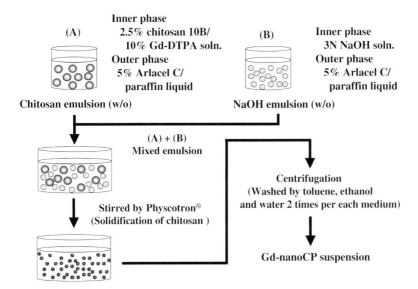

Fig. 3.7. Preparation process of Gd-nanoCPs using an emulsion droplet coalescence technique.

Tab. 3.5. Mean particle size and Gd content of Gd-nanoCP.

	No. of batch	Mean particle size (nm)[a]	Gd content (% w/w)[a] [Gd-DTPA content (%)]
Chitosan 10B			
5% Gd-DTPA soln	3	461 ± 15	7.7 ± 1.7 [26.9 ± 5.9]
10% Gd-DTPA soln	6	426 ± 28	9.3 ± 3.2 [32.4 ± 11.0]
15% Gd-DTPA soln	3	452 ± 25	13.0 ± 1.8 [45.3 ± 6.2]
Chitosan 9B			
10% Gd-DTPA soln	3	594 ± 96	4.1 ± 1.0 [14.2 ± 3.4]
Chitosan 8B			
10% Gd-DTPA soln	3	750 ± 77	3.3 ± 0.8 [11.6 ± 2.7]

[a] Value shows average ± S.D. of 3–6 batches.

successively. Finally, the Gd-nanoCPs were obtained as water suspensions or powders lyophilized after suspending in isotonic mannitol solution.

This technique utilized the fact that, when two emulsions of the same type with the same continuous phase are mixed and stirred vigorously, droplets of each emulsion collide at random, coalesce and split, and all of the droplets finally become uniform in content. Thus, Gd-nanoCP generation was triggered by neutralization of the acidic chitosan-dissolving droplets of emulsion A with sodium hydroxide in the droplets of emulsion B. Nanoparticle generation consequently occurred within the emulsion droplets. The size of nanoparticles did not reflect the droplet size.

Chitosans of grade 10B (100% deacetylated), 9B (91.4% deacetylated) and 8B (84.9% deacetylated) (Katokichi Bio, Japan) were used. Table 3.5 shows the mean particle diameter and gadolinium content in Gd-nanoCPs for each chitosan grade. When chitosan 9B and 8B were used, the mean particle diameters were 594 and 750 nm, respectively, and the gadolinium contents were 4.1% and 3.3% (corresponding to 14.2% and 11.6% as Gd-DTPA), respectively. Namely, as the deacetylation degree of chitosan decreased, the particle size increased gradually and, in contrast, the Gd content decreased markedly. With chitosan 10B, Gd-nanoCPs prepared using 5% and 15% Gd-DTPA solution had mean particle diameters of 461 and 452 nm, respectively, and the gadolinium contents were 7.7% and 13.0% (corresponding to 27% and 45% as Gd-DTPA), respectively. Thus, as the Gd-DTPA concentration in the solution of chitosan 10B increased, the gadolinium content increased, but the particle size was not significantly influenced.

The crosslinking [110–115] and electrostatic interaction by anionic materials such as alginate [116, 117] have been generally used to prepare solid chitosan particles, since it is difficult to form microparticles using only chitosan and drug. In addition, the precipitation technique [118, 119] and block copolymerization [119] have been studied to produce chitosan nanoparticulate carriers in recent years. A major problem with the clinical application of nanoparticles is that the drug load is too low to deliver an effective dose; the performance of the above chitosan particles in past studies was also unexceptional. In the present emulsion-droplet coalescence technique, the Gd-DTPA appears to strongly interact electrostatically with the amino groups of chitosan in the deposition of Gd-nanoCPs. This would contribute to the extraordinarily high Gd-DTPA content (45.3%) and their small particle size, which was reduced to i.t. injectable size (452 nm), in Gd-nanoCPs prepared using chitosan 10B and 15% Gd-DTPA solution.

3.4.3.2 Gd-DTPA Release Property of Gd-nanoCPs

Gd-DTPA release from Gd-nanoCPs that were prepared using chitosan 10B and 10% Gd-DTPA solution has been examined *in vitro* [109]. As test media, an isotonic phosphate-buffered saline solution (PBS) of pH 7.4 and human plasma, as a biological medium, were used, and the method was based on a dynamic dialysis. The gadolinium release behavior was significantly different between Gd-nanoCPs in PBS and those in the human plasma. Gd-nanoCPs released only 1.8% up to 7 d in PBS, whereas 67.9 and 91.5% of gadolinium were eluted from Gd-nanoCPs in human plasma for 6 and 24 h, respectively. This again suggested strong complex formation of Gd-DTPA with chitosan in a simple aqueous medium, because highly water-soluble Gd-DTPA was hardly eluted for a long time in PBS. This releasing property might be advantageous to GdNCT trial by i.t. injection into a solid tumor. The mechanism of fast release of gadolinium from Gd-nanoCPs in human plasma was not clear.

3.4.3.3 Gd-DTPA Retention in Tumor Tissue after Intratumoral Injection

The quantity of gadolinium in the melanoma tissue on mice after i.t. injection of each Gd-DTPA dosage form containing 1200 µg as gadolinium has been determined [109]. In an *in vivo* experiment, the B16F10 malignant melanoma cell suspension (0.1 mL) in an isotonic PBS containing 3×10^6 cells was carefully inoculated subcutaneously (s.c.) into the posterior flank of a six-week-old C57BL/6 mouse. At 10 d after tumor implantation by the above procedure, 200 µL of the Gd-nanoCP (prepared with chitosan 10B and 10% Gd-DTPA solution) suspension in isotonic mannitol solution (Gd 6000 ppm) was injected gently into a block of grown tumor that was about 10 mm in diameter (Gd dose, 1200 µg per mouse). In parallel, the dilute Magnevist® solution (Gd 6000 ppm) was injected in the same manner. At 5 min or 24 h after i.t. injection, mice were sacrificed and tumoral blocks were excised. The amount of gadolinium in the tumor tissue was analyzed by ICP-AES after incineration.

When a dilute solution of Magnevist® was administered, the quantities of gadolinium in a tumor block were 452 µg (37.6% of dose) and 5.3 µg (0.4%) 5 min and

24 h after injection, respectively. In contrast, 892 µg (74.3% of dose) and 821 µg (68.4%) of gadolinium remained in the tumor block 5 min and 24 h after administration of Gd-nanoCP suspension, respectively. Thus, since the gadolinium hardly leaked to the surrounding normal tissues, in particular to the subcutaneous space on tumor tissue, over 24 h, damage beyond the tumor part by GdNCT would be kept to a minimum. No Gd-DTPA release from Gd-nanoCPs in the aqueous medium might contribute to prolonged retention in the tumor tissue. This extended retention was thought to lead to greater enhancement of the antitumor effect as compared with past GdNCT trials using Magnevist® [52].

3.4.3.4 *In vivo* Growth Suppression of Experimental Melanoma Solid Tumor

A GdNCT trial has been carried out *in vivo* by i.t. injection using Gd-nanoCPs prepared with chitosan 10B and 10% Gd-DTPA solution [120]. The Gd-nanoCPs were 430 nm in mean particle diameter and the content of the natural form of gadolinium was 9.3 w/w %. Therefore, the content of ^{157}Gd, 15.6% of the natural form, corresponded to 1.45 w/w %. The radioresistive B16F10 malignant melanoma was selected as tumor model [121, 122] to demonstrate a potential for GdNCT.

When the s.c. B16F10 melanoma tumor in the five-week-old male C57BL/6 mice (body weight, 21–27 g) grew to about 10 mm in diameter 10 d after tumor implantation by the above procedure, i.t. administration of the Gd-nanoCP suspension in isotonic mannitol solution or Magnevist® solution as a control was started. Each gadolinium dosage form (corresponding to natural Gd 6000 µg mL), 200 µL, was injected twice into the block of tumor, 24 and 8 h before neutron irradiation (total natural Gd dose, 2400 µg per mouse). The source of the thermal neutron beams was obtained from the Kyoto University Research Reactor Institute, Japan (the Heavy Water Facility; operating power, 5 MW; irradiation time, 60 min; operating mode, OO-0011-F) [123, 124]. The measured average fluence on the tumor surface was 6.32×10^{12} neutrons cm^{-2}. Neutron irradiation was performed only once.

In the gadolinium-loaded nanoparticle-administered and neutron-irradiated (Gd-P, N+) group, the tumor growth was significantly suppressed despite the radioresistance of melanoma model [121, 122] and the long interval (8 h) until neutron irradiation after second gadolinium administration. Its mean tumor volume was less than 15%, compared with that in the non-gadolinium-administered and neutron-irradiated (Gd−, N+) group, 14 d after neutron irradiation. The mean time taken to reach a tumor volume ratio of 10 in the [Gd-P, N+] group was prolonged to 23.2 d, which was 227% and 374% of that in the [Gd−, N+] group and the non-gadolinium-administered and non-neutron-irradiated (Gd−, N−) group, respectively. In addition, the survival time of mice in the [Gd-P, N+] group was also significantly prolonged, to 22.2 d as the mean time. Three of the six mice in the [Gd-P, N+] group were alive at 28 d after neutron irradiation, while all mice in the other groups were dead by 21 d.

Conversely, no GdNCT effect was observed in the Magnevist® solution-administered and neutron-irradiated (Gd-S, N+) group, since the change in the tumor volume ratio and the survival time of mice did not differ from those in the [Gd−, N+] group. When the gadolinium nanoparticle or solution was administered and there was no neutron irradiation (Gd-P or Gd-S, N−), tumor growth was not inhibited at all: the mean time to reaching a tumor volume ratio of 10 was about 6–7 d, which is no different from that in the [Gd−, N−] group.

The content of gadolinium in the tumor just at the starting point of neutron irradiation was estimated to explain the GdNCT effect enhanced by Gd-nanoCPs. When Gd-nanoCPs were administered in the same manner as that in the GdNCT trial, the gadolinium content in melanoma tissue in mice was 1766 ± 96 μg, corresponding to 74% of dose. However, following the administration of Magnevist® solution, the gadolinium content in the tumoral block was 16 ± 7 μg, corresponding to only 0.7% of dose. Clearly, the strong suppression of tumor growth observed in the present GdNCT trial with Gd-nanoCPs resulted from the excellent Gd-DTPA retention in the tumor tissue of chitosan nanoparticles after i.t. injection.

No skin damage over the tumor was apparent in [Gd−, N+] and [Gd-S, N+] groups. However, the skin over the tumor in the [Gd-P, N+] group became red for a few days after neutron irradiation and, later, ulcer formation was observed. This severe side effect demonstrated that the photons and/or electrons emitted by the frequent Gd-NCR would have a sufficient destructive effect for the tumor if selective gadolinium distribution in the tumor tissue and neutron fluence were controlled as optimally as possible in GdNCT.

In one report on a GdNCT trial, a neutral macrocyclic gadolinium complex (Gadobutrol, Gadovist®) was used as the gadolinium source and excellent results were obtained by i.t. injection [53]. The biological half-life of the wash-out from the tumor tissue after i.t. injection of Gadobutrol was estimated to be 50–151 min (average, 115 min) and might contribute to the assured antitumor effect, because it was longer than that of Magnevist® [53, 54]. In the present study, the outstanding gadolinium retention in the tumor tissue following i.t. injection of Gd-nanoCPs clearly led to the potent GdNCT effect even though the tumors were irradiated at an exceptionally longer interval after gadolinium administration compared with the usual GdNCTs with Magnevist® and Gadovist®. This property will also contribute to flexible adaptation in duration and frequency of neutron irradiation in future GdNCT trials.

In the present GdNCT trial, the dose of natural gadolinium per a tumor block with the nanoparticle formulation (2400 μg-Gd per tumor) was considerably smaller than that in past trials [53, 54]. This gadolinium dose indicated that if gadolinium enriched to 100% of ^{157}Gd was used the corresponding level of growth inhibition could be gained by about 16% of gadolinium dose (460 μg) against B16F10 melanoma block about 10 mm in diameter [53].

3.4.3.5 Bioadhesion and Uptake of Gd-nanoCP in Three Different Cell Lines

The extracellular matrix is composed of sulfated glycosaminoglycans and polysaccharide acids, which form hydrophilic, negatively charged gels over the cell membrane. Membrane glycoproteins, most often bearing sialic acid residues, also contribute to the negative charge of the cell surface. Chitosan, however, has the characteristics of a cationic polyelectrolyte, thereby providing a strong electrostatic interaction with negatively charged cell surfaces. Therefore, the use of cationic polymers, such as chitosan, in preparing particulate carriers would give rise to higher drug retention in cells and tissues because of the electrostatic interaction between particle surfaces and cell surfaces. Indeed, many workers have reported that the bioadhesive properties of chitosan originate from electrostatic interaction between positively charged amino groups in chitosan and negatively charged sialic acid residues on the cell surface [125–127]. In addition, Chatelet et al. have demonstrated the role of deacetylation degree of chitosan on bioadhesive properties using chitosan film [128]. They showed that the adhesion of chitosan on fibroblasts isolated from foreskins of children was increased as the deacetylation degree of chitosan was increased. Therefore, the use of 100% deacetylated chitosan in the present Gd-nanoCPs was expected to lead to such a bioadhesive property.

L929 mouse fibroblast cells, B16F10 melanoma cells and SCC-VII squamous cell carcinoma were employed to evaluate the bioadhesion and uptake of Gd-nanoCPs in cultured cells in order to clarify the mechanism of high tumor-killing effects observed in our GdNCT trials [129]. Gd-DTPA incorporated in Gd-nanoCPs was hardly released in culture medium during the experiments.

Transmission Electron Microscopy Using a transmission electron microscope (TEM), L929 cells after exposure to Gd-nanoCPs were observed. Following exposure to the autoclaved Gd-nanoCP suspension in the culture medium at 37 °C for 12 h, L929 cells were washed with phosphate-buffered saline solution (PBS) to remove the free Gd-nanoCPs not adhered on and not endocytosed into the cells. The cells were then fixed with formaldehyde solution and epoxy resin, thinly sliced with a microtome, and observed by TEM.

Transmission electron micrographs of L929 cells incubated with Gd-nanoCPs for 12 h at 37 °C indicated that Gd-nanoCPs were not so stably dispersed in the culture medium because they loosely aggregated (Fig. 3.8). However, a considerable number of Gd-nanoCPs adhered to L929 cells and some of these were being endocytosed or incorporated in the cells in the form of particles.

Effect of Gd Concentration, Incubation Time and Temperature The amount of Gd in and on the cells after 12 h incubation was increased with increasing feed concentration of Gd (ranging from 0 to 20 ppm). At 20 ppm of applied Gd concentration, the amount of Gd reached 18 µg-Gd per 10^6 cells and this value was unchanged up to 40 ppm. However, it was experimentally confirmed that no further increase in cellular uptake and adhesion was observed beyond 40 ppm. Cationic macromolecules such as polylysine affect, in general, cell viability [130]. However, the cytotoxicity of Gd-nanoCPs against L929 cells was negligible in the range 20–40 ppm. Therefore, the following studies were performed at a Gd feed concentration of 40 ppm.

The Gd accumulation in L929 cells, B16F10 melanoma cells and SCC-VII squamous cell carcinoma incubated with Gd-nanoCPs at 37 and 4 °C was determined as a function of incubation time. The accumulation at 37 °C seemed to level off after 12 h. When incubated with Gd-nanoCPs for 12 h at 37 °C, the total Gd amounts were 18, 27 and 60 µg Gd per 10^6 cells with L929 fibroblast cells,

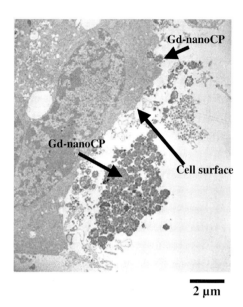

Fig. 3.8. Transmission electron micrograph of L929 fibroblast cells after exposure to Gd-nanoCPs for 12 h.

B16F10 cells and SCC-VII cells, respectively (Fig. 3.9). In contrast, the Gd accumulation at 4 °C where endocytic activity would not exist rapidly leveled off and was far lower, about 5 or less µg Gd per 10^6 cells in all three cell lines. Thus, this significantly higher cellular-accumulation at 37 °C, which would be attributed to endocytosis that was active at 37 °C but suppressed at 4 °C, indicated that endocytic

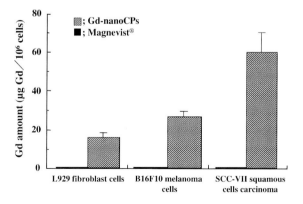

Fig. 3.9. Uptake and adhesion of Gd-nanoCPs or Magnevist at Gd-dose of 40 ppm in three different cell lines 12 h after exposure to the cells under a 5% CO_2 atmosphere at 37 °C.

uptake of Gd-nanoCPs made a major contribution to the total accumulation. The notably varied accumulation behavior of Gd in these three cell lines is attributed to their differing endocytic activity.

The size and bioadhesive property of Gd-nanoCPs is related to the endocytic uptake by the cells. Indeed, Green et al. have examined the particle size dependence of endocytosis of polyethylene particles in C3H murine peritoneal macrophages [131]. They found that polyethylene particles in the range 0.3–10 μm in diameter were endocytosable. The size of the present Gd-nanoCPs, 430 nm, is included in this range. In addition, Lee et al. have demonstrated that a high level uptake of liposomes in J774 cells, a murine macrophage-like cell line, could be obtained by altering the lipid formulation to provide an adhesive property on their surfaces, as might be achieved by the present Gd-nanoCPs [132]. These results support the idea that the adhesion of optimally sized Gd-nanoCPs on the cell surface might be an important step in the subsequent uptake by endocytosis.

The cellular accumulation behavior observed *in vitro* may also be related to the tumor-killing effects *in vivo*. B16F10 cell-bearing mice were employed in our GdNCT trials *in vivo* as described earlier [120]. The tumor-killing effects observed in SCC-VII cell-bearing mice [133] were nearly identical to those achieved with B16F10 cell-bearing mice [120] at the same i.t. Gd dose of 2400 μg per tumor even though the thermal neutron fluence in the GdNCT trials using SCC-VII cell-bearing mice (3.31×10^{12} neutrons cm^{-2}) was only a half of that in the cases of B16F10 cell-bearing mice (6.32×10^{12} neutrons cm^{-2}). From these results, at least one reason for the relatively higher tumor-killing effects obtained with SCC-VII cell-bearing mice *in vivo* would be their higher endocytic activity. Further, since the accumulation was almost saturated at 40 ppm of Gd in the *in vitro* studies described here, which was far lower than the Gd level in the previous GdNCT trials at a dose of 2400 μg per tumor, it was suggested that most of Gd i.t. administered as Gd-nanoCPs might exist in tumor tissue extracellularly, as might be estimated from the TEM photograph given in Fig. 3.8, which shows the presence of agglomerates in the cultured cells.

Comparative Studies with Magnevist® In a comparative study using the solution system, the most important point to be emphasized again was that Gd accumulation in all cell lines utilized in the present study was mostly achieved in the form of particles by cell-surface adhesion and endocytosis of Gd-nanoCPs, though the activity more or less varied. In fact, Magnevist® solution scarcely showed intra- and extracellular Gd accumulation in any of the cell lines (Fig. 3.9). The Gd amount detected in the Magnevist® group after 12 h at 37 °C was less than 1% of that in Gd-nanoCP group (Fig. 3.9), which is probably related to the well-known fact that Magnevist® was eliminated from the tumor tissue immediately after i.t. injection [120]. Conversely, bioadhesion, endocytosis and strong Gd-DTPA-binding of optimally sized Gd-nanoCPs would significantly extend the elimination half-life of Gd i.t. administered [120]. The present results evidenced that Gd-nanoCPs have great potential to accumulate Gd into tumor tissue and/or cells, consequently leading to improved therapeutic efficiency in our GdNCT trials [120].

A major concern in past GdNCT trials using cultured cells and solution systems had been the necessity of a large amount of Gd for cell-growth suppression. Akine et al. reported that 10% survival level of the cultured Chinese hamster cells was obtained with a fluence of 1.55×10^{12} neutrons cm^{-2} in the presence of Gd of 5000 ppm [58]. Tokuuye et al. have evaluated the effect of radiation released during neutron capture reaction by ^{157}Gd of 800 ppm, ^{10}B of 51 ppm or their combination, using Magnevist® and BSH, on cell survival [134] – fluences of 1.13, 1.69 and 0.95×10^{12} neutrons cm^{-2} were required for 10% survival levels, respectively. Further, Hofmann et al. have demonstrated, using the Gadovist®, that at a fluence of 3.6×10^{12} neutrons cm^{-2} a Gd concentration of 10 mmol-Gd L^{-1}, equivalent to 1570 ppm, was required to obtain about a 52% survival level of Sk-Mel-28 cells, a melanoma cell line of human origin [53]. A possible problem of these studies was that they had to apply high Gd concentrations, because only Gd-agents that show no affinity to the tumor cells, possibly including Gadovist®, were available. As a result, cells were unable to suffer sufficient neutron-irradiation because of obstruction by the presence of excess neutron capture elements in the medium. Unlike such a solution-based formulation, Gd would be concentrated on and in cells when using the present Gd-nanoCPs. This situation is clearly favorable for obtaining effective cell-growth suppression.

It has been believed that a contact between the cells and Gd is not always necessary for the tumor cell inactivation in GdNCT, because of the long-range γ-rays emitted as a result of Gd-NCR. Akine and coworkers, however, have proposed that the electrons might also play an important role in the tumor-killing effect in GdNCT [58]. Therefore, the presence of Gd in the intracellular space may be rather desirable for GdNCT, since the electrons, especially Auger electrons, have a short-range and a high linear energy transfer. As shown in the present study, Gd-nanoCPs could bind to the cell surfaces, owing to their cationic nature, and subsequently be endocytosed. These intra- and extracellular accumulation behaviors would provide a compatible way to accomplish the high suppression of tumor growth in GdNCT.

3.5
Conclusions

Recently, clinical BNCT trials have often been carried out, mainly for brain tumors in Japan. It is hopeful that BNCT has been extended to other types of cancer such as head-neck and tongue cancers, with excellent treatment results being reported. In the near future, lung cancer and hepatoma are to be treated by BNCT. Proton accelerators as a source of neutron beams are also going to become a reality, instead of the nuclear reactors currently used. Success in NCT, essentially, depends on the selective tumor-accumulation of boron or gadolinium. Only BPA and BSH are now used clinically as solutions, but their selectivity in the accumulation is not necessarily high. We believe that applications of nanoparticulate "atom" delivery technologies can contribute to NCT.

References

1 CODERRE J. A., MORRIS G. M., The radiation biology of boron neutron capture therapy, *Radiat. Res.*, **1999**, 151, 1–18.
2 BARTH R. F., SOLOWAY A. H., Boron neutron capture therapy of primary and metastatic brain tumors, *Mol. Chem. Neuropathol.*, **1994**, 21, 139–154.
3 BARTH R. F., SOLOWAY A. H., FAIRCHILD R. G., Boron neutron capture therapy of cancer, *Cancer Res.*, **1990**, 50, 1061–1070.
4 GREENWOOD R. C., REICH C. W., BAADER H. A., KOCH H. R., BREITIG D., SCHULT O. W. B., FOGELBERG B., BÄCKLIN A., MAMPE W., VON EGIDY T., SCHRECKENBACH K., Collective and two-quasiparticle states in 158Gd observed through study of radiative neutron capture in 157Gd, *Nucl. Phys.*, **1978**, A304, 327–428.
5 BRUGGER R. M., SHIH J. A., Evaluation of gadolinium-157 as a neutron capture therapy agent, *Strahlenther. Onkol.*, **1989**, 165, 153–156.
6 ALLEN B. J., MCGREGOR B. J., MARTIN R. F., Neutron capture therapy with gadolinium-157. *Strahlenther. Onkol.*, **1989**, 165, 156–157.
7 GARBER D. I., KINSEY R. R., in *Neutron Cross Sections Report*, BNL-325, 3rd edn., Brookhaven National Laboratory, New York, **1976**.
8 SHIH J. A., BRUGGER R., Gadolinium as a neutron capture therapy agent, *Med. Phys.*, **1992**, 19, 733–744.
9 MARTIN R. F., D'CUNHA G., PARDEE M., ALLEN B. J., Induction of double-strand breaks following neutron capture by DNA-bound 157Gd, *Int. J. Radiat. Biol.*, **1988**, 54, 205–208.
10 WEINMANN H. J., BRASCH R. C., PRESS W. R., WESBEY G. E., Characteristics of gadolinium-DTPA complex: A potential NMR contrast agent, *Am. J. Rad.*, **1984**, 142, 619–624.
11 GOLDSTEIN H. A., KASHANIAN F. K., BLUMETTI R. F., HOLYOAK W. L., HUGO F. P., BLUMENFIELD D. M., Safety assessment of gadopentetate dimeglumine in U.S. clinical trials, *Radiology*, **1990**, 174, 17–23.
12 LOCHER G. L., Biological effects and therapeutic possibilities of neutron, *Am. J. Roentgenol. Radium Ther.*, **1936**, 36, 1–13.
13 HATANAKA H., Boron neutron capture therapy for tumors, in *Boron Neutron Capture Therapy for Tumors* (H. HATANAKA, ed.), Nishimura, Niigata, **1986**, pp. 1–28.
14 HATANAKA H., NAKAGAWA Y., Clinical results of long-surviving brain tumor patients who underwent boron neutron capture therapy, *Int. J. Radiat. Oncol. Biol. Phys.*, **1994**, 28, 1061–1066.
15 NAKAGAWA Y., HATANAKA H., Boron neutron capture therapy: Clinical brain tumor studies, *J. Neuro-Oncol.*, **1997**, 33, 105–115.
16 MISHIMA Y., ICHIHASHI M., NAKANISHI T., TSUJI M., UEDA M., NAKAGAWA T., SUZUKI T., Cure of malignant melanoma by single thermal neutron capture treatment using melanoma seeking compounds: [10]B/melanogenesis interaction to in vitro/in vivo radiobiological analysis to preclinical studies, in *Neutron Capture Therapy* (FAIRCHILD R. G., BROWNNELL G., eds.), Brookhaven National Laboratory, Upton, NY, **1983**, pp. 355–364.
17 MISHIMA Y., ICHIHASHI M., HATTA S., HONDA C., YAMAMURA K., NAKAGAWA T., New thermal neutron capture therapy for malignant melanoma: Melanogenesis-seeking 10B molecule-melanoma cell interaction from in vitro to first clinical trial, *Pigment Cell Res.*, **1989**, 2, 226–234.
18 SOLOWAY A. H., TJARKS W., BARNUM B. A., RONG F. G., BARTH R. F., CODOGNI I. M., WILSON J. G., The chemistry of neutron capture therapy, *Chem. Rev.*, **1998**, 98, 1515–1562.
19 FUKUDA H., HONDA C., WADABAYASHI N., KOBAYASHI T., YOSHINO K., HIRATSUKA J., TAKAHASHI J.,

Akaizawa T., Abe Y., Ichihashi M., Mishima Y., Pharmacokinetics of 10B-p-boronophenylalanine in tumours, skin and blood of melanoma patients: A study of boron neutron capture therapy for malignant melanoma, *Melanoma Res.*, **1999**, 9(1), 75–83.
20 Rossi S., Schinazi R. F., Martini G., ESR as a valuable tool for the investigation of the dynamics of EPC and EPC/cholesterol liposomes containing a carboranyl-nucleoside intended for BNCT, *Biochim. Biophys. Acta*, **2005**, 1712(1), 81–91.
21 Ristori S., Oberdisse J., Grillo I., Donati A., Spalla O., Structural characterization of cationic liposomes loaded with sugar-based carboranes, *Biophys. J.*, **2005**, 88(1), 535–547.
22 Martini S., Ristori S., Pucci A., Bonechi C., Becciolini A., Martini G., Rossi C., Boronphenylalanine insertion in cationic liposomes for Boron Neutron Capture Therapy, *Biophys. Chem.*, **2004**, 111(1), 27–34.
23 Nakamura H., Miyajima Y., Takei T., Kasaoka S., Maruyama K., Synthesis and vesicle formation of a nido-carborane cluster lipid for boron neutron capture therapy, *Chem. Commun.*, **2004**, (17), 1910–1911.
24 Koning G. A., Fretz M. M., Woroniecka U., Storm G., Krijger G. C., Targeting liposomes to tumor endothelial cells for neutron capture therapy, *Appl. Radiat. Isot.*, **2004**, 61(5), 963–967.
25 Morandi S., Ristori S., Berti D., Panza L., Becciolini A., Martini G., Association of sugar-based carboranes with cationic liposomes: An electron spin resonance and light scattering study, *Biochim. Biophys. Acta*, **2004**, 1664(1), 53–63.
26 Wei Q., Kullberg E. B., Gedda L., Trastuzumab-conjugated boron-containing liposomes for tumor-cell targeting; development and cellular studies, *Int. J. Oncol.*, **2003**, 23(4), 1159–1165.
27 Stephenson S. M., Yang W., Stevens P. J., Tjarks W., Barth R. F., Lee R. J., Folate receptor-targeted liposomes as possible delivery vehicles for boron neutron capture therapy, *Anticancer Res.*, **2003**, 23(4), 3341–3345.
28 Kullberg E. B., Carlsson J., Edwards K., Capala J., Sjoberg S., Gedda L., Introductory experiments on ligand liposomes as delivery agents for boron neutron capture therapy, *Int. J. Oncol.*, **2003**, 23(2), 461–467.
29 Carlsson J., Kullberg E. B., Capala J., Sjoberg S., Edwards K., Gedda L., Ligand liposomes and boron neutron capture therapy, *J. Neurooncol.*, **2003**, 62(1–2), 47–59.
30 Kullberg E. B., Nestor M., Gedda L., Tumor-cell targeted epiderimal growth factor liposomes loaded with boronated acridine: Uptake and processing, *Pharm. Res.*, **2003**, 20(2), 229–236.
31 Peacock G. F., Ji B., Wang C. K., Lu D. R., Cell culture studies of a carborane cholesteryl ester with conventional and PEG liposomes, *Drug. Deliv.*, **2003**, 10(1), 29–34.
32 Sudimack J. J., Adams D., Rotaru J., Shukla S., Yan J., Sekido M., Barth R. F., Tjarks W., Lee R. J., Folate receptor-mediated liposomal delivery of a lipophilic boron agent to tumor cells in vitro for neutron capture therapy, *Pharm. Res.*, **2002**, 19(10), 1502–1508.
33 Pan X. Q., Wang H., Lee R. J., Boron delivery to a murine lung carcinoma using folate receptor-targeted liposomes, *Anticancer Res.*, **2002**, 22(3), 1629–1633.
34 Kullberg E. B., Bergstrand N., Carlsson J., Edwards K., Johnsson M., Sjoberg S., Gedda L., Development of EGF-conjugated liposomes for targeted delivery of boronated DNA-binding agents, *Bioconj. Chem.*, **2002**, 13(4), 737–743.
35 Pan X. Q., Wang H., Shukla S., Sekido M., Adams D. M., Tjarks W., Barth R. F., Lee R. J., Boron-containing folate receptor-targeted liposomes as potential delivery agents for neutron capture therapy, *Bioconj. Chem.*, **2002**, 13(3), 435–442.
36 Ji B., Chen W., Lu D. R., Halpern D. S., Cell culture and animal studies

for intracerebral delivery of borocaptate in liposomal formulation, **2001**, *Drug Deliv.*, 8(1), 13–17.

37 PAVANETTO F., PERUGINI P., GENTA I., MINOIA C., RONCHI A., PRATI U., ROVEDA L., NANO R., Boron-loaded liposomes in the treatment of hepatic metastases: Preliminary investigation by autoradiography analysis, *Drug Deliv.*, **2000**, 7(2), 97–103.

38 MORAES A. M., SANTANA M. H., CARBONELL R. G., Preparation and characterization of liposomal systems entrapping the boronated compound o-carboranylpropylamine, *J. Microencapsul.*, **1999**, 16(5), 647–664.

39 ZHOU R., BALASUBRAMANIAN S. V., KAHL S. B., STRAUBINGER R. M., Biopharmaceutics of boronated radiosensitizers: Liposomal formulation of MnBOPP (manganese chelate of 2,4-(alpha, beta-dihydroxyethyl) deuterioporphyrin IX) and comparative toxicity in mice, *J. Pharm. Sci.*, **1999**, 88(9), 912–917.

40 SHELLY K., FEAKES D. A., HAWTHORNE M. F., SCHMIDT P. G., KRISCH T. A., BAUER W. F., Model studies directed toward the boron neutron-capture therapy of cancer: Boron delivery to murine tumors with liposomes, *Proc. Natl. Acad. Sci. U.S.A.*, **1992**, 89, 9039–9043.

41 FEAKES D. A., SHELLY K., KNOBLER C. B., HAWTHORNE M. F., Na3[B20H17NH3]: Synthesis and liposomal delivery to murine tumors, *Proc. Natl. Acad. Sci. U.S.A.*, **1994**, 91, 3029–3033.

42 FEAKES D. A., SHELLY K., HAWTHORNE M. F., Selective boron delivery to murine tumors by lipophilic species incorporated in the membranes of unilamellar liposomes, *Proc. Natl. Acad. Sci. U.S.A.*, **1995**, 92, 1367–1370.

43 FEAKES D. A., WALLER R. C., HATHAWAY D. K., MORTON V. S., Synthesis and in vivo murine evaluation of Na4[1-(1*-B10H9)-6-SHB10H8] as a potential agent for boron neutron capture therapy, *Proc. Natl. Acad. Sci. U.S.A.*, **1999**, 96, 6406–6410.

44 HAWTHORNE M. F., SHELLY K., Liposomes as drug delivery vehicles for boron agents, *J. Neurooncol.*, **1997**, 33(1–2), 53–58.

45 WATSON-CLARK R. A., BANQUERIGO M. L., SHELLY K., HAWTHORNE M. F., BRAHN E., Model studies directed toward the application of boron neutron capture therapy to rheumatoid arthritis: Boron delivery by liposomes in rat collagen-induced arthritis, *Proc. Natl. Acad. Sci. U.S.A.*, **1998**, 95, 2531–2534.

46 YANAGIE H., FUJII Y., TAKAHASHI T., TOMITA T., FUKANO Y., HASUMI K., NARIUCHI H., YASUDA T., SEKIGUCHI M., UCHIDA H., Boron neutron capture therapy using 10B entrapped anti-CEA immunoliposome, *Hum. Cell*, **1989**, 2(3), 290–296.

47 YANAGIE H., TOMITA T., KOBAYASHI H., FUJII Y., TAKAHASHI T., HASUMI K., NARIUCHI H., SEKIGUCHI M., Application of boronated anti-CEA immunoliposome to tumour cell growth inhibition in in vitro boron neutron capture therapy model, *Br. J. Cancer*, **1991**, 63(4), 522–526.

48 YANAGIE H., TOMITA T., KOBAYASHI H., FUJII Y., NONAKA Y., SAEGUSA Y., HASUMI K., ERIGUCHI M., KOBAYASHI T., ONO K., Inhibition of human pancreatic cancer growth in nude mice by boron neutron capture therapy, *Br. J. Cancer*, **1997**, 75(5), 660–665.

49 YANAGIE H., KOBAYASHI H., TAKEDA Y., YOSHIZAKI I., NONAKA Y., NAKA S., NOJIRI A., SHINNKAWA H., FURUYA Y., NIWA H., ARIKI K., YASUHARA H., ERIGUCHI M., Inhibition of growth of human breast cancer cells in culture by neutron capture using liposomes containing 10B, *Biomed. Pharmacother.*, **2002**, 56(2), 93–99.

50 YANAGIE H., OGURA K., TAKAGI K., MARUYAMA K., MATSUMOTO T., SAKURAI Y., SKVARC J., ILLIC R., KUHNE G., HISA T., YOSHIZAKI I., KONO K., FURUYA Y., SUGIYAMA H., KOBAYASHI H., ONO K., NAKAGAWA K., ERIGUCHI M., Accumulation of boron compounds to tumor with polyethylene-glycol binding liposome by using neutron capture

autoradiography, *Appl. Radiat. Isot.*, **2004**, 61(4), 639–646.
51 MARUYAMA K., ISHIDA O., KASAOKA S., TAKIZAWA T., UTOGUCHI N., SHINOHARA A., CHIBA M., KOBAYASHI H., ERIGUCHI M., YANAGIE H., Intracellular targeting of sodium mercaptoundecahydrododecaborate (BSH) to solid tumors by transferrin-PEG liposomes for boron neutron-capture therapy (BNCT), *J. Controlled Release*, **2004**, 98(2), 195–207.
52 LASTER B. H., KAHL S. B., POPENOE E. A., PATE D. W., FAIRCHILD R. G., Biological efficacy of boronated low-density lipoprotein for boron neutron capture therapy as measured in cell culture, *Cancer Res.*, **1991**, 51(17), 4588–4593.
53 HOFMANN B., FISCHER C. O., LAWACZECK R., PLATZEK J., SEMMLER W., Gadolinium neutron capture therapy (GdNCT) of melanoma cells and solid tumors with the magnetic resonance imaging contrast agent Gadobutrol, *Invest. Radiol.*, **1999**, 34, 126–133.
54 KHOKHLOV V. F., YASHKIN P. N., SILIN D. I., DJOROVA E. S., LAWACZECK R., Neutron capture therapy with Gd-DTPA in tumor-bearing rats, in *Cancer Neutron Capture Therapy* (Y. MISHIMA, ed.), Plenum Press, New York, **1996**, pp. 865–869.
55 LAWACZECK R., FISCHER C. O., KRÜGER U., LEUTHER W., MENRAD J., Gadolinium neutron capture therapy (GdNCT) with MRI contrast media. In vitro study, in *Cancer Neutron Capture Therapy* (Y. MISHIMA, ed.), Plenum Press, New York, **1996**, pp. 859–864.
56 AKINE Y., TOKITA N., TOKUUYE K., SATOH M., CHUREI H., PECHOUX C. L., KOBAYASHI T., KANDA K., Suppression of rabbit VX-2 subcutaneous tumor growth by gadolinium neutron capture therapy, *Jpn. J. Cancer Res.*, **1993**, 84, 841–843.
57 AKINE Y., TOKITA N., TOKUUYE K., SATOH M., KOBAYASHI T., KANDA K., Electron-equivalent dose for the effect of gadolinium neutron capture therapy on the growth of subcutaneously-inoculated Ehrlich tumor cells in mice, *Jpn. J. Clin. Oncol.*, **1993**, 23, 145–148.
58 AKINE Y., TOKITA N., MATSUMOTO T., OYAMA H., EGAWA S., AIZAWA O., Radiation effect of gadolinium-neutron capture reactions on survival of Chinese hamster cells, *Strahlenther. Onkol.*, **1990**, 166, 831–833.
59 TAKAGAKI M., ODA T., MIYATAKE S., KIKUCHI H., KOBAYASHI T., KANDA K., UJENO Y., Killing effects of gadolinium neutron capture reactions on brain tumors, in *Progress in Neutron Capture Therapy for Cancer* (ALLEN B. J. et al. eds.), Plenum Press, New York, **1992**, pp. 407–410.
60 MATSUMOTO T., Transport calculations of depth-dose distributions for gadolinium neutron capture therapy, *Phys. Med. Biol.*, **1992**, 37, 155–162.
61 FUKUMORI Y., ICHIKAWA H., TOKUMITSU H., MIYAMOTO M., JONO K., KANAMORI R., AKINE Y., TOKITA N., Design and preparation of ethyl cellulose microcapsules of gadopentetate dimeglumine for neutron-capture therapy using the Wurster process, *Chem. Pharm. Bull.*, **1993**, 41, 1144–1148.
62 AKINE Y., TOKITA N., TOKUUYE K., SATOH M., FUKUMORI Y., TOKUMITSU H., KANAMORI R., KOBAYASHI T., KANDA K., Neutron-capture therapy of murine ascites tumor with gadolinium-containing microcapsules, *J. Cancer Res. Clin. Oncol.*, **1992**, 119, 71–73.
63 KOBAYASHI H., KAWAMOTO S., SAGA T., SATO N., ISHIMORI T., KONISHI J., ONO K., TOGASHI K., BRECHBIEL M. W., Avidin-dendrimer-1B4M-Gd(254): A tumor-targeting therapeutic agent for gadolinium neutron capture therapy of intraperitoneal disseminated tumor which can be monitored by MRI, *Bioconj. Chem.*, **2001**, 12, 587–593.
64 DE STASIO G., CASALBORE P., PALLINI R., GILBERT B., SANITA F., CIOTTI M. T., ROSI G., FESTINESI A., LAROCCA L. M., RINELLI A., PERRET D., MOGK D. W., PERFETTI P., MEHTA M. P., MERCANTI D., Gadolinium in human glioblastoma cells for gadolinium

cancer by a novel emulsion-droplet coalescence technique, *Chem. Pharm. Bull.*, **1999**, 47, 838–842.
109 TOKUMITSU H., ICHIKAWA H., FUKUMORI Y., Chitosan-gadopentetic acid complex nanoparticles for gadolinium neutron-capture therapy of cancer: Preparation by novel emulsion-coalescence technique and characterization, *Pharm. Res.*, **1999**, 16, 1830–1835.
110 NISHIOKA Y., KYOTANI S., OKAMURA M., MIYAZAKI M., OKAZAKI K., OHNISHI S., YAMAMOTO Y., ITO K., Release characteristics of cisplatin chitosan microspheres and effect of containing chitin, *Chem. Pharm. Bull.*, **1990**, 38, 2871–2873.
111 OHYA Y., TAKEI T., KOBAYASHI H., OUCHI T., Release behavior of 5-fluorouracil from chitosan-gel microspheres immobilizing 5-fluorouracil derivative coated with polysaccharides and their cell specific recognition, *J. Microencapsulat.*, **1993**, 10, 1–9.
112 AKBUGA J., DURMAZ G., Preparation and evaluation of crosslinked chitosan microspheres containing furosemide, *Int. J. Pharm.*, **1994**, 111, 217–222.
113 HASSAN E. E., PARISH R. C., GALLO J. M., Optimized formulation of magnetic chitosan microspheres containing the anticancer agent, oxantrazole, *Pharm. Res.*, **1992**, 9, 390–397.
114 THANOO B. C., SUNNY M. C., JAYAKRISHNAN A., Cross-linked chitosan microspheres: Preparation and evaluation as a matrix for the controlled release of pharmaceuticals, *J. Pharm. Pharmacol.*, **1992**, 44, 283–286.
115 JAMEELA S. R., JAYAKRISHNAN A., Glutaraldehyde cross-linked chitosan microspheres as a long acting biodegradable drug delivery vehicle: Studies on the in vitro release of mitoxantrone and in vivo degradation of microspheres in rat muscle, *Biomaterials*, **1995**, 16, 769–775.
116 POLK A., AMSDEN B., DE YAO K., PENG T., GOOSEN M. F. A., Controlled release of albumin from chitosan-alginate microcapsules, *J. Pharm. Sci.*, **1994**, 83, 178–185.
117 LIU L. S., LIU S. Q., NG S. Y., FROIX M., OHNO T., HELLER J., Controlled release of interleukin-2 for tumour immunotherapy using alginate/chitosan porous microspheres, *J. Controlled Release*, **1997**, 43, 65–74.
118 BERTHOLD A., CREMER K., KREUTER J., Preparation and characterization of chitosan microspheres as drug carrier for prednisolone sodium phosphate as model for anti-inflammatory drugs, *J. Controlled Release*, **1996**, 39, 17–25.
119 CALVO P., REMUÑAN-LÓPEZ C., VILA-JATO J. L., ALONSO M. J., Chitosan and chitosan/ethylene oxide-propylene oxide block copolymer nanoparticles as novel carriers for proteins and vaccines, *Pharm. Res.*, **1997**, 14, 1431–1436.
120 TOKUMITSU H., HIRATSUKA J., SAKURAI Y., KOBAYASHI T., ICHIKAWA H., FUKUMORI Y., Gadolinium neutron capture therapy using novel gadopentetic acid chitosan complex nanoparticles: In vivo growth suppression of experimental melanoma solid tumor, *Cancer Lett.*, **1999**, 150, 177–182.
121 ICHIHASHI M., SASASE A., HIRAMOTO T., FUNASAKA Y., HATTA S., MISHIMA Y., KOBAYASHI T., FUKUDA H., YOSHINO K., Relative biological effectiveness (RBE) of thermal neutron capture therapy of cultured B-16 melanoma cells preincubated with 10B-paraboronophenylalanine, *Pigment Cell Res.*, **1989**, 2, 325–329.
122 FUKUDA H., KOBAYASHI T., MATSUZAWA T., KANDA K., ICHIHASHI M., MISHIMA Y., RBE of a thermal neutron beam and the 10B (n,α) 7Li reaction on cultured B-16 melanoma cells, *Int. J. Radiat. Biol.*, **1987**, 51, 167–175.
123 HIRATSUKA J., FUKUDA H., KOBAYASHI T., KARASHIMA H., YOSHINO K., IMAJO Y., MISHIMA Y., The relative biological effectiveness of 10B-neutron capture therapy for early skin reaction in the hamster, *Radiat. Res.*, **1991**, 128, 186–191.
124 KOBAYASHI T., KANDA K., FUJITA Y.,

SAKURAI Y., ONO K., The upgrade of the heavy water facility of the Kyoto University Reactor for neutron capture therapy, in *Advances in Neutron Capture Therapy* (LARSSON B. et al. eds.), Elsevier Science, Amsterdam, 1997, vol. I, pp. 321–325.

125 HE P., DAVIS S. S., ILLUN L., *Int. J. Pharm.*, 1998, 166, 75–88.

126 LUEßEN H. L., LEEUW B. J. D., LANGEMEYER M. W. E., BOER A. G., VERHOEF J. C., JUNGINGER H. E., *Pharm. Res.*, 1996, 13(11), 1668–1672.

127 DESAI M. P., LABHASETWAR V., WALTER E., LEVY R. J., AMIDON G. L., *Pharm. Res.*, 1997, 14(11), 1568–1573.

128 CHATELET C., DAMOUR O., DOMARD A., *Biomaterials*, 2001, 22(3), 261–268.

129 SHIKATA F., TOKUMITSU H., ICHIKAWA H., FUKUMORI Y., In vitro cellular accumulation of gadolinium incorporated into chitosan nanoparticles designed for neutron-capture therapy of cancer, *Eur. J. Pharm. Biopharm.*, 2002, 53, 57–63.

130 CHOKSAKULNIMITR S., MASUDA S., TOKUDA H., TAKAKURA Y., HASHIDA M., *J. Controlled Release*, 1995, 34(3), 233–241.

131 GREEN T. R., FISHER J., STONE M., WROBLEWSKI B. M., INGHAM E., *Biomaterials*, 1998, 19, 2297–2302.

132 LEE K. D., HONG K., PAPAHADJOPOULOS D., *Biochem. Biophys. Acta*, 1992, 1103, 185–197.

133 TOKUMITSU H., HIRATSUKA J., SAKURAI Y., SHIKATA F., ICHIKAWA H., FUKUMORI Y., KOBAYASHI T., Gadolinium neutron-capture therapy trial using gadopentetic acid-chitosan complex nanoparticles: Effect of gadolinium dose on tumor growth suppression in vivo, *Kyoto University Research Reactor Progress Report*, Section I, 1998, p. 179.

134 TOKUUYE K., TOKITA N., AKINE Y., NAKAYAMA H., SAKURAI Y., KOBAYASHI T., KANDA K., Comparison of radiation effects of gadolinium and boron neutron capture reactions, *Strahlenther. Onkol.* 2000, 176, 81–83.

4
Nanovehicles and High Molecular Weight Delivery Agents for Boron Neutron Capture Therapy[1]

Gong Wu, Rolf F. Barth, Weilian Yang, Robert Lee, Werner Tjarks, Marina V. Backer, and Joseph M. Backer

4.1
Introduction

4.1.1
Overview

Boron neutron capture therapy (BNCT) is based on the nuclear capture and fission reactions that occur when non-radioactive boron-10 is irradiated with low energy thermal neutrons to yield high linear energy transfer (LET) alpha particles (^{4}He) and recoiling lithium-7 (^{7}Li) nuclei. For BNCT to be successful, a sufficient number of ^{10}B atoms ($\sim 10^9$ atoms per cell) must be selectively delivered to the tumor and enough thermal neutrons must be absorbed by them to sustain a lethal ^{10}B(n, α) ^{7}Li capture reaction. BNCT primarily has been used to treat patients with brain tumors, and more recently those with head and neck cancer. Two low molecular weight (LMW) boron delivery agents are currently used clinically, sodium borocaptate and boronophenylalanine. However, various high molecular weight (HMW) agents consisting of macromolecules and nanovehicles have been developed. This chapter focuses on the latter, which include monoclonal antibodies, dendrimers, liposomes, dextrans, polylysine, folate receptor targeting agents, epidermal and vascular endothelial growth factors (EGF and VEGF). Procedures for introducing boron atoms into these HMW agents and their chemical properties will be discussed. *In vivo* studies on their biodistribution will be described, and the efficacy of a subset of them, which have been used for BNCT of tumors in experimental animals, will be discussed. Since brain tumors currently are the primary candidates for treatment by BNCT, delivery of these HMW agents across the blood–brain barrier presents a special challenge. Various routes of administration will be discussed, including receptor-facilitated transcytosis following intravenous admin-

---

[1] We thank Bentham Science Publishers, Ltd for granting copyright release to republish the text and figures of this article.

istration, direct intratumoral injection and convection-enhanced delivery by which a pump is used to apply a pressure gradient to establish bulk flow of the HMW agent during interstitial infusion. Finally, we conclude with a discussion relating to issues that must be addressed if these HMW agents are to be used clinically.

4.1.2
General Background

After decades of intensive research, high grade gliomas, and specifically glioblastoma multiforme (GBM), are still extremely resistant to all current forms of therapy, including surgery, chemotherapy, radiotherapy, immunotherapy and gene therapy [1–5]. The five-year survival rate of patients diagnosed with GBM in the United States is less than a few percent [6, 7] despite aggressive treatment using combinations of therapeutic modalities. This is due to the infiltration of malignant cells beyond the margins of resection and their spread into both gray and white matter by the time of surgical resection [8, 9]. High grade gliomas are histologically complex and heterogeneous in their cellular composition. Recent molecular genetic studies of gliomas have shown how complex the development of these tumors is [10]. Glioma cells and their neoplastic precursors have biologic properties that allow them to evade a tumor-associated host immune response [11], and biochemical properties that allow them to invade the unique extracellular environment of the brain [12, 13]. Consequently, high grade supratentorial gliomas must be regarded as a whole brain disease [14]. The inability of chemo- and radiotherapy to cure patients with high grade gliomas is due to their failure to eradicate microinvasive tumor cells within the brain. To successfully treat these tumors, therefore, strategies must be developed that can selectively target malignant cells with little or no effect on normal cells and tissues adjacent to the tumor.

BNCT is based on nuclear capture and fission reactions that occur when nonradioactive ^{10}B is irradiated with low energy thermal neutrons to yield high LET alpha particles (^{4}He) and recoiling ^{7}Li nuclei. In order to be successful BNCT, sufficient ^{10}B atoms ($\sim 10^9$ atoms per cell) must be delivered selectively to the tumor and enough thermal neutrons must be absorbed by them to sustain a lethal $^{10}B(n,\alpha)$ ^{7}Li capture reaction. The destructive effects of these high-energy particles are limited to boron-containing cells. BNCT primarily has been used to treat high grade gliomas [15, 16], and either cutaneous primaries [17] or cerebral metastases of melanoma [18]. More recently, it also has been used to treat patients with head and neck [19, 20] and metastatic liver cancer [21, 22]. BNCT is a biologically rather than physically targeted type of radiation treatment. If sufficient amounts of ^{10}B and thermal neutrons can be delivered to the target volume, the potential exists to destroy tumor cells dispersed in the normal tissue parenchyma. Readers interested in more in-depth coverage of other topics related to BNCT are referred to several recent reviews and monographs [15, 23–25]. The present chapter focuses on boron-containing macromolecules and nanovehicles as boron delivery agents.

4.2
General Requirements for Boron Delivery Agents

A successful boron delivery agent should have (1) no or minimal systemic toxicity with rapid clearance from blood and normal tissues; (2) high tumor (~20 μg-^{10}B g^{-1}) and low normal tissue uptake; (3) high tumor:brain (T:Br) and tumor: blood (T:Bl) concentration ratios (>3–4:1); and (4) persistence in the tumor for a sufficient period of time to carry out BNCT. At this time *no* single boron delivery agent fulfills all these criteria. However, as a result of new synthetic techniques and increased knowledge of the biological and biochemical requirements for an effective agent, several promising new boron agents have emerged, and these are described in a special issue of Anti-Cancer Agents in Medicinal Chemistry (6, 2, 2006). The major challenge in their development has been the requirement for specific tumor targeting to achieve boron concentrations sufficient to deliver therapeutic doses of radiation to the tumor with minimal normal tissue toxicity. The selective destruction of GBM cells in the presence of normal cells represents an even greater challenge than malignancies at other anatomic sites.

4.3
Low Molecular Weight Delivery Agents

In the 1950s and early 1960s clinical trials of BNCT were carried out using boric acid and some of its derivatives as delivery agents. These simple chemical compounds had poor tumor retention, attained low T:Br ratios and were non-selective [26, 27]. Among the hundreds of low-molecular weight boron-containing compounds that were synthesized, two appeared to be promising. One, based on arylboronic acids [28], was L-4-dihydroxyborylphenylalanine, referred to as boronophenylalanine or BPA (Fig. 4.1, **1**). The second, a polyhedral borane anion, was

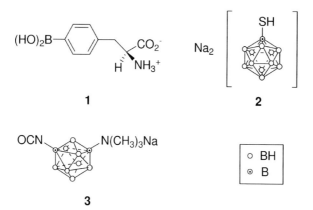

Fig. 4.1. Structure of two compounds used clinically for BNCT, dihydroxyborylphenylalanine or boronophenylalanine (BPA, **1**) and disodium undecahydro-*closo*-dodecaborate or sodium borocaptate (BSH, **2**), and the isocyanato polyhedral borane (**3**), which has been used to heavily boronate dendrimers.

sodium mercaptoundecahydro-*closo*-dodecaborate [29], more commonly known as sodium borocaptate or BSH (**2**). These two compounds persisted longer in animal tumors than did related molecules, attained T:Br and T:Bl boron ratios > 1 and had low toxicity. ^{10}B-enriched BPA, complexed with fructose to improve its water solubility, and BSH have been used clinically for BNCT of brain, as well as extracranial tumors. Although their selective accumulation in tumors is not ideal, the safety of these two drugs following i.v. administration has been well established [30, 31].

4.4
High Molecular Weight Boron Delivery Agents

High molecular weight (HMW) delivery agents usually contain a stable boron group or cluster linked via a hydrolytically stable bond to a tumor-targeting moiety, such as monoclonal antibodies (mAbs) or low molecular weight receptor targeting ligands. Examples of these include epidermal growth factor (EGF) or the mAb cetuximab (IMC-C225) to target the EGF receptor or its mutant isoform EGFRvIII, which are overexpressed in various malignant tumors, including gliomas and squamous cell carcinomas of the head and neck [32]. Agents that are to be administered systemically should be water soluble, but lipophilicity is important in order to cross the blood–brain barrier (BBB) and diffuse within the brain and the tumor. There should be a favorable differential in boron concentrations between tumor and normal brain, thereby enhancing their tumor specificity. Their amphiphilic character is not as crucial for LMW agents that target specific biological transport systems and/or are incorporated into nanovehicles such as liposomes. Molecular weight also is an important factor, since it determines the rate of diffusion both within the brain and the tumor. Detailed reviews of the state-of-the-art of compound development for BNCT have been published [33, 34]. The present chapter focuses on boron-containing macromolecules and liposomes as delivery agents for BNCT, and how they can be most effectively administered.

4.5
Dendrimer-related Delivery Agents

4.5.1
Properties of Dendrimers

Dendrimers are synthetic polymers with a well-defined globular structure. They are composed of a core molecule, repeat units that have three or more functionalities, and reactive surface groups (Fig. 4.2) [35, 36]. Two techniques have been used to synthesize these macromolecules: divergent growth outwards from the core [37], or convergent growth from the terminal groups inwards towards the core [36, 38]. Regular and repeated branching at each monomer group gives rise to a symmetric structure and pattern to the entire globular dendrimer. Dendrimers are an attrac-

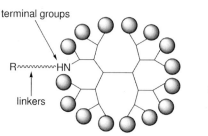

Fig. 4.2. Structure of a boronated PAMAM dendrimer that has been linked to targeting moieties. PAMAM dendrimers consist of a core, repeating polyamido amino units, and reactive terminal groups. Each successively higher generation of PAMAM dendrimer has a geometrically incremental number of terminal groups. Dendrimers have been boronated by reaction with water-soluble isocyanato polyhedral boranes and subsequently attached to targeting moieties by means of heterobifunctional linkers.

tive platform for macromolecular imaging and gene delivery because of their low cytotoxicity and their multiple types of reactive terminal groups [36, 39–44].

4.5.2
Boronated Dendrimers Linked to Monoclonal Antibodies

4.5.2.1 Boron Clusters Directly Linked to mAb

Monoclonal antibodies have been attractive targeting agents for delivering radionuclides [45], drugs [46–50], toxins [51] and boron to tumors [52–55]. Before the introduction of dendrimers as boron carriers, boron compounds were directly attached to mAbs [53, 54]. Approximately ~10^9 ^{10}B atoms per cell (~20 µg g^{-1} tumor) must be delivered to kill tumor cells [55, 56]. Based on the assumption of 10^6 antigenic receptor sites per cell, ~50–100 boron cage structures of carboranes, or polyhedral borane anions and their derivatives must be linked to each mAb molecule to deliver the required amount of boron for NCT. The attachment of such a large number of boron cages to a mAb may result in precipitation of the bioconjugate or a loss of its immunological activity. Solubility can be improved by inserting a water-soluble gluconamide group into the protein-binding boron cage compounds, thereby enhancing their water solubility [57]. This modification makes it possible to incorporate up to 1100 boron atoms into a human gamma globulin (HGG) molecule without any precipitation. Other approaches to enhance solubility include the use of negatively charged carboranes [58] or polyhedral borane anions [59], as well as the insertion of carbohydrate groups [60, 61]. A major limitation of using an agent containing a single boron cage is that many sites must be modified to deliver 10^3 boron atoms per molecule of antibody and this can reduce its immunoreactivity activity. Alam et al. have shown that attachment of an average of

1300 boron atoms to mAb 17-1A, which is directed against human colorectal carcinoma cells, resulted in a 90% loss of immunoreactivity [62].

4.5.2.2 Attachment of Boronated Dendrimers to mAb

Dendrimers are one of the most attractive polymers that have been used as boron carriers due to their well-defined structure and multiple reactive terminal groups. Depending on the antigen site density, ~1000 boron atoms need to be attached per molecule of dendrimer and subsequently linked to the mAb. In our first study, second- and fourth-generation polyamido amino (PAMAM or "starburst") dendrimers, which have 12 and 48 reactive terminal amino groups, respectively, were reacted with the water-soluble isocyanato polyhedral borane [$Na(CH_3)_3NB_{10}H_8NCO$] (3, Fig. 4.1) [63, 64]. The boronated dendrimer then was linked to the mAb IB16-6, which is directed against the murine B16 melanoma, by means of two heterobifunctional linkers, m-maleimidobenzoyl-N-hydroxysulfosuccinimide ester (sulfo-MBS) and N-succinimidyl 3-(2-pyridyldithio)propionate (SPDP) [63, 65]. However, following intravenous (i.v.) administration, large amounts of the bioconjugate accumulated in the liver and spleen and it was concluded that random conjugation of boronated dendrimers to a mAb could alter its binding affinity and biodistribution. To minimize the loss of mAb reactivity, a 5th generation PAMAM dendrimer was boronated with the same polyhedral borane anion, and more recently it was site-specifically linked to the anti-EGFR mAb cetuximab (or IMC-C225) or the EGFRvIII specific mAb L8A4 (Fig. 4.3). Cetuximab was linked via glycosidic moieties in the Fc region by means of two heterobifunctional reagents, SPDP and N-(κ-maleimidoundecanoic acid) hydrazide (KMUH) [66, 67]. The resulting bioconjugate, designated C225-G5-B_{1100}, contained ~1100 boron atoms per molecule of cetuximab and retained its aqueous solubility in 10% DMSO and its *in vitro* and *in vivo* immunoreactivity. As determined by a competitive binding assay, there was a <1 log unit decrease in affinity for EGFR(+) glioma cell line $F98_{EGFR}$, compared to that of unmodified cetuximab [66]. In vivo biodistribution studies, carried out 24 h after intratumoral (i.t.) administration of the bioconjugate, demonstrated that 92.3 µg g^{-1} of boron was retained in rats bearing $F98_{EGFR}$ gliomas compared to 36.5 µg g^{-1} in EGFR(−) F98 parental tumors and 6.7 µg g^{-1} in normal brain [67] thereby indicating specific molecular targeting of the receptor.

4.5.3
Boronated Dendrimers Delivered by Receptor Ligands

4.5.3.1 Epidermal Growth Factors (EGF)

Due to its increased expression in various human tumors, including high grade gliomas, and its low or undetectable expression in normal brain, EGFR is an attractive target for cancer therapy [68–70]. As described above, targeting of EGFR has been carried out using either mAbs or alternatively, as described in this section, EGF, which is a single-chain, 53-mer heat- and acid-stable polypeptide. It binds to a transmembrane glycoprotein with tyrosine kinase activity, which triggers dimeri-

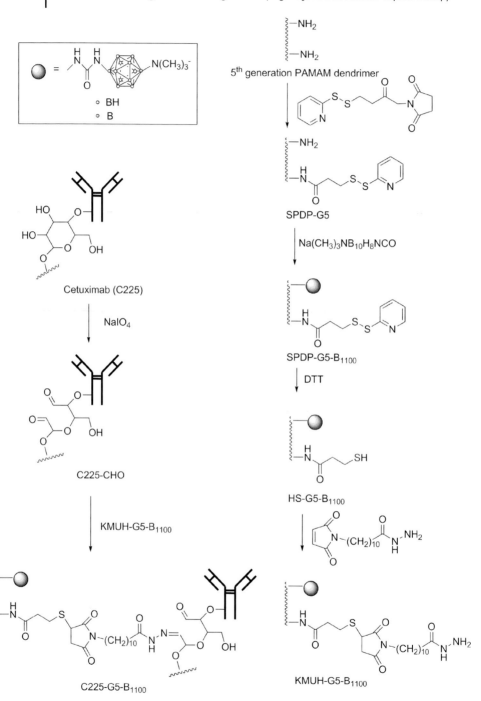

Fig. 4.3. Conjugation scheme for linking a boron-containing dendrimer to the monoclonal antibody C225 (cetuximab), which is directed against EGFR [66, 69].

zation and internalization [71, 72]. Since the EGF boron bioconjugates have a much smaller MW than mAb conjugates, they should be capable of more rapid and effective tumor targeting than mAbs [69, 73].

The procedure used to conjugate EGF to a boronated dendrimer was slightly different from that used to boronate mAbs. A fourth-generation PAMAM dendrimer was reacted with the isocyanato polyhedral borane $Na(CH_3)_3NB_{10}H_8NCO$. Next, reactive thiol groups were introduced into the boronated dendrimer using SPDP, and EGF was derivatized with m-maleimidobenzoyl-N-hydroxysulfosuccinimide ester (MBS). Reaction of the thiol groups of the derivatized, boronated dendrimer with maleimide groups produced stable BSD-EGF bioconjugates, which contained ~960 atoms of boron per molecule of EGF [74]. The BSD-EGF initially was bound to the cell surface membrane and then was endocytosed, which resulted in accumulation of boron in lysosomes [74]. Subsequently, *in vitro* and *in vivo* studies were carried out to evaluate the potential efficacy of the bioconjugate as a boron delivery agent for BNCT [73]. As will be described in more detail later below (Section 4.9.3), therapy studies demonstrated that $F98_{EGFR}$ glioma bearing rats that received either boronated EGF or mAb by either direct i.t. injection or convection-enhanced delivery into the brain had a longer mean survival time than animals bearing F98 parental tumors following BNCT [75–77].

4.5.3.2 Folate Receptor Targeting Agents

Folate receptor (FR) is overexpressed on various human cancers, including those originating in ovary, lung, breast, endometrium and kidney [78–80]. Folic acid (FA) is a vitamin that is transported into cells via FR-mediated endocytosis. The attachment of FA via its γ-carboxylic function to other molecules does not alter its endocytosis by FR-expressing cells [81]. FR targeting has been used successfully to deliver protein toxins, chemotherapeutic, radio-imaging, therapeutic and MRI contrast agents [82], liposomes [83], gene transfer vectors [84], antisense oligonucleotides [85], ribozymes and immunotherapeutic agents to FR-positive cancers [86]. To deliver boron compounds, FA was conjugated to heavily boronated 3rd generation PAMAM dendrimers containing poly(ethylene glycol) (PEG) [87]. PEG was introduced into the bioconjugate to reduce its uptake by the reticuloendothelial system (RES) and, more specifically, the liver and spleen. Folate linked to 3rd generation PAMAM dendrimers containing 12–15 decaborate clusters and 1–1.5 PEG_{2000} units had the lowest hepatic uptake in C57Bl/6 mice (7.2–7.7% injected dose [ID] per gram of liver). *In vitro* studies using FR (+) KB cells have demonstrated receptor-dependent uptake of the bioconjugate. Biodistribution studies with this conjugate, carried out in C57Bl/6 mice bearing subcutaneous (s.c.) implants of the FR (+) murine sarcoma 24JK-FBP, demonstrated selective tumor uptake (6.0% ID per g tumor), but there was high hepatic (38.8% ID per g) and renal (62.8% ID per g) uptake [87].

4.5.3.3 Vascular Endothelial Growth Factor (VEGF)

There is preclinical and clinical evidence indicating that angiogenesis plays a major role in the growth and dissemination of malignant tumors [88, 89]. Inhibition of angiogenesis has yielded promising results in several experimental animal tumor

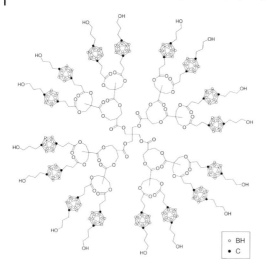

Fig. 4.5. Structure of a carborane-containing aliphatic polyester dendrimer. Carborane cages were incorporated into the interior of the dendrimer structure and the peripheral hydrophilic groups improved water solubility and were available for modification.

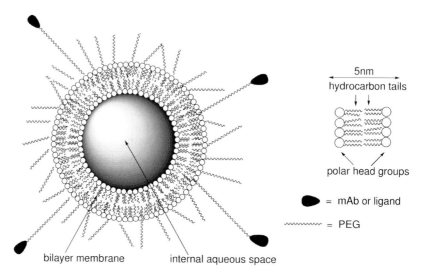

Fig. 4.6. Schematic structure of a liposome that has an aqueous core and a lipid bilayer membrane. The latter is composed of polar head groups with hydrocarbon tails. The liposomal surface can be modified by PEGylation to prolong its circulation time and can be linked to either a mAb or a ligand for targeting.

("stealth") liposomes can increase the amounts of anticancer drugs that can be delivered to solid tumors by passive targeting. Rapidly growing solid tumors have increased permeability to nanoparticles due to increased capillary pore size. These can range from 100 to 800 nm compared to 60–80 nm for those in normal tissues, which are impermeable to liposomes. In addition, tumors lack efficient lymphatic drainage, and, consequently, clearance of extravasated liposomes is slow [98]. Modification of the liposomal surface by PEGylation or attachment of antibodies or receptor ligands will improve their selective targeting and increase their circulation time [98].

4.6.2
Liposomal Encapsulation of Sodium Borocaptate and Boronophenylalanine

4.6.2.1 Boron Delivery by Non-targeted Liposomes

Liposomes have been extensively evaluated as nanovehicles for the delivery of boron compounds for NCT [99, 100]. *In vitro* and *in vivo* studies have demonstrated that they can effectively and selectively deliver large quantities of boron to tumors and that the compounds delivered by liposomes have a longer tumor retention time. BPA is an amino acid analogue that is preferentially taken up by cells with increased metabolic activity, such as tumor cells of varying histopathologic types, including melanomas [31, 101], gliomas [102] and squamous cell carcinomas [103, 104]. Because of its low aqueous solubility, BPA has been used as a fructose complex, which has permitted it to be administered i.v. rather than orally [105, 106]. Following i.v. administration of BPA, which had been incorporated into conventional liposomes, there was rapid elimination by the reticuloendothelial system (RES) with very low blood and liver boron concentrations at 3 h. In contrast, if BPA was incorporated into liposomes composed of DSPE-PEG, therapeutically useful tumor boron concentrations ($>20\ \mu g\ g^{-1}$) were seen at 3 and at 6 h, indicating that PEG-liposomes had evaded the RES [107]. In addition, BPA has been incorporated into the lipid bilayer of liposomes, composed of either the positively charged lipid 1,2-dioleoyl-3-trimethylammonium propane (DOTAP) or the zwitterionic lipid, 1,2-dioleoyl-sn-glycerol-3-phosphoethanolamine (DOPE) [108]. Cationic liposomes have been widely used as carriers of biomolecules that specifically target the cell nucleus [109], which would be advantageous for BNCT. Another clinically used drug, sodium borocaptate (BSH), has been incorporated into liposomes composed of DPPC/Chol in a 1:1 molar ratio with or without PEG stabilization [110]. The average diameter of liposomes containing BSH was in the range 100–110 nm. Both types of liposomes resulted in a significant improvement in their circulation time compared with that of free BSH. At 24 h following i.v. injection of PEG-liposomes, 19% of the injected dose of boron was in the blood compared with 7% following formulation of BSH in conventional liposomes. The mean percent uptake by the liver and spleen was not significantly different for the two types of liposomes. However, the blood:RES ratios were higher for PEG-liposomes at all time points, indicating that a higher fraction of the injected dose of BSH was still in the blood. Ji et al. have reported that there were no significant differences in the *in vitro* up-

active NH^{3-} substituted $[n-B_{20}H_{18}]^{2-}$ electrophilic anion, $[B_{20}H_{17}NH_3]^-$. Another anion, $[ae-B_{20}H_{17}NH_3]^{3-}$, also was encapsulated into liposomes prepared with 5% PEG-2000-distearoyl phosphatidylethanolamine as a constituent of the membrane. These liposomes had longer *in vivo* circulation times, which resulted in continued accumulation of boron in the tumor over the entire 48 h, and reached a maximum concentration of 47 µg-B per g tumor.

$[B_{20}H_{17}SH]^{4-}$ (7), a thiol derivative of $[B_{20}H_{18}]^{4-}$, possesses a reactive thiol substituent that can be oxidized to give the more reactive $[B_{20}H_{17}SH]^{2-}$ anion. Both species were considered to be essential for high tumor boron retention [120] and they have been encapsulated into small, unilamellar liposomes. Biodistribution was determined after i.v. injection into BALB/c mice bearing EMT6 tumors. At low injected doses, tumor boron concentrations increased throughout the experiment, resulting in a maximum concentration of 47 µg-B per g tumor at 48 h, which corresponded to 22.2% ID g^{-1} and a T:Bl ratio of 7.7. This was the most promising of the polyhedral borane anions that had been investigated for liposomal delivery. Although they were able to deliver adequate amounts of boron to tumor cells, their application to BNCT has been limited due to their low incorporation efficiency (~3%).

Lipophilic boron compounds incorporated into the lipid bilayer would be an alternative approach. Small unilamellar vesicles composed of a 3:3:1 ratio of distearoylphosphatidylcholine, cholesterol and $K[nido-7-CH_3(CH_2)_{15}-7,8-C_2B_9H_{11}]$ (8) in the lipid bilayer and $Na_3[a2-B_{20}H_{17}NH_2CH_2CH_2NH_2]$ (9) in the aqueous core were produced as a delivery agents for NCT-mediated synovectomy [121]. Biodistribution studies were carried out in Louvain rats that had a collagen-induced arthritis. The maximum synovial boron concentration was 29 µg per gram of tissue at 30 h and this had only decreased to 22 µg g^{-1} at 96 h following i.v. administration. The prolonged retention by synovium provided sufficient time for extensive clearance of boron from other tissues so that at 96 h the synovium:blood (Syn:Bl) ratio was 3.0. To accelerate blood clearance, serum stability of the liposomes was lowered by increasing the proportion of compound 8 embedded in the lipid bilayer. Liposomes were formulated with a 3:3:2 ratio DSPC:Ch:8 in the lipid bilayer and 9 was encapsulated in the aqueous core. The boron concentration in the synovium reached a maximum of 26 µg g^{-1} at 48 h with a Syn:Bl ratio of 2, following which it slowly decreased to 14 µg g^{-1}g at 96 h, at which time the Syn:Bl ratio was 7.5 [121].

Another method to deliver hydrophilic boron-containing compounds would be to incorporate them into cholesterol to target tumor cells expressing amplified low density lipoprotein (LDL) receptors [122–124]. Glioma cells, which absorb more cholesterol, have been reported to take up more LDL than the corresponding normal tissue cells [125–127]. The cellular uptake of liposomal cholesteryl 1,12-dicarba-*closo*-dodecaboranel-carboxylate (10) by two fast-growing human glioma cell lines, SF-763 and SF-767, was mediated via the LDL receptor and was much higher than that of human neurons. The cellular boron concentration was ~10–11× greater than that required for BNCT [128].

4.6.3
Boron Delivery by Targeted Liposomes

To improve the specificity of liposomally encapsulated drugs and to increase the amount of boron delivered, targeting moieties have been attached to the surface of liposomes. These could be any molecules that selectively recognized and bound to target antigens or receptors that were overexpressed on neoplastic cells or tumor-associated neovasculature. These have included either intact mAb molecules or fragments, and low molecular weight, naturally occurring or synthetic ligands such as peptides or receptor-binding ligands such as EGF. To date, liposomes linked to mAbs or their fragments [129], EGF [130], folate [131] and transferrin [112] have been the most extensively studied as targeting moieties (Fig. 4.8).

4.6.3.1 Immunoliposomes

The murine anti-carcinoembryonic antigen (CEA) mAb 2C-8 has been conjugated to large multilamellar liposomes containing ^{10}B compounds [132, 133]. The maximum number of ^{10}B atoms attached per molecule of mAb was $\sim 1.2 \times 10^4$. These immunoliposomes bound selectively to the human pancreatic carcinoma cell line, AsPC-1, that overexpressed CEA. Incubating the immunoliposomes with either MRKnu/nu-1 or AsPC-1 tumor cells suppressed *in vitro* tumor cell growth follow-

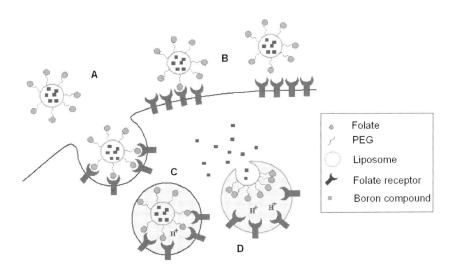

Fig. 4.8. Proposed mechanism of intracellular boron delivery based on receptor-mediated tumor cell targeting. Liposomes loaded with a boron compound (A) are conjugated to a targeting ligand (e.g., folate, transferrin, or anti-EGFR antibody). These bind to receptors on the cell surface (B), and are then internalized by receptor-mediated endocytosis (C) into the acidified endosomal/lysosomal compartment. The boron compound is then released (D) into the cytosol by liposomal degradation, endosome/lysosome disruption or liposome–endosome/lysome membrane fusion.

ing thermal neutron irradiation [134]. This was dependent upon the liposomal concentration of the ^{10}B-compound and on the number of molecules of mAb conjugated to the liposomes. Immunoliposomes containing either $(Et_4N)_2B_{10}H_{10}$ and linked to the mAb MGb 2, directed against human gastric cancer [135, 136] or water-soluble boronated acridine (WSA, 11; Fig. 4.7) linked to trastuzumab, directed against HER-2, have been prepared and evaluated *in vitro* [129]. There was specific binding and high uptake of these immunoliposomes, which delivered a sufficient amount of ^{10}B to produce a tumoricidal effect following thermal neutron irradiation.

4.6.3.2 Folate Receptor-targeted Liposomes

A highly ionized boron compound, $Na_3B_{20}H_{17}NH_3$, has been incorporated into liposomes by passive loading [131, 137, 138]. This showed high *in vitro* uptake by the FR-expressing human cell line KB (American Type Culture Collection CCL 17), which originally was thought to be derived from a squamous cell carcinoma of the mouth, and subsequently was shown to be identical to HeLa cells, as determined by isoenzyme markers, DNA fingerprinting and karyotypic analysis. KB-tumor-bearing mice that received either FR-targeted or non-targeted control liposomes had equivalent tumor boron values ($\sim$85 µg g^{-1}), which attained a maximum at 24 h, while the T:Bl ratio reached a maximum at 72 h. Additional studies were carried out with the lipophilic boron compound K[*nido*-7-CH$_3$(CH$_2$)$_{15}$-7,8-C$_2$B$_9$H$_{11}$] (8). This was incorporated into large unilamellar vesicles, $\sim$200 nm in diameter, composed of egg PC/chol/8 in a 2:2:1 mol/mol ratio, and an additional 0.5 mol% of folate-PEG-DSPE or PEG-DSPE for the FR-targeted or non-targeted liposomal formulations [139]. Boron uptake by FR-overexpressing KB cells, treated with these targeted liposomes, was $\sim$10$\times$ greater compared with those treated with control liposomes. In addition, BSH and five weakly basic boronated polyamines were evaluated (Fig. 4.9). Two of these were the spermidine derivatives N^5-(4-carboranylbutyl)spermidine · 3HCl (12) and N^5-[4-(2-aminoethyl-*o*-carboranyl)butyl]-spermidine · 4HCl (13). Three were the spermine derivatives N^5-(4-*o*-carboranylbutyl)spermine · 4HCl (14), N^5-[4-(2-aminoethyl-*o*-carboranyl)butyl]spermine · 5HCl (15), and N^5,N^{10}-bis(4-*o*-carboranylbutyl)spermine · 4HCl (16). These were incorporated into liposomes by a pH-gradient-driven remote-loading method with varying loading efficiencies, which were influenced by the specific trapping agent and the structure of the boron compound. Greater loading efficiencies were obtained with lower molecular weight boron derivatives, using ammonium sulfate as the trapping agent, compared with those obtained with sodium citrate.

4.6.3.3 EGFR Targeted Liposomes

Acridine is a water-soluble (WS) DNA-intercalator. Its boronated derivative WSA has been incorporated into liposomes composed of EGF-conjugated lipids. Their surface contained $\sim$5 mol% PEG and 10–15 molecules of EGF, and 10^4–10^5 of the WSA molecules were encapsulated. These liposomes had EGFR-specific cellular binding to cultured human glioma cells [130, 140] and were internalized following specific binding to the receptor. Following internalization, WSA, primarily, was

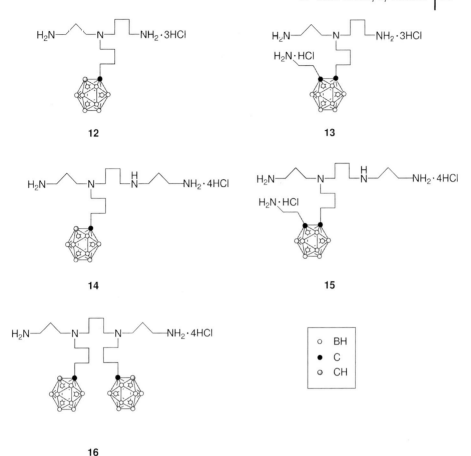

Fig. 4.9. Structures of five weakly basic boronated polyamines encapsulated in FR targeting liposomes. The two spermidine (**12**, **13**) and three spermine derivatives (**14–16**), which contain hydrophilic amine groups and lipophilic carboranyl cages, had DNA-binding properties.

localized in the cytoplasm, and had high cellular retention, with 80% of the boron remaining cell-associated after 48 h [141].

4.7
Boron Delivery by Dextrans

Dextrans are glucose polymers that consist mainly of a linear α-1,6-glucosidic linkage with some degree of branching via a 1,3-linkage [142, 143]. Dextrans have been used extensively as drug and protein carriers to increase drug circulation time [144, 145]. In addition, native or chemically-modified dextrans have been used for passive targeting to tumors, the RES or active receptor-specific cellular targeting. To link boron compounds to dextrans [146], β-decachloro-o-carborane derivatives, in which one of the carbon atoms is substituted by -CH$_2$CHOHCH$_2$-O-CH$_2$CH=CH$_2$, have been epoxidized and subsequently bound to dextran, with a resulting boron content of 4.3% (w/w) [147]. The modified dextran could then be attached to tumor-specific antibodies [147–150]. BSH has been covalently coupled to dextran derivatives by two methods [151]. In the first, dextran was activated with 1-cyano-4-(dimethylamino)pyridine (CDAP) and subsequently coupled with 2-aminoethyl pyridyl disulfide. Then, thiolated dextran was linked to BSH in a disulfide exchange reaction. A total of 10–20 boron cages were attached to each dextran chain. In the second method, dextran was derivatized to a multiallyl derivative (Fig. 4.10, 17), which was reacted with BSH in a free-radical-initiated addition reaction. Using this method, 100–125 boron cages could be attached per dextran chain, suggesting

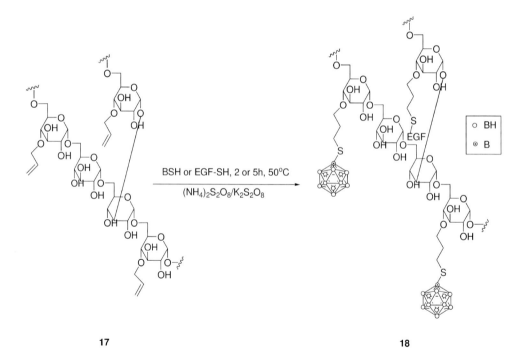

Fig. 4.10. Preparation of EGF-targeted, boronated dextrans. The bioconjugate was prepared by a free-radical-initiated addition reaction between multiallyl dextran derivatives and BSH or thiolated EGF at 50 °C using K$_2$S$_2$O$_8$ as an initiator.

that this derivative might be a promising template for the development of other HMW delivery agents. By the second method, designed to target EGFR overexpressing cells, EGF and BSH have been covalently linked to a 70 kDa dextran (**18**) [152–154]. Bioconjugates having a few BSH molecules attained maximum *in vitro* binding at 4 h with the human glioma cell line U-343 MGaC12:6. In contrast, there was a slow increase of binding over 24 h for those having many BSH molecules. Although most of the bioconjugates were internalized, *in vitro* retention was low, as was *in vivo* uptake following i.v. injection into nude mice bearing s.c. implants of Chinese hamster ovary (CHO) cells transfected with the human gene encoding EGFR (designated CHO-EGFR). However, following i.t. injection, boron uptake was higher with CHO-EGFR(+) tumors compared to wild-type EGFR(−) CHO tumors [155].

4.8
Other Macromolecules used for Delivering Boron Compounds

Polylysine is another polymer having multiple reactive amino groups that has been used as a platform for the delivery of boron compounds [53, 156]. The protein-binding polyhedral boron derivative isocyanatoundecahydro-*closo*-dodecaborate ($B_{12}H_{11}NCO^{2-}$) has been linked to polylysine and subsequently to the anti-B16 melanoma mAb IB16-6 using two heterobifunctional linkers, SPDP and sulfo-MBS. The bioconjugate had an average of 2700 boron atoms per molecule and retained 58% of the immunoreactivity of the native antibody, as determined by a semiquantitative immunofluorescent assay or by ELISA. Other bioconjugates prepared by this method had >1000 boron atoms per molecule of antibody and retained 40–90% of the immunoreactivity of the native antibody [53]. Using another approach, site-specific linkage of boronated polylysine to the carbohydrate moieties of anti-TSH antibody resulted in a bioconjugate that had ~6×10^3 boron atoms with retention of its immunoreactivity [156].

A streptavidin–biotin system has also been developed to specifically deliver boron to tumors. Biotin was linked to a mAb and streptavidin was attached to the boron-containing moiety. The indirect linking of boron to the mAb minimized loss of its immunoreactivity. BSH was attached to poly-(D-glutamate D-lysine) (poly-GL) via a heterobifunctional agent [157]. This boronated poly-GL then was activated by a carbodiimide reagent and in turn reacted with streptavidin. Another approach employed a streptavidin mutant that had 20 cysteine residues per molecule. BSH was conjugated via sulfhydryl-specific bifunctional reagents to incorporate ~230 boron atoms per molecule [158]. A closomer species with an icosahedral dodecaborate core and twelve pendant anionic *nido*-7,8-carborane groups has been developed as a new class of unimolecular nanovehicles for evaluation as a delivery agent for BNCT [159].

4.9
Delivery of Boron-containing Macromolecules to Brain Tumors

4.9.1
General Considerations

Drug delivery to brain tumors is dependent upon (1) the plasma concentration profile of the drug, which depends upon the amount and route of administration, (2) the ability of the agent to traverse the blood brain barrier (BBB), (3) blood flow within the tumor, and (4) the lipophilicity of the drug. In general, a high steady-state blood concentration will maximize brain uptake, while rapid clearance will reduce it, except in the case of intra-arterial (i.a.) drug administration. Although the i.v. route is currently being used clinically to administer both BSH and BPA, this may not be ideal for boron-containing macromolecules, and other strategies must be employed to improve their delivery.

4.9.2
Drug-transport Vectors

One approach to improve brain tumor uptake of boron compounds has been to conjugate them to a drug-transport vector by means of receptor-specific transport systems [160, 161]. Proteins such as insulin, insulin-like growth factor (IGF), transferrin (TF) [162], and leptin can traverse the BBB. BSH encapsulated in TF-PEG-liposomes had a prolonged residence time in the circulation and low RES uptake in tumor-bearing mice, resulting in enhanced extravasation of the liposomes into the tumor and concomitant internalization by receptor-mediated endocytosis [163, 164]. Mice that received BSH-containing TF-liposomes followed by BNCT had significantly longer survival times than those that received PEG-liposomes, bare liposomes and free BSH, thereby establishing *proof-of-principle* for transcytosis of a boron-containing nanovehicle [112].

4.9.3
Direct Intracerebral Delivery

Studies carried out by us have clearly demonstrated that the i.v. route of administration is not suitable for delivery of boronated EGF or mAbs to glioma bearing rats [75, 165]. Intravenous injection of technetium-99m labeled EGF to rats bearing intracerebral implants of the C6 rat glioma, which had been genetically engineered to express the human EGFR gene, resulted in 0.14% ID localizing in the tumor. Intracarotid (i.c.) injection with or without BBB disruption increased the tumor uptake to 0.34 to 0.45% ID g^{-1}, but based even on the most optimistic assumptions the amount of boron that could be delivered to the tumor by i.v. injection would be inadequate for BNCT [165]. Direct i.t. injection of boronated EGF (BSD-EGF), however, resulted in tumor boron concentrations of 22 µg g^{-1} compared to 0.01 µg g^{-1} following i.v. injection and almost identical boron uptake values were obtained using the F98$_{EGFR}$ glioma model [77]. This was produced by transfecting F98 glioma cells with the gene encoding human EGFR. Based on our biodistribution results, therapy studies were initiated with the F98$_{EGFR}$ glioma in syngeneic Fischer rats. F98$_{EGFR}$ glioma bearing rats that received BSD-EGF i.t. had a mean survival time (MST) of 45 ± 1 d compared to 33 ± 2 d in animals that had EGFR-(−) wild-type F98 gliomas. Since it is unlikely that any single boron delivery agent will be able to target all tumor cells, the combination of i.t. administration of BSD-EGF with i.v. injection of BPA was evaluated. This furthered increased the MST to 57 ± 8 d compared to 39 ± 2 d for i.v. BPA alone [73]. These data provide *proof-of-principle* for the idea of using a combination of low and HMW boron delivery agents.

4.9.4
Convection-enhanced Delivery (CED)

CED, by which therapeutic agents are directly infused into the brain, is an innovative method to increase their uptake and distribution [166–168]. Under normal physiological conditions, interstitial fluids move through the brain by both convection and diffusion. Diffusion of a drug in tissue depends upon its molecular weight, ionic charge and its concentration gradient within normal tissue and the tumor. The higher the molecular weight of the drug and the more positively charged the ionic species, the lower its concentration, then the slower its diffusion. For example, diffusion of antibody into a tumor requires three days to diffuse 1 mm from the point of origin. Unlike diffusion, however, convection or "bulk" flow results from a pressure gradient that is independent of the molecular weight of the substance. CED, potentially, can improve the targeting of both low and HMW molecules, as well as liposomes, to the CNS by applying a pressure gradient to establish bulk flow during interstitial infusion. The volume of distribution (V_d) is a linear function of the volume of the infusate (V_i). CED has been used to efficiently deliver drugs and HMW agents such as mAbs and toxin fusion proteins to brain tumors [168–170]. CED can provide a more homogenous dispersion of the agent and at higher concentrations than otherwise would be attainable by i.v. injection [165]. For example, in our own studies, CED of ^{125}I-labeled EGF to F98$_{EGFR}$ glioma bearing rats resulted in 47% ID g^{-1} of the bioconjugate localizing in the tumor compared to 10% ID g^{-1} in normal brain at 24 h following administration. The corresponding boron values were 22 and 2.9–4.9 µg g^{-1}, respectively [76]. Based on these results, therapy studies were initiated. F98$_{EGFR}$-glioma-bearing rats that received BD-EGF by CED had a MST of 53 ± 13 d compared to 40 ± 5 d for animals that received BPA i.v. [73]. Similar studies have been carried out using either boronated cetuximab (IMC-C225) or the mAb L8A4 [171, 172], which is specifically directed against the tumor-specific mutant isoform EGFRvIII; comparable results were obtained [173]. Direct intracerebral administration of these and other HMW agents by CED has opened up the possibility that they actually could be used clinically, since CED is being used to administer radiolabeled antibodies, toxin fusion proteins and gene vectors to patients with GBM [168–170, 175–179].

It is only a matter of time before this approach is also used to deliver both low and HMW boron-containing agents for NCT.

4.10
Clinical Considerations and Conclusions

In this chapter we have focused on HMW boron delivery agents and nanovehicles that potentially could be used clinically for targeting intra- and extracranial tumors. Animal studies, carried out in glioma-bearing rats, have demonstrated that boronated EGF and the mAb cetuximab, both of which bind to EGFR, selectively target receptor (+) tumors following direct i.c. delivery. Furthermore, following BNCT, a significant increase in MST was observed, and this was further enhanced if BPA was administered in combination with the HMW agents. These studies provide *proof-of-principle* first for the potential utility of HMW agents and, second, the therapeutic gain associated with combining them with LMW boron delivery agents. A major question is whether any of these agents will ever be used clinically? There are several critical issues that must be addressed if BNCT is to ever become a useful modality for the treatment of cancer. *First*, large clinical trials, preferably randomized, must be carried out to convincingly demonstrate the efficacy of the two drugs that currently are being used, BPA and BSH. Once this has been established, studies with HMW EGFR targeting agents could move forward. Both direct i.t. injection [174, 175] and CED [170, 176–179] have been used clinically to deliver mAbs and toxin fusion proteins to patients who have had surgical resection of their brain tumors. These studies provide a strong clinical rationale for the direct intracerebral delivery of HMW agents. Initially, the primary focus should be on determining the safety of administering them to patients prior to surgical resection of their brain tumors. Once this has been established, then biodistribution studies could be carried out in patients who were going to have surgical resection of their brain tumors. Tumor and normal tissues would be analyzed for their boron content, and if there was evidence of preferential tumor localization with boron concentrations in the range 10–20 $\mu g\ g^{-1}$ and normal brain concentrations of <5 $\mu g\ g^{-1}$ then therapy studies could be undertaken. Since there is considerable variability in EGFR expression in gliomas, it is highly unlikely that any single agent will be able to deliver the requisite amount of boron to all tumor cells, and they would be used in combination with BPA/BSH. This general plan would also be applicable to the other HMW delivery agents and nanovehicles that have been discussed here. The joining together of chemistry and nanotechnology [180, 181] represents a major step towards the development of effective boron delivery agents for NCT. Nanovehicles offer the possibility of tumor targeting with enhanced boron payloads. Potentially, this could solve the central problem of how to selectively deliver a large number of boron atoms to individual cancer cells.

The preceding discussion shows that the development of HMW boron delivery agents must proceed in step with strategies to optimize their delivery and an appreciation as to how they would be used clinically. Intracerebral delivery has been used in clinically advanced settings, but nuclear reactors, which currently are the only source of neutrons for BNCT, would not be conducive to this. Therefore, the development of accelerator neutron sources [24], which could be easily sited in hospitals, is especially important. This would also facilitate the initiation of large-scale clinical trials at selected centers that treat large numbers of patients with brain tumors and would permit evaluation of new boron delivery agents. In conclusion, as should be apparent from this review, a plethora of HMW boron delivery agents have been designed and synthesized. The challenge is to move from experimental animal studies to clinical biodistribution studies, a step that has yet to be taken.

Acknowledgments

We thank Dr. Achintya Bandyopadhyaya for suggestions on figures and Mrs. Beth Kahl for secretarial assistance. Experimental studies described in this report were supported by N.I.H. grants 1 R01 CA098945 (R.F. Barth), 1 R01 CA79758 (W. Tjarks), and Department of Energy grants DE-FG02-98ER62595 (R.F. Barth) and DE-FG-2-02FR83520 (J.M. Backer).

References

1 Berger M. S., Malignant astrocytomas: Surgical aspects, *Semin. Oncol.*, **1994**, 21, 172–185.
2 Gutin P. H., Posner J. B., Neuro-oncology: Diagnosis and management of cerebral gliomas – past, present, and future, *Neurosurgery*, **2000**, 47, 1–8.
3 Parney I. F., Chang S. M., Current chemotherapy for glioblastoma, *Cancer J.*, **2003**, 9, 149–156.
4 Paul D. B., Kruse C. A., Immunologic approaches to therapy for brain tumors, *Curr. Neurol. Neurosci. Rep.*, **2001**, 1, 238–244.
5 Rainov N. G., Ren H., Gene therapy for human malignant brain tumors, *Cancer J.*, **2003**, 9, 180–188.
6 Curran W. J., Jr., Scott C. B., Horton J., Nelson J. S. et al., Recursive partitioning analysis of prognostic factors in three Radiation Therapy Oncology Group malignant glioma trials, *J. Natl. Cancer Inst.*, **1993**, 85, 704–710.
7 Lacroix M., Abi-Said D., Fourney D. R., Gokaslan Z. L. et al., A multivariate analysis of 416 patients with glioblastoma multiforme: Prognosis, extent of resection, and survival, *J. Neurosurg.*, **2001**, 95, 190–198.
8 Hentschel S. J., Lang F. F., Current surgical management of glioblastoma, *Cancer J.*, **2003**, 9, 113–125.
9 Laws E. R., Jr., Shaffrey M. E., The inherent invasiveness of cerebral gliomas: Implications for clinical management, *Int. J. Dev. Neurosci.*, **1999**, 17, 413–420.
10 Ware M. L., Berger M. S., Binder D. K., Molecular biology of glioma tumorigenesis, *Histol. Histopathol.*, **2003**, 18, 207–216.
11 Parney I. F., Hao C., Petruk K. C., Glioma immunology and

immunotherapy, *Neurosurgery*, **2000**, 46, 778–792.

12 KACZAREK E., ZAPF S., BOUTERFA H., TONN J. C. et al., Dissecting glioma invasion: Interrelation of adhesion, migration and intercellular contacts determine the invasive phenotype, *Int. J. Dev. Neurosci.*, **1999**, 17, 625–641.

13 NUTT C. L., MATTHEWS R. T., HOCKFIELD S., Glial tumor invasion: A role for the upregulation and cleavage of BEHAB/brevican, *Neuroscientist*, **2001**, 7, 113–122.

14 HALPERIN E. C., BURGER P. C., BULLARD D. E., The fallacy of the localized supratentorial malignant glioma, *Int. J. Radiat. Oncol. Biol. Phys.*, **1988**, 15, 505–509.

15 BARTH R. F., A critical assessment of boron neutron capture therapy: An overview, *J. Neurooncol.*, **2003**, 62, 1–5.

16 NAKAGAWA Y., POOH K., KOBAYASHI T., KAGEJI T. et al., Clinical review of the Japanese experience with boron neutron capture therapy and a proposed strategy using epithermal neutron beams, *J. Neurooncol.*, **2003**, 62, 87–99.

17 WADABAYASHI N., HONDA C., MISHIMA Y., ICHIHASHI M., Selective boron accumulation in human ocular melanoma vs surrounding eye components after 10B1-p-boronophenylalanine administration. Prerequisite for clinical trial of neutron-capture therapy, *Melanoma Res.*, **1994**, 4, 185–190.

18 BUSSE P. M., HARLING O. K., PALMER M. R., KIGER W. S. 3rd et al., A critical examination of the results from the Harvard-MIT NCT program phase I clinical trial of neutron capture therapy for intracranial disease, *J. Neurooncol.*, **2003**, 62, 111–121.

19 KATO I., ONO K., SAKURAI Y., OHMAE M. et al., Effectiveness of BNCT for recurrent head and neck malignancies, *Appl. Radiat. Isot.*, **2004**, 61, 1069–1073.

20 RAO M., TRIVILLIN V. A., HEBER E. M., CANTARELLI MDE L. et al., BNCT of 3 cases of spontaneous head and neck cancer in feline patients, *Appl. Radiat. Isot.*, **2004**, 61, 947–952.

21 KOIVUNORO H., BLEUEL D. L., NASTASI U., LOU T. P. et al., BNCT dose distribution in liver with epithermal D-D and D-T fusion-based neutron beams, *Appl. Radiat. Isot.*, **2004**, 61, 853–859.

22 PINELLI T., ZONTA A., ALTIERI S. et al., TAOrMINA: From the first idea to the application to the human liver, in *Research and Development in Neutron Capture Therapy, Proceedings of the International Congress on Neutron Capture Therapy*, SAUERWEIN M. W., MOSS R., WITTIG A. eds., Monduzzi Editore, Bologna, Italy, **2002**, pp. 1065–1072.

23 CODERRE J. A., TURCOTTE J. C., RILEY K. J., BINNS P. J. et al., Boron neutron capture therapy: Cellular targeting of high linear energy transfer radiation, *Technol. Cancer Res. Treat.*, **2003**, 2, 355–375.

24 BARTH R. F., CODERRE J. A., VICENTE M. G., BLUE T. E., Boron neutron capture therapy of cancer: Current status and future prospects, *Clin. Cancer Res.*, **2005**, 11, 3987–4002.

25 ZAMENHOF R. G., CODERRE J. A., RIVARD M. J., PATEL H., Eleventh world congress on neutron capture therapy, *Appl. Radiat. Isot.*, **2004**, 61, 731–1130.

26 FARR L. E., SWEET W. H., ROBERTSON J. S., FOSTER C. G. et al., Neutron capture therapy with boron in the treatment of glioblastoma multiforme, *Am. J. Roentgenol. Radium Ther. Nucl. Med.*, **1954**, 71, 279–293.

27 GOODWIN J. T., FARR L. E., SWEET W. H., ROBERTSON J. S., Pathological study of eight patients with glioblastoma multiforme treated by neutron-capture therapy using boron 10, *Cancer*, **1955**, 8, 601–615.

28 SNYDER H. R., REEDY A. J., LENNARZ W. J., Synthesis of aromatic boronic acids. Aldehydo boronic acids and a boronic acid analog of tyrosine, *J. Am. Chem. Soc.*, **1958**, 80, 835–838.

29 SOLOWAY A. H., HATANAKA H., DAVIS M. A., Penetration of brain and brain tumor. VII. Tumor-binding sulfhydryl boron compounds, *J. Med. Chem.*, **1967**, 10, 714–717.

30 Hatanaka H., Nakagawa, Y., Clinical results of long-surviving brain tumor patients who underwent boron neutron capture therapy, *Int. J. Radiat. Oncol. Biol. Phys.*, **1994**, 28, 1061–1066.

31 Mishima Y., Selective thermal neutron capture therapy of cancer cells using their specific metabolic activities – melanoma as prototype, in *Cancer Neutron Capture Therapy*, Y. Mishima, ed., Plenum Press, New York, **1996**, pp. 1–26.

32 Ang K. K., Berkey B. A., Tu X., Zhang H. Z. et al., Impact of epidermal growth factor receptor expression on survival and pattern of relapse in patients with advanced head and neck carcinoma, *Cancer Res.*, **2002**, 62, 7350–7356.

33 Hawthorne M. F., Lee M. W., A critical assessment of boron target compounds for boron neutron capture therapy, *J. Neurooncol.*, **2003**, 62, 33–45.

34 Soloway A. H., Tjarks W., Barnum B. A., Rong F. G. et al., The chemistry of neutron capture therapy, *Chem. Rev.*, **1998**, 98, 1515–1562.

35 Gillies E. R., Frechet J. M. J., Dendrimers and dendritic polymers in drug delivery, *Drug. Discovery Today*, **2005**, 10, 35–43.

36 Esfand R., Tomalia D. A., Poly(amidoamine) (PAMAM) dendrimers: From biomimicry to drug delivery and biomedical applications, *Drug. Discov Today*, **2001**, 6, 427–436.

37 Tomalia D. A., Naylor A. M., Goddard W. A. III, Starburst dendrimers: Control of size, shape, surface chemistry, topology and flexibility in the conversion of atoms to macroscopic materials, *Angew. Chem.*, **1990**, 102, 119–157.

38 Verheyde B., Maes W., Dehaen W., The use of 1,3,5-triazines in dendrimer synthesis, *Mater. Sci. Eng., C: Biomim. Supramol. Systems*, **2001**, C18, 243–245.

39 McCarthy T. D., Karellas P., Henderson S. A., Giannis M. et al., Dendrimers as drugs: Discovery and preclinical and clinical development of dendrimer-based microbicides for HIV and STI prevention, *Mol. Pharm.*, **2005**, 2, 312–318.

40 Venditto V. J., Regino C. A., Brechbiel M. W., PAMAM dendrimer based macromolecules as improved contrast agents, *Mol. Pharm.*, **2005**, 2, 302–311.

41 Ambade A. V., Savariar E. N., Thayumanavan S., Dendrimeric micelles for controlled drug release and targeted delivery, *Mol. Pharm.*, **2005**, 2, 264–272.

42 Majoros I. J., Thomas T. P., Mehta C. B., Baker, Jr., J. R., Poly(amidoamine) dendrimer-based multifunctional engineered nanodevice for cancer therapy, *J. Med. Chem.*, **2005**, 48, 5892–5899.

43 Boas U., Heegaard P. M., Dendrimers in drug research, *Chem. Soc. Rev.*, **2004**, 33, 43–63.

44 Klajnert B., Bryszewska M., Dendrimers: Properties and applications, *Acta Biochim. Pol.*, **2001**, 48, 199–208.

45 Sharkey R. M., Goldenberg D. M., Perspectives on cancer therapy with radiolabeled monoclonal antibodies, *J. Nucl. Med.*, **2005**, 46(Suppl 1), 115S–127S.

46 Jaracz S., Chen J., Kuznetsova L. V., Ojima I., Recent advances in tumor-targeting anticancer drug conjugates, *Bioorg. Med. Chem.*, **2005**, 13, 5043–5054.

47 Garnett M. C., Targeted drug conjugates: Principles and progress, *Adv. Drug Deliv. Rev.*, **2001**, 53, 171–216.

48 Chari R. V., Targeted delivery of chemotherapeutics: Tumor-activated prodrug therapy, *Adv. Drug Deliv. Rev.*, **1998**, 31, 89–104.

49 Hamblett K. J., Senter P. D., Chace D. F., Sun M. M. et al., Effects of drug loading on the antitumor activity of a monoclonal antibody drug conjugate, *Clin. Cancer Res.*, **2004**, 10, 7063–7070.

50 Trail P. A., King H. D., Dubowchik G. M., Monoclonal antibody drug immunoconjugates for targeted

treatment of cancer, *Cancer Immunol. Immunother.*, **2003**, 52, 328–337.

51 FRACASSO G., BELLISOLA G., CASTELLETTI D., TRIDENTE G., COLOMBATTI M., Immunotoxins and other conjugates: Preparation and general characteristics, *Mini Rev. Med. Chem.*, **2004**, 4, 545–562.

52 LIU L., BARTH R. F., ADAMS D. M., SOLOWAY A. H., REISFELD R. A., Bispecific antibodies as targeting agents for boron neutron capture therapy of brain tumors, *J. Hematother.*, **1995**, 4, 477–483.

53 ALAM F., SOLOWAY A. H., BARTH R. F., MAFUNE N. et al., Boron neutron capture therapy: Linkage of a boronated macromolecule to monoclonal antibodies directed against tumor-associated antigens, *J. Med. Chem.*, **1989**, 32, 2326–2330.

54 BARTH R. F., JOHNSON C. W., WEI W. Z., CAREY W. E. et al., Neutron capture using boronated monoclonal antibody directed against tumor-associated antigens, *Cancer Detect. Prev.*, **1982**, 5, 315–323.

55 ALAM F., BARTH R. F., SOLOWAY A. H., Boron containing immunoconjugates for neutron capture therapy of cancer and for immunocytochemistry, *Antibody, Immunoconjugates, and Radiopharmaceuticals*, **1989**, 2, 145–163.

56 TOLPIN E. I., WELLUM G. R., DOHAN F. C. JR., KORNBLITH P. L., ZAMENHOF R. G., Boron neutron capture therapy of cerebral gliomas. II. Utilization of the blood-brain barrier and tumor-specific antigens for the selective concentration of boron in gliomas, *Oncology*, **1975**, 32, 223–246.

57 SNEATH R. L. JR., WRIGHT J. E., SOLOWAY A. H., O'KEEFE S. M. et al., Protein-binding polyhedral boranes, *J. Med. Chem.*, **1976**, 19, 1290–1294.

58 VARADARAJAN A., SHARKEY R. M., GOLDENBERG D. M., HAWTHORNE M. F., Conjugation of phenyl isothiocyanate derivatives of carborane to antitumor antibody and in vivo localization of conjugates in nude mice, *Bioconj. Chem.*, **1991**, 2, 102–110.

59 TAKAHASHI T., FUJII Y., FUJII G., NARIUCHI H., Preliminary study for application of anti-alpha-fetoprotein monoclonal antibody to boron-neutron capture therapy, *Jpn. J. Exp. Med.*, **1987**, 57, 83–91.

60 COMPOSTELLA F., MONTI D., PANZA L., POLETTI L., PROSPERI D., Synthesis of glycosyl carboranes with different linkers between the sugar and the boron cage moieties, *Research and Development in Neutron Capture Therapy, Proceedings of the International Congress on Neutron Capture Therapy*, SAUERWEIN M.W., MOSS R., WITTIG A. eds., Monduzzi Editore, Bologna, Italy, **2002**, 81–84.

61 GIOVENZANA G. B., LAY L., MONTI D., PALMISANO G., PANZA L., Synthesis of carboranyl derivatives of alkynyl glycosides as potential BNCT agents, *Tetrahedron*, **1999**, 55, 14 123–14 136.

62 ALAM F., SOLOWAY A. H., MCGUIRE J. E., BARTH R. F. et al., Dicesium N-succinimidyl 3-(undecahydro-closo-dodecaboranyldithio)propionate, a novel heterobifunctional boronating agent, *J. Med. Chem.*, **1985**, 28, 522–525.

63 BARTH R. F., ADAMS D. M., SOLOWAY A. H., ALAM F., DARBY M. V., Boronated starburst dendrimer-monoclonal antibody immuno-conjugates: Evaluation as a potential delivery system for neutron capture therapy, *Bioconj. Chem.*, **1994**, 5, 58–66.

64 LIU L., BARTH R. F., ADAMS D. M., SOLOWAY A. H., REISFELD R. A., Critical evaluation of bispecific antibodies as targeting agents for boron neutron capture therapy of brain tumors, *Anticancer Res.*, **1996**, 16, 2581–2588.

65 BARTH R. F., ADAMS D. M., SOLOWAY A. H., DARBY M. V., In vivo distribution of boronated monoclonal antibodies and starburst dendrimers, *Adv. Neutron Capture Ther.*, [*Proc. 5th Int. Symp.*], **1993**, 351–355.

66 WU G., BARTH R. F., YANG W., CHATTERJEE M. et al., Site-specific conjugation of boron-containing dendrimers to anti-EGF receptor

monoclonal antibody cetuximab (IMC-C225) and its evaluation as a potential delivery agent for neutron capture therapy, *Bioconj. Chem.*, **2004**, 15, 185–194.
67 ARTEAGA C. L., Overview of epidermal growth factor receptor biology and its role as a therapeutic target in human neoplasia, *Semin. Oncol.*, **2002**, 29, 3–9.
68 MENDELSOHN J., Targeting the epidermal growth factor receptor for cancer therapy, *J. Clin. Oncol.*, **2002**, 20, 1S–13S.
69 BARTH R. F., WU G., YANG W., BINNS P. J. et al., Neutron capture therapy of epidermal growth factor (+) gliomas using boronated cetuximab (IMC-C225) as a delivery agent, *Appl. Radiat. Isot.*, **2004**, 61, 899–903.
70 PAL S. K., PEGRAM M., Epidermal growth factor receptor and signal transduction: Potential targets for anti-cancer therapy, *Anticancer Drugs*, **2005**, 16, 483–494.
71 NORMANNO N., BIANCO C., STRIZZI L., MANCINO M. et al., The ErbB receptors and their ligands in cancer: An overview, *Curr. Drug Targets*, **2005**, 6, 243–257.
72 JORISSEN R. N., WALKER F., POULIOT N., GARRETT T. P. et al., Epidermal growth factor receptor: Mechanisms of activation and signalling, *Exp. Cell Res.*, **2003**, 284, 31–53.
73 YANG W., BARTH R. F., WU G., BANDYOPADHYAYA, A. K. et al., Boronated epidermal growth factor as a delivery agent for neutron capture therapy of EGF receptor positive gliomas, *Appl. Radiat. Isot.*, **2004**, 61, 981–985.
74 CAPALA J., BARTH R. F., BENDAYAN M., LAUZON M. et al., Boronated epidermal growth factor as a potential targeting agent for boron neutron capture therapy of brain tumors, *Bioconj. Chem.*, **1996**, 7, 7–15.
75 YANG W., BARTH R. F., ADAMS D. M., SOLOWAY A. H., Intratumoral delivery of boronated epidermal growth factor for neutron capture therapy of brain tumors, *Cancer Res.*, **1997**, 57, 4333–4339.
76 YANG W., BARTH R. F., ADAMS D. M., CIESIELSKI M. J. et al., Convection-enhanced delivery of boronated epidermal growth factor for molecular targeting of EGF receptor-positive gliomas, *Cancer Res.*, **2002**, 62, 6552–6558.
77 BARTH R. F., YANG W., ADAMS D. M., ROTARU J. H. et al., Molecular targeting of the epidermal growth factor receptor for neutron capture therapy of gliomas, *Cancer Res.*, **2002**, 62, 3159–3166.
78 REDDY J. A., ALLAGADDA V. M., LEAMON C. P., Targeting therapeutic and imaging agents to folate receptor positive tumors, *Curr. Pharm. Biotechnol.*, **2005**, 6, 131–150.
79 LEAMON C. P., REDDY J. A., Folate-targeted chemotherapy, *Adv. Drug Deliv. Rev.*, **2004**, 56, 1127–1141.
80 SUDIMACK J., LEE R. J., Targeted drug delivery via the folate receptor, *Adv. Drug Deliv. Rev.*, **2000**, 41, 147–162.
81 PAULOS C. M., REDDY J. A., LEAMON C. P., TURK M. J., LOW P. S., Ligand binding and kinetics of folate receptor recycling in vivo: Impact on receptor-mediated drug delivery, *Mol. Pharmacol.*, **2004**, 66, 1406–1414.
82 REDDY J. A., LOW P. S., Folate-mediated targeting of therapeutic and imaging agents to cancers, *Crit. Rev. Ther. Drug Carrier Syst.*, **1998**, 15, 587–627.
83 STEPHENSON S. M., YANG W., STEVENS P. J., TJARKS W. et al., Folate receptor-targeted liposomes as possible delivery vehicles for boron neutron capture therapy, *Anticancer Res.*, **2003**, 23, 3341–3345.
84 WARD C. M., Folate-targeted non-viral DNA vectors for cancer gene therapy, *Curr. Opin. Mol. Ther.*, **2000**, 2, 182–187.
85 GOTTSCHALK S., CRISTIANO R. J., SMITH L. C., WOO S. L., Folate receptor mediated DNA delivery into tumor cells: Protosomal disruption results in enhanced gene expression, *Gene Ther.*, **1994**, 1, 185–191.
86 HILGENBRINK A. R., LOW P. S., Folate receptor-mediated drug targeting:

From therapeutics to diagnostics, *J. Pharm. Sci.*, **2005**, 94, 2135–2146.
87 SHUKLA S., WU G., CHATTERJEE M., YANG W. et al., Synthesis and biological evaluation of folate receptor-targeted boronated PAMAM dendrimers as potential agents for neutron capture therapy, *Bioconj. Chem.*, **2003**, 14, 158–167.
88 BENOUCHAN M., COLOMBO B. M., Anti-angiogenic strategies for cancer therapy (Review), *Int. J. Oncol.*, **2005**, 27, 563–571.
89 TORTORA G., MELISI D., CIARDIELLO F., Angiogenesis: A target for cancer therapy, *Curr. Pharm. Des.*, **2004**, 10, 11–26.
90 BREKKEN R. A., LI C., KUMAR S., Strategies for vascular targeting in tumors, *Int. J. Cancer*, **2002**, 100, 123–130.
91 GAYA A. M., RUSTIN G. J., Vascular disrupting agents: A new class of drug in cancer therapy, *Clin. Oncol. (R. Coll. Radiol.)*, **2005**, 17, 277–290.
92 BERGSLAND E. K., Vascular endothelial growth factor as a therapeutic target in cancer, *Am. J. Health Syst. Pharm.*, **2004**, 61, S4–11.
93 HICKLIN D. J., ELLIS L. M., Role of the vascular endothelial growth factor pathway in tumor growth and angiogenesis, *J. Clin. Oncol.*, **2005**, 23, 1011–1027.
94 BACKER M. V., GAYNUTDINOV T. I., PATEL V., BANDYOPADHYAYA A. K. et al., Vascular endothelial growth factor selectively targets boronated dendrimers to tumor vasculature, *Mol. Cancer Ther.*, **2005**, 4, 1423–1429.
95 BARTH R. F., YANG W., CODERRE J. A., Rat brain tumor models to assess the efficacy of boron neutron capture therapy: A critical evaluation, *J. Neurooncol.*, **2003**, 62, 61–74.
96 PARROTT M. C., MARCHINGTON E. B., VALLIANT J. F., ADRONOV A., Synthesis and properties of carborane-functionalized aliphatic polyester dendrimers, *J. Am. Chem. Soc.*, **2005**, 127, 12 081–12 089.
97 PARK J. W., BENZ C. C., MARTIN F. J., Future directions of liposome- and immunoliposome-based cancer therapeutics, *Semin. Oncol.*, **2004**, 31, 196–205.
98 SAPRA P., ALLEN T. M., Ligand-targeted liposomal anticancer drugs, *Prog. Lipid Res.*, **2003**, 42, 439–462.
99 CARLSSON J., KULLBERG E. B., CAPALA J., SJOBERG S. et al., Ligand liposomes and boron neutron capture therapy, *J. Neurooncol.*, **2003**, 62, 47–59.
100 HAWTHORNE M. F., SHELLY K., Liposomes as drug delivery vehicles for boron agents, *J. Neurooncol.*, **1997**, 33, 53–58.
101 MISHIMA Y., HONDA C., ICHIHASHI M., OBARA H. et al., Treatment of malignant melanoma by single thermal neutron capture therapy with melanoma-seeking ^{10}B-compound, *Lancet*, **1989**, 2, 388–389.
102 CODERRE J. A., GLASS J. D., FAIRCHILD R. G., MICCA P. L. et al., Selective delivery of boron by the melanin precursor analogue p-borono-phenylalanine to tumors other than melanoma, *Cancer Res.*, **1990**, 50, 138–141.
103 ONO K., MASUNAGA S., SUZUKI M., KINASHI Y. et al., The combined effect of boronophenylalanine and borocaptate in boron neutron capture therapy for SCCVII tumors in mice, *Int. J. Radiat. Oncol. Biol. Phys.*, **1999**, 43, 431–436.
104 OBAYASHI S., KATO I., ONO K., MASUNAGA S. et al., Delivery of (10)boron to oral squamous cell carcinoma using boronophenylalanine and borocaptate sodium for boron neutron capture therapy, *Oral Oncol.*, **2004**, 40, 474–482.
105 YOSHINO K., SUZUKI A., MORI Y., KAKIHANA H. et al., Improvement of solubility of p-boronophenylalanine by complex formation with monosaccharides, *Strahlenther Onkol.*, **1989**, 165, 127–129.
106 RYYNANEN P. M., KORTESNIEMI M., CODERRE J. A., DIAZ A. Z. et al., Models for estimation of the ^{10}B concentration after BPA-fructose complex infusion in patients during epithermal neutron irradiation in BNCT, *Int. J. Radiat. Oncol. Biol. Phys.*, **2000**, 48, 1145–1154.

107 PAVANETTO F., PERUGINI P., GENTA I., MINOIA C. et al., Boron-loaded liposomes in the treatment of hepatic metastases: Preliminary investigation by autoradiography analysis, *Drug Deliv.*, **2000**, 7, 97–103.

108 MARTINI S., RISTORI S., PUCCI A., BONECHI C. et al., Boronphenylalanine insertion in cationic liposomes for Boron Neutron Capture Therapy, *Biophys. Chem.*, **2004**, 111, 27–34.

109 SMYTH TEMPLETON N., Cationic liposomes as in vivo delivery vehicles, *Curr. Med. Chem.*, **2003**, 10, 1279–1287.

110 MEHTA S. C., LAI J. C., LU D. R., Liposomal formulations containing sodium mercaptoundecahydrododecaborate (BSH) for boron neutron capture therapy, *J. Microencapsul.*, **1996**, 13, 269–279.

111 JI B., CHEN W., LU D. R., HALPERN D. S., Cell culture and animal studies for intracerebral delivery of borocaptate in liposomal formulation, *Drug Deliv.*, **2001**, 8, 13–17.

112 MARUYAMA K., ISHIDA O., KASAOKA S., TAKIZAWA T. et al., Intracellular targeting of sodium mercaptoundecahydrododecaborate (BSH) to solid tumors by transferrin-PEG liposomes, for boron neutron-capture therapy (BNCT), *J. Controlled Release*, **2004**, 98, 195–207.

113 VALLIANT J. F., GUENTHER K. J., KING A. S., MOREL P. et al., The medicinal chemistry of carboranes, *Coord. Chem. Rev.*, **2002**, 232, 173–230.

114 HAWTHORNE M. F., The role of chemistry in the development of cancer therapy by the boron-neutron capture reaction, *Angew. Chem.*, **1993**, 105, 997–1033 (See also *Angew. Chem., Int. Ed. Engl.*, **1993**, 950–984).

115 MORAES A. M., SANTANA M. H. A., CARBONELL R. G., Preparation and characterization of liposomal systems entrapping the boronated compound o-carboranylpropylamine, *J. Microencapsulat.*, **1999**, 16, 647–664.

116 MORAES A. M., SANTANA M. H. A., CARBONELL R. G., Characterization of liposomal systems entrapping boron-containing compounds in response to pH gradients, *Biofunctional Membranes, Proceedings of the International Conference on Biofunctional Membranes*, BUTTERFIELD D. A. ed., Plenum Press, New York, **1996**, 259–275.

117 SHELLY K., FEAKES D. A., HAWTHORNE M. F., SCHMIDT P. G. et al., Model studies directed toward the boron neutron-capture therapy of cancer: Boron delivery to murine tumors with liposomes, *Proc. Natl. Acad. Sci. U.S.A.*, **1992**, 89, 9039–9043.

118 FEAKES D. A., SHELLY K., KNOBLER C. B., HAWTHORNE M. F., $Na_3[B_{20}H_{17}NH_3]$: Synthesis and liposomal delivery to murine tumors, *Proc. Natl. Acad. Sci. U.S.A.*, **1994**, 91, 3029–3033.

119 HAWTHORNE M. F., SHELLY K., LI F., The versatile chemistry of the $[B_{20}H_{18}]^{2-}$ ions: Novel reactions and structural motifs, *Chem. Commun. (Camb)*, **2002**, 547–554.

120 FEAKES D. A., WALLER R. C., HATHAWAY D. K., MORTON V. S., Synthesis and in vivo murine evaluation of $Na_4[1-(1'-B_{10}H_9)-6-SHB_{10}H_8]$ as a potential agent for boron neutron capture therapy, *Proc. Natl. Acad. Sci. U.S.A.*, **1999**, 96, 6406–6410.

121 WATSON-CLARK R. A., BANQUERIGO M. L., SHELLY K., HAWTHORNE M. F., BRAHN E., Model studies directed toward the application of boron neutron capture therapy to rheumatoid arthritis: Boron delivery by liposomes in rat collagen-induced arthritis, *Proc. Natl. Acad. Sci. U.S.A.*, **1998**, 95, 2531–2534.

122 FEAKES D. A., SPINLER J. K., HARRIS F. R., Synthesis of boron-containing cholesterol derivatives for incorporation into unilamellar liposomes and evaluation as potential agents for BNCT, *Tetrahedron*, **1999**, 55, 11 177–11 186.

123 JI B., PEACOCK G., LU D. R., Synthesis of cholesterol-carborane conjugate for targeted drug delivery, *Bioorg. Med. Chem. Lett.*, **2002**, 12, 2455–2458.

124 PAN G., OIE S., LU D. R., Uptake of

the carborane derivative of cholesteryl ester by glioma cancer cells is mediated through LDL receptors, *Pharm. Res.*, **2004**, 21, 1257–1262.
125 MALETINSKA L., BLAKELY E. A., BJORNSTAD K. A., DEEN D. F. et al., Human glioblastoma cell lines: Levels of low-density lipoprotein receptor and low-density lipoprotein receptor-related protein, *Cancer Res.*, **2000**, 60, 2300–2303.
126 NYGREN C., VON HOLST H., MANSSON J. E., FREDMAN P., Increased levels of cholesterol esters in glioma tissue and surrounding areas of human brain, *Br. J. Neurosurg.*, **1997**, 11, 216–220.
127 LEPPALA J., KALLIO M., NIKULA T., NIKKINEN P. et al., Accumulation of ^{99m}Tc-low-density lipoprotein in human malignant glioma, *Br. J. Cancer*, **1995**, 71, 383–387.
128 PEACOCK G., SIDWELL R., PAN G., OLE S., LU D. R., In vitro uptake of a new cholesteryl carborane ester compound by human glioma cell lines, *J. Pharm. Sci.*, **2004**, 93, 13–19.
129 WEI Q., KULLBERG E. B., GEDDA L., Trastuzumab-conjugated boron-containing liposomes for tumor-cell targeting, development and cellular studies, *Int. J. Oncol.*, **2003**, 23, 1159–1165.
130 BOHL KULLBERG E., BERGSTRAND N., CARLSSON J., EDWARDS K. et al., Development of EGF-conjugated liposomes for targeted delivery of boronated DNA-binding agents, *Bioconj. Chem.*, **2002**, 13, 737–743.
131 STEPHENSON S. M., YANG W., STEVENS P. J., TJARKS W. et al., Folate receptor-targeted liposomes as possible delivery vehicles for boron neutron capture therapy, *Anticancer Res.*, **2003**, 23, 3341–3345.
132 YANAGIE H., FUJII Y., TAKAHASHI T., TOMITA T. et al., Boron neutron capture therapy using ^{10}B entrapped anti-CEA immunoliposome, *Hum. Cell*, **1989**, 2, 290–296.
133 YANAGIE H., TOMITA T., KOBAYASHI H., FUJII Y. et al., Application of boronated anti-CEA immunoliposome to tumour cell growth inhibition in in vitro boron neutron capture therapy model, *Br. J. Cancer*, **1991**, 63, 522–526.
134 YANAGIE H., KOBAYASHI H., TAKEDA Y., YOSHIZAKI I. et al., Inhibition of growth of human breast cancer cells in culture by neutron capture using liposomes containing ^{10}B, *Biomed. Pharmacother.*, **2002**, 56, 93–99.
135 XU L., Boron neutron capture therapy of human gastric cancer by boron-containing immunoliposomes under thermal neutron irradiation (in Chinese), *Zhonghua Yi Xue Za Zhi*, **1991**, 71, 568–571.
136 XU L., ZHANG X. Y., ZHANG S. Y., In vitro and in vivo targeting therapy of immunoliposomes against human gastric cancer (in Chinese), *Zhonghua Yi Xue Za Zhi*, **1994**, 74, 83–86, 126.
137 PAN X. Q., WANG H., LEE R. J., Boron delivery to a murine lung carcinoma using folate receptor-targeted liposomes, *Anticancer Res.*, **2002**, 22, 1629–1633.
138 PAN X. Q., WANG H., SHUKLA S., SEKIDO M. et al., Boron-containing folate receptor-targeted liposomes as potential delivery agents for neutron capture therapy, *Bioconj. Chem.*, **2002**, 13, 435–442.
139 SUDIMACK J. J., ADAMS D., ROTARU J., SHUKLA S. et al., Folate receptor-mediated liposomal delivery of a lipophilic boron agent to tumor cells in vitro for neutron capture therapy, *Pharm. Res.*, **2002**, 19, 1502–1508.
140 BOHL KULLBERG E., CARLSSON J., EDWARDS K., CAPALA J. et al., Introductory experiments on ligand liposomes as delivery agents for boron neutron capture therapy, *Int. J. Oncol.*, **2003**, 23, 461–467.
141 KULLBERG E. B., NESTOR M., GEDDA L., Tumor-cell targeted epiderimal growth factor liposomes loaded with boronated acridine: Uptake and processing, *Pharm. Res.*, **2003**, 20, 229–236.
142 MEHVAR R., Dextrans for targeted and sustained delivery of therapeutic and imaging agents, *J. Controlled Release*, **2000**, 69, 1–25.

143 MEHVAR, R., Recent trends in the use of polysaccharides for improved delivery of therapeutic agents: Pharmacokinetic and pharmacodynamic perspectives, *Curr. Pharm. Biotechnol.*, **2003**, 4, 283–302.

144 CHAU Y., TAN F. E., LANGER R., Synthesis and characterization of dextran-peptide-methotrexate conjugates for tumor targeting via mediation by matrix metalloproteinase II and matrix metalloproteinase IX, *Bioconj. Chem.*, **2004**, 15, 931–941.

145 ZHANG X., MEHVAR R., Dextran-methylprednisolone succinate as a prodrug of methylprednisolone: Plasma and tissue disposition, *J. Pharm. Sci.*, **2001**, 90, 2078–2087.

146 LARSSON B., GABEL D., BORNER H. G., Boron-loaded macromolecules in experimental physiology: Tracing by neutron capture radiography, *Phys. Med. Biol.*, **1984**, 29, 361–370.

147 GABEL D., WALCZYNA R., B-Decachloro-o-carborane derivatives suitable for the preparation of boron-labeled biological macromolecules, *Z. Naturforsch. Teil C*, **1982**, 37, 1038–1039.

148 PETTERSSON M. L., COUREL M. N., GIRARD N., GABEL D., DELPECH B., In vitro immunological activity of a dextran-boronated monoclonal antibody, *Strahlenther Onkol.*, **1989**, 165, 151–152.

149 UJENO Y., AKABOSHI M., MAKI H., KAWAI K. et al., The enhancement of thermal-neutron induced cell death by 10-boron dextran, *Strahlenther Onkol.*, **1989**, 165, 201–203.

150 PETTERSSON M. L., COUREL M. N., GIRARD N., ABRAHAM R. et al., Immunoreactivity of boronated antibodies, *J. Immunol. Methods*, **1990**, 126, 95–102.

151 HOLMBERG A., MEURLING L., Preparation of sulfhydrylborane-dextran conjugates for boron neutron capture therapy, *Bioconj. Chem.*, **1993**, 4, 570–573.

152 CARLSSON J., GEDDA L., GRONVIK C., HARTMAN T. et al., Strategy for boron neutron capture therapy against tumor cells with over-expression of the epidermal growth factor-receptor, *Int. J. Radiat. Oncol. Biol. Phys.*, **1994**, 30, 105–115.

153 GEDDA L., OLSSON P., PONTEN J., CARLSSON J., Development and in vitro studies of epidermal growth factor-dextran conjugates for boron neutron capture therapy, *Bioconj. Chem.*, **1996**, 7, 584–591.

154 MEHTA S. C., LU D. R., Targeted drug delivery for boron neutron capture therapy, *Pharm. Res.*, **1996**, 13, 344–351.

155 OLSSON P., GEDDA L., GOIKE H., LIU L. et al., Uptake of a boronated epidermal growth factor-dextran conjugate in CHO xenografts with and without human EGF-receptor expression, *Anticancer Drug Des.*, **1998**, 13, 279–289.

156 NOVICK S., QUASTEL M. R., MARCUS S., CHIPMAN D. et al., Linkage of boronated polylysine to glycoside moieties of polyclonal antibody; boronated antibodies as potential delivery agents for neutron capture therapy, *Nucl. Med. Biol.*, **2002**, 29, 159–167.

157 FERRO V. A., MORRIS J. H., STIMSON W. H., A novel method for boronating antibodies without loss of immunoreactivity, for use in neutron capture therapy, *Drug Des. Discov.*, **1995**, 13, 13–25.

158 SANO T., Boron-enriched streptavidin potentially useful as a component of boron carriers for neutron capture therapy of cancer, *Bioconj. Chem.*, **1999**, 10, 905–911.

159 THOMAS J., HAWTHORNE M. F., Dodeca(carboranyl)-substituted closomers: Toward unimolecular nanoparticles as delivery vehicles for BNCT, *Chem. Commun. (Camb)*, **2001**, 1884–1885.

160 PARDRIDGE W. M., Vector-mediated drug delivery to the brain, *Adv. Drug Deliv. Rev.*, **1999**, 36, 299–321.

161 PARDRIDGE W. M., Blood-brain barrier biology and methodology, *J. Neurovirol.*, **1999**, 5, 556–569.

162 Hatakeyama H., Akita H., Maruyama K., Suhara T., Harashima H., Factors governing the in vivo tissue uptake of transferrin-coupled polyethylene glycol liposomes in vivo, *Int. J. Pharm.*, **2004**, 281, 25–33.

163 Yanagie H., Ogura K., Takagi K., Maruyama K. et al., Accumulation of boron compounds to tumor with polyethylene-glycol binding liposome by using neutron capture autoradiography, *Appl. Radiat. Isot.*, **2004**, 61, 639–646.

164 Maruyama K., Takizawa T., Yuda T., Kennel S. J. et al., Targetability of novel immunoliposomes modified with amphipathic poly(ethylene glycol)s conjugated at their distal terminals to monoclonal antibodies, *Biochim. Biophys. Acta*, **1995**, 1234, 74–80.

165 Yang W., Barth R. F., Leveille R., Adams D. M. et al., Evaluation of systemically administered radiolabeled epidermal growth factor as a brain tumor targeting agent, *J. Neurooncol.*, **2001**, 55, 19–28.

166 Bobo R. H., Laske D. W., Akbasak A., Morrison P. F. et al., Convection-enhanced delivery of macromolecules in the brain, *Proc. Natl. Acad. Sci. U.S.A.*, **1994**, 91, 2076–2080.

167 Groothuis D. R., The blood-brain and blood-tumor barriers: A review of strategies for increasing drug delivery, *Neurooncol*, **2000**, 2, 45–59.

168 Vogelbaum M. A., Convection enhanced delivery for the treatment of malignant gliomas: Symposium review, *J. Neurooncol.*, **2005**, 73, 57–69.

169 Husain S. R., Puri R. K., Interleukin-13 receptor-directed cytotoxin for malignant glioma therapy: From bench to bedside, *J. Neurooncol.*, **2003**, 65, 37–48.

170 Kunwar S., Convection enhanced delivery of IL13-PE38QQR for treatment of recurrent malignant glioma: Presentation of interim findings from ongoing phase 1 studies, *Acta Neurochir. Suppl.*, **2003**, 88, 105–111.

171 Wikstrand C. J., Hale L. P., Batra S. K., Hill M. L. et al., Monoclonal antibodies against EGFRvIII are tumor specific and react with breast and lung carcinomas and malignant gliomas, *Cancer Res.*, **1995**, 55, 3140–3148.

172 Wikstrand C. J., McLendon R. E., Friedman A. H., Bigner D. D., Cell surface localization and density of the tumor-associated variant of the epidermal growth factor receptor, EGFRvIII, *Cancer Res.*, **1997**, 57, 4130–4140.

173 Yang W., Barth R. F., Wu G., Ciesielski M. J. et al., Development of a syngeneic rat brain tumor model expressing EGFRvIII and its use for molecular targeting studies with monoclonal antibody L8A4, *Clin. Cancer Res.*, **2005**, 11, 341–350.

174 Cokgor I., Akabani G., Kuan C. T., Friedman H. S. et al., Phase I trial results of iodine-131-labeled antitenascin monoclonal antibody 81C6 treatment of patients with newly diagnosed malignant gliomas, *J. Clin. Oncol.*, **2000**, 18, 3862–3872.

175 Akabani G., Reardon D. A., Coleman R. E., Wong T. Z. et al., Dosimetry and radiographic analysis of 131I-labeled anti-tenascin 81C6 murine monoclonal antibody in newly diagnosed patients with malignant gliomas: A phase II study, *J. Nucl. Med.*, **2005**, 46, 1042–1051.

176 Laske D. W., Youle R. J., Oldfield E. H., Tumor regression with regional distribution of the targeted toxin TF-CRM107 in patients with malignant brain tumors, *Nat. Med.*, **1997**, 3, 1362–1368.

177 Sampson J. H., Akabani G., Archer G. E., Bigner D. D. et al., Progress report of a Phase I study of the intracerebral microinfusion of a recombinant chimeric protein composed of transforming growth factor (TGF)-alpha and a mutated form of the Pseudomonas exotoxin termed PE-38 (TP-38) for the treatment of malignant brain tumors, *J. Neurooncol.*, **2003**, 65, 27–35.

178 Weber F., Asher A., Bucholz R., Berger M. et al., Safety, tolerability,

and tumor response of IL4-Pseudomonas exotoxin (NBI-3001) in patients with recurrent malignant glioma, *J. Neurooncol.*, **2003**, 64, 125–137.

179 WEBER F. W., FLOETH F., ASHER A., BUCHOLZ R. et al., Local convection enhanced delivery of IL4-Pseudomonas exotoxin (NBI-3001) for treatment of patients with recurrent malignant glioma, *Acta Neurochir. Suppl.*, **2003**, 88, 93–103.

180 FERRARI M., Nanovector therapeutics, *Curr. Opin. Chem. Biol.*, **2005**, 9, 343–346.

181 FERRARI M., Cancer nanotechnology: Opportunities and challenges, *Nat. Rev. Cancer*, **2005**, 5, 161–171.

5
Local Cancer Therapy with Magnetic Drug Targeting using Magnetic Nanoparticles

Christoph Alexiou and Roland Jurgons

5.1
Introduction

A continual problem in cancer therapy is the fact that applied drugs do not have a selective site of action. To achieve a therapeutic concentration in the region of interest using systemic administration it is necessary to administer high doses of drugs. Frequently, the dose of therapeutic agents is limited by their body distribution and subsequently their negative side effects and toxicity. During the last three decades many approaches and techniques have been developed in medicine for diagnosis and therapy both *in vivo* and *in vitro* to obtain a more site-specific transport of therapeutic agents, especially anticancer drugs in *local chemotherapy*, so as to increase the agent's local concentration in specific body compartments without harming healthy tissue [1–5]. Targeted drug delivery systems for accumulating pharmaceuticals in specific areas can be passive, based on specific properties of pathological tissues or specific characteristics of targeted organs [6], or active, often magnetically directed as *magnetic targeting* or *magnetic drug delivery*, based on various carrier systems [7]. Inflamed tissues or tumors, for example, differ from healthy tissue concerning pH, temperature or permeability of the vascular endothelium [8]. This enables the use of drug loaded pH-sensitive or thermosensitive liposomes to achieve a high amount of the therapeutic agent in the respective tissue [9–11]. Among liposomes, various substances such as peptides or proteins can be used for targeted drug delivery. Drug-containing monoclonal antibodies can be used for thrombolysis [12] or tumor-specific antibodies can be employed in cancer treatment [13, 14]. In addition to reviewing magnetic nanoparticles in biomedicine, this chapter overviews local chemotherapies, focusing especially on regional cancer therapy with magnetic nanoparticles.

5.2
Local Chemotherapy

The therapeutic effectiveness of drugs, especially of chemotherapeutic agents, is connected with their cell toxic properties [15]. This cell toxicity can cause a cell-

Tab. 5.1. Application techniques of local chemotherapy.

1. Intratumoral (intralesional) therapy
2. Intracavital therapy
3. Intra-arterial infusion
4. Intraportal infusion
5. Regional perfusion of extremities
6. Chemoembolization

depression of bone marrow or the gastrointestinal tract. Furthermore, it may have serious consequences such as mutagenic, teratogenic or cancerogenic effects. The occurrence of secondary tumors induced by chemotherapy is evaluated at about 3% [16]. Targeted drug delivery may increase the therapeutic effects, reduce the costs of therapies and, especially for patients, would mean less negative side effects.

The easiest way of drug targeting involves local injection directly in the region of interest, which in cancer therapy is the tumor region (Table 5.1) [17, 18].

Another possibility of local chemotherapy is the intracavital application of cytostatics. After systemic administration, chemotherapeutic agents reach the tumor by the vascular system; given intracavitally the site of action is penetration of the tumor tissue. The effectiveness of such local chemotherapy depends on a high plasma clearance and a small cavital clearance of the chemotherapeutic agent [19], on a good penetration of the tumor tissue and on the tumor size. With tumors larger than 2 cm in diameter, penetration is less and there is no advantage over systemic administration [20]. Application can be performed intraperitoneally [21], intrapleurally [22] or intravesically [23].

Secondary liver tumors can be treated by the injection of cytostatics in the portal vein [24] or in most cases in the hepatic artery [25].

Actually, regional perfusion with chemotherapeutic agents in an isolated circulation is used in treating liver tumors. Furthermore, isolated perfusion is performed in malignant melanoma of extremities [26].

A very effective palliative treatment of non-resectable hepatocellular carcinomas is chemoembolization. The combination of embolization and application of chemotherapeutics induces a high accumulation and a long residence time of the drug in the tumor [27].

Intraarterial infusion of chemotherapeutics bypasses a possible so-called first-pass-effect through liver and spleen before reaching the tumor. This results in a higher concentration of the drug compared with systemic administration. This effect can be increased by diminishing the arterial flow rate, which can be accomplished by the use of balloon-catheters [28], by administration of vasoconstrictive drugs [29, 30] or embolization [31, 32].

5.3
Magnetic Drug Delivery

Drug targeting with magnetic carrier systems enables an active transport of drugs into the respective region. Therefore, therapeutic agents such as drugs [33, 34] or radioisotopes [35] bound to a magnetic carrier will be injected in the vascular system. Under the influence of an external magnetic field this compound will be held in the targeted area and concentrated in the region of interest. Magnetic particles have been used in medicine for about 40 years and different sophisticated biomedical applications have been developed. Magnetic particles are frequently used *in vitro* and *in vivo*, depending on their size, surface and coating.

5.3.1
In Vitro Applications

For *in vitro* applications several biotechnological approaches have been developed. A molecular-biological diagnostic procedure that has become increasingly important is biomagnetic separation. Magnetic nanoparticles are bound to a ligand (i.e., antibodies) that can be used to target cells, DNA or bacteria. After binding of the ligands to the targets they can be separated by the use of an external magnetic field. Magnetic cell separation is a medical approach of this biomagnetic separation technique that has become standard for *in vitro* diagnosis in cancer patients [36, 37]. By the use of superparamagnetic nanospheres (Fig. 5.1) and an automated magnetic cell separation (MACS®, Miltenyi Biotech®) this system allows the detection of disseminated tumor cells in the peripheral blood of cancer patients.

Another approach for *in vitro* application of magnetic nanoparticles is Magnetofection™. Gene vectors bound to starch-coated magnetic nanoparticles are attracted by an external magnetic field and used to transfect cells *in vitro*. The respective cells show greater transfection than without the influence of an external magnetic field [38].

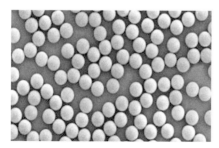

Fig. 5.1. Magnetic particles with homogenous size for the magnetic cell separation system (Dynabeads®).

5.3.2
In Vivo Applications

For *in vivo* applications the magnetic particles have to be biocompatible. They can be incorporated in, or coated with, different biological materials. Therefore, different carriers have been used, such as magnetic microparticles, magnetoliposomes or magnetic iron oxide nanoparticles and also ferromagnetic substances. Alksne et al. (1966) have used carbonated iron and the influence of an external magnetic field to occlude intracranial aneurysms [39].

McNeil et al. (1995) have described a magnetic guidance system for a promising use in neurosurgery. Six external electromagnetic coils are able to guide magnetic neodymium-iron-bor capsules in the brain [40].

Another drug delivery system that can target chemotherapeutic agents is magnetic microspheres [41, 42]. Widder et al. have used magnetic albumin microspheres to increase the concentration of drugs in the tumor-region after applying an external magnetic field to this region [43–46]. Tumor remissions in rats could be shown [45]. These particles were about 1–7 µm in size. Because of enzymatic breakdown of the albumin, a controlled release of the incorporated drug could not be achieved.

Another approach is the use of activated carbonated iron [47, 48]. Allen et al. (1997) have formulated Magnetically Targetable Carriers (MTC^{TM}) made of iron and activated carbon bound to the chemotherapeutic agent paclitaxel for targeted drug delivery [49].

A magnetically targeted carrier bound to doxorubicin (MTC-Dox, FerX®, CA, USA) has been used to treat hepatocellular carcinomas. In human patients, MTC-Dox was administered in the hepatic artery while an external magnetic field was focused to the tumor region. In one case, reduction of tumor size could be shown [50]. However, another study has described a necrosis of the liver corresponding to embolization after treatment with these particles [51]. Recently, a global multicenter phase II/III study using MTC-Dox in human patients had to be stopped because "the clinical endpoints of the trial could not be met with statistical significance with the product as currently manufactured" (FerX®, San Diego, April 30, 2004).

Magnetoliposomes are also used for magnetically targeted drug delivery. These colloidal particles are composed of a lipid layer and qualified as drug carriers [52]. The chemotherapeutic agent Methotrexat encapsulated in thermosensitive liposomes can be accumulated in the tumor region after intravenous administration and local hyperthermia [53]. The use of magnetic thermosensitive liposomes and the influence of an external magnetic field placed over the tumor-region may increase the concentration of the encapsulated drug [54]. Furthermore, thermosensitive magnetoliposomes can also be used for local hyperthermia in cancer treatment [55]. Whole body hyperthermia in combination with chemotherapy is a very common approach in cancer treatment. A possibility to achieve local hyperthermia with magnetic particles is the use of magnetic iron oxide particles, which can be heated to over 40 °C by an alternating magnetic field after intralesional injection [56].

The size of albumin-microspheres or magnetoliposomes is in the range of μms. As a consequence there is a risk of embolization in the capillary system because of these particles. Also useful for magnetically targeted drug delivery and smaller in size are magnetic nanoparticles. On nanoparticles, the surface-to-volume ratio is much higher than on larger particles. Therefore, nanomaterials show completely new characteristics and can be used for various new approaches. After application in an organism, different components such as material, size, charge and coating of the nanoparticles and the bound chemotherapeutic agent have an influence on biocompatibility and the biodistribution as well as the applied magnetic field and the respective gradient [57–68]. Nanoparticles have to be small enough not to occlude the capillary system and cause embolization, but they have to be large enough to be attractable by an external magnetic field. Furthermore, nanoparticles over 100 nm in diameter are more readily cleared by cells of the MPS (mononuclear phagocyte system) than smaller particles [57]. Research on biocompatible magnetic nanoparticles has focused on superparamagnetic iron oxide nanoparticles (Fe_3O_4). These particles have been used as a contrast agent in magnetic resonance imaging for lymphography [69] (Fig. 5.2).

A first clinical trial with superparamagnetic iron oxide nanoparticles bound to a chemotherapeutic agent was performed in 1996 by Lübbe et al. [33]. For experimental studies on rats and in a clinical trial on 14 human patients they used

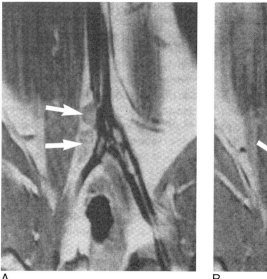
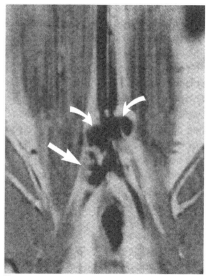

A   B

Fig. 5.2. MR-Imaging of iliac lymph nodes in tumor-bearing rabbits. MRI before (A) and after (B) administration of superparamagnetic iron oxide particles. There is a homogeneous signal in healthy lymph nodes (curved arrows) and an inhomogeneous signal in metastatic lymph nodes [69].

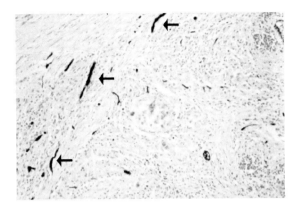

Fig. 5.3. Histological cross section of a rabbit's VX2–squamous cell carcinoma after treatment with intraarterial magnetic drug targeting (MDT). Arrows show the ferrofluids in tumor vessels. Staining: Prussian blue; magnification: 200× [72].

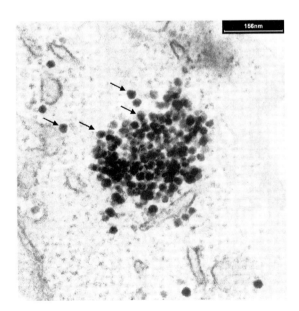

Fig. 5.4. Typical multidomain-particle in the intracellular space of tumor cells. Single particles (marked with arrows) are surrounded by starch polymers (not visible in the electron-microscopic picture) [73].

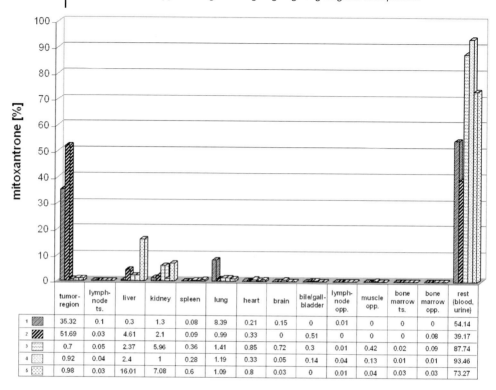

Fig. 5.5. Concentration of mitoxantrone in different body compartments 60 min after the respective injection in percent of the applied dose. Experiments were performed on tumor-bearing rabbits (i.a. = intraarterial, i.v. = intravenous).
1 i.a. application of 5 mg m^{-2} ferrofluid-bound mitoxantrone with an external magnetic field [$n = 1$, magnetic drug targeting (MDT)].
2 i.a. application of 2 mg m^{-2} ferrofluid-bound mitoxantrone with an external magnetic field ($n = 1$, MDT).
3 i.v. application of 10 mg m^{-2} mitoxantrone without an external magnetic field ($n = 1$, application of the regular systemic dose).
4 i.a. application of 5 mg m^{-2} ferrofluid-bound mitoxantrone without an external magnetic field ($n = 1$).
5 i.a. application of 10 mg m^{-2} mitoxantrone without an external magnetic field ($n = 1$).
ts. = tumor side; opp. = opposite; tumor region = tumor and surrounding area (<1 cm) [72].

starch-coated magnetic nanoparticles ionically bound to epidoxorubicin. It could be demonstrated that these particles are biocompatible. In another preclinical study, intraarterial Magnetic Drug Targeting (MDT) on tumor-bearing rabbits was performed using a powerful external magnetic field (1.7 Tesla) placed over the tumor region. Starch-coated superparamagnetic iron oxide nanoparticles (about 100–150 nm in diameter) ionically bound to the chemotherapeutic agent mitoxantrone were injected in the tumor supplying artery.

Through this intraarterial MDT, complete tumor remissions could be shown only by the use of 50% and 20% of the regular systemic chemotherapeutic dose without any negative side effects [34, 70, 71]. It could be demonstrated histologically that the nanoparticles were enriched in the tumor tissue (Figs. 5.3 and 5.4). Radioactive ^{59}Fe-nanoparticles showed 114× more activity in the tumor region after MDT than the control without magnetic field [72]. Furthermore, with this system a high and specific enrichment of the bound chemotherapeutic agent in a desired body compartment (i.e. the tumor) is possible. HPLC-analysis of the chemotherapeutic agent after MDT revealed a 75× higher concentration of the administered chemotherapeutic agent in the tumor region compared with the regular systemic administration (Fig. 5.5) [72, 73]. These encouraging data will lead to further studies on cancer patients.

References

1 COLLINS, J.M., Pharmacological rationale for regional drug delivery, *J. Clin. Oncol.* **1984**, 2, 498–505.
2 GUPTA, P.K., HUNG, C.T., Comparative disposition of adriamycin delivered via magnetic albumin microspheres in presence and absence of magnetic fields in rats, *Life Sci.* **1990**, 46, 471–484.
3 GUPTA, P.K., HUNG, C.T., Targeted delivery of low dose doxorubicin hydrochloride administered via magnetic albumin mirospheres in rats, *J. Microencap.* **1990**, 7, 85–92.
4 GUPTA, P.K., Drug targeting in chemotherapy: A clinical perspective, *J. Pharm. Sci.*, **1990**, 79, 949–962.
5 TORCHILIN, V.P., Drug targeting, *Eur. J. Pharm. Sci.*, **2000**, 11, 81–91.
6 ARAP, W., HAEDICKE, W., BERNASCONIE, M., KAIN, R., RAJOTTE, D., KRAJEWSKI, S., ELLERBY, H.M., BREDESEN, D.E., PASQUALINI, R., RUOSLATHI, E., Targeting the prostate for destruction through a vascular address, *Proc. Natl. Acad. Sci. U.S.A.*, **2002**, 99, 1527–1531.
7 ZACHARSKI, L.R., ORNSTEIN, D.L., GABAZZA, E.C., D'ALESSANDRO-GABAZZA, C.N., BRUGAROLAS, A., SCHNEIDER, J., Treatment of malignancy by activation of the plasminogen system, *Semin. Thromb. Hemost.*, **2002**, 28, 5–18.
8 GERLOWSKI, L.E., JAIN, R.K., Microvascular permeability of normal and neoplastic tissues, *Microvasc. Res.*, **1986**, 31, 288–299.
9 TORCHILIN, V.P., ZHOU, F., HUANG, L., pH-sensitive liposomes, *Liposome Res.*, **1993**, 3, 201–205.
10 YATVIN, M.B., KREUTZ, W., HORWITZ, B.A., SHINITZKY, M., pH-sensitive liposomes: Possible clinical implications, *Science* **1980**, 210, 1253–1255.
11 SULLIVAN, S.M., HUANG, L., Preparation and characterization of heat-sensitive immunoliposomes, *Biochem. Biophys. Acta*, **1985**, 812, 116–126.
12 HABER, E., Antibody targeting as a strategy in thrombolysis, in KAW, B.A., NARULA, J., STRAUSS, H.W. (eds.), *Monoclonal Antibodies in Cardiovascular Diseases*, Lea and Febiger, Malvern, PA, U.S.A. **1994**, pp. 187–197.
13 SACHDEVA, M.S., Drug targeting systems for cancer therapy, *Expert Opin. Investig. Drugs*, **1998**, 7, 1849–1864.
14 LEVEUGLE, B., MANN, D., MADIYALAKAN, R., NOUJAIM, A.A., Therapeutic antibodies for prostate cancer, *IDrugs*, **2000**, 3(10), 1191–1198.
15 POSTE, G., KIRSH, R., Site specific (targeted) drug delivery in cancer therapy, *Biotechnology*, **1983**, 10, 869–885.
16 NEGLIA, J.P., FRIEDMANN, D.L., YASUI, Y., Second malignant neoplasms in five-year survivors of childhood cancer:

Childhood cancer survivor study, *J. Natl. Cancer Inst.*, **2001**, 93, 618.
17 WALTER, K.A., TAMARGO, R.J., OLIVI, A., BURGER, P.C., BREM, H., Intratumoral chemotherapy, *Neurosurgery* **1995**, 37, 1128–1145.
18 WERNER, J.A., KEHRL, W., PLUZANSKA, A., ARNDT, O., LAVERY, K.M., GLAHOLM, J., DIETZ, A., DYCKHOFF, A., MAUNE, S., STEWART, M.E., ORENBERG, E.K., LEAVITT, R.D., A phase III placebo-controlled study in advanced head and neck cancer using intratumoural cisplatin/epinephrine gel, *Br. J. Cancer*, **2002**, 87, 938–944.
19 PREIß, J., LINK, K.H., SCHMOLL, E., Regionale chemotherapie, in SCHMOLL, H.J., HÖFFKEN, K., POSSINGER, K. (eds.), *Kompendium Internistische Onkologie*, Teil 1, Springer Verlag, Berlin, Heidelberg, New York, **1999**, pp. 1755–1763.
20 KEMENY, N.E., Is hepatic infusion of chemotherapy effective treatment for liver metastases? Yes. In, DEVITA, V.T. JR., HELLMAN, S., ROSENBERG, S.A. (eds.), *Important Advances in Oncology*, Lippincott, Philadelphia, **1992**, pp. 207–227.
21 DUFOUR, P., BERGERAT, J.P., BARATS, J.C., GIRON, C., DUCLOS, B., DELLENBACH, P., RITTER, J., RENAUD, R., AUDHUY, B., OBERLING, F., Intraperitonel mitoxantrone as consolidation treatment of patients with ovarian carcinoma in pathologic complete remission, *Cancer*, **1994**, 73, 1865–1869.
22 KIKUMORI, T., HAYASHI, H., SHIBATA, A., SEKIYA, M., ITOH, T., MASE, T., OIWA, M., IMAI, T., FUNAHASHI, H., Administration of docetaxel in cases of recurrent breast carcinoma with malignant pleural effusion controlled by intrapleural administration of OK-432, *Gan To Kagaku Ryoho.*, **1999**, 26, 2091–2094.
23 RASSWEILER, J., Neoadjuvant chemotherapy of invasive bladder cancer, *Urologe A*, **1994**, 33, 576–581.
24 LORENZ, M., ROSSION, I., Adjuvante und palliative regionale therapie von lebermetastasen kolorektaler tumoren, *Dtsch. Med. Wschr.*, **1995**, 120, 690–697.
25 OETTLE, H., KRATSCHMER, B., LÖFFEL, J., VOGL, T.J., RIESS, H., Regionale chemotherapie von lebermetastasen bei kolorektalen karzinomen, *Onkologe*, **1998**, 4, 67–78.
26 SCOTT, R.N., KERR, D.J., BLACKIE, R., HUGHES, J., BURNSIDE, G., MACKIE, R.M., BYRNE, D.S., MCKAY, A.J., The pharmacokinetic advantages of isolated limb perfusion with melphalan for malignant melanoma, *Br. J. Cancer*, **1992**, 66, 159–166.
27 BERGER, H., BAETHGE, I., RUDOLPHI, A., BOOS, K., STÄBLER, A., REISER, M., SEIDEL, D., Transcatheter chemoembolisation of hepatocellular carcinoma: A study of the pharmacokinetics of epirubicin with Lipiodol® or starch microspheres as embolizing agent, *Reg. Cancer Treat.*, **1996**, 9, 181–185.
28 BENGMARK, S., Regional chemotherapy of liver and hepatic artery occlusion, in BERGER, H.G., BÜCHLER, M., REISFELD, R.A., SCHULZ, G. (eds.), *Cancer Therapy Monoclonal Antibodies, Lymphokines, New Development in Surgical Oncology and Chemo- and Hormonal Therapy*, Springer, Berlin, Heidelberg, New York, **1989**, pp. 201–215.
29 HEMINGWAY, D.M., ANDERSON, W.J., ANDERSON, J.H., GOLDBERG, J.A., MCARDLE, C.S., COOKE, T.G., Monitoring blood flow to colorectal liver metastases using laser Doppler flowmetry: The effect of angiotensin II, *Br. J. Cancer*, **1992**, 66, 958–960.
30 NOGUCHI, S., MIYAUCHI, K., NISHIZAWA, Y., SASAKI, Y., IMAOKA, S., IWANAGA, T., KOYAMA, H., TERASAWA, T., Augmentation of anticancer effect with angiotensin II in intraarterial infusion chemotherapy for breast cancer, *Cancer*, **1988**, 62, 467–473.
31 LEYLAND-JONES, B., Targeted drug delivery, *Semin. Oncol.*, **1993**, 20, 12–17.
32 SCHULTHEIS, K.H., PLIES, M., GENTSCH, H.H., Chemoembolization of liver tumors, in BERGER, H.G., BÜCHLER, M., REISFELD, R.A., SCHULZ,

G. (eds.), *Cancer Therapy With Monoclonal Antibodies, Lymphokines, New Development in Surgical Oncology and Chemo and Hormonal Therapy*, Springer, Heidelberg, New York, Tokyo, **1989**, pp. 201–215.

33 LÜBBE, A.S., BERGEMANN, Ch., RIESS, H., SCHRIEVER, F., REICHARDT, P., POSSINGER, K., MATTHIAS, M., DORKEN, B., HERRMANN, F., GURTLER, R., HOHENBERGER, P., HAAS, N., SOHR, R., SANDER, B., LEMKE, A.J., OHLENDORF, D., HUHNT, W., HUHN, D., Clinical experiences with magnetic drug targeting: A phase I study with 4′-epidoxorubicin in 14 patients with advanced solid tumors, *Cancer Res.*, **1996**, 56, 4686–4693.

34 ALEXIOU, Ch., ARNOLD, W., KLEIN, R., PARAK, F., HULIN, P., BERGEMANN, C., ERHARDT, W., WAGENPFEIL, S., LÜBBE, A.S., Locoregional cancer treatment with magnetic drug targeting, *Cancer Res.*, **2000**, 60, 6641–6648.

35 HÄFELI, U.O., Magneto- and radio-pharmaceuticals, radioactive magnetic microspheres, in ARSHADY, R. (ed.), *Microspheres, Microcapsules Liposomes*, **2001**, 3, 559–584.

36 BILKENROTH, U., TAUBERT, H., RIEMANN, D., REBMANN, U., HEYNEMANN, H., MEYE, A., Detection and enrichment of disseminated renal carcinoma cells from peripheral blood by immunomagnetic cell separation., *Int. J. Cancer*, **2001**, 92, 577–582.

37 TRELEAVEN, J.G., GIBSON, F.M., UGELSTAD, J., REMBAUM, A., PHILIP, T., CAINE, G.D., KEMSHEAD, J.T., Removal of neuroblastoma cells from bone marrow with monoclonal antibodies conjugated to magnetic microspheres, *The Lancet*, **1984**, 70–73.

38 SCHERER, F., ANTON, M., SCHILLINGER, U., HENKE, J., BERGEMANN, C., KRUGER, A., GÄNSBACHER, B., PLANK, C., Magnetofection enhancing and targeting gene delivery by magnetic force in vitro and in vivo, *Gene Ther.*, **2002**, 9, 102–109.

39 ALKSNE, J.F., FINGERHUT, A., RAND, R., Magnetically controlled metallic thrombosis of intracranial aneurysms, *Surgery*, **1966**, 60, 212–218.

40 MCNEIL, R.G., RITTER, R.C., WANG, B., LAWSON, M.A., GILLIES, G.T., WIKA, K.G., QUATE, E.G., HOWARD, M.A., GRADY, M.S., Functional design features and initial performance characteristics of a magnetic-implant guidance system for stereotactic neurosurgery, *IEEE Trans. Biomed. Eng.*, **1995**, 42, 793–801.

41 DEVINENI, D., BLANTON, C.D., GALLO, J.M., Preparation and in vitro evaluation of magnetic microsphere-methotrexate conjugate drug delivery systems, *Bioconj. Chem.*, **1995**, 6, 203–210.

42 DRISCOLL, C.F., MORRIS, R.M., SENYEI, A.E., WIDDER, K.J., HELLER, G.S., Magnetic targeting of microspheres in blood flow, *Microvascular Res.*, **1984**, 27, 353–369.

43 WIDDER, K.J., SENYEI, A.E., SCARPELLI, D.G., Magnetic microspheres: A model system for site specific drug delivery in vivo, *Proc. Soc. Exp. Biol. Med.*, **1978**, 158, 141–146.

44 WIDDER, K.J., SENYEI, A.E., RANNEY, D.F., Magnetically responsive microspheres and other carriers for the biophysical targeting of antitumor agents, in GAVATTINI, S., GOLDIN, A., HOWKING, F., KOPIN, I.J., SCHNITZER, R.J. (eds.), *Advances in Pharmacology and Chemotherapy*, Academic Press, New York, **1979**, pp. 213–239.

45 WIDDER, K.J., MORRIS, R.M., POORE, G.A., HOWARD, D.P., SENYEI, A.E., Selective targeting of magnetic albumin microspheres containing low-dose doxorubicin: Total remission in yoshida sarcoma bearing rats, *Eur. Cancer Clin. Oncol.*, **1983**, 19, 135–199.

46 WIDDER, K.J., MANRINO, P.A., MORRIS, R.M., HOWARD, D.P., POORE, G.A., SENYEI, A.E., Selective targeting of magnetic albumin microspheres to the yoshida sarcoma: Ultrastructural evaluation of microsphere distribution, *Eur. J. Cancer. Clin. Oncol.*, **1993**, 19, 141–147.

47 LEAKAKOS, T., JI, C., LAWSON, G., PETERSON, C., GOODWIN, S., Intravesical administration of

doxorubicin to swine bladder using magnetically target carriers, *Cancer Chemother. Parmacol.*, **1996**, 51, 445–450.

48 RUDGE, S., PETERSON, C., VESSELY, C., KODA, J., STEVENS, S., CATTERALL, L., Adsorption and desorption of chemotherapeutic drugs from a magnetically targeted carrier (MTC), *J. Controlled Release*, **2001**, 74, 335–340.

49 ALLEN, L.M., KENT, J., WOLFE, C., FICCO, C., JOHNSON, J., MTC™: A magnetically targetable drug carrier for paclitaxel, in HÄFELI, U., SCHÜTT, W., TELLER, J., ZBOROWSKI, M. (eds.), *Scientific and Clinical Applications of Magnetic Carriers*, Plenum Press, New York and London, **1997**, pp. 481–494.

50 WILSON, M.W., KERLAN, R.K., FIDELMAN, N.A., VENOOK, A.P., LA BERGE, J.M., KODA, J., GORDON, R.L., Hepatocellular carcinoma: Regional therapy with a magnetic targeted carrier bound to doxorubicin in a dual MR-imaging/conventional angiography suite – initial experience with four patients, *Radiology*, **2004**, 230, 287–293.

51 GOODWIN, S.C., BITTNER, C.A., PETERSON, C.L., WONG, G., Single dose toxicity study of hepatic intra-arterial infusion of doxorubicin coupled to a novel magnetically targeted drug carrier, *Toxicol. Sci.*, **2001**, 60, 177–183.

52 KUBO, T., SUGITA, T., SHIMOSE, S., NITTA, Y., IKUTA, Y., MURAKAMI, T., Targeted delivery of anticancer drugs with intravenously administered magnetic liposomes in osteosarcoma-bearing hamsters, *Int. J. Oncol.*, **2000**, 17, 309–315.

53 WEINSTEIN, J.N., MAGIN, R.L., YATVIN, M.B., ZAHARKO, D.S., Liposomes and local hyperthermia: Selective delivery of methotrexate to heated tumors, *Science*, **1979**, 13, 188–191.

54 VIROONCHATAPAN, E., SATO, H., UENO, M., ADACHI, I., TAZAWA, K., HORKOSHI, I., Magnetic targeting of thermosensitive magnetoliposomes to mouse livers in an in situ on-line perfusion system, *Life Sci.*, **1996**, 58, 2251–2261.

55 SHINKAI, M., SUZUKI, M., IIJIMA, S., KOBAYASHI, T., Antibody-conjugated magnetoliposomes for targeting cancer cells and their application in hyperthermia, *Biotechnol. Appl. Biochem.*, **1994**, 21, 125–137.

56 HILGER, I., HIERGEIST, R., HERGT, R., WINNEFELD, K., SCHUBERT, H., KAISER, W.A., Thermal ablation of tumors using magnetic nanoparticles: An in vivo feasibility study, *Invest. Radiol.*, **2002**, 37, 580–586.

57 STORM, G., BELLIOT, S.O., DAEMEN, T., LASIC, D.D., Suface modification of nanoparticles to oppose uptake by the mononuclear phagocyte system, *Adv. Drug Deliv. Rev.*, **1995**, 17, 31–48.

58 TORCHILIN, V.P., TRUBETSKOY, V.S., Which polymer can make nanoparticulated drug carriers long circulating?, *Adv. Drug Deliv. Rev.*, **1995**, 16, 141–155.

59 ZHAO, A., YAO, P., KANG, C., YUAN, X., CHANG, J., PU, P., Synthesis and characterization of tat-mediated O-CMC magnetic nanoparticles having anticancer function, *J. Magn. Magn. Mater.*, **2005**, 295, 37–43.

60 MYKHAYLYK, O., DUDCHENKO, N., DUDCHENKO, A., Doxorubicin magnetic conjugate targeting upon intravenous injection into mice: High gradient magnetic field inhibits the clearance of nanoparticles from the blood, *J. Magn. Magn. Mater.*, **2005**, 293, 473–482.

61 NEUBERGER, T., SCHÖPF, B., HOFMANN, H., HOFMANN, M., VON RECHENBERG, B., Superparamagnetic nanoparticles for biomedical applications: Possibilities and limitations of a new drug delivery system, *J. Magn. Magn. Mater.*, **2005**, 293, 483–496.

62 XU, H., SONG, T., BAO, X., HU, L., Site-directed research of magnetic nanoparticles in magnetic drug targeting, *J. Magn. Magn. Mater.*, **2005**, 293, 514–519.

63 ASMATULU, R., ZALICH, M.A., CLAUS, R.O., RIFFLE, J.S., Synthesis, characterization and targeting of

biodegradable magnetic nanocomposite particles by external magnetic fields, *J. Magn. Magn. Mater.*, **2005**, 292, 108–119.

64 ITO, A., KUGA, Y., HONDA, H., KIKKAWA, H., HORIUCHI, A., WATANABE, Y., KOBAYASHI, T., Magnetite nanoparticle-loaded anti-HER2 immunoliposomes for combination of antibody therapy with hyperthermia, *Cancer Lett.*, **2004**, 212, 167–175.

65 SON, S.J., REICHEL, J., HE, B., SCHUCHMAN, M., LEE, S.B., Magnetic nanotubes for magnetic-field-assisted bioseparation, biointeraction, and drug delivery, *J. Am. Chem. Soc.*, **2005**, 127, 7316–7317.

66 JAIN, T.K., MORALES, M.A., SAHOO, S.K., LESLIE-PELECKY, D.L., LABHASETWAR, V., Iron oxide nanoparticles for sustained delivery of anticancer agents, *Mol. Pharm.*, **2005**, 2, 194–205.

67 KOHLER, N., SUN, C., WANG, J., ZHANG, M., Methotrexate-modified superparamagnetic nanoparticles and their intracellular uptake into human cancer cells, *Langmuir*, **2005**, 21, 8858–8864.

68 VEISEH, O., SUN, C., GUNN, J., KOHLER, N., GABIKIAN, P., LEE, D., BHATTARAI, N., ELLENBOGEN, R., SZE, R., HALLAHAN, A., OLSON, J., ZHANG, M., Optical and MRI multifunctional nanoprobe for targeting gliomas, *Nano Lett.*, **2005**, 5, 1003–1008.

69 TAUPITZ, M., WAGNER, S., HAMM, B., DIENEMANN, D., LAWACZECK, R., WOLF, K.J., MR lymphography using iron oxide particles. Detection of lymph node metastases in the VX2 rabbit tumor model, *Acta Radiol.*, **1993**, 34, 10–15.

70 ALEXIOU, Ch., ARNOLD, W., HULIN, P., KLEIN, R.J., RENZ, H., PARAK, F.G., BERGEMANN, Ch., LÜBBE, A.S., Magnetic mitoxantrone nanoparticle detection by histology, x-ray and MRI after magnetic drug targeting, *J. Magn. Magn. Mater.*, **2001**, 225, 187–193.

71 ALEXIOU, Ch., SCHMIDT, A., HULIN, P., KLEIN, R.J., BERGEMANN, Ch., ARNOLD, W., Magnetic drug targeting: Biodistribution and dependency on magnetic field strength, *J. Magn. Magn. Mater.*, **2002**, 252, 363–366.

72 ALEXIOU, Ch., JURGONS, R., SCHMID, R.J., BERGEMANN, Ch., HENKE, J., ERHARDT, W., HUENGES, E., PARAK, F.G., Magnetic drug targeting-biodistribution of the magnetic carrier and the chemotherapeutic agent Mitoxantron after locoregional cancer treatment, *J. Drug Targeting*, **2003**, 11, 139–149.

73 ALEXIOU, Ch., JURGONS, R., SCHMID, R.J., HILPERT, A., BERGEMANN, Ch., PARAK, F.G., IRO, H., In vitro and in vivo investigations of targeted chemotherapy with magnetic nanoparticles, *J. Magn. Magn. Mater.*, **2005**, 293, 389–393.

6
Nanomaterials for Controlled Release of Anticancer Agents

Do Kyung Kim, Yun Suk Jo, Jon Dobson, Alicia El Haj, and Mamoun Muhammed

6.1
Introduction

The enormous advances made in basic cancer research have not been paralleled by similar improvements in treatment results. Metastatic breast cancer, late-stage colon cancer, malignant melanoma and other forms of cancer are still essentially incurable in most cases. Enormous efforts have been made in recent years to develop "smart drugs" that are directed to tumor cell-specific enzymes and surface receptors. Almost without exception, these drugs have encountered problems in clinical trials. Today's cancer treatment is therefore primarily based on the use of conventional cytotoxic drugs that have adverse side effects and only limited effectiveness. The gains in health that would be achieved by efficient tumor treatment with fewer side effects (owing to specific targeting to tumor tissue) are therefore enormous. Cancer is a group of diseases characterized by uncontrolled growth and spread of abnormal cells. If the growth and spread of cancer cells is not controlled, the disease is fatal. Though in many cases the specific cause of cancer is unknown, in general it can be caused by several external factors (tobacco, chemicals, radiation, and infectious organisms) and internal factors (inherited mutations, hormones, immune conditions and mutations that occur from metabolism). Theses factors may act together or in sequence to initiate or promote carcinogenesis. Generally, cancer is treated by surgery, radiation, chemotherapy, hormones and immunotherapy [1]. Cancer therapy can be successful in treating solid tumors/cancers in lesions that can be removed by surgery or treated by radiotherapy/chemotherapy, the major treatment modalities for primary cancer/tumor and metastases. Each of these cancer treatments has advantages and disadvantages, thus the combination with other treatments is recommended to achieve the optimum outcome.

Chemotherapy for cancer is a whole body treatment that is administered either orally or intravenously. This results in systemic distribution of cytotoxic chemotherapeutic compounds that can be more effective for the treatment of micrometastases. However, the systemic distribution of cytotoxic compounds results in more severe side effects (some of which can be life-threatening) compared to surgery or radiotherapy. The major goal of targeted therapies is to reduce the side effects of cytotoxic drugs, resulting in more effective control of cell growth or tumor angiogenesis.

Presently, numerous anticancer agents are available in the clinic. These anticancer agent/drugs have an elimination half-life that results in a decrease of therapeutic potential and side effects such as bone marrow depression and gastrointestinal damage [2]. For example, poly(alkyl cyanoacrylate) nanoparticles as drug delivery systems (DDSs) play an important role in the incorporation of anticancer drugs as they can enhance the drug's concentration in the tumor and reduce drug levels in the heart, thus avoiding some side effects [3]. Targeting toxic therapeutics to tumors through binding to receptors overexpressed on the surface of cancer cells can also reduce systemic toxicity and increase the effectiveness of the targeted compounds. Small molecule-targeted therapeutics have several advantages over toxic immunoconjugates, including better tumor penetration, lack of neutralizing host immune response and superior flexibility in the selection of drug components with optimal specificity, potency and stability in circulation [4].

Novel strategies have been developed to decrease the toxicity of active molecules by targeting the specific tumor site, where the drug can selectively bind to the targeted tissue at cellular and/or sub-cellular level to influence its therapeutic effects. The chemotherapeutic activity can be enhanced by using macromolecules as a vector to control the release rate of anticancer agents. The use of polymeric nanoparticles is mainly focused on controlling the loaded anticancer agent/drug at the targeted lesion. Polymeric nanoparticles are a form of core–shell structure/ nanocapsule that can be loaded with therapeutic agents. Several kinds of inorganic or metallic nanoparticles can be used due to their intrinsic physical properties even without a core–shell structure. Superparamagnetic iron oxide nanoparticles (SPION) in particular can be used as a transducer for active targeting by responding to both external ac and dc magnetic fields [5].

Despite major advances in the development of small-scale devices, however, most DDSs still use small molecules administrated orally, transdermally, parenterally, or through the nose or lung. The emergence of novel, biologically targeted anticancer agents such as gefitinib ("Iressa", ZD1839) has raised the question of how the dose for later-stage clinical development and clinical use is best determined. For cytotoxic drugs, because toxic effects and antitumor activity often fall within the same dose range and are dose dependent, the clinically used dose will depend on the therapeutic window [6]. Cytostatic drug handling in hospitals is not unproblematic. Large doses are given of compounds that are highly mutagenic. Administration of the drugs and also disposal of patient urine etc. is therefore potentially hazardous. Target-oriented, controlled drug delivery will lead to decreases in the quantities of drugs used and will, therefore, help to reduce these problems. The development of novel smart biomaterials is already having an enormous effect on nanomedicine [7].

6.2
Nanoparticles for Biomedical Applications

Modern technological realizations are concerned with the development of advanced, multifunctional, and even more "smart" materials for specific applications in highly integrated biomedical approaches (Table 6.1). These novel interdisciplinary concepts are recently emerging at the intersection of material science and molecular biotechnology. They are closely associated with surface chemistry and the physical properties of inorganic nanoparticles, the topics of bioorganic and bioinorganic chemistry, and various aspects of molecular biology, recombinant DNA technology and protein expression, and immunology.

Tab. 6.1. Examples of nano-sized inorganic components in biomedical applications.

Particle composition	Particle size (nm)	Applications	Ref.
Metals			
Au	2–150	Drug and gene delivery	8
Ag	1–80	Antibody tagged marker	9
Pt	1–20	Sensors and electrodes	10
Co	1–50	Magnetic separation, drug targeting	11
Semiconductors			
CdX (X = S, Se, Te)	1–20	Fluorescent labeling	12
ZnX (X = S, Se, Te)	1–20	Fluorescent labeling	13
PbS	2–18	Photoluminescence	14
TiO_2	3–50	Biomedical devices for nerve tissue monitoring	15
ZnO	1–30	Photoluminescence	16
CaAs, InP	1–15	Nonlinear optics	17
Ge	6–30	Photoluminescence	18
Magnetic			
Fe-O	6–40	MR contrast agent, drug delivery	[19]
Fe-Pt	2–10	MR contrast agent, drug delivery	[20]

6.2.1
First Generation Nanoparticles

First generation nanoparticles have been available for several years. Colloids are representative of nanoparticles stabilized in solution to prevent uncontrolled growth, aggregation, and flocculation of the nanoparticles. Utilization of colloidal processing leads to attractive new concepts for the fabrication of advanced nanostructured materials. For this reason, many investigations have focused on colloidal processing of inorganic materials through chemical methods.

6.2.2
Second Generation Nanoparticles

As a result of an increasing degree of complexity and sophistication needed for the engineering of nanostructures for advanced applications, second generation nanoparticles have emerged. A key aspect here is the need for multifunctionality of these materials in which several properties are combined to achieve a specific function. For example, certain properties could be achieved through the reduction of building blocks to the nanometer regime, e.g., magnetite (Fe_3O_4)/maghemite (γFe_2O_3) nanoparticles become superparamagnetic at sizes below about 30 nm. Ferrofluids – which have the fluid properties of a liquid but are strongly magnetized in applied fields – can be produced by suspending these sub-30 nm nanoparticles in a suitable media. Ferrofluids are useful as active components for enhancing the performance of many devices, e.g., mechanical (seals, bearings and dampers) or electromechanical (loudspeakers, stepper motors and sensors, etc.). However, by combining the superparamagnetic properties of ferrofluids with functional chemical groups on the particles' surface there are opportunities for advanced applications in magnetically targeted drug delivery.

For second generation nanoparticles, the surface layer (a few or several monolayers) is distinctly different from that of the core material (composition or structure). Again, such particles are categorized as core–shell structures. The surface layer may be thin or thick, depending on the functionality required. Figure 6.1 shows a schematic representation of different types of second generation nanoparticles with surface modifications and nanoparticles with a core–shell structure. Broadly speaking, these particles can also be considered as composite nanoparticles. However, the term nanocomposite generally refers to materials consisting of a dispersion of nanoparticles within a suitable matrix. The most common example of nanocomposites is the precipitation of inorganic (often metal) nanoparticles within a nanoporous polymer structure. Interestingly, the fundamental properties of the polymeric materials can be dramatically altered as a result of the dispersion of few percent of inorganic nanoparticles – particularly with the addition of magnetic iron oxides.

172 | 6 Nanomaterials for Controlled Release of Anticancer Agents

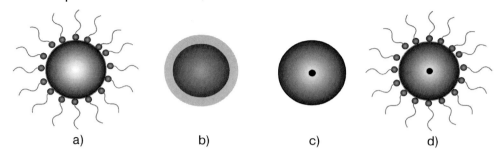

Fig. 6.1. Evolution of second generation nanoparticles: (a) Nanoparticle coated with surfactant to form a stable suspension, (b) nanoparticle coated with a thin metallic layer, (c) small nanoparticle coated with a porous ceramic layer and (d) dispersion of core–shell combination of (a) and (c) for stable suspension.

6.2.3
Advanced Generation Nanoparticles

Recently, advanced generation nanoparticles have emerged to meet the need for the fabrication of more complex nanoparticles. Further to the core–shell structure, nanoparticles with structures similar to nanocomposites have been fabricated (nanobeads). In these nanobeads, a single bead consists of a nanocomposite core, where one or more smaller nanoparticles are dispersed into the matrix. Several possible combinations of organic and inorganic particles can be dispersed within the matrix of the core structure. Each dispersed component can be selected to achieve a specific function or properties of the particle. The surface layer can combine both physical (e.g., diffusion control) and chemical (e.g., allowing certain conjugation chemistries) functionality to the particles. In this way, it is possible to "program" the nanobeads with multiple functionalities, suitable for performing certain tasks, that can be triggered under specific conditions. For example, it is possible to fabricate such nanobeads that can be magnetically moved or localized for controlled drug release. The release of the drug can be controlled by diffusion control of the matrix of the bead or through the control of the porosity of a suitable shell layer on the surface of the bead. Nanobeads can be programmed to be responsive to the environment, e.g., small variations in temperature or pH. Fabrication of these advanced generation nanoparticles requires the use of comprehensive and detailed procedures.

The design and fabrication of biochemically functionalized superparamagnetic iron oxide nanoparticles and near-infrared light absorbing nanoparticles is of particular interest for cancer targeting and therapy applications. Figure 6.2 illustrates a strategy to construct magnetic drug carriers in combination with thermosensitive polymeric materials. Target-oriented release of drugs encapsulated in polymeric nanocapsules is presently the most active research area in this field. Processing of

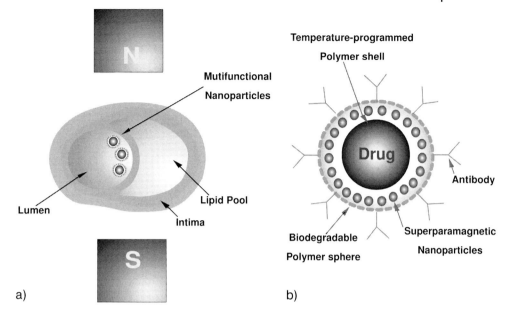

Fig. 6.2. Functional nanoparticles: (A) Coronary plaque removal by nanofluid of magnetic nanoparticles under RF magnetic field and (b) multipurpose nanovectors for target oriented controlled drug release.

nanoparticles with controlled properties, such as chemical properties (composition of the bulk, interaction between the particles, and surface charge) and structural properties (crystalline or amorphous structure, size, and morphology), is the main feature in designing the nanoprecursors (nanoparticles/nanotube/nanolayer). The development of supramolecular, biomolecular, and dendrimer chemistries for engineering substances of ångström and nanoscale dimensions has been encouraged for requirements in nanotechnology. The emerging disciplines of nanoengineering, nanoelectronics, and nanobioelectronics require suitably sized and functional building blocks to construct their architecture and devices [21].

Nanoparticles are in the solid phase and may be either amorphous or crystalline. They can be constructed to absorb, conjugate and encapsulate therapeutic agents inside or outside. Several parameters of colloidal systems developed have been considered, such as the temperature, osmolality and pH of the polymerization medium, that could influence the characteristics (morphology and morphometry, drug content, melting point transition or the enthalpy of transition) and stability of nanospheres. On the other hand, based on their unique mesoscopic physical, chemical, thermal and mechanical properties, nanoparticles offer great potential for many biomedical applications, including bioanalysis and bioseparation, tissue-specific drug therapeutic applications, gene and radionuclide delivery [5, 22].

To be used effectively in fighting disease, the specific surface chemistry of the nanoparticles must be tailored for the desired biomedical applications. Magnetic nanoparticles are also of particular interest as inhomogeneous external magnetic fields exert a force on them, and thus they can be manipulated or transported to a specific diseased tissue by a magnetic field gradient. They also have controllable sizes, so that their dimensions can match either that of a virus (20–500 nm), a protein (5–50 nm) or a gene (2 nm wide and 10–100 nm long). In addition, superparamagnetic particles are of interest because they do not retain any magnetism after removal of the magnetic field.

6.3
Polymer Materials for Drug Delivery Systems

The past few decades has seen considerable interest in developing biodegradable nanoparticles as effective drug delivery devices [23]. Biodegradable polymers are polymers that can be degraded and/or catabolized, eventually to carbon dioxide and water, by microorganisms (bacteria, fungi, etc.) under natural environments [24]. However, due to the development of a wide variety of synthetic biocompatible polymers, the definition has been altered to include many artificially synthesized polymeric materials. Needless to say, degraded components of the polymers should not be toxic and should not promote the generation of harmful substances. Biodegradable polymers can be classified into three major categories:

1. Polyesters produced by microorganisms.
2. Natural polysaccharides (i.e., chitosan [25–29], dextran [30]).
3. Artificially synthesized polymers, especially aliphatic polyesters, {i.e., polylactide (PLA) [31, 32], poly(lactide-r-glycolide) (PLGA) [33], and poly(ε-caprolactone) (PCL) [34]}, polyamide (i.e., poly L-lysine [35]), and others such as poly(methyl methacrylate) (PMMA) [36] and poly(ethyl-2-cyanoacrylate) (PECA) [37], which have also been developed as nanoparticles for the same purpose.

Biodegradable polymers are not only limited to medical devices and wound dressing, but are also used for the fabrication of scaffolds in tissue engineering [38], and as DDSs for controlled release of 5-fluorouracil [39], cisplatin [40], lidocaine [41–43], indomethacin [32, 34], taxol [44], 4-nitroanisole [45], dexamethasone [46], radioactive compound [47], peptides [48], and proteins [49–54] at characteristic rates and specific target sites. For DDS, interest has focused on the use of particle formations prepared from aliphatic polyesters due to their biocompatibility and resorbability. In terms of these required characteristics, numerous workers use polyesters produced from glycolic acid and lactic acid polymers, which are approved by the FDA. These polymers do not require surgical removal after the completion of drug release [55].

6.4
Design of Drug Delivery Vectors and Their Prerequisites

The development of micro/nanospheres for novel drug delivery systems has become and important area of research as such systems enable the controlled release of toxic drugs into the target organs. They also make it possible to deliver useful drugs into sites of inflammation or tumor cells. One of the most important characteristics required by these materials to be suitable for biomedical applications is their biocompatibility, especially with respect to their surface chemistry.

Biocompatibility of nanospheres can be further improved by modifying the terminal groups located on the polymer surface as well as its structure. Obviously, it is important to stabilize the nanospheres sterically by a coating process or chemical modification so as to minimize recognition by phagocytic cells in the reticuloendothelial system (RES). The most promising materials for this purpose are polymeric drug carriers. In general, a polymer that tends to lose mass over time within a living organism is called an absorbable, resorbable, or bioabsorbable, as well as a biodegradable polymer. In comparison with the strict definition, biodegradable polymers require enzymes of microorganisms for natural hydrolytic or oxidative degradation. Regardless of its degradation behavior, this terminology applies to both enzymatic and non-enzymatic hydrolysis.

The physiochemical properties of the materials should be considered to develop nanoparticles as vectors for controlled DDS. Toxins must be removed from the drugs in the patient's body as quickly as possible. Engineered nanoparticles are strong candidates for drug detoxification because the particle size is the key to preventing further damage to the patient's healthy organs [33].

6.4.1
Polymeric Nanoparticles

Numerous factors should be considered when designing a DDS. The first generation of polymeric vectors is simple core–shell structures/capsules within which the therapeutic agents are loaded. The second generation of polymeric vectors aims to enhance degradation rates by synthesizing block copolymers with more than two different species and by varying the molecular weight of the polymer components. Obviously, the release time of the therapeutic agents can be effectively improved in this way by understanding the physiochemical properties of the human body. Even though the drugs are effectively incorporated into the vectors, the nanoparticles can be easily agglomerated after dosing via various routes of administration (i.e., oral, intravenous, intramuscular, subcutaneous, etc.).

A particular consideration to take into account when designing nanoparticles for these applications is that the body's fluids are composed of quite complex compositions such as water, hormones, plasma (e.g., erythrocytes, leukocytes, platelets), fats, protein (e.g., albumin, globulin, fibrinogen), and numerous ions etc. For example, stable colloidal suspensions of DDS can be prepared simply in water-based solvent. However, the nanoparticles, once administrated into the blood stream, typically create a serious blockage of blood flow due to their agglomeration and particle growth through interactions with compounds in the blood. Moreover, many kinds of biological substances are hydrophilic and water soluble, and cannot therefore pass through the hydrophobic lipid bilayer membranes. For example, MPEG has uncharged hydrophilic residues and a very high surface mobility, leading to high steric exclusion. Therefore, it is expected to effectively improve the biocompatibility of nanoparticles and to possibly avoid accumulation in the RES or the mononuclear phagocyte system. Obviously, it is important to stabilize the nanospheres sterically by a coating process or chemical modification so as to minimize recognition by phagocyte cells in the RES.

The most advanced generation of polymeric DDSs has been developed to resolve the above-mentioned problems for multifunctional applications. This can be achieved via the use of nano-biotechnology to fabricate nanovectors with a higher degree of complexity based on the design and fabrication of bio-active and biocompatible, functionalized nanofluids using nanoparticles with several chemistries relevant to specific biological/medical activities.

Surface functionalization of nanovectors is designed to perform the following tasks:

1. Carrier for certain functional compound, e.g., drug(s) and other materials.
2. Formation of stable suspension in physiologically compatible solutions.
3. Controlled targeting to an organ or tissue within the body of a living animal or human subject.
4. Keeping the particles in a given location for a desired period.
5. Controlled release of drugs or chemicals through the pores of the shell according to defined conditions.

As one of the stimuli-sensitive polymers (SSPs), PNIPAAm is well known as a thermosensitive polymer due to its distinct phase transition at a specific lower critical solution temperature (LCST) of 32 °C in water [56–60]. PNIPAAm is hydrophilic below the LCST but becomes hydrophobic when it is heated up above the LCST. PNIPAAm has been consistently investigated as it has "smart" characteristics and is being developed for biomedical applications in the form of micelles [56], tablets [59], and hydrogels.

A new class of temperature-programmed "shell-in-shell" structures with two different copolymers synthesized by a modified-double-emulsion method (MDEM) has been reported as an advanced generation nanovector [31]. In this approach, thermosensitive inner shells composed of poly(N-isopropylacrylamide-co-D,L-lactide) (PNIPAAm-PDLA) with a lower critical solution temperature (LCST) can be fabricated. This novel concept can effectively load any hydrophilic proteins into a polymeric DDS and construct an adequate vector together with the programmable release rate. The release rates are governed by several key parameters, which only involve the PLLA-PEG outer shell, such as the volumetric ratio between the organic phase and aqueous phase, the interaction parameter between the therapeutic agents and the core domain, tacticity of the copolymer, the encapsulation efficiency, etc. Figure 6.3 shows a schematic representation of "shell-in-shell"

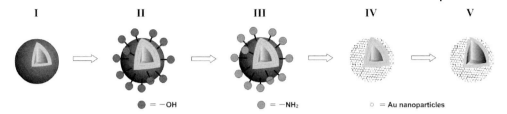

Fig. 6.3. Schematic of the fabrication of Au@PLLA-PEG@PNIPAAm-PDLA: (I) Formation of the hydrophilic protein-loaded PNIPAAm-PDLA sphere; (II) construction of PLLA-PEG@PNIPAAm-PDLA dual-shell structure via a MDEM; (III) functionalization of PLLA-PEG@PNIPAAm-PDLA dual-shell structure with 3-aminopropyltrimethoxysilane (APTMS); (IV) *in situ* reduction of Au^{3+} for the self-assembly of Au nanoparticles on the surface of PLLA-PEG@PNIPAAm-PDLA; (V) completion to load a protein in Au@PLLA-PEG@PNIPAAm-PDLA by elimination of the PNIPAAm-PDLA inner shell.

structures produced by a MDEM. Generally, a hydrophilic protein such as bovine serum albumin (BSA) can be encapsulated in the polymeric spheres using a double-emulsion method (DEM), a so-called "water-in-oil-in-water" (w/o/w) emulsion method [51, 60, 61]. However, the DEM has a disadvantage with respect to the stability of amphiphilic polymer spheres because both the inner shells and the outer shells are composed of the same species of copolymer. Instead, as for the MDEM, two different kinds of copolymers are sequentially incorporated in the organic phase to promote enhanced stability of the spheres. In this way, the inner shells can be prepared with PNIPAAm-PDLA diblock copolymers and the outer shells can be prepared with PLLA-PEG diblock copolymers.

For another class of thermosensitive nanocarriers, poly[(NIPAAm-r-AAm)-*co*-lactic acid] (PNAL) has been reported [62]. As schematically illustrated in Fig. 6.4(b), Au nanoparticles can be directly self-assembled on the surface of PNAL nanospheres by virtue of primary amino groups coming from acrylamide (AAm) molecules of the PNAL diblock terpolymer. The primary amino groups can be strongly bound to noble metals such as gold or silver. Therefore, the "shell" domain of Au@PNAL becomes an affinity site for biomolecules to be conjugated. Furthermore, the LCST of poly(*N*-isopropylacrylamide-*r*-acrylamide) (PNA) was modulated from 32 up to approximately 36 °C through the manipulation of the ratio between N-isopropylacrylamide (NIPAAm) and AAm units. This nanostructure is expected to serve as a synchronous delivery system by virtue of its Au-modified surface and hydrophobic inner core site (Fig. 6.4).

Figure 6.5 shows TEM images of the "shell-in-shell" spherical structures of PLLA-PEG@PNIPAAm-PDLA (parts a and b) and Au@PLLA-PEG@PNIPAAm-PDLA (c and d). The well-defined PLLA-PEG@PNIPAAm-PDLA can be prepared by MDEM and the Au nanoparticles are further deposited by self-assembly, resulting in hybrid nanosphere, i.e., Au@PLLA-PEG@PNIPAAm-PDLA. The bright contrast in the TEM images shown in Fig. 6.5(b) can be identified by a distinct difference in gray scale, which infers the "shell-in-shell" structures are properly fabricated. By increasing the temperature above the LCST, the inner shell of

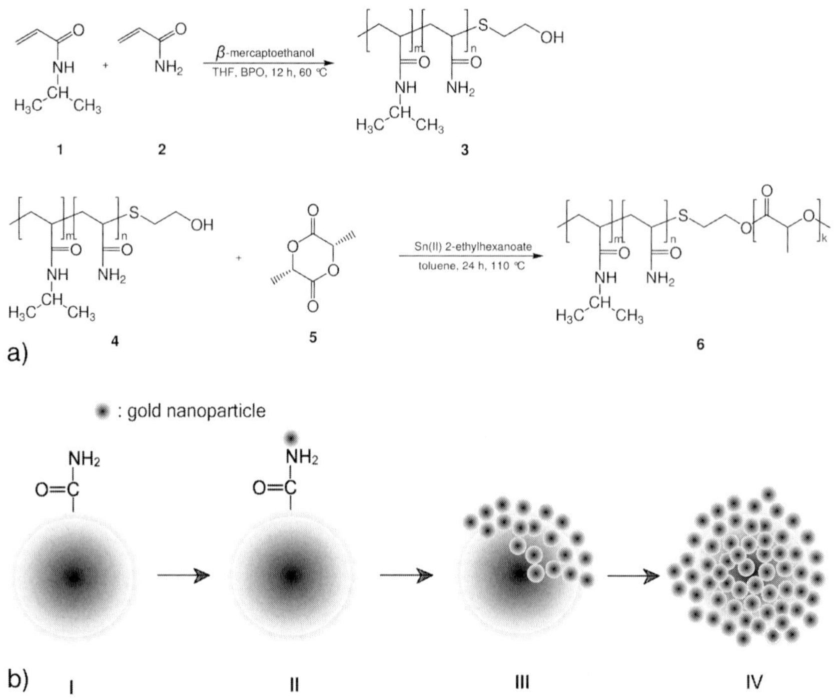

Fig. 6.4. Strategy to fabricate Au@PNAL spheres: (a) Synthetic pathway to PNAL diblock terpolymer: (**1**) NIPAAm; (**2**) AAm; (**3**) PNA; (**4**) hydroxyl-terminated PNA; (**5**) L,L-lactide; (**6**) PNAL. (b) Schematic of direct self-assembly of Au nanoparticles on PNAL nanospheres: step (I) PNAL nanospheres; (II) direct self-assembly of Au nanoparticles to the primary amide groups of Aam; (III and IV) completion of the self-assembly of Au nanoparticles.

PLLA-PEG@PNIPAAm-PDLA undergoes the phase transition, resulting in elimination of the inner shell. Finally, PLLA-PEG@PNIPAAm-PDLA dual-shell structures are changed to the simple core–shell structure.

Hydrolyzable diblock copolymers of poly(ethylene glycol)-poly(L-lactic acid) (PEG-PLA) or poly(ethylene glycol)-poly(caprolactone) (PEG-PCL) have been prepared and loaded with doxorubicin for controlled release of the anticancer agent. The release rates of doxorubicin from the hydrolyzable vector can be modulated by increasing the amount of PEG in the polymeric systems, and also increased linearly with the molar ratio of degradable copolymer blended into the nondegradable membranes. In both nano- (100 nm) and micro-size vectors, the average release time reflects a highly quantized process in which any given vector is either intact or retains its encapsulant. Poration occurs as the hydrophobic PLA or PCL block is hydrolytically scissioned, progressively generating an increasing number of pore-preferring copolymers in the membrane. The kinetics of this evolving detergent mechanism underlies the phase behavior of amphiphiles, with transitions from membranes to micelles allowing controlled release [63].

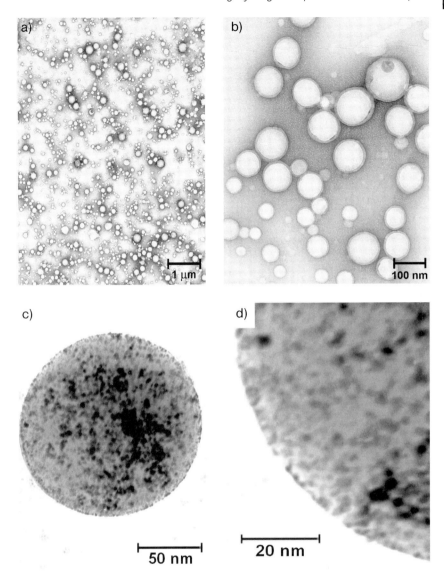

Fig. 6.5. TEM images of PLLA-PEG@PNIPAAm-PDLA (negatively stained with 2 wt.% ammonium molybdate aqueous solution for 2 min) (a and b) and Au@PLLA-PEG@PNIPAAm-PDLA (without staining) (c and d).

Chitosan is a polysaccharide obtained by alkaline deacetylation of the naturally occurring abundant polysaccharide chitin. It is biodegradable, biocompatible, non-immunogenic, and non-carcinogenic, making it suitable for pharmaceutical applications [64]. Chitosan tripolyphosphate (TPP) microspheres have been prepared by spray-drying methods using acetaminophen as a model drug substance. Such ionically crosslinked chitosan-TPP microspheres afforded slower release rates. The vectors can be prepared using a higher concentration of chitosan, higher volume of TPP solution, a higher molecular weight chitosan and/or higher drug loading. In this study, acetaminophen release rates were mainly controlled by the chitosan-TPP matrix density and by the degree of swelling of the hydrogel matrix. The overall release trend of acetaminophen from spray-dried chitosan-TPP microspheres is a two-step biphasic process, with an initial burst followed by subsequent slower release [65]. A complex composite of proteins glycoproteins and proteoglycans has provided an important model for the design of biomaterials [66, 67]. Recent progress in the development of methods for incorporating non-natural amino acids into recombinant proteins using DNA technology points the way to an alternative strategy for preparing new types of drug deliver systems (DDSs).

6.4.2
Inorganic Nanoparticles

Inorganic porous materials are emerging as novel host systems. Owing to some interesting features, such as their biological stability and their drug-releasing properties, there is a significant and increasing interest in these potential vectors. Several porous materials have been used, including synthetic zeolites and silica xerogel materials. MCM 41 is typical mesoporous, templated silica that has been widely investigated. This material presents nanosized pores that allow the incorporation of therapeutic agents. The release properties of ibuprofen-loaded MCM 41 show the feasibility of such systems as vectors for DDS [68].

Porous calcium hydroxyapatite is also used as a vector for anticancer drugs (cis-platinum). The slow release of cis-platinum from the inorganic vector has been confirmed by *in vitro* experiments [69]. When the drug-loaded vector was implanted into normal back muscle, or the tibia, sustained release of cis-platinum was observed during the subsequent 12 weeks. The diffusion rate of cis-platinum into the blood and other organs (liver, kidney, brain) was <10% of that at the implanted site. The vector administered into tumors of mice also showed a steady release of cis-platinum for more than three months. Inhibition of tumor growth was more obvious after local implantation of the cis-platinum-loaded inorganic vector than after intraperitoneal administration of cis-platinum. Based on these results, this novel mesoporous vector shows great potential as controlled DDS of anticancer agents. It is more attractive in bone tumors because the mechanical strength of calcium hydroxyapatite permits partial surgical excision and replacement of the bone defect at the same time.

Porous CuX zeolite has been synthesized as an inorganic vector to incorporate cyclophosphamide (CP). Biochemical and anatomopathological evaluations of antitumoral effects by oral administration of the CP-loaded CuX zeolite show that the intensity of the antitumoral effects is similar to that with normal administration of CP. An advantage of the CP-loaded CuX zeolite is the maintenance of a CP concentration ranging between 100 and 1000 ng per mL of plasma in the blood [70].

6.4.3
Metallic Nanoparticles

Recently, the immobilization of biomolecules, ligands and therapeutic agents onto the surface of metallic nanoparticles has been the focus of intense activity in biological engineering and biotechnology [71–77]. Especially, amine groups and cysteine residues in the proteins can be bound onto the surface of Au nanoparticles and can be stabilized electrostatically [71, 75]. In parallel with this work, several groups have focused on the fabrication of Au nanoparticles-organic/inorganic hybrid structures with polyurethane (PU) [71], silica [78–80], polystyrene (PS) [81], and stimuli-sensitive polymers (SSPs) [31, 62, 82].

Colloidal Au can be used as a vector for therapeutic agents as well as an indicator for immunodiagnostics. However, the use of these Au nanoparticles for *in vivo* DDS was not well established. A colloidal gold (cAu) nanoparticle vector has been used to target the delivery of tumor necrosis factor (TNF) to a solid tumor growing in mice [83]. The optimal vector, designated PT-cAu-TNF, consists of molecules of thiol-derivatized PEG (PT) and recombinant human TNF that can be directly bound onto the surface of the Au nanoparticles. Following intravenous administration, PT-cAu-TNF rapidly accumulates in MC-38 colon carcinoma tumors and shows little to no accumulation in the liver, spleen (i.e., the RES) or other healthy organs of animals. PT-cAu-TNF was less toxic and more effective in reducing tumor burden than native TNF since maximal antitumor responses were achieved at lower drug doses. Svarovsky et al. have described the synthesis of Au nanoshells encapsulated with up to 90 units of the Thomsen-Friedenreich (TF) tumor-associated carbohydrate antigen (TACA) disaccharide (Galbeta1-3GaINAc-alpha-O-Ser/Thr) as well as the assembly of a suitably linked designer glycopeptide as a precursor to similar multivalent presentations on Au. The TF-coated Au nanoparticles are highly stable, water soluble, and easily handled. Improvements in the linker technology used to attach the disaccharide to the Au nanoparticles led to a robust multivalent platform. The antigen retains all recognition characteristics while displayed on this template, as shown by several *in vitro* assays. This approach can be used to develop novel therapeutic agents that inhibit protein–carbohydrate interactions [84].

6.5
Kinetics of the Controlled Release of Anticancer Agents

In vivo treatment requires the release of therapeutic agents into the body; this is followed by absorption, metabolism, distribution, and elimination of the therapeutic agents that are administered, and subsequent target organ effects, both therapeutic and toxic. The process of release followed by absorption, distribution, metabolism, and elimination is referred to as "pharmacokinetics" – a process commonly represented by a mathematical description of the behavior of a drug, and possibly its metabolites, in the system. Pharmacokinetics are frequently described as what the body does to the drug. A major goal of clinical pharmacology is to integrate pharmacokinetics and pharmacodynamics so that their relationships can be understood, and so that drug treatment can be optimized based upon such an understanding [85].

The release rate of DDS predicted by the diffusion- and the dissolution-based release systems can be applicable for the controlled-release of drugs during circulation in the blood stream or localization at the target site. Diffusion models are defined literally as a mass transfer process of the individual substance, brought about by random molecular motion and associated with a concentration gradient. In monolithic devices, the drug is uniformly mixed within a polymeric matrix and is present either in a dissolved or dispersed structure.

Generally, the release model of the devices where the drug is dissolved follows Fickian kinetics. When the drug is dispersed in a polymeric matrix, the rate of release follows the square root of time kinetics until the concentration of the drug decreases below the saturation value. In addition, the preferred release profiles of the drug in bulk degrading systems can be manipulated by adjusting the molecular weight of the polymer, copolymer composition, crystallinity, loading amounts of the drug, and interactions between polymer and drug, etc. Generally, drug release may be diffusion controlled or dissolution controlled, depending on parameters such as the permeability of the polymer to water, the solubility of the drug in the polymer and in the water phase, the molecular weight of the drug, etc. With methotrexate-loaded gelatin nanoparticles, the drug release follows a diffusion-controlled mechanism [86].

Polakovič et al. have made a significant contribution to the investigation of the release profiles of model drugs for spherical shapes of DDS [42]. They suggested that two main models, consisting of the diffusion and the dissolution, should be considered to determine the release rate of the drug from polymeric spheres. Generally, drug release mechanisms from micro/nanoparticles should be assumed based on (a) surface desorption, (b) diffusion through particle pores, (c) diffusion through intact polymers, (d) diffusion through water swollen polymers and (e) surface or bulk erosion of polymeric matrix [87]. The last phenomenon partly represents a mechanism of drug-release. However, other factors will change the morphologies of the platforms, resulting in changes to the rate of diffusion for drug release [42].

6.5.1
Diffusion Model

The release of a drug from a polymeric matrix generally follows Fick's second law. The concentration gradient of the spherical particles follows the form given by Eq. (1).

$$\frac{\partial c}{\partial t} = D\left(\frac{\partial^2 c}{\partial r^2} + \frac{2}{r}\frac{\partial c}{\partial r}\right) \tag{1}$$

where c is the local drug concentration at time t and the distance r from the center of the particle and D is the diffusion coefficient of the drug in the polymeric matrix.

Therefore,

$$\frac{c_1}{c_{1\infty}} = 1 - \sum_{n=1}^{\infty} \frac{6(\alpha+1)\alpha}{(9+3\alpha+q_n^2\alpha^2)} e^{-(q_n^2/R^2)Dt} \qquad (2)$$

$\alpha = V/(V_s K_p)$, where V is the bulk liquid volume of the surrounding medium, V_s is the total volume of the particles $q_n = \lambda R$ where λ is eigen value and R is radius of particles.

6.5.2
Dissolution Model

The dissolution model can be expected when a solid drug is dissolved in media. The dissolution rate of a drug can often be the rate-determining step when the absorption rates are faster than the dissolution rates (e.g., as with steroids). If the drugs are not dissolved before they are removed from the intestinal absorption site, the proper effects can not be expected due to a limited residence time at the absorption site. Therefore, the rate of the dissolution should be considered rather than the diffusion model for drugs that are poorly soluble. For this reason, the dissolution rate of drugs in solid dosage forms is an important parameter in the design of proper DDSs.

Drug release by the dissolution model can be defined by Eq. (3).

$$r_d = -\frac{dc}{dt} = k(c - K_p c_1) \qquad (3)$$

where r_d is the rate of drug dissolution and k is the dissolution coefficient. The drug concentration is eliminated after introducing the mass balance and its derivative in Eq. (3). Integration of the rearranged equation provides the relationship of c_1 with time given in Eq. (4).

$$c_1 = \frac{c_0}{K_p(\alpha+1)} \left[1 - \exp\left(-\frac{\alpha+1}{\alpha} kt\right)\right] \qquad (4)$$

$K_p = c_\infty/c_{1\infty}$ is the partition coefficient characterized by the concentration ratio of the concentration inside the particles to the bulk liquid drug concentration in thermodynamic equilibrium.

6.5.3
Kinetics of the Indomethacin (IMC, 1-[p-chlorobenzoyl]-2-methyl-5-methoxy-3-indoleacetic acid) Release

The release kinetics of IMC as a model drug have been examined with parameters obtained by comparing the diffusion and dissolution kinetic models. Table 6.2

Tab. 6.2. Different series of IMC-loaded PLA-PEO spheres.

Code	Ratio (o/w)	Ratio (drug/polymer)	Water phase (mL)	Organic phase	
				PLA soln (mL)	IMC soln (mL)
LL_1	0.11	1	3	0.165	0.165
LL_2	0.22	1	3	0.330	0.330
LL_3	0.33	1	3	0.660	0.660
DL_1	0.11	1	3	0.165	0.165

summarizes the codes of IMC-loaded PLA-PEO spheres for *in vitro* experiments on the controlled-release of IMC. Figure 6.6(a–d), representing LL_1, LL_2, LL_3, and DL_1 respectively, shows four different kinetic behaviors. The results of LL_1 and DL_1 are in a good agreement with the diffusion model only. However, the release profiles gradually shift from the diffusion model towards the dissolution model as $c_{1\infty} V_r/c_0$ decreases. Figure 6.6(b) shows the release profile of LL_2, which is more dissolution-dependent than that of LL_1.

A similar phenomenon can be observed in LL_3, which shows the greatest dissolution-dependent characteristics with the lowest $c_{1\infty} V_r/c_0$. Consequently, LL_2 has intermediate behavior between the two different models, showing that the *in vitro* release profile is located between the diffusion and the dissolution profiles. This demonstrates that the kinetic models are strongly dependent upon $c_{1\infty} V_r/c_0$. A high $c_{1\infty} V_r/c_0$ can directly affect the diffusion kinetics. Therefore, a high apparent efficiency (ζ) can result in an intermediate characteristic between the diffusion- and dissolution-dependent mechanisms. In addition to the parameters mentioned above, other parameters should be considered during the fabrication of nanospheres for DDS. Yang et al. have reported a correlation between the release behavior and surface porosity of microspheres [61]. They also reported that the organic phase used for the fabrication of the nanospheres by emulsion/evaporation techniques plays a critical role in determining the overall release characteristics of the drugs. The fabrication temperature also becomes a critical factor because the shell of the nanospheres is vulnerable to evaporating solvents with a low-boiling point such as chlorine-based solvents (i.e., methylene chloride and/or chloroform). The release rate is increased owing to a larger pore size on the surface of the spheres when the DDS has a lower V_r. The release profiles of LL_1 and DL_1 show initial fast release rates. In contrast, the initial "bursting" behaviors of LL_2 and LL_3 are less pronounced. This phenomenon explains that V_r can affect the release rates at the initial stage, especially within a short period, i.e., as soon as exposed to the surrounding medium. From the mathematical modeling and *in vitro* experiments, we should not use only one model to predict the release behavior of therapeutic agents because the releasing environments are usually more complicated than ex-

6.5 Kinetics of the Controlled Release of Anticancer Agents

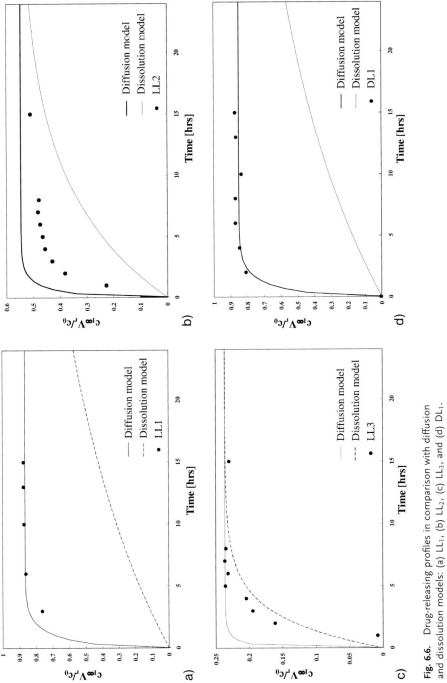

Fig. 6.6. Drug-releasing profiles in comparison with diffusion and dissolution models: (a) LL_1, (b) LL_2, (c) LL_3, and (d) DL_1.

pected. Thus, proper assumptions and possible variables should be considered for both theoretical and practical applications.

6.6
Controlled Release of Anticancer Agents

Although biomaterials (biologically derived components) are useful for new medical treatments, critical problems in biocompatibility, mechanical properties, degradation and numerous other issues remain. Stealth properties and responsiveness to factors such as pH, temperature, specificity and other critical problems have to be resolved. The possibility of delivering cytotoxic agents directly into tumor cells has several advantages: drug losses in the bloodstream and upon liposome–cell interaction are minimized and the preparation of drug-loaded nanoparticles becomes simpler. To be effective, a material must possess several attributes, including the ability to condense therapeutic molecules to a size of less than 150 nm so that it can be taken up by receptor-meditated endocytosis, the ability to be taken up by endosomes in the cell and to allow therapeutic molecules to be released in active form, and to enable it to travel to cell's nucleus. Moreover, gene therapy is gaining in popularity as a medical treatment for cancer, tumors, Alzheimer's diseases, diabetes etc.; however, the clinical efficacy is lower than expected due to the detergent effect. When administered directly into the blood vessel or lesion, the therapeutic molecules are taken up by other healthy organs/cells and the residual time in biological systems is less than 2 h, and thus the pharmacological action is diminished. In addition, it has been reported that therapeutic molecules taken up by healthy organs/cells undergo mutation and may cause other, more serious diseases, including cancer. Also, most antitumor agents are hydrophilic compounds and, therefore, cannot be retained within the membrane. Thus, the use of prodrug forms of anticancer agents to alter the phase behavior of the chemicals is becoming more popular.

6.6.1
Alkylating Agents

Alkylating reagents are chemical reagents that have an alkyl group such as propyl in place of a nucleophilic group. They include several cytotoxic drugs, some of which react specifically with N7 of the purine ring, resulting in depurination of DNA. These alkylating drugs interact with DNA and prevent the division of the cells.

The alkylation of DNA bases can disrupt the replication mechanism of the cell. The nitrogen bases in DNA molecules are nucleophilic and can be easily alkylated. If the N–H groups are replaced by N–R groups then the DNA base pairing is disrupted and can lead to cellular dysfunction. This should have an effect on the replication of cancerous cells, thus leading to a slow-down or stoppage of growth of the cancer.

6.6.1.1 Chlorambucil

Chlorambucil (pronounced "klor-AM-byoo-sill") is a well-known anticancer agents for blood cancers and acts to reduce the number of blood cells. It is also used to treat other cancers such as lymphomas. Chlorambucil is an aromatic derivative of mechlorethamine and is closely related in structure to melphalan. The therapeutic effects are the slowest acting and generally least toxic among the alkylating agents. Alkylation of DNA results in breaks in the DNA molecule as well as crosslinking of the twin strands, thus interfering with DNA replication and transcription of RNA. Like other alkylators, chlorambucil is cell cycle phase-nonspecific [88, 89].

Leroux et al. have demonstrated that polymeric nanoparticles can be loaded with chlorambucil (8.52% m/m) with an entrapment efficiency of 60%. Polymeric nanospheres have been prepared by emulsification of a benzyl alcohol solution of a polymer in a hydrocolloid-stabilized aqueous solution followed by dilution of the emulsion with water. Nanoparticles as small as 70 nm in diameter can be produced by increasing the percentage of poly(vinyl alcohol) to 27.5% in the external phase. The particle size can be controlled by using gelatin instead of poly(vinyl alcohol) and the smallest nanoparticles, with an average size of 70 nm, can be obtained [90].

Chitin-based biodegradable microspheres have also been investigated for their ability to encapsulate chlorambucil as a model drug. The polymer sphere can be prepared by directly blending chitin with different contents of poly(D,L-lactide-*co*-glycolide 50:50) (PLGA 50/50) in dimethylacetamide–lithium chloride solution, followed by coagulating in water via wet phase inversion. Chlorambucil-loaded chitin/PLGA (50/50) has a two-step release mechanism. In the initial stage, the drug release rate increases with increased chitin content due to hydration and surface erosion of the hydrophilic chitin phase; however, the subsequent slow release is sustained for several days, mainly due to bulk hydrolysis of the hydrophobic PLGA phase [91].

6.6.1.2 Cyclophosphamide

Cyclophosphamide is a cyclic phosphamide ester of mechlorethamine. It is transformed via hepatic and intracellular enzymes into active alkylating metabolites, acrolein and phosphoramide mustard. Cyclophosphamide prevents cell division primarily by crosslinking DNA strands. This anticancer agent is applicable to breast cancer, lung cancer, multiple myeloma, mycosis fungoides, neuroblastoma and retinoblastoma etc. It must be handled carefully as it is considered to be highly carcinogenic in humans. Cyclophosphamide-loaded poly(butyl cyanoacrylate) nanospheres have been investigated to obtain a suitable and tolerated ocular delivery device for therapeutic applications, involving treatment of severe ocular inflammatory processes that localize in the anterior chamber of the eye [92, 93].

Local delivery of 4-hydroperoxycyclophosphamide (4HC derived from cyclophosphamide) has been carried out via a controlled-release biodegradable polymer to determine whether the use of a polymer vector can enhance efficacy. Ninety Fischer 344 rats implanted with 9L or F98 gliomas were treated with an intracranial polymer implant containing 0–50% loaded 4HC in the polymer. The 20%

4HC-loaded polymers caused minimum local brain toxicity and maximum survival. These polymers were then used to compare the *in vivo* efficacy of 4HC to BCNU in rats implanted with 9L glioma. Animals with brain tumors treated with 4HC had a median survival of 77 days compared to that of 21 days in BCNU-treated animals and 14 days in untreated animals. Long-term survival for more than 80 days was 40% in the 4HC-treated rats versus 30% in the BCNU-treated rats.

In conclusion, 4HC-impregnated polymers provide an effective, safe local treatment for rat glioma [94].

6.6.1.3 Carmustine

Carmustine [BCNU, 1,3-bis(2-chloroethyl)-l-nitrosourea] is a highly lipophilic nitrosourea compound that undergoes hydrolysis *in vivo* to form reactive metabolites. These metabolites cause alkylation and crosslinking of DNA. Nitrosoureas generally lack cross-resistance with other alkylating agents [88, 89]. The US Food and Drug Administration (FDA) approval of Gliadel® in 1996 represented the first new treatment approved for brain tumors in over 20 years. It has also been approved by numerous regulatory agencies worldwide.

Gliadel® is a polymer–drug combination that delivers the chemotherapeutic agent carmustine directly to the site of a brain tumor via controlled release from a biodegradable matrix [95]. To compare the effectiveness of lipid microspheres with Gliadel®, Takenaga has incorporated a BCNU into lipid microspheres by homogenizing a soybean oil solution of BCNU with egg yolk lecithin. Compared with the corresponding conventional dose of BCNU, the lipid microsphere-encapsulated BCNU significantly enhanced antitumor activity with reduced toxicity in mice with L1210 leukemia. Lipid nanospheres with an average size of 50 nm also showed a similar level of *in vivo* antitumor activity. In this report, [^{14}C]triolein uptake by L1210 leukemia cells was increased by incorporation into microspheres. The nanospheres showed a longer *in vivo* half-life due to the avoidance of cellular uptake by the RES, resulting in higher accumulation at the tumor sites [96].

6.6.2
Antimetabolic Agent

6.6.2.1 Cytarabine

Cytarabine is metabolized intracellularly into its active triphosphate form (cytosine arabinoside triphosphate). This metabolite then damages DNA by multiple mechanisms, including the inhibition of α-DNA polymerase, inhibition of DNA repair through an effect on β-DNA polymerase, and incorporation into the DNA. The latter mechanism is probably the most important. Cytotoxicity is highly specific for the S phase of the cell cycle [88, 89]. Ellena et al. have investigated the distribution of phospholipid and triglyceride molecules in the membranes forming the nonconcentric vesicular network within a multivesicular lipid particle (MLP). MLP formulations exhibited controlled release of encapsulated pharmaceuticals on time scales of a few days to a few weeks. The MLP can be synthesized by a double emulsification process with a neutral lipid such as a triglyceride. MLP formulations with the antineoplastic agent cytarabine encapsulated in the aqueous compartments have been prepared that further contained [^{13}C]carbonyl-enriched triolein. This rational approach can be used to develop MLP formulations with variable rates of sustained release, modulated by changes in the distribution of various phospholipids and triglycerides [97].

6.6.2.2 Fluorouracil (FU)

Fluorouracil was developed in 1957 based on the observation that tumor cells utilized the base pair uracil for DNA synthesis more efficiently than did normal cells of the intestinal mucosa. It is a fluorinated pyrimidine that is metabolized intracellularly to its active form, fluorodeoxyuridine monophosphate (FdUMP). The active form inhibits DNA synthesis by inhibiting the normal production of thymidine. Fluorouracil is cell cycle phase-specific (S phase) [88]. 5-Fluorouracil (5-FU)-loaded poly(L-lactide) (PLLA) or its carbonate copolymer microspheres have been prepared by a modified oil-in-oil (o/o) emulsion solvent evaporation technique. The dispersed phase was a solvent mixture of N,N-dimethylformamide (DMF) and acetonitrile, and the continuous phase was liquid paraffin containing 1–10% (w/v) Span 80(R). Using this modified process, microspheres with various particle sizes can be prepared with high 5-FU entrapment efficiency (~80%). *In vitro* drug release experiments showed a burst release of 5-FU from PLLA microspheres, followed by a sustained release over 50 days. With other vectors, poly(L-lactide-*co*-1,3-trimethylene carbonate) (PLTMC) and poly(L-lactide-*co*-2,2-dimethyl-1,3-trimethylene carbonate) (PLDTMC), the drug release rate can be prolonged to over 60 days [98].

Roullin et al. have developed 5-FU-loaded poly(L-lactide-*co*-glycolide) (PLGA) microspheres to deliver therapeutic agents into the CNS for stereotactic intracerebral implantation [39]. *In vivo* experiments with C6 glioma-bearing rats showed promising results – the median survival time was doubled [99]. A phase I–II pilot study was conducted on eight patients with high-grade glioma who underwent surgical removal before 5-FU-loaded microspheres were implanted. After 18 months the patients' survival rate and welfare was improved [100]. Microsphere fate and the 5-FU diffusion area from these particles in the brain was also investigated, depending on the inserted locations of the drug-loaded microspheres. [3H]5-FU microspheres were used to evaluate diffusion areas from the implantation site [39].

Another approach for controlled DDSs into the brain has also been developed using implantable, biodegradable microspheres. The strategy was evaluated initially to provide localized and sustained delivery of the radiosensitizer 5-FU after patients underwent surgical resection of malignant glioma [101].

6.6.2.3 Methotrexate

Methotrexate and its active metabolites compete for folate-binding sites of the enzyme dihydrofolate reductase. Folic acid must be reduced to tetrahydrofolic acid by this enzyme for DNA synthesis and cellular replication to occur. Competitive inhibition of the enzyme leads to blockage of tetrahydrofolate synthesis, depletion of

novel techniques for the effective loading of active molecules and surface activation (i.e., antibodies and functional groups) for active targeting are essential to improve the therapeutic effectiveness by reducing both the dose and the side effects. Several functions should be considered, such as effective targeting, maximum uptake and retention at the target sites, and rapid clearance after finishing the mission. For this reason, the generation of intelligent smart biomaterials (ISB) based on nanotechnology is being intensively investigated for controlled DDS of anticancer agents and is already having an enormous effect on nanomedicine as a new research field.

While many of nanoparticle synthesis techniques have focused on the empirical basis for controlled release of anticancer agents, increasing demand for multifunctional vectors represents a major fabrication challenge. These vectors should be designed to integrate several aspects: (a) theoretical and practical considerations of the evolution of novel phenomena coming from their composition and size, (b) design of complex or composite structures with given morphology required for multifunctionality, (c) generation and assembly of new molecular and macromolecular structures using suitable processing routes, (d) incorporation of drugs to be delivered to the target cells, and (e) modification of the nanoparticulate surfaces and interfaces, rendering them suitable for interaction with the target.

To develop artificial synthetic biomaterials based on nanotechnology for cancer-oriented drug delivery systems (CoDDS), we should consider the extracellular matrix biology, cell receptors and immunology, and how the body responds to specific materials. These novel vectors should incorporate hydrophobic/hydrophilic drugs with active surface modifications by the attachment of active functional ligands for passive targeting at specific target organs, receptors, etc.

Moreover, novel concepts of ISB for CoDDS can be developed based on core–shell, mesoporous nanotechnology. These vectors can be programmed to respond to external environmental conditions, e.g., pH, ionic strength, temperature, ultrasound, radiation, magnetic fields, UV-light, etc. The development of suitable nanostructures, methodologies for drug incorporation, methodologies for controlling releasing rates, and acceptable biological activity should be considered in the design process.

Targeting is also an important concept for CoDDS. Passive and active targeting to specific lesions, degradation/release rates, and surface activation/functionalization should be considered for optimization of these systems. Toxicology and pharmacological evaluation are also important factors. Thus, the therapeutic compounds, loading methods, quantity and efficacy, etc. should all be carefully considered during the design proper vectors.

References

1 American Cancer Society, http://www.cancer.org.

2 F. L. MI, S. S. SHYU, C. Y. KUAN, S. T. LEE, K. T. LU, S. F. JANG, Chitosan-polyelectrolyte complexation for the preparation of gel beads and con-

trolled release of anticancer drug. I. Effect of phosphorous polyelectrolyte complex and enzymatic hydrolysis of polymer, *J. Appl. Polym. Sci.* **1999**, 74, 1868–1879.

3 S. GOLUBOVIC, B. Z. RADMANOVIC, Increase of corneal graft-survival by use of topically immunosuppressive agents in rabbits, *Graefes Arch. Clin. Exp. Ophthalmol.* **1988**, 226, 288–290.

4 M. DYBA, N. I. TARASOVA, C. J. MICHEJDA, Small molecule toxins targeting tumor receptors, *Curr. Pharm. Des.* **2004**, 10, 2311–2334.

5 Q. A. PANKHURST, J. CONNOLLY, S. K. JONES, J. DOBSON, Applications of magnetic nanoparticles in biomedicine, *J. Phys. D: Appl. Phys.* **2003**, 36, R167–R181.

6 M. WOLF, H. SWAISLAND, S. AVERBUCH, Development of the novel biologically targeted anticancer agent gefitinib: Determining the optimum dose for clinical efficacy, *Clin. Cancer Res.* **2004**, 10, 4607–4613.

7 M. P. LUTOLF, J. A. HUBBELL, Synthetic biomaterials as instructive extracellular microenvironments for morphogenesis in tissue engineering, *Nat. Biotechnol.* **2005**, 23, 47–55.

8 L. REN, G. M. CHOW, Synthesis of nir-sensitive Au-Au2S nanocolloids for drug delivery, *Mater. Sci. Eng., C* **2003**, 23, 113–116.

9 M. MORAWSKI, T. REINERT, C. MEINECKE, T. ARENDT, T. BUTZ, Antibody meets the microbeam – Or how to find neurofibrillary tangles, *Nucl. Instrum. Methods Phys. Res., Sect. B* **2005**, 231, 229–233.

10 K. SIVAKUMAR, B. PANCHAPAKESAN, Electric field-assisted deposition of nanowires on carbon nanotubes for nanoelectronics and sensor applications, *J. Nanosci. Nanotechnol.* **2005**, 5, 313–318.

11 J. CONNOLLY, T. G. ST PIERRE, M. RUTNAKORNPITUK, J. S. RIFFLE, Cobalt nanoparticles formed in polysiloxane copolymer micelles: Effect of production methods on magnetic properties, *J. Phys. D: Appl. Phys.* **2004**, 37, 2475–2482.

12 C. BARGLIK-CHORY, D. BUCHOLD, M. SCHMITT, W. KIEFER, C. HESKE, C. KUMPF, O. FUCHS, L. WEINHARDT, A. STAHL, E. UMBACH, M. LENTZE, J. GEURTS, G. MULLER, Synthesis, structure and spectroscopic characterization of water-soluble CdS nanoparticles, *Chem. Phys. Lett.* **2003**, 379, 443–451.

13 L. MENG, Z. X. SONG, Applications of quantum dots to biological medicine, *Prog. Biochem. Biophys.* **2004**, 31, 185–187.

14 A. MARTUCCI, J. FICK, S. E. LEBLANC, M. LOCASCIO, A. HACHE, Optical properties of PbS quantum dot doped sol-gel films, *J. Non-Cryst. Solids* **2004**, 345–346, 639–642.

15 D. S. KOKTYSH, X. R. LIANG, B. G. YUN, I. PASTORIZA-SANTOS, R. L. MATTS, M. GIERSIG, C. SERRA-RODRIGUEZ, L. M. LIZ-MARZAN, N. A. KOTOV, Biomaterials by design: Layer-by-layer assembled ion-selective and biocompatible films of TiO_2 nano-shells for neurochemical monitoring, *Adv. Funct. Mater.* **2002**, 12, 255–265.

16 T. HIRAI, Y. ASADA, Preparation of ZnO nanoparticles in a reverse micellar system and their photo-luminescence properties, *J. Colloid Interface Sci.* **2005**, 284, 184–189.

17 R. A. GANEEV, M. BABA, A. I. RYASNYANSKY, M. SUZUKI, H. KURODA, Laser ablation of GaAs in liquids: Structural, optical, and nonlinear optical characteristics of colloidal solutions, *Appl. Phys. B* **2005**, 80, 595–601.

18 H. YANG, R. YANG, X. WAN, W. WAN, Structure and photoluminescence of Ge nanoparticles with different sizes embedded in SiO_2 glasses fabricated by a sol–gel method, *J. Cryst. Growth* **2004**, 261, 549–556.

19 M. MIKHAYLOVA, D. K. KIM, N. BOBRYSHEVA, M. OSMOLOWSKY, V. SEMENOV, T. TSAKALAKOS, M. MUHAMMED, Superparamagnetism of magnetite nanoparticles: Dependence on surface modification, *Langmuir* **2004**, 20, 2472–2477.

20 D. K. KIM, D. KAN, T. VERES, F. NORMADIN, J. K. LIAO, H. H. KIM, S.-H. LEE, M. ZAHN, M. MUHAMMED,

Monodispersed Fe–Pt nanoparticles for biomedical applications, *J. Appl. Phys.* **2005**, 97, 10Q918.

21 A. S. EDELSTEIN, R. C. CAMMARATA, Nanomaterials: Synthesis, Properties and Applications, Institute of Physics Publishing, London, **1998**.

22 A. K. GUPTA, M. GUPTA, Synthesis and surface engineering of iron oxide nanoparticles for biomedical applications, *Biomaterials* **2005**, 26, 3995–4021.

23 K. S. SOPPIMATH, T. M. AMINABHAVI, A. R. KULKARNI, W. E. RUDZINSKI, Biodegradable polymeric nanoparticles as drug delivery devices, *J. Controlled Release* **2001**, 70, 1–20.

24 M. OKADA, Chemical syntheses of biodegradable polymers, *Prog. Polym. Sci.* **2002**, 27, 87–133.

25 Z. S. MA, H. H. YEOH, L. Y. LIM, Formulation pH modulates the interaction of insulin with chitosan nanoparticles, *J. Pharm. Sci.* **2002**, 91, 1396–1404.

26 E. RUEL-GARIEPY, A. CHENITE, C. CHAPUT, S. GUIRGUIS, J. C. LEROUX, Characterization of thermosensitive chitosan gels for the sustained delivery of drugs, *Int. J. Pharm.* **2000**, 203, 89–98.

27 K. A. JANES, M. P. FRESNEAU, A. MARAZUELA, A. FABRA, M. J. ALONSO, Chitosan nanoparticles as delivery systems for doxorubicin, *J. Controlled Release* **2001**, 73, 255–267.

28 A. M. DE CAMPOS, A. SANCHEZ, M. J. ALONSO, Chitosan nanoparticles: A new vehicle for the improvement of the delivery of drugs to the ocular surface. Application to cyclosporin A, *Int. J. Pharm.* **2001**, 224, 159–168.

29 Y. HU, X. Q. JIANG, Y. DING, H. X. GE, Y. Y. YUAN, C. Z. YANG, Synthesis and characterization of chitosan-poly(acrylic acid) nanoparticles, *Biomaterials* **2002**, 23, 3193–3201.

30 S. MITRA, U. GAUR, P. C. GHOSH, A. N. MAITRA, Tumour targeted delivery of encapsulated dextran-doxorubicin conjugate using chitosan nanoparticles as carrier, *J. Controlled Release* **2001**, 74, 317–323.

31 Y. S. JO, D. K. KIM, Y. K. JEONG, K. J. KIM, M. MUHAMMED, Encapsulation of bovine serum albumin in temperature-programmed "shell-in-shell" structures, *Macromol. Rapid Commun.* **2003**, 24, 957–962.

32 S. Y. KIM, I. G. SHIN, Y. M. LEE, Preparation and characterization of biodegradable nanospheres composed of methoxy poly(ethylene glycol) and DL-lactide block copolymer as novel drug carriers, *J. Controlled Release* **1998**, 56, 197–208.

33 D. KIM, H. EL-SHALL, D. DENNIS, T. MOREY, Interaction of PLGA nanoparticles with human blood constituents, *Colloids Surf., B* **2005**, 40, 83–91.

34 S. Y. KIM, I. L. G. SHIN, Y. M. LEE, C. S. CHO, Y. K. SUNG, Methoxy poly(ethylene glycol) and epsilon-caprolactone amphiphilic block copolymeric micelle containing indomethacin. II. Micelle formation and drug release behaviours, *J. Controlled Release* **1998**, 51, 13–22.

35 P. CALVO, J. L. VILAJATO, M. J. ALONSO, Evaluation of cationic polymer-coated nanocapsules as ocular drug carriers, *Int. J. Pharm.* **1997**, 153, 41–50.

36 P. AHLIN, J. KRISTL, A. KRISTL, F. VRECER, Investigation of polymeric nanoparticles as carriers of enalaprilat for oral administration, *Int. J. Pharm.* **2002**, 239, 113–120.

37 J. L. ARIAS, V. GALLARDO, S. A. GOMEZ-LOPERA, R. C. PLAZA, A. V. DELGADO, Synthesis and characterization of poly(ethyl-2-cyanoacrylate) nanoparticles with a magnetic core, *J. Controlled Release* **2001**, 77, 309–321.

38 N. KUMAR, M. N. V. RAVIKUMAR, A. J. DOMB, Biodegradable block copolymers, *Adv. Drug Deliv. Rev.* **2001**, 53, 23–44.

39 V. G. ROULLIN, J. R. DEVERRE, L. LEMAIRE, F. HINDRE, M. C. VENIER-JULIENNE, R. VIENET, J. P. BENOIT, Anti-cancer drug diffusion within living rat brain tissue: An experimental study using [H-3](6)-5-fluorouracil-loaded PLGA microspheres, *Eur. J. Pharm. Biopharm.* **2002**, 53, 293–299.

40 K. Avgoustakis, A. Beletsi, Z. Panagi, P. Klepetsanis, A. G. Karydas, D. S. Ithakissios, PLGA-mPEG nanoparticles of cisplatin: In vitro nanoparticle degradation, in vitro drug release and in vivo drug residence in blood properties, *J. Controlled Release* **2002**, 79, 123–135.

41 T. Gorner, R. Gref, D. Michenot, F. Sommer, M. N. Tran, E. Dellacherie, Lidocaine-loaded biodegradable nanospheres. I. Optimization of the drug incorporation into the polymer matrix, *J. Controlled Release* **1999**, 57, 259–268.

42 M. Polakovic, T. Gorner, R. Gref, E. Dellacherie, Lidocaine loaded biodegradable nanospheres. II. Modelling of drug release, *J. Controlled Release* **1999**, 60, 169–177.

43 M. T. Peracchia, R. Gref, Y. Minamitake, A. Domb, N. Lotan, R. Langer, PEG-coated nanospheres from amphiphilic diblock and multiblock copolymers: Investigation of their drug encapsulation and release characteristics, *J. Controlled Release* **1997**, 46, 223–231.

44 L. Mu, S. S. Feng, A novel controlled release formulation for the anticancer drug paclitaxel (Taxol (R)): PLGA nanoparticles containing vitamin E TPGS, *J. Controlled Release* **2003**, 86, 33–48.

45 M. S. Romero-Cano, B. Vincent, Controlled release of 4-nitroanisole from poly(lactic acid) nanoparticles, *J. Controlled Release* **2002**, 82, 127–135.

46 S. Ghassabian, T. Ehtezazi, S. M. Forutan, S. A. Mortazavi, Dexamethasone-loaded magnetic albumin microspheres: Preparation and in vitro release, *Int. J. Pharm.* **1996**, 130, 49–55.

47 J. F. W. Nijsen, M. J. van Steenbergen, H. Kooijman, H. Talsma, L. M. J. Kroon-Batenburg, M. van de Weert, P. P. van Rijk, A. de Witte, A. D. V. Schip, Characterization of poly(L-lactic acid) microspheres loaded with holmium acetylacetonate, *Biomaterials* **2001**, 22, 3073–3081.

48 W. I. Li, K. W. Anderson, P. P. DeLuca, Kinetic and thermodynamic modeling of the formation of polymeric microspheres using solvent extraction/evaporation method, *J. Controlled Release* **1995**, 37, 187–198.

49 J. C. Gayet, G. Fortier, High water content BSA-PEG hydrogel for controlled release device: Evaluation of the drug release properties, *J. Controlled Release* **1996**, 38, 177–184.

50 T. Verrecchia, G. Spenlehauer, D. V. Bazile, A. Murrybrelier, Y. Archimbaud, M. Veillard, Non-stealth (poly(lactic acid albumin)) and stealth (poly(lactic acid-polyethylene glycol)) nanoparticles as injectable drug carriers, *J. Controlled Release* **1995**, 36, 49–61.

51 P. Quellec, R. Gref, L. Perrin, E. Dellacherie, F. Sommer, J. M. Verbavatz, M. J. Alonso, Protein encapsulation within polyethylene glycol-coated nanospheres. I. Physicochemical characterization, *J. Biomed. Mater. Res.* **1998**, 42, 45–54.

52 J. Slager, A. J. Domb, Biopolymer stereocomplexes, *Adv. Drug Deliv. Rev.* **2003**, 55, 549–583.

53 J. M. Bezemer, R. Radersma, D. W. Grijpma, P. J. Dijkstra, C. A. van Blitterswijk, J. Feijen, Microspheres for protein delivery prepared from amphiphilic multiblock copolymers 1. Influence of preparation techniques on particle characteristics and protein delivery, *J. Controlled Release* **2000**, 67, 233–248.

54 T. Morita, Y. Horikiri, T. Suzuki, H. Yoshino, Applicability of various amphiphilic polymers to the modification of protein release kinetics from biodegradable reservoir-type microspheres, *Eur. J. Pharm. Biopharm.* **2001**, 51, 45–53.

55 H. Y. Kwon, J. Y. Lee, S. W. Choi, Y. S. Jang, J. H. Kim, Preparation of PLGA nanoparticles containing estrogen by emulsification-diffusion method, *Colloids Surf., A* **2001**, 182, 123–130.

56 S. H. Yuk, S. H. Cho, S. H. Lee, pH/temperature-responsive polymer composed of poly((N,N-dimethylamino)ethyl methacrylate-co-

ethylacrylamide), *Macromolecules* **1997**, 30, 6856–6859.
57 S. R. SERSHEN, S. L. WESTCOTT, N. J. HALAS, J. L. WEST, Temperature-sensitive polymer-nanoshell composites for photothermally modulated drug delivery, *J. Biomed. Mater. Res.* **2000**, 51, 293–298.
58 F. EECKMAN, A. J. MOES, K. AMIGHI, Surfactant induced drug delivery based on the use of thermosensitive polymers, *J. Controlled Release* **2003**, 88, 105–116.
59 J. E. CHUNG, M. YOKOYAMA, K. SUZUKI, T. AOYAGI, Y. SAKURAI, T. OKANO, Reversibly thermo-responsive alkyl-terminated poly(N-isopropylacrylamide) core-shell micellar structures, *Colloids Surf., B* **1997**, 9, 37–48.
60 S. B. ZHOU, X. M. DENG, H. YANG, Biodegradable poly(epsilon-caprolactone)-poly(ethylene glycol) block copolymers: Characterization and their use as drug carriers for a controlled delivery system, *Biomaterials* **2003**, 24, 3563–3570.
61 Y. Y. YANG, T. S. CHUNG, X. L. BAI, W. K. CHAN, Effect of preparation conditions on morphology and release profiles of biodegradable polymeric microspheres containing protein fabricated by double-emulsion method, *Chem. Eng. Sci.* **2000**, 55, 2223–2236.
62 Y. S. JO, D. K. KIM, M. MUHAMMED, Synchronous delivery systems composed of Au nanoparticles and stimuli-sensitive diblock terpolymer, *J. Mater. Sci.: Mater. Med.* **2004**, 15, 1291–1295.
63 F. AHMED, D. E. DISCHER, Self-porating polymersomes of PEG-PLA and PEG-PCL: Hydrolysis-triggered controlled release vesicles, *J. Controlled Release* **2004**, 96, 37–53.
64 R. HEJAZI, M. AMIJI, Chitosan-based gastrointestinal delivery systems, *J. Controlled Release* **2003**, 89, 151–165.
65 K. G. H. DESAI, H. J. PARK, Preparation and characterization of drug-loaded chitosan-tripolyphosphate microspheres by spray drying, *Drug Dev. Res.* **2005**, 64, 114–128.

66 R. LANGER, Drug delivery and targeting, *Nature* **1998**, 392, 5–10.
67 D. G. ANDERSON, J. A. BURDICK, R. LANGER, Materials science – Smart biomaterials, *Science* **2004**, 305, 1923–1924.
68 C. CHARNAY, S. BEGU, C. TOURNE-PETEILH, L. NICOLE, D. A. LERNER, J. M. DEVOISSELLE, Inclusion of ibuprofen in mesoporous templated silica: Drug loading and release property, *Eur. J. Pharm. Biopharm.* **2004**, 57, 533–540.
69 A. UCHIDA, Y. SHINTO, N. ARAKI, K. ONO, Slow release of anticancer drugs from porous calcium hydroxyapatite ceramic, *J. Orthop. Res.* **1992**, 10, 440–445.
70 C. V. UGLEA, I. ALBU, A. VATAJANU, M. CROITORU, S. ANTONIU, L. PANAITESCU, R. M. OTTENBRITE, Drug-delivery systems based on inorganic materials. 1. Synthesis and character-ization of a zeolite-cyclophosphamide system, *J. Biomater. Sci. Polym. Ed.* **1994**, 6, 633–637.
71 S. PHADTARE, A. KUMAR, V. P. VINOD, C. DASH, D. V. PALASKAR, M. RAO, P. G. SHUKLA, S. SIVARAM, M. SASTRY, Direct assembly of gold nanoparticle "shells" on polyurethane microsphere "cores" and their application as enzyme immobilization templates, *Chem. Mater.* **2003**, 15, 1944–1949.
72 A. GOLE, C. DASH, V. RAMAKRISHNAN, S. R. SAINKAR, A. B. MANDALE, M. RAO, M. SASTRY, Pepsin-gold colloid conjugates: Preparation, characteriza-tion, and enzymatic activity, *Langmuir* **2001**, 17, 1674–1679.
73 A. GOLE, C. DASH, C. SOMAN, S. R. SAINKAR, M. RAO, M. SASTRY, On the preparation, characterization, and enzymatic activity of fungal protease-gold colloid bioconjugates, *Bioconjug. Chem.* **2001**, 12, 684–690.
74 A. GOLE, S. VYAS, S. PHADTARE, A. LACHKE, M. SASTRY, Studies on the formation of bioconjugates of endoglucanase with colloidal gold, *Colloids Surf., B* **2002**, 25, 129–138.
75 A. CSAKI, G. MAUBACH, D. BORN, J. REICHERT, W. FRITZSCHE, DNA-based

molecular nanotechnology, *Single Mol.* **2002**, 3, 275–280.

76 A. Schroedter, H. Weller, Ligand design and bioconjugation of colloidal gold nanoparticles, *Angew. Chem. Int. Ed.* **2002**, 41, 3218–3221.

77 R. C. Mucic, J. J. Storhoff, C. A. Mirkin, R. L. Letsinger, DNA-directed synthesis of binary nanoparticle network materials, *J. Am. Chem. Soc.* **1998**, 120, 12674–12675.

78 V. G. Pol, A. Gedanken, J. Calderon-Moreno, Deposition of gold nanoparticles on silica spheres: A sonochemical approach, *Chem. Mater.* **2003**, 15, 1111–1118.

79 M. S. Fleming, D. R. Walt, Stability and exchange studies of alkanethiol monolayers on gold-nanoparticle-coated silica microspheres, *Langmuir* **2001**, 17, 4836–4843.

80 K. S. Mayya, B. Schoeler, F. Caruso, Preparation and organization of nanoscale polyelectrolyte-coated gold nanoparticles, *Adv. Funct. Mater.* **2003**, 13, 183–188.

81 Z. J. Liang, A. Susha, F. Caruso, Gold nanoparticle-based core-shell and hollow spheres and ordered assemblies thereof, *Chem. Mater.* **2003**, 15, 3176–3183.

82 N. Nath, A. Chilkoti, Creating "smart" surfaces using stimuli responsive polymers, *Adv. Mater.* **2002**, 14, 1243–1247.

83 G. F. Paciotti, L. Myer, D. Weinreich, D. Goia, N. Pavel, R. E. McLaughlin, L. Tamarkin, Colloidal gold: A novel nanoparticle vector for tumor directed drug delivery, *Drug Deliv.* **2004**, 11, 169–183.

84 S. A. Svarovsky, Z. Szekely, J. J. Barchi, Synthesis of gold nanoparticles bearing the Thomsen-Friedenreich disaccharide: A new multivalent presentation of an important tumor antigen, *Tetrahedron: Asymmetry* **2005**, 16, 587–598.

85 M. J. Egorin, Overview of recent topics in clinical pharmacology of anticancer agents, *Cancer Chemother. Pharmacol.* **1998**, 42, S22–S30.

86 M. G. Cascone, L. Lazzeri, C. Carmignani, Z. H. Zhu, Gelatin nanoparticles produced by a simple W/O emulsion as delivery system for methotrexate, *J. Mater. Sci.: Mater. Med.* **2002**, 13, 523–526.

87 R. Jalil, J. R. Nixon, Biodegradable poly(lactic acid) and poly(lactide-co-glycolide) microcapsules – Problems associated with preparative techniques and release properties, *J. Microencapsul.* **1990**, 7, 297–325.

88 C. M. Haskell, *Cancer Treatment*, WB Saunders Co., Philadelphia, **1990**.

89 B. A. Chabner, C. E. Myers, *Cancer: Principles and Practice of Oncology, Clinical Pharmacology of Cancer Chemotherapy*, JB Lippincott Co, Philadelphia, **1989**.

90 J. C. Leroux, E. Allemann, E. Doelker, R. Gurny, New approach for the preparation of nanoparticles by an emulsification-diffusion method, *Eur. J. Pharm. Biopharm.* **1995**, 41, 14–18.

91 F. L. Mi, Y. M. Lin, Y. B. Wu, S. S. Shyu, Y. H. Tsai, Chitin/PLGA blend microspheres as a biodegradable drug-delivery system: Phase-separation, degradation and release behavior, *Biomaterials* **2002**, 23, 3257–3267.

92 A. Salgueiro, F. Gamisans, M. Espina, X. Alcober, M. L. Garcia, M. A. Egea, Cyclophosphamide-loaded nanospheres: Analysis of the matrix structure by thermal and spectroscopic methods, *J. Microencapsul.* **2002**, 19, 305–310.

93 A. Salgueiro, M. A. Egea, M. Espina, O. Valls, M. L. Garcia, Stability and ocular tolerance of cyclophosphamide-loaded nanospheres, *J. Microencapsul.* **2004**, 21, 213–223.

94 K. D. Judy, A. Olivi, K. G. Buahin, A. Domb, J. I. Epstein, O. M. Colvin, H. Brem, Effectiveness of controlled-release of a cyclophosphamide derivative with polymers against rat gliomas, *J. Neurosurg.* **1995**, 82, 481–486.

95 C. Guerin, A. Olivi, J. D. Weingart, H. C. Lawson, H. Brem, Recent advances in brain tumor therapy: Local intracerebral drug delivery by polymers, *Invest. New Drugs* **2004**, 22, 27–37.

96 M. TAKENAGA, Application of lipid microspheres for the treatment of cancer, *Adv. Drug Deliv. Rev.* **1996**, 20, 209–219.

97 J. F. ELLENA, M. LE, D. S. CAFISO, R. M. SOLIS, M. LANGSTON, M. B. SANKARAM, Distribution of phospholipids and triglycerides in multivesicular lipid particles, *Drug Deliv.* **1999**, 6, 97–106.

98 K. J. ZHU, J. X. ZHANG, C. WANG, H. YASUDA, A. ICHIMARU, K. YAMAMOTO, Preparation and in vitro release behaviour of 5-fluorouracil-loaded microspheres based on poly(L-lactide) and its carbonate copolymers, *J. Microencapsul.* **2003**, 20, 731–743.

99 P. MENEI, M. BOISDRONCELLE, A. CROUE, G. GUY, J. P. BENOIT, Effect of stereotactic implantation of biodegradable 5-fluorouracil-loaded microspheres in healthy and C6 glioma-bearing rats, *Neurosurgery* **1996**, 39, 117–123.

100 P. MENEI, M. C. VENIER, E. GAMELIN, J. P. SAINT-ANDRE, G. HAYEK, E. JADAUD, D. FOURNIER, P. MERCIER, G. GUY, J. P. BENOIT, Local and sustained delivery of 5-fluorouracil from biodegradable microspheres for the radiosensitization of glioblastoma – A pilot study, *Cancer* **1999**, 86, 325–330.

101 P. MENEI, E. JADAUD, N. FAISANT, M. BOISDRON-CELLE, S. MICHALAK, D. FOURNIER, M. DELHAYE, J. P. BENOIT, Stereotaxic implantation of 5-fluorouracil-releasing microspheres in malignant glioma – A phase I study, *Cancer* **2004**, 100, 405–410.

102 D. S. FISCHER, M. T. KNOBF, H. J. DURIVAGE, N. BEAULIEU, *The Cancer Chemotherapy Handbook*, Mosby-Year Book, Inc., New York, **2003**.

103 Y. ZHANG, R. X. ZHUO, Synthesis and drug release behavior of poly(trimethylene carbonate)-poly(ethylene glycol)-poly(trimethylene carbonate) nanoparticles, *Biomaterials* **2005**, 26, 2089–2094.

104 L. MANIL, J. C. DAVIN, C. DUCHENNE, C. KUBIAK, J. FOIDART, P. COUVREUR, P. MAHIEU, Uptake of nanoparticles by rat glomerular mesangial cells in-vivo and in-vitro, *Pharm. Res.* **1994**, 11, 1160–1165.

105 F. BRASSEUR, P. COUVREUR, B. KANTE, L. DECKERSPASSAU, M. ROLAND, C. DECKERS, P. SPEISER, Actinomycin-D adsorbed on polymethylcyanoacrylate nanoparticles – Increased efficiency against an experimental tumor, *Eur. J. Cancer* **1980**, 16, 1441–1445.

106 B. KANTE, P. COUVREUR, V. LENAERTS, P. GUIOT, M. ROLAND, P. BAUDHUIN, P. SPEISER, Tissue distribution of [actinomycin]-H-3 D adsorbed on polybutylcyanoacrylate nanoparticles, *Int. J. Pharm.* **1980**, 7, 45–53.

107 R. H. BLUM, S. K. CARTER, K. AGRE, Clinical review of bleomycin – New antineoplastic agent, *Cancer* **1973**, 31, 903–914.

108 I. F. UCHEGBU, A. G. SCHATZLEIN, L. TETLEY, A. I. GRAY, J. SLUDDEN, S. SIDDIQUE, E. MOSHA, Polymeric chitosan-based vesicles for drug delivery, *J. Pharm. Pharmacol.* **1998**, 50, 453–458.

109 A. HAGIWARA, T. TAKAHASHI, O. KOJIMA, K. KITAMURA, C. SAKAKURA, S. SHOUBAYASHI, K. OSAKI, A. IWAMOTO, M. LEE, K. FUJITA, Endoscopic local injection of a new drug-delivery format of peplomycin for superficial esophageal cancer – A pilot-study, *Gastroenterology* **1993**, 104, 1037–1043.

110 K. G. LAU, S. CHOPRA, Y. MAITANI, Entrapment of bleomycin in ultra-deformable liposomes, *Stp Pharma Sci.* **2003**, 13, 237–239.

111 K. G. LAU, Y. HATTORI, S. CHOPRA, E. A. O'TOOLE, A. STOREY, T. NAGAI, Y. MAITANI, Ultra-deformable liposomes containing bleomycin: In vitro stability and toxicity on human cutaneous keratinocyte cell lines, *Int. J. Pharm.* **2005**, 300, 4–12.

112 R. S. BENJAMIN, Clinical-pharmacology of daunorubicin, *Cancer Treat. Rep.* **1981**, 65, 109–110.

113 E. A. FORSSEN, The design and development of DaunoXome(R) for solid tumor targeting in vivo, *Adv. Drug Deliv. Rev.* **1997**, 24, 133–150.

7
Critical Analysis of Cancer Therapy using Nanomaterials

Lucienne Juillerat-Jeanneret

7.1
Introduction

The treatment of diseases such as cancer is challenging because these pathologies involve dysregulation of endogenous and often essential cellular processes. Cancer cells replicate faster than most non-tumoral cells, and the vast majority of presently used therapies capitalize on these differences. More selective therapies, such as anti-angiogenic therapies, are under development and/or clinical evaluation, whereas a few targeted therapies are in clinical use, such as antiestrogen therapies in estrogen receptor-positive breast cancer. With targeted approaches, not only patient survival will increase due to improved treatment efficiency, but also the quality of life of patients will improve by decreasing side effects to normal cells. Most solid tumors possess unique features, such as extensive angiogenesis, defective vascular architecture, increased vascular permeability, impaired lymphatic drainage, which can also be used as therapeutic targets. Nanoparticles can take advantage of these features and act as a vehicle to selectively and specifically deliver anticancer drugs to tumors, either by using passive mechanisms such as increased vascular permeability or acting as drug reservoir in a defined location, or by using active targeting. These combined approaches would result not only in increased efficacy but also in decreased collateral side effects. However, targeting strategies and chemical synthesis routes need to be improved, and the mechanisms of interactions of these functionalized nanostructures with living materials need to be better understood.

This chapter describes not only the tools of nanoparticle technology that can be used to treat cancer, as many excellent reviews (indicated below the section headings) have been published, but rather critically reviews and discusses the advantages and drawbacks of nanoparticles for the *targeted* delivery of anticancer agents to defined cells of human cancers. Approaches that have been developed or are under development to achieve improved cancer therapy using nanoparticles as targeting delivery agents for anticancer drugs are reviewed, and the problems and issues that need to be answered to validate these approaches are summarized. Section 7.2 reviews the characteristics of human tumors that can be used to develop anticancer treatments. Section 7.3 covers the characteristics of already developed nanoparticles, considering the chemical, biophysical and biological demands of such devices. Information from the previous two sections is combined in Section 7.4 to review the characteristics of already developed, or under development, nanoparticles for anticancer treatments using drugs employed in clinics or under development, photodynamic and gene therapy approaches, or magnetically controlled delivery. Section 7.5 describes the defects and characteristics of human tumors that can be used to develop targeted anticancer treatments using nanoparticles, via cancer-associated cells, cancer cell molecules, or by achieving intracellular delivery, and also describes the chemical challenges involved in preparing these nanovectors. Section 7.6 hypothesizes how nanoparticles may result in the delivery of anticancer drugs in drug-resistant human cancers. Issues not yet resolved issues concerning the potential toxicity of nanoparticulate vectors to patients and the general population are covered in Section 7.7. Finally, Section 7.8 describes what might be the ideal nanoparticle in the context of targeted treatment of human cancer using nanodevices.

7.2
Anticancer Therapies

For a more extensive review see Ref. [1] and references herein. Effective therapies of cancer capitalize on differences between diseased and healthy tissues that can be targeted with drugs. Cancer drugs used in patients target the cell cycle, DNA replication, and cytoskeletal assembly. The general toxicity to the whole body of current anticancer chemotherapeutic treatments, resulting in important side effects such as sterility, loss of digestive capacities and appetite, loss of hair, defects in immune functions, etc., is a challenge facing the development of new modalities of cancer treatment. In addition, combination therapies or scheduled therapies improved patient response to chemotherapy regimen. One of the main challenges in cancer treatment is no longer the development of efficient drugs but the improvement of drug selectivity. More recently, the targeting of growth receptors and cellular signaling pathways, "targeted therapeutics", has became a new approach in cancer treatment. However, even targeted and combined therapies suffer from side effects that result from imperfect selectivity for diseased tissue. The availability of novel molecular targets and drug delivery systems that distinguish diseased from healthy cells could vastly amplify therapeutic opportunities. The identification of new disease-associated changes in cellular biology, and the development of associated tools, that may be used to improve selectivity in diagnosis, treatment and evaluation of response to treatment will be the next challenges in cancer therapy. Most cancer-related deaths occur as a consequence of metastasis, and the major problem facing oncologists treating cancer is metastatic disease. Metastasis of tumor cells to organs distant from the original primary tumor site involves about half of all cancers and is generally detected only at an advanced stage of metastatic disease. Successful eradication of metastatic lesions still depends on the early detection of metastases, which only rarely happens. Therefore, devising new means to detect metastatic tumor lesions at the earliest possible stage, and at the same time to be able to treat them, would be an important breakthrough in cancer treatment. Magnetic nanoparticles can be used for this purpose. Therefore, one goal in the field of cancer is to develop chemically derivatized nanoparticles able to target tumor cells and tumor-associated stromal cells via specific recognition mechanisms, and dual cancer detection such as magnetic resonance imaging and cancer therapy using cell-directed drugs.

Neoplastic tissue can be divided into three compartments, vascular, interstitial and cellular, and cancers are constituted of several cell types:

- Endothelial cells and pericytes, either overnumbered or undernumbered;
- immune/inflammatory cells, including macrophages and lymphocytes, and fibroblasts and myofibroblasts;
- normal cells and tumor cells derived from these normal cells;
- and by the absence of a well-defined lymphatic network.

Tumor vasculature is highly abnormal, proliferating, activated, tortuous, and presenting increased permeability and gaps, with pores between 350 and 800 nm, and a cutoff around 400 nm. Tumor vascularization is generally poorly perfused. Tumor interstitium is predominantly constituted of a protein network, including collagens, elastin, proteoglycans and glycoproteins, forming a hydrophilic gel and producing high interstitial osmotic pressure, leading to an outward convective fluid flow. The tumor environment is oxidative and acidic, and thus ionization of basic drugs may decrease their interstitial transport and oxidation destroy their anticancer properties. The transport of drugs in the interstitium will thus be governed by interstitial osmotic pressure and the relative chemical composition and characteristics of drugs and the interstitium.

Therefore, the delivery of a therapeutic agent to tumor must:

1. Resist hydrostatic, hydrophilic/hydrophobic and biophysical/biochemical barriers;
2. resist cellular resistance to treatment;
3. resist biotransformation, degradation and clearance mechanisms;
4. reach its treatment target: extracellular or intracellular compartments, tumor cells or vascular cells, etc.;
5. achieve distribution in all tumor areas even with low vascularization, or poor perfusion;
6. and be active in tumors at efficient concentrations, without unacceptable side effects to non-tumoral cells.

Both tissue- and cell-distribution of anticancer drugs can be controlled and improved by their entrapment in colloidal nanoparticles, increasing anti-tumor efficacy, reversing resistance mechanisms and decreasing side effects. Active tumor targeting with long-circulating nanoparticles decorated with targeting agents is the main development required to achieve these goals. The association of drugs with colloidal nanoparticles may be a way to overcome many of these resistance mechanisms of tumors to treatment and increase selectivity. In this context, nanoparticles are defined as submicroscopic colloidal systems, which may act as a drug vehicle either as nanospheres (a matrix system in which the drug is dispersed) or nanocapsules, which are reservoirs in which the drug is confined in a hydrophobic or hydrophilic core surrounded by a single polymeric membrane. The structure of the polymer and the method of trapping the drugs in the nanoparticles will define the drug release kinetics and characteristics. The necessary characteristics to be useful in cancer treatment, and the drug–nanoparticle systems that have reached clinical use, are under clinical evaluation or are under development will be reviewed.

7.3
Characteristics of Nanoparticles for Cancer Therapy

For more extensive reviews see Refs. [2–6] and references herein. The treatment of cancer is limited by the inability to deliver therapeutic agents in such a way that most drug molecules will selectively reach the desired targets, with only marginal collateral damage. To achieve such efficient treatments, two main goals must be met:

- Increasing targeting selectivity for defined organ, tissue or cells.
- Devising a therapeutic formulation able to overcome the biological barriers that prevent drugs efficiently reaching their targets.

However, the realization of such a system faces formidable challenges, which include:

- Identification of neoplastic biomarkers as biological targets, and their evolution over time.
- Development of biotechnologies to develop biomarker-targeted delivery of multiple therapeutic agents, coupled to the possibility to avoid biological barriers and various resistance mechanisms.

Nanoparticles are interesting for medical application since they present a large surface for functionalization with drugs compared to larger particles made of the same materials, and hopefully achieve targeted drug delivery. They can pass epithelial and vascular barriers. Thus, nanoparticles have the potential to provide opportunities to meet the challenges of cancer therapy, and also of therapeutic approaches for other disorders.

The dawning era of polymer therapeutics started with improved knowledge of polymer characteristics and the development of polymer chemistry (reviewed in Ref. [3]). Initially, polymer–drugs, polymer–proteins, and, in particular, PEGylated derivatives and HPMA [N-(hydroxypropyl)-methacrylamide] copolymers have been used in the context of anticancer therapy [7], and are the precursors of nanoparticulate systems. Self-assembling block-copolymers were then evaluated to deliver drugs to cancer. A pluronic block copolymer-doxorubicin was able to circumvent Pgp [8, 9], and PEG-polyAsp-doxorubicin accumulated preferentially in tumors due to vascular leakage [10, 11], while the increased size prevented back-diffusion and renal clearance when evaluated in clinical trials. These approaches opened the way to nanotechnology for drug delivery.

7.3.1
Nanovectors

Nanotechnology implies that the drug delivery device is man-made, and of dimensions in the nm range (sub-cellular size), which includes nanovectors such as liposomes or monomeric or block copolymeric nanoparticles for the (targeted) delivery of anticancer drugs, imaging contrast agents such as gadolinium or iron-oxide nanosized magnetic resonance imaging contrast agents (cf., for example, Chapters 3 and 5), or quantum dots. First, basic definitions for the various nanoparticulate systems in the context of cancer are given as they will be developed in this chapter. Nanoparticles are non-viral solid nanovectors made of one or several different materials, including water-soluble polymers, whose upper size limit is <1 µm, generally <100 nm. These nanostructures have unique properties, such as modification of the properties, spacing and arrangement of surface atoms, and physics and chemistry compared to larger particles of the same material, and they also have a large surface area to volume ratio. Nanoparticles may be built of polymeric drugs (nanosuspensions), polymer–drug conjugates, polymer–protein conjugates, polymeric drug-micelles, etc. Micelles are self-assembling colloidal aggregates of amphipathic molecules–polymeric block copolymers to give polymeric micelles, which occurs when the concentration reaches the crucial micelle concentration (c.m.c.). Polyplexes are polyelectrolyte complexes formed by a polycation and an anionic molecule, generally an oligonucleotide. Dendrimers are macromolecule that contain symmetrically arranged branches arising from a multifunctional core, to which a precise number of terminal groups are added stepwise. Drug nanosuspensions are insoluble nanocrystals of drugs, generally coated with a surfactant.

Rational approaches in design and surface engineering for site-specific delivery of drugs, genetic material and diagnostic agents, to tissues, cells and intracellular cell compartments, after intravascular, parenteral, intraperitoneal, etc., administration, include:

- Liposomes: closed vesicles formed by hydratation of phospholipids above their transition temperature. Nanoliposomes are bilayer structures of less than 100 nm, surrounding the drug entrapped in the aqueous space. Drugs can also be contained in the lipid space between bilayers. Surface modification is possible, and nucleic acids are adsorbed on cationic liposomes by ion-pairing. These structures have been the first to reach clinical use.
- Micelles: amphiphilic aggregates, <50 nm, made of hydrophilic (A) and hydro-

phobic (B) block copolymers (AB or ABA), in which hydrophobic or hydrophilic drugs are physically trapped or covalently bound. Drugs may include nucleic acids for transfection and gene therapy.
- Nanospheres: made of polymer (synthetic or natural) aggregates (tens to hundreds nm size) in which the drug is either dissolved, entrapped, encapsulated or covalently attached. Surface modification is possible.
- Superparamagnetic iron oxide crystals: made of an iron oxide core (5–10 nm) obtained by coprecipitation of Fe^{2+} and Fe^{3+}, and coated with a polymer [dextran, poly(ethylene glycol), poly(vinyl alcohol), etc.]. Surface modification is possible by covalent links or adsorption (drug, antibodies, nucleic acid, targeting agents). These nanostructures have mainly been used for cancer detection, and are now developed to couple detection with drug delivery.
- Carbohydrates-ceramic nanoparticles: core composed of calcium phosphate or ceramic, surrounded by a polyhydroxyl oligomeric film, on which drugs are adsorbed.
- Dendrimers: highly branched three-dimensional macromolecules that grow by outward–controlled polymerization; drugs are covalently bound at the surface.

Therapeutic agents (chemically synthesized therapeutic small or large drugs, therapeutic peptides or proteins, nucleic acids for gene therapy) can be entrapped, encapsulated, adsorbed, covalently bound either to the surface or at the interior of biodegradable polymeric nanoparticles (Table 7.1). These approaches can improve drug solubility, and also achieve better drug selectivity. Drugs can be made to form small aggregates, surrounded by a water- and bio-compatible, biodegradable poly-

Tab. 7.1. Biological characteristics of the polymers.

Polymer	Efficient for
PEG	Increases biocompatibility, increases circulating time, decreases uptake by macrophages
Dextrans	Increases circulating time, decreases aggregation and opsonization, decreases uptake by macrophages
Poly(vinylpyrrolidone)	Increases circulating time, decreases aggregation and opsonization, decreases uptake by macrophages
PVA	Decreased aggregation/coagulation
Polyacrylates	Biostabilization, biocompatibility, increases bioadhesion
Polypeptides	Targeting
Poly(DL-lactide)	Increases biocompatibility, decreases cytotoxicity
Chitosan	Increases biocompatibility, increases hydrophilicity

meric thin layer, improving the biodistribution and bioavailability of drugs. Surface or polymers functionalization with appropriate ligands can allow the targeting of these nanostructures to defined cells, tissues or body locations, depending on the chosen specificity and selectivity of the ligand, improving the therapeutic effectiveness and decreasing side effects of drugs. Finally, the polymer properties may also be defined to respond to changes in pH or redox state, chemical environment, heat (either internal or external), or an external physical stimulus, therefore allowing choice for the rate and location of drug release, e.g., acidic intracellular organelles such as the lysosomes.

The development of nanovectors has given birth to what has been called "nanomedicine": the applications of nanotechnology for the rational delivery and targeting of pharmaceuticals, therapeutics and diagnostic agents. The next challenges include:

- Identification of precise targets for selective delivery.
- Choice of appropriate nano-carriers.
- Avoidance of mononuclear phagocytes and the reticuloendothelial system, to selectively target either angiogenic cancer-associated endothelial cells or cancer cells, which are key targets in cancer.

7.3.2
Biological Issues

Injection in the blood and lymphatic vessels, but also inhalation or intraperitoneal injection, etc., are possible routes of administration for the delivery of therapeutic nanoparticles. The stability, extracellular or cellular distribution of nanoparticles depend on their surface properties, chemical composition, morphology and size. The main challenges for intravenously injected nanoparticles are rapid opsonization and clearance by the reticuloendothelial system (RES) of the liver and the spleen or excretion by the kidneys (Fig. 7.1). Opsonization by complement proteins, vitronectin, fibronectin, immunoglobulins, lipoproteins, etc. renders nanoparticles recognizable by the major defense systems of the body, the RES and the mononuclear phagocyte system, depending on the surface properties of nanoparticles, their size ($<$ or $>$200 nm) and surface characteristics. Nanoparticles with a largely hydrophobic surface are efficiently coated with plasma components, trapped in the liver and rapidly removed from circulation, while smaller particles can stay in circulation. More hydrophilic nanoparticles can resist coating process to a variable extent and are more slowly cleared from the blood stream [2]. Therefore, clearance kinetics depend on the chemical and physical properties of the nanoparticles: surface charges and charge density, lipophilic/hydrophilic area ratio, presence of functional and chemically reactive groups. Consequently, successful drug delivery requires careful control of the physicochemical properties (size and surface) of nanovectors. Suppression of opsonization will increase the retention of nanoparticles in locations other than macrophages and so afford a longer circulatory time. In addition, macrophages are heterogeneous, in different tissues and within a tis-

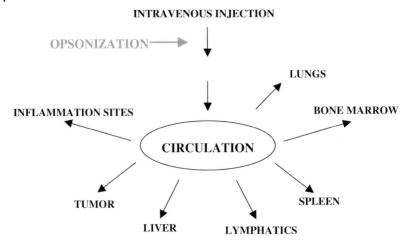

Fig. 7.1. Fate and distribution in the body of intravenously injected nanoparticles – their uptake by different organs, according to the particles' size and surface.

sue. PEGylation [the coating of the surface by poly(ethylene glycols)] of nanoparticles has been the main approach up to now to solve opsonization problems and short circulation times. PEG coating decreases liver, spleen, lung, or kidney clearance. A large surface area is also an issue in the aggregation of nanoparticles in a biological environment, determining the effective clearance rates and mechanisms.

7.3.3
Nanoparticle Targeting: Passive or Active

Long circulation times are efficient in both treating circulatory disorders and vascular imaging. Passive targeting depends on vascular leakage and passive diffusion [enhanced permeability and retention (EPR) effect] of nanovectors to achieve drug delivery. Nanoparticles escape from the vasculature compared with virus behavior and is restricted to endothelial fenestration (between 150 and 300 nm) of leaky areas in inflammation or tumors, or splenic filtration for non-deformable nanoparticles (200–250 nm). Initially, targeting was passively achieved at the organ/tissue level by virtue of particle size, 50–200 nm carriers restricting the distribution volume to the blood compartment.

Active (ligand-targeting, cationic lipids or polymers cytotoxic and low circulating times) delivery using selective recognition mechanisms can be achieved with more recent nanoparticles, and is presently under development and evaluation. Therefore, the surface engineering of nanoparticles is a crucial determinant of their biological behavior, and much effort is presently undertaken to improve these modifications (see below). A thorough understanding of the elaborate cell transport machinery as well as an understanding and the finding of targets to achieve selective delivery will be necessary to fulfill the potential of nanobiotechnologies in cancer.

Many drugs are agonists and antagonists of chemicals inappropriately produced by diseased cells. Intelligent systems should respond to differences in concentration of these chemicals or changes in external biological conditions, e.g., by modification of the polymer lattices, allowing increased or decreased release of drugs, dependent on a biosensor to achieve "intelligent therapeutics". Delivery of therapeutic agents precisely where and when they are needed in the human body is becoming realistic due to rapid, tremendous progress in physiology, nanoparticle and nanobiotechnology. Nanoparticles can improve the targeting of cytotoxic drugs to cancer only if they can be directed to cancer areas and maintained there for long time periods with their drug cargo for selective and local release of drugs such as alkylating agents, 5-fluorouracyl, platinum derivatives, taxol derivatives, and more selective kinase inhibitors. Chemical bonding, ionic or hydrophobic adsorption or embedding of drugs into nanoparticles have been devised to be sensitive to cancer tissue properties, including high proteolytic or glycolytic activity, high metabolism, low pH, and high oxidative environments in tumors, to further ensure selective release of the drug in the tumor area. The discovery and design of intelligent material wills be the next drug-delivery system generation for chronic diseases. These approaches will now be reviewed.

7.4 Nanovectors in Biomedical Applications: Drug Delivery Systems (DDS) for Cancer

For general reviews see Refs. [6, 12] and references herein. Present treatments for cancer include various unique or combined approaches, encompassing total or partial surgical excision of tumor tissue, chemotherapy and radiotherapy. Anticancer chemotherapeutic drugs are generally administered intravenously, leading to general systemic distribution. As drugs used in cancer chemotherapy are mostly non-selective for tumor or tumor-associated cells, important and deleterious side effects result from their use [2, 13, 14]. These secondary effects in patients result in loss of quality of life and necessitate drugs to alleviate them. Therefore, means are needed to deliver drugs to specific areas of the body, maximizing drug action (exclusively in diseased cells), together with minimizing side effects, and consequently increasing treatment efficiency. Selective or targeted DDS, and in particular nanoparticles, have the potential to achieve the goals of drug-targeted delivery:

- Delivery to a particular organ;
- to a specific cell type (differentiation from tumor and normal cells);
- to a structure within a cell (such as the nucleus in gene therapy, where targeting to the nucleus is a prerequisite of gene expression).

They use (a) physicochemical, (b) biological or (c) chemical methods to control the distribution of drugs to improve the outcome of chemotherapy.

7.4.1
Physicochemical Drug Delivery

Polymers and colloidal nanocarriers (nanoparticles) can be used for passive or active drug delivery. Passive targeting implies a physiological uptake mechanism (filtration or macrophage sequestration), while active methods involve the use of a recognition ligand.

7.4.2
Biological Drug Delivery

With biological targeting a specific marker (a target) is selectively expressed on diseased cells and not, or at a much lower level, in normal cells. The targeting agent, such as antibodies, a ligand for a receptor or a lectin, is covalently conjugated via an appropriate spacer to the nanoparticle. It can be directed towards an antigen or receptor residing on or within the target tissue.

7.4.3
Chemical Drug Delivery

Chemical methods involve the use of modified forms of active drugs, e.g., prodrugs, by exploiting differences in pathophysiological conditions within target tissues (e.g., pH, redox state, enzyme content) and normal tissues.

The efficiency of drug delivery depends strongly on the nanovector size and surface characteristics, controlling the fate of a drug in the organism, and the selective delivery of drugs. The size, zeta potential (surface charge, coating), release characteristics (polymers and linkers), biodegradability and cytotoxicity and encapsulation efficiency are the main factors determining efficient drug delivery. Expression at the surface of nanovectors of cell-specific ligands can further increase selectivity. Control of the release of a drug, in a defined localization in an organism, organ, tissue, at the cell surface or intracellularly, and the kinetics of drug release must also be characterized. Finally, the stability, cytotoxicity, mechanisms of cell uptake and the biodegradation of the nanovectors are also important. Passive (EPR effect) versus active (ligand-targeting, cationic lipids or polymer cytotoxicity and poor circulating times) delivery is also a choice to consider for efficient delivery, and must be dictated by the characteristics of the cancer to be treated, whether a marked angiogenesis is present or not.

Various nanovectors are available to achieve drug delivery, such as liposomes, micro/nanospheres, nanoemulsions and micro/nanocapsules. They can be used to deliver hydrophilic drugs, hydrophobic drugs, proteins, nucleic acids, vaccines, biological macromolecules, etc. Nanovectors, such as liposomes, protect drugs from degradation and biological metabolism; however, liposomes have a low encapsulation efficiency, poor storage stability, and rapid leakage of water-soluble drugs in the blood. As such, their ability to control the release of many drugs is not optimal. Nanoparticles made of colloidal suspensions and biodegradable polymers offer better stability than liposomes and allow controlled release. Finally, the efficiency of drug delivery is increased by using magnetic vectors, which can be targeted with an external gradient magnetic field; such vectors also have the potential to record the drug delivery sites as contrast agents in magnetic resonance imaging (MRI) for diagnostics (see below). The characteristics, composition, etc. of nanoparticles (which are in clinical use, in clinical trials or under development) necessary to achieve targeted delivery of anticancer agents to treat human cancers will be reviewed as a function of their physicochemical properties and the chemical and biological properties of the drug, or the cancer type characteristics. However, this chapter is not an extensive review of all the published information available.

7.4.4
Nanoparticles for Anticancer Drug Delivery

For more extensive reviews see Refs. [6, 12, 15]. What has been attempted and what are the next challenges of the nanoparticulate approach to drug delivery in cancer?

7.4.4.1 Existing Systems

Colloidal Delivery Systems Encapsulation of therapeutic agents in colloidal carriers, including liposomes, emulsion, solid lipid nanoparticles, polymeric particle and polymeric micelles, form colloidal delivery systems.

Liposome Nanoparticles These nanoparticles are biodegradable and flexible, have an aqueous core containing the drug, and are bilayer amphipathic lipids. Drugs encapsulated in liposomes under evaluation or in clinics include paclitaxel, lurtotecan, platinum derivatives, vincristine, doxorubicin (see below).

Emulsion/Solid Lipid Nanoparticles Here an oily core, either liquid or solid lipids at body temperature, or a monolayer of amphipathic lipids contains the drug. Drugs under evaluation include protoporphyrin IX, for photodynamic therapy, and taxol (see below).

Polymeric Nanospheres Polymers [such as poly(lactide-co-glycolide), poly(vinylpyrrolidone), poly(ε-caprolactone)] and entrapped drugs allow controlled drug release from the polymer.

Polymeric Micelles Biodegradable polymeric micelles with the drug in the core have been prepared from di-block copolymers (one core-forming segment, one shell-forming segment).

Drug-conjugated Delivery Systems In these systems drugs are covalently bound to the polymer and are either active when coupled to the polymer or after release (acid-sensitive, enzyme-sensitive bonds) or polymer degradation.

Lipid-based or polymer–drug conjugate-based nanoparticles (<200 nm) can improve the pharmacological properties, pharmacokinetics and biodistribution, and sustained release of free drugs. But they also present new challenges and issues [12] that need to be taken into account for the DDS to reach clinical use, including potency (the fewer the carrier can carry, the more potent the drug must be), the stability of the drug carrier (either shelf or biological stability), solubility of the drug, size of the carrier and the cargo (e.g., proteins and small carriers), charge, carrier biocompatibility, cytotoxicity, degradation products, drug survival to chemical procedures and coupling routes, rates and efficacy of drug release in the tumor space (e.g., for schedule-dependent anticancer drug therapies drugs must stay above minimal efficacy levels for several hours or days, for schedule-independent drugs a large burst is more important than constant release), hypersensitivity reaction to the carrier–drug conjugates. A drug linked to a carrier may have the same activity as the free drug or, alternatively, its pharmacological properties may be modified, e.g., the toxicity of liposomal vincristine is similar to that of the free drug, but its potency is augmented, liposomal topotecan is protected from biodegradation, and doxorubicin-linked *N*-2-hydroxypropyl methacrylamide copolymer displays a slow release and an increased maximal tolerated dose.

7.4.4.2 Systems under Development and Challenges

Targeted Drug Delivery Nanoparticulate Systems These are the most recent development, presently only at the initial stages. Drugs are conjugated to ligands/antibodies or incorporated in carriers bearing ligands/antibodies for recognition by cell surface receptors expressed by target cells, e.g., doxorubicin-nanoparticles targeting HerB2/neu for breast cancer. Major obstacles for the delivery of anticancer drugs include the definition of selective cell-specific targets and physiological barriers, in particular the epithelial and blood–brain barriers. For example, approaches using LDL-mimic nanoparticles targeting the LDL receptors on brain endothelial cells [16] or galactose-HPMA copolymer bearing doxorubicin for the asialoglycoprotein receptors in liver tumors, which is under clinical trial [17], have been attempted. One major problem in these approaches is to identify relevant targeting entities in cancers compared with normal cells of the whole body. A few have been identified and evaluated in preliminary trials, e.g., folate receptors and PSA-doxorubicin conjugates for PSA-positive prostate cancer.

Intracellular Delivery For maximal efficacy, cancer drugs must reach their appropriate targets, in the appropriate location within cells, which are mainly located either in the cytosol, the nucleus, and more recently in cell organelles such as mitochondria. Therefore, either drug-loaded carriers must be transported intact inside cells, then carriers must release their cargo, or carriers must release their cargo at the targeted-cell surface, and free drug must be transported inside cells. Colloids and nanoparticles are mostly taken up by endocytosis in cancer cells, via the endosome/lysosome pathway. Therefore, the drug must be released intact from the lysosomal compartment. Most approaches have used the acidic characteristics of this compartment to dissociate the drug from its carrier, e.g., HPMA copolymer, liposomes, polymeric micelles, cationic lipids and photosensitizers [18–22], and release into the cytosol; more recently, the presence of lysosomal enzymes have been used to release covalently bound drug from a carrier system. Nuclear delivery is also necessary for many drugs, mainly nucleic acid or protein drugs. In this case tagging with a nuclear location signal is necessary.

Avoiding Drug Resistance Mechanisms Resistance of tumor cells to chemotherapeutic agents is a major cause of treatment failure in cancer therapy, mainly mediated by families of energy-dependent ATP-driven efflux pumps (MDR, Pgp). Liposomes, polymer conjugates, polymeric micelles have the potential to overcome resistance mechanisms, and positive results in this direction have been obtained (reviewed in Ref. [23], and see below).

Methods of delivery and tumor targeting are key areas for the future of nanotechnology in anticancer therapy. For this, innovative nanotechnological methodologies have been initiated, are under active development and will be further improved, to achieve better, more efficient and less aggressive therapies of cancer for the whole organism.

7.4.5
Nanoparticles for Drug Delivery in Clinical Use or under Clinical Evaluation

More extensive reviews are given in Refs. [6, 12, 24]. Several polymer–drug conjugates are in clinical use or under clinical evaluation as anticancer agents (Table 7.2).

Hydrophilic drugs can be easily entrapped with high efficiency in the aqueous core of liposomes whereas hydrophobic weak bases, such as doxorubicin or vincristine, are loaded by pH and chemical gradients across the liposome bilayer. Consequently, as many of the agents active against cancer are hydrophobic molecules, most presently used DDSs in clinics are liposomes, and many have been decorated with PEG to increase their bioavailability (decreased opsonization, decreased clearance by the RES, increased circulation time), e.g., PEGylated doxorubicin-liposomes.

7.4.5.1 Doxorubicin Family

Liposomes encapsulating doxorubicin are the archetypal, simplest and first form of nanoparticles used for cancer therapy. Doxorubicin-liposomes were the first to be used clinically and ameliorated versions, such as PEGylated doxorubicin liposomes, have been in clinical use for breast and other cancers for several years.

Doxorubicin belongs to the anthracyclines and is used to treat breast, ovarian, bronchial cancers by inhibiting the synthesis of nucleic acids in cancer cells, but at the price of cardiotoxicity and myelosuppression and a very narrow therapeutic index. Doxorubicin-liposomes of phosphatidylcholinum-carbamoyl-cholesterol coated (glycosylated) with methoxypoly(ethylene glycol) (MPEG) (100 nm particles) to form an hydrophilic layer that protects against phagocytosis by macrophages and increases the circulation half-life, extravasation via defective tumor vessels and release of doxorubicin chlorhydrate selectively in tumor vicinity are in clinical use.

Tab. 7.2. Some nanoparticulate drugs under clinical use or evaluation. (Adapted from Ref. [12] with modifications.)

Nanoparticulate drug	Clinical use or evaluation
Doxorubicin-liposome/PEG	Breast and ovarian carcinoma
Zinostatin-styrene	Hepatocellular carcinoma
Liposomal-vincristine	non-Hodgkin's lymphoma
Liposomal all-trans-retinoic acid	non-Hodgkin's lymphoma
Polyglutamate-paclitaxel	non-Small cell lung carcinoma
Liposomal paclitaxel	Advanced solid tumors
Liposomal oxaliplatin	Colorectal cancer
Liposomal lurtotecan or irinotecan	Solid tumors, ovarian, small cell lung cancer
N-(2-Hydroxypropyl)methacrylamide copolymer doxorubicin	Breast, colon, lung cancer
N-(2-Hydroxypropyl)methacrylamide copolymer doxorubicin-galactosamine	Liver cancer

Conjugates of doxorubicin and dextran have been encapsulated with chitosan (100 nm size particles) or conjugated to PLGA, PCAA or poly(γ-benzyl-L-glutamate)/poly(ethylene oxide) nanoparticles (200–250 nm), resulting in long-term *in vitro* release of the drug [25] and *in vivo* suppression of tumor growth in *in vivo* experimental models [26], proving efficacy. Efficacy was also suggested to be macrophage-mediated [27]. Brain delivery of doxorubicin has also been obtained by biodegradable poly(butyl cyanoacrylate)-polysorbate 80-coated nanoparticles [28]. Recent developments of nanoparticles for doxorubicin include the development of lecithin lipid core-drug/pluronic [poly(ethylene oxide)-poly(propylene oxide)-poly(ethylene oxide) triblock copolymer]-shell nanoparticles, obtained by a freeze-drying procedure [29]. In addition, solid lipid nanoparticles of cholesteryl butyrate of doxorubicine with paclitaxel, which had additive effects [30], and nanoparticles of poly(isohexyl cyanoacrylate) able to overcome MDR and increase sensitivity to doxorubicin [31], have also been developed recently.

7.4.5.2 Paclitaxel (Taxol)

Paclitaxel (taxol) is a microtubule-stabilizing agent that causes polymerization of tubulin and cell death. Paclitaxel is used in ovarian, breast, colon, non-small cell lung carcinomas. Paclitaxel is poorly aqueous soluble, but soluble in organic solvents, and is presently formulated in Cremophor EL (polyoxyethyleneglycerol triricinoleate 35) or polysorbate (Tween) 80 (polyoxyethylenene-sorbitan-20-monooleate), which have important side effects and drawbacks. Strategies have been developed to formulate taxanes in Cremophor and Tween 80 based on pharmaceuticals such as albumin nanoparticles, emulsions, liposomes, polyglutamates, and prodrugs strategies [32]. New formulations that have been developed include biodegradable nanoparticle formulations (140 nm mean diameter) using poly(lactic-*co*-glycolic) and nanoprecipitation in acetone, showing efficiency and high incorporation loading [33, 34], poly(ethylene glycol)-poly(lactide) (PLGA) nanoparticles (<100 nm) [35–38] and d-α-tocopheryl PEG as an emulsifier. In addition, the use of methoxy poly(ethylene glycol)-poly(lactide) (MPEG-PLA) results in a slow and long-lasting release after an initial burst [39]. More recently, entrapment of paclitaxel by hydrophobic forces in micelles of block copolymers (NK105) led to increased blood stability, bioavailability and anticancer drug efficacy, extended *in vivo* antitumor activity and reduced the neurotoxicity of paclitaxel [40].

7.4.5.3 5-Fluorouracil

5-Fluorouracil is used in the treatment of rectum, colon, breast, stomach, pancreas, liver, uterus, ovarian and bladder cancers. Intravenous injection of an aqueous solution of 5-fluorouracil inhibits cell growth by blocking thymine (5-methyluracile) formation, and DNA synthesis, and the formation of aberrant RNA; however, the drug is very short-lived under this route of application. For clinical formulation, 5-fluorouracil has been incorporated in dendrimers of poly(amidoamine) modified with mPEG-500 by simple incubation due to the hydrophilicity of the drug, achieving longer-lasting release than the free drug. The *in vivo* half-life of 5-fluorouracil, a clinical problem necessitating continuous infusion for several hours, has been increased by drug formulation by a diafiltration procedure in biodegradable amphiphilic PEG-poly(γ-benzyl-l-Glu) micelles (180–250 nm) and an hydrophilic shell of 30 nm [41].

7.4.5.4 Tamoxifen

Tamoxifen is a non-steroid inhibitor of estrogen receptors mainly used to treat breast cancer. For clinical use, tamoxifen is encapsulated within PEG-coated nanospheres and located at the nanosphere surface, resulting in immediate drug release [23]. More recently, tamoxifen was encapsulated in poly(ethylene oxide)-modified poly(ε-caprolactone) nanoparticles of an average diameter of 150–250 nm, for targeted delivery of tamoxifen to breast cancer. The primary site of accumulation *in vivo* was the liver; however, 26% of the total drug-loaded nanoparticles were recovered in the tumor at 6 h post-injection, increasing further with time. Extended presence in the circulation was also observed. Therefore, these nanoparticles achieved preferential tumor-targeting and a circulating drug reservoir [42]. Antiestrogens were also incorporated at high amounts in nanocapsules [polymers with an oily core and coupling with PEG] displaying enhanced anti-tumoral activity toward breast cancer cells [43].

7.4.5.5 Cisplatin

Platinum derivatives (cisplatin and carboplatin) are inorganic metallic complexes of the Pt^{2+} cation and pairs of chloride and amino ligands in the cis position for cisplatin. Their mechanism of action resembles that of alkylating agents active on DNA, independently of the cell cycle. Cisplatin, alone or in combination with other drugs, is used to treat testicular, ovarian, bladder, prostate, respiratory carcinomas, and lymphoma, sarcomas and melanoma. Platinum derivatives have a short half-life in the biological environment and are frequently associated with drug resistance in many cancers and, therefore, their entrapment in liposomes has been attempted to enhance efficacy – these formulations are presently under clinical evaluation [44–47].

7.4.5.6 Campthotecins

Campthotecins (irinotecan and topotecan) are cytostatic drugs that act as specific inhibitors of DNA topoisomerase I, active in S phase and on P-glycoprotein (MDR)-positive cells, with activity that depends on exposure time; they induce single-strand lesions in DNA and inhibition of replication. They are in clinical use mainly for advanced colorectal cancer in conjunction with 5-fluorouracil, but are extremely hydrophobic. Lipid-based nanoparticles (100–375 nm) of irinotecan have been prepared and their activity was size-dependent: 375 nm irinotecan-nanoparticles > irinotecan > 100 nm irinotecan-nanoparticles > no treatment [24].

7.4.5.7 Methotrexate

Methotrexate is a cytotoxic antagonist (antimetabolite) of folic acid during the S phase of the cell cycle of actively proliferating cells, acting by competitive inhibition of dihydrofolate reductase and blocking the reduction of dihydrofolic acid (FH_2) into tetrahydrofolic acid (FH_4), and hence the synthesis of pyrimidine and purine bases and amino acids, and of DNA, RNA and protein synthesis. Methotrexate has been entrapped in triblock poly(trimethylene carbonate)-PEG-poly(trimethylene carbonate) copolymer core–shell type nanoparticles (50–160 nm) by ring-opening polymerization [48]. Very recently, in a very elegant approach, polyamidoamine (PAMAM) dendrimers (<5 nm size), conjugated to folic acid for tumor targeting and methotrexate as anticancer agent for tumor treatment, injected i.v. into folate receptor-positive human KB tumor-bearing immunodeficient mice displayed increased anti-tumor activity of methotrexate and markedly decreased toxicity, allowing a tumor drug dosage not possible with the free drug [49].

7.4.6
New Experimental Drugs and Therapies

7.4.6.1 Proteins, Peptides, their Inhibitors and Antagonists

- The cysteine protease cathepsin B inhibitor cystatin, potentially active as an anticancer drug, has been incorporated in poly(lactide-co-glycolide) nanoparticles (300–350 nm size), preserving its cathepsin B inhibitory activity; it was internalized and was cytotoxic for tumor cells whereas free cystatin was not [50, 51].
- Conjugation of an anti-EGF receptor monoclonal antibody to colloidal gold nano-

particles increased their selectivity toward cancer cells versus non-cancer cells, suggesting their utility in both detection and targeted delivery [52].
- Magnetic nanoparticle-loaded anti-Her2 immunoliposomes have been developed for a combination of antibody therapy with hyperthermia [53].
- Increased transmembrane transport has been observed using positively charged, membrane translocation-inducing poly-D-Arg and Tat-peptide crosslinked to dextran nanoparticles (30 nm) [54].

7.4.6.2 New Drugs

- All-trans retinoic acid incorporated in solid lipid nanoparticle powder displayed increased stability, decreased hemolytic toxicity, and efficacy was maintained against human cell lines [55].
- Clinical application of cucurbitacin in poly(lactic acid) nanoparticles (85 nm diameter), as anticancer targeting against metastasis foci of cervical lymph nodes in patients with oral carcinoma, displayed enhanced efficacy, decreased side effects and a long-lasting high concentration of the drug in lymph nodes [56].

7.4.6.3 New Therapeutic Approaches: Photodynamic Therapy (PDT)

Ceramic-based nanoparticles entrapping water-insoluble photosensitizing anticancer drugs have been developed as a novel drug-carrier system for the PDT of tumor cells [57]. Methylene blue, as a photosensitizer and a source of singlet oxygen in PDT, has been encapsulated in polyacrylamide, sol–gel silica or organically modified silicate sub-200 nm nanoparticles. Of these three matrices, polyacrylamide was the most efficient delivery system but its content in methylene blue was low, whereas the opposite was true for the sol–gel nanoparticles. PDT treatment of cells was demonstrated to be efficient [58]. PDT protocols have also been developed using photosensitizers incorporated in PLA-PEG nanospheres and tumor irradiation (EMT-6 cells). However, the large (>900 nm) particle size did not allow high enough intra-tumoral accumulation [59]. Therefore, the nanoplatform represents a functional system for decreasing the side effects in detection and therapy of cancer in PDT protocols.

7.4.7
Gene Therapy

Viral vectors are very efficient but present the inherent risk that the inactivated virus reverts to wild-type and that viruses induce immunogenic reaction. Therefore, attempts have been made to replace viruses by synthetic vectors with improved safety, greater flexibility and easier manufacturing. Synthetic polymers (mainly cyclodextrin-modified, branched polyethyleneimine) are generally cationic molecules to which negatively charged DNA or RNA molecules electrostatically bind, forming polyplexes of a few tens to hundreds of nm in diameter, but with excess cationic charges maintained to favor cell uptake. Serum stability, aggregation, and clearance of polyplexes depend on the packaging polymer and surface modification. Surface modification with a hydrophilic polymer, such as PEG, N-2(hydroxypropyl)methacrylamide, oligosaccharides and sugars or proteins, can increase serum compatibility and provides a basis for chemically grafting target addresses, which include membrane receptors, lectins, transferrin, antibodies, lectins. Targeting efficacy depends on the conjugation chemistry, the length and chemical composition of the spacer between the polyplex and the ligand, the kinetics characteristics and the number of interactions between the ligand and receptor. Intracellular trafficking and endosomal escape are also important factors for targeting approaches.

Polymers used for gene delivery include polylysine, branched polyethyleneimine, poly(amidoamine) dendrimers, membrane-disrupting pH-responsive or pH-degradable acrylate-based polymers, allowing lysosomal escape, peptides, cyclodextrins and non-cytotoxic biodegradable polymers of L-glycolic acid (for a discussion of the advantages/disadvantages of the various polymers see the reviews in Refs. [60–62]). However, whereas substantial progress has been made *in vitro*, their utilization in clinics remains to be evaluated. Moreover, the much lower efficacy of all synthetic vectors than that of viruses makes them generally considered as unacceptable for clinical applications.

7.4.7.1 Nanoparticle for Gene Delivery: Non-chitosan and Chitosan-type Polymers

Progress in gene therapy has relied on the emergence of polymeric and non-polymeric nanoparticles that have been investigated for their ability to deliver genes, mainly encompassing two families: chitosan-related and chitosan-unrelated materials.

Non-chitosan Polymers

- Polyethyleneimine-DNA complexes (39–1200 nm) linked to PEG coated with transferrin or EGF have achieved 10–100× higher tumor selectivity over other organs in mice [24].
- Dextran-SPION plasmid as a model for gene carrier (59 nm effective diameter) has achieved efficient transfer in a human bladder cancer cells line [63].
- Positively charged calcium phosphate nanoparticles (30 nm) were able to transfer foreign DNA with high transfection efficiency and were less cytotoxic than DNA liposomes [64].

Chitosan-type Polymers The inclusion of poly(propyl acrylic acid), which disrupts lipid bilayer membranes at defined pH incorporated in chitosan-DNA plasmid, increased delivery from the endosomal to cytoplasmic compartment for non-viral gene delivery [65].

7.4.8
New Approaches

Recent, new and innovative approaches have been aimed mainly at (a) improving existing systems, by enhancing biocompatibility, biodegradability and biological characteristics of existing systems, and (b) developing novel nanotechnological approaches for drug–nanoparticle preparation as anticancer agents (for more detailed reviews see Refs. [6, 23]).

7.4.8.1 Improvement of Biological Characteristics

Conventional Drug-loaded Nanoparticles These nanoparticles are rapidly opsonized in the circulation and cleared by the RES, mainly macrophages, in the liver, spleen, lungs, bone marrow where drug accumulates. Conventional nanoparticles induce some cytotoxic effects against phagocytes, and drugs accumulate in bone marrow and result in myelosuppression, an unfavorable event. Nephrotoxicity has also been observed. However, the drug-nanoparticle toxicity profile is more favorable than the free drug toxicity profile, in particular for the cardiotoxicity of doxorubicin. The accumulation of nanoparticle-doxorubicin in the lysosomes of Kupffer cells in the liver, but not in tumor cells, acts as a long-term active drug delivery system [66]. For example, in mice treated with doxorubicin-poly(isohexyl cyanoacrylate) (PIHCA) nanospheres [67] the drug accumulates in the liver, spleen and lungs, but a reduction of hepatic metastasis and a longer life span were also observed [68] compared with free doxorubicin. The same was seen for actinomycin D adsorbed on poly(methyl cyanoacrylate) (PMCA) and lung accumulation [69], and for actinomycin D adsorbed on the more slowly degradable poly(ethyl cyanoacrylate) (PECA) and small intestine accumulation [70] or vinblastine incorporated into PECA and accumulation in the spleen [70]. Therefore, as both the polymer composition and the drug chemical characteristics, as well as the location in the nanoparticle (adsorbed or incorporated), are important for tissue localization these factors have been modified.

An ideal drug carrier should be a system that can reside *in vivo* for long periods, targeting particular cell types, compartmentalizing a large set of molecules and releasing them in the appropriate environment at the appropriate rate and dose. Creating a toolbox of molecules that hierarchically assemble in ordered structures, spatially and chemically controlled, is requisite to making them attractive and efficient for encapsulating and delivering drugs. The chemistry used to achieve biomimetic assemblies, which consist of the polymer, a linker and a bioactive molecule, has been reviewed [71]. Di- or tri-block copolymers with low polydispersity can be obtained via anionic polymerization. Subsequent manipulation involves self-assembly at concentrations favoring spherical micelles, controlled crosslinking using radical chemistry to obtain a hydrogel shell and polymer micelle architecture, followed by conjugation with a biological molecule to achieve targeting, such as peptides via a carboxyl end group on the polymer. The most commonly incorporated polymer is PEG [poly(ethylene glycol)], a flexible water-soluble molecule that can be end-functionalized to obtain aldehyde, methacryloyl, hydroxyl, primary amine, acetal, mercapto groups, and copolymerization with other polymers [72–74]. The chemical procedures available have been outlined [71].

Factors controlling the rate of drug release are not well understood but depend on the assembly morphology, drug molecular weight, chemical composition, etc. These factors have also been addressed. Biodegradable polymers are preferred for controlled drug delivery systems, and are made of natural or synthetic polymers, which have the advantage over natural polymers in that they can be tailored to obtain defined properties. They must match the mechanical properties and degradation rates required for the application. Commonly used biodegradable polymers for biomedical applications include polyglycolic acid, L-, D-, DL-polylactic acid, poly(caprolactone), poly(DL-lactide-co-glycolide), and poly(vinyl alcohol). These polymers have features such as controllable mechanical properties, controllable degradation rates, minimal toxicity and immune responses [75].

Long-circulating Biocompatible and Biodegradable Drug-loaded Nanoparticles To escape the RES, stealth nanoparticles that are "invisible" to macrophages, and which are long-lasting in the blood compartment, need to be designed. For this, modification of the surface (hydrophilic/hydrophobic) and a small size (<100 nm) of the colloidal carrier are essential [23]. The hydrophilicity of the carrier and sustained drug release are also important. A previous breakthrough was the use of PEG or polysaccharides to coat the nanoparticle surface and, thereby, repel plasma proteins. Hydrophilic polymers are either adsorbed at the surface of nanoparticles or are block copolymers. Finally, covalent linkage of amphiphilic copolymers is the preferred way to obtain a protective hydrophilic coat, using poly(lactic) (PLA), poly(caprolactone), or poly(cyanoacrylate) copolymers chemically coupled to PEG. Controlled drug-delivery systems are generally based on either diffusion or degradation of the polymer. The choice of polymer depends on potential interactions with the drugs, surface characteristics, the hydrophilic–hydrophobic balance, surface charges, and the biological properties of the target. Novel approaches to solve the drug release problems have been addressed.

The preparation and characterization of biodegradable/bioerodible polymers for the controlled targeted release of proteic drugs (interferon or growth factors for tissue engineering) have been described. *Bioerodible polymeric matrices* are hemiesters of alternating copolymers of maleic anhydride with alkyl vinyl ethers of oligo(ethylene glycol). Hydrophilic shell coating to minimize opsonization was achieved with grafted β-cyclodextrins. Coprecipitation was used for formulation, based on dropwise addition of synthetic polymer in water-miscible organic solvent to aqueous protein solution under stirring, followed by the addition of the glycolipid. Particles 130–150 nm diameter were obtained, with β-galactose residues exposed at the surface. *Biodegradable polymer matrices* solve the issue of removal of the device after drug delivery. Among them, poly(malolactonates) are biocompatible, degrading to malic acid, and contain reactive side-chain carboxyl groups that can be esterified [76] for adjustment of the hydrophilic–hydrophobic balance [76–81]. The starting components are commercially available and nanoparticles of 100–150 nm can be obtained by coprecipitation (reviewed in Ref. [82]).

The effects of the following surface modifications have also been evaluated:

- Poly(lactic-co-glycolic acid) nanoparticles coated with vitamin E TPGS [82, 83], or polysaccharide-decorated nanoparticles [84].
- Poly(alkyl cyanoacrylate) nanoparticles [85].
- Amphiphilic poly(lactic)-pluronic block copolymers, which can release either hy-

drophobic anthracene and hydrophilic procaine drugs due to slow hydrolytic degradation of poly(lactic acid) [86].
- Poly(ethylene oxide)-β-poly(N-isopropylacrylamide) nanoparticles with crosslinked cores as hydrophobic drug carriers [87].

7.4.8.2 New Technological Approaches

Local Administration and Long-time Delivery Small particles injected locally in the tumor vicinity infiltrate gradually the lymphatics, representing a tool for lymph node metastases, and locoregional adjuvant therapy, for breast cancer for example. In this model the size and choice of the carrier are of fundamental importance to avoid clogging of small lymphatic vessels. For this purpose magnetic systems would be a good choice (see below).

Polymer gels have been designed for controlled release and as systems for modulated delivery. The hydrophobic–hydrophilic balance of a gel carrier can be modulated to provide useful diffusion characteristics for periods up to months. Current polymer network drug delivery systems incorporate the pharmaceutical agent by imbibition, equilibrium partitioning after the network is formed, before or after polymerization depending on the drug stability to UV light. Reference [88] gives a more extensive review of the features of these devices.

New Materials Evaluated as Carriers: Thermoresponsive or Oxidation-sensitive Materials The preparation and characterization of "intelligent" core–shell nanoparticles has been described. Thermoresponsive, pH-responsive and biodegradable nanoparticles of poly(D,L-lactide)-*graft*-poly(N-isopropyl acrylamide-*co*-methacrylic acid) have been developed as a core–shell type nanoparticle with high drug loading capacity. A hydrophilic outer shell and a hydrophobic inner core, with a phase transition above 37 °C, have been prepared. Heating above the phase transition temperature or pH modification caused leakage of the drugs, demonstrating the potential of these nanoparticles as drug carriers for intracellular delivery of anticancer drugs [89]. Monodisperse polymeric nanoparticles of polyacrylic/isopropyl acrylamide, prepared by seed-and-feed method, displayed inverse thermogellation at 33 °C. Therefore, drug–polymer mixtures that are liquids at room temperature become a gel at body temperature without chemical reaction, allowing the sustained release of drugs [90]. Oxidation-sensitive polymeric nanoparticles of crosslinked polysulfides have also been developed as oxidation-sensitive vehicles for hydrophilic or hydrophobic drugs able to release drugs in the more oxidative environment of cancer and inflammation [91].

7.4.9
Superparamagnetic Iron Oxide Nanoparticles (SPIONs) as Magnetic Drug Nanovectors

Nanovectors able to both detect cancer and deliver drugs have also been developed, mainly using the outstanding superparamagnetic properties of iron-oxide nanoparticles. Superparamagnetic iron oxide nanoparticles (SPIONs) are usually used as magnetic nanovectors since their superparamagnetic properties, depending on their nanoscale size, make them magnetic in the presence of an external magnetic field, but no magnetic remanence remains after removal of this external field. Magnetic nanoparticles offer additional attractive possibilities in biomedicine since their magnetic properties allow them to be manipulated by an external magnetic field gradient, further improving chemical and biological drug delivery strategies, in addition to magnetic resonance imaging (MRI) contrast enhancement detection of their sites of localization. Previous experiments using this approach have been performed by some groups with magnetic particles or magnetic liposomes, ranging from 10 µm to 100 nm, loaded with chemotherapeutic agents [92–94]. Some positive effects were obtained, in particular a decreased general toxicity, due to lower dosages, of the agents. This approach has the advantage of being non-invasive and efficient.

Biocompatible superparamagnetic nanoparticles have been developed for *in vivo* biomedical application, mainly in magnetic resonance imaging [63, 95, 96], and have been only preclinically evaluated in the tissue-specific release of therapeutic agents [97]. Dextran-SPIONs have been used in gene transfer experiments in a human bladder cancer cell line [98]. SPIONs combined with albumin [99] or as liposomes with antibodies targeting Her2 have been used for combination of antibody therapy with hyperthermia [53]. Surface modification of the PEG film of SPIONs with folic acid decreased their uptake by mouse macrophages and increased their uptake by human cancer cells [100]. Anionic SPIONs, coated with albumin, were taken up by cells [101], and anionic to cationic surface charge in the acidic lysosomal compartment may allow escape from the endo-lysosomal compartment to the cytosol [102]. Dextran-coated SPIONs of 150 nm, but not of 10 nm, were taken up by macrophages involving type I and II scavenger receptor SR-A-mediated endocytosis [103]. For the latter application, the internalization of nanoparticles into specific cells is the critical step, but is limited by non-specific targeting and a low efficiency of internalization of the endocytosed ligands grafted on the nanoparticles [104]. Clearly, this step needs to be improved to enhance efficiency. Few published studies have addressed the uptake by cells or the cell-surface binding of functionalized superparamagnetic nanoparticles [105]. Therefore, a firm theoretical foundation of magnetic drug targeting is still lacking.

Interactions of magnetic nanoparticles with cells may have an impact on cell functions. Macaque T cells labeled with monocrystalline SPIONs following adsorptive pinocytosis or receptor-mediated endocytosis and which localized in the cytoplasm did not cause any measurable effects on T cell functions [106]. Conversely, transferrin magnetic nanoparticles localized to the cell membrane without inducing endocytosis, but this surface localization induced the expression of several genes of the cytoskeleton and cell signaling pathways [107].

7.5
Targeting

Detecting and treating cancer by targeted means at the earliest possible stage of cancer will improve the quality of life and the life expectancy of patients. The use of nanoparticles has a bright future towards achieving these goals. Probably the most promising features of nanovectors are related to the possibility of modifying their properties, mainly their surface, to achieve organ-, tissue- and cell-specific and -selective delivery of anticancer drugs, to increase drug efficacy and decrease side effects, and ultimately improve the quality of life and survival of patients. Technical problems of developing targeting nanoparticles include the increased complexity/size of the nanoparticles, as well as the increased risks of adverse reactions, while advantages include the fact that more drug will reach the target, selectivity is increased, the delivery of multiple agents at the same site will become possible for targeted combination therapies, and the ability to bypass biological barriers.

Cancer targeting using drug-loaded nanovectors may be achieved in three ways:

1. Using the tissue characteristics of cancer, mainly the leaky vessels associated with many human cancers – what has been called passive targeting.
2. Using the characteristics of cancer cells, mainly increased metabolism and active cell proliferation, and *active circumstantial targeting*.
3. Using molecular markers overexpressed by either cancer cells or cancer-associated cells, mainly the neo-angiogenic endothelium of cancer tissue, and *active molecular targeting*.

Active targeting of drugs to cancer is probably the most challenging bio-nanotechnological approach. To achieve these goals it will be necessary to:

1. Define valid targets: markers of tumor and tumor-associated cells and vasculature.
2. Develop the necessary chemistry to bind drugs to nanovectors: validate synthetic routes.
3. Release the drugs at the right place and time from the carrier: develop intelligent linkers.
4. Achieve intracellular delivery: define cell penetration means and organelle addresses.

The greatest challenge in designing strategies for targeting is to define the optimal targeting system, able to selective bind to tumor or tumor-associated cells, then to trigger internalization of the drug cargo (for a more extensive review see Ref. [24]). Most studies have used a modification of the nanoparticle surface, e.g.,

- covalently-linked antibodies,
- tumor vascular markers (RGD and integrins, VEGF-R, EGF-R),
- and permeation-enhancing agents.

The endothelium is a very attractive tissue for drug targeting using nanoparticles in cancer. Surface properties of the nanocarrier may provide passive–active targeting to endothelium, e.g., polystyrene nanoparticles arrest in bone marrow endothelium, followed by receptor-mediated endocytosis [108], while polysorbate 80-coated nanoparticles arrest in the blood–brain barrier vasculature [16, 28]. Cationic lipo-

Tab. 7.3. Some examples of linkers.

Linker	Ref.
Peptides	
Peptidyl spacers designed for cleavage by the lysosomal thiol-dependent (cysteine) proteases (Gly-Phe-Leu-Gly)	153, 154
Glu-Lys (cleaved with bovine trypsin)	155
Leu, Ala or Leu-Ala	154
Trp-Ser-Gln	156
Valine (less prone to ester hydrolysis)	157
Chemical	
4-Acyl-oxy-3-carboxybenzyloxy moiety (recognized by a specific lipase)	155
C4-Chain spacer tetramethylenediamine (TMDA, MW = 88)	158
Diaminated poly(ethylene glycol) (PEG, MW 3400)	158
Ester linker, formed between a hydroxyl-containing drug and a mercaptoacid	159
Ethylene glycobis(succinimidylsuccinate) (EGS): used to link particles having amino groups with each other via an ester linkage	160
Glutaraldehyde (crosslinker)	161
N-Hydroxysuccinimidyl active ester linker (synthesis with dianhydride)	74
Phenylacetyl derivatives, phenylacetamide moiety	162, 155
Scissile linkers	155
Succinic acid, glutaric acid	163
3-Sulfanylpropionyl	159
4-Sulfanylbutyryl	159
Triazene	164, 165
Dual linker systems	166
Coupling agents for peptide synthesis: 1-ethyl-3-(3-dimethylaminopropyl)carbodiimide (EDCI) N-succinimidyl 3-(2-pyridyldithio) propionate (SPDP) N-hydroxysuccinimide N,N'-methylene bis-acrylamide	115

the drug. Thus it is necessary to maintain the drugs for a long period at the tumor site. Nanoparticles with long, controlled drug release may thus ameliorate this defect in anticancer treatment.

A second important mechanism of resistance of cancer cells to therapeutic drugs is the simultaneous resistance mechanisms linked to ATP-driven efflux pumps (MDR, Pgp) able to efflux positive drugs (such as doxorubicin) out of cells. Resistance may appear at the onset of therapy, or later during treatment. It is mainly linked to ATP-driven cell membrane efflux pumps on the tumor cells or the tumor-associated vascular cells, or as detoxifying enzymes expressed by these cells. Increased efficacy of PIHCA-doxorubicin was dependent on an increased adsorption of nanoparticles to the cell surface, and to increased diffusion of doxorubicin forming ion pairs between the drug and cyanoacrylic monomer degradation product [139]. This effect was initially observed with cyanoacrylate polymers. Inhibitors of Pgp (verapamil, amiodarone, cyclosporine), or the encapsulation of drugs into colloidal carriers, shield drugs from Pgp, which recognizes drugs associated to the membrane but not in the cytoplasm. Inhibitors for pumps, or for detoxifying enzymes, may be co-administered, or anticancer drugs delivered using nanoparticles able to overcome the resistance of pumps. Since the amino sugar of doxorubicin is a substrate for Pgp, doxorubicin was covalently coupled to PEG-poly(Asp) (MW 14 400 and 3500 kDa, respectively) block copolymer before micelle formation or to gelatin, but without success, probably due to poor diffusion of the drug out of the carrier. Co-encapsulation of the Pgp inhibitor cyclosporin with doxorubicin into PICBA inhibited efficiently growth of doxorubicin-resistant tumor cells, by continuous and simultaneous release of both drugs by polymer biodegradation [140, 141]. Doxorubicin encapsulated in poly(alkyl cyanoacrylate) PIHCA nanoparticles was able to kill Pgp-positive tumor cells, but not PIBCA nanospheres, which are more rapidly degraded and do not enter cells [139, 142–144]. This effect requires direct contact between the carrier and tumor cells and was mediated by poly(alkyl cyanoacrylate) degradation products, which formed ion-pairs with the drug, and the doxorubicin concentration was increased in tumor cells. Pgp remains active. The same effect was obtained by encapsulating doxorubicin in the core and expressing cyclosporin A at the surface of the nanoparticles. Pluronic block copolymer may inhibit the efflux pumps (MDR/P-glycoproteins), and their gene expression, on the endothelial cells forming the blood–brain barrier (BBB) [145, 146] that protects the brain from unwanted chemicals, but which are also involved in the resistance of brain tumors to chemotherapeutic agents. Therefore, they may be of interest in the treatment of brain cancer only if they can be made selective to tumor-associated endothelium.

7.7
Toxicity Issues

For more extensive reviews see Refs. [5, 6]. Nanoparticles interact differently with organs, tissues and cells than do larger particles made of the same components.

Therefore, the evaluation of toxicity performed with larger particles cannot be extrapolated to nanoparticles without control. The hazards of inhaled micro- and nanoparticles in air pollution are well established, epidemiological and toxicological studies have coherently demonstrated pro-inflammatory and prothrombotic adverse effects in diverse organs. There is virtually no toxicological data available, for patients, researchers or medical workers concerning the new types of nanoparticles under development for drug delivery. Size and surface modification may modify biocompatibility and biodistribution, and a combination of drugs, devices and biological agents may behave differently than each agent separately – therefore combination approval must be obtained from drug control agencies. The exact mechanisms of interaction of nanovectors and cells have been determined in very few situations. Therefore, coordinated studies will be rapidly needed to address these issues. Nanostructures can minimize solubility and stability problems, and improve the negative impact of drugs to collateral non-tumoral tissues and organs. However, nanomaterials themselves may be cytotoxic [147, 148] or induce and/or potentiate cell death [149, 150] or immunogenic reactions, or nanoparticle aggregates may clog small blood vessels. For example, micelles of cisplatin differently induced gene expression than cisplatin alone [149], and degradation products from poly(L-lactic acid) were cytotoxic for immune cells [150].

Gene therapy with viruses has had poor success and many problems have been linked to immunogenic reaction to the viral vector constituents, and random integration in the genome. Therefore, cationic nanoparticle vectors have been designed to complete viral vectors. Polycations are cytotoxic for cells, inducing mitochondria-mediated necrosis and/or apoptosis or membrane destabilization and pore-formation meditated by interactions of polycations with negatively-charged cell-surface glycoproteins or actin [151]. The interactions of cationic polymers with mitochondria need to be better understood, in particular with the proteins of the bcl-2 family. Hypersensitivity reactions secondary to complement activation and induced by infusion of PEG-modified liposomes [152] may be a potential problem with these materials. Careful design of the polymer formulation and surface functionalization is needed to reduce these side effects. Therefore, nanoparticle design and polymer formulation and functionalization for gene therapy *in vivo* must be carefully optimized before such treatments can be envisioned.

To gain wide acceptance of nanovectors as anticancer delivery agents, the following toxicity issues need to be addressed:

- The ultimate biological fate of nanomaterials, and their degradation products, particularly non-biodegradable nanomaterials such as functionalized poly(ethylene glycol).
- The immunological and pharmacological activities and toxicities.
- Possible interferences with cellular machineries, gene expression, protein processing.
- Short- and long-term consequences of exposure to nanovectors.
- How *in vitro* studies translate to *in vivo* application.

7.8
Conclusions

In summary, medical benefits can be expected within the next 10 years from the branch of nanotechnology named "Nanomedicine". However, for drug delivery, nanocarriers have to become smarter, and the biophysical, biochemical and biological processes and the mechanisms associated with the interaction of nanoparticles with living tissue, at large, must be understood in detail to design optimized materials for defined therapeutic goals. This includes carrier stability, targeting, extracellular and intracellular drug release mechanisms, overcoming mechanisms of resistance to biological barriers and to anticancer agents, and understanding the immunogenic reaction mechanisms of the nanodevices, and toxicity issues in great detail. The future of nanomedicine will in part depend on the rational design of smart nanoparticles.

7.8.1
Opportunities and Challenges of Nanomedicine in Cancer

1. Passive targeting effects can enhance the amounts of drug at tumor sites, but circumstantial or active targeting can be designed to improve the selectivity and specificity of drug delivery; therefore, combining these approaches may greatly increase the quality of cancer treatment.
2. Targeted drug delivery systems can improve the therapeutic index of approved drugs.
3. New drugs that are macromolecules, proteins, peptides, oligonucleotides or plasmids will need such delivery systems to achieve an efficient local concentration.
4. Externally activated (ultrasound, irradiation, magnetic field, photodynamic therapy, etc.) drug delivery of therapeutic agents can be designed to control the timing of drug delivery and administration schedules, without multiple injections.
5. Overcoming of the biological barriers:
   - Including tight junctions (BBB) and epithelial barriers.
   - Using permeation enhancers: tight junction opening agents, osmotic shock.
   - Avoiding the RES: using surface modification of nanoparticles, e.g., by poly(ethylene glycol)s.
   - Controlling sensitization reactions: protein-dendrimers are highly immunogenic.
   - Overpassing intratumoral increased osmotic barriers: limit to the diffusion of nanoparticles and drugs.
6. Challenges also include the control of drug release and bioavailability at the tumor site, enhancing selective targeting, and achieving controlled and efficient intracellular delivery, together with acceptable toxicological hazards.

Therefore, an ideal nanovector should look as shown in Fig. 7.2. One representation of the ideal goal is an injectable nanoparticle for targeted drug delivery, without, or with highly reduced, collateral toxicity, and with a reporter device.

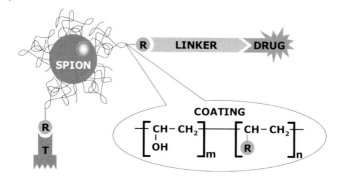

Fig. 7.2. Ideal drug delivery nanovector, containing a reporter system (SPION), a targeting agent (T) linked to the polymer, and a drug linked to the polymer via a releasing linker.

References

1 Gordon, E. M., Hall, F. L., Nanotechnology blooms, at last, *Oncol. Rep.*, **2005**, 13, 1003–1007.
2 Davis, S. S., Biomedical applications of nanotechnology – Implications for drug targeting and gene therapy, *Trends Biotechnol.*, **1997**, 15, 217–224.
3 Duncan, R., The dawning era of polymer therapeutics, *Nat. Rev. Drug Discov.*, **2003**, 2, 347–360.
4 Rabinow, B. E., Nanosuspensions in drug delivery. *Nat. Rev. Drug Discov.*, **2004**, 3, 785–796.
5 Moghimi, S. M., Hunter, A. C., Murray, J. C., Nanomedicine: Current status and future prospects, *FASEB J.*, **2005**, 19, 311–330.
6 Ferrari, M., Cancer nanotechnology: Opportunities and challenges, *Nat. Rev., Cancer*, **2005**, 5, 161–171.
7 Satchi-Fainaro, R., Puder, M., Davies, J. W., Tran, H. T., Sampson, D. A., Greene, A. K., Corfas, G., Folkman, J., Targeting angiogenesis with a conjugate of HPMA copolymer and TNP-470, *Nat. Med.*, **2004**, 10, 255–261.
8 Batrakova, E. V., Dorodnych, T. Y., Klinskii, E. Y., Kliushnenkova, E. N., Shemchukova, O. B., Goncharova, O. N., Arjakov, S. A., Alakhov, V. Y., Kabanov, A. V., Anthracyclin antibiotics noncovalently incorporated into block copolymer micelles: In vivo evaluation of anticancer activity, *Br. J. Cancer*, **1996**, 74, 1545–1552.
9 Danson, S., Ferry, D., Alakhov, V., Margison, J., Kerr, D., Jowle, D., Brampton, M., Halbert, G., Ranson, M., Phase I dose escalation and pharmacokinetic study of pluronic polymer-bound doxorubicin (SP1049C) in patients with advanced cancer, *Br. J. Cancer*, **2004**, 90, 2085–2091.
10 Yasugi, K., Nagasaki, Y., Kato, M., Kataoka, K., Preparation and characterization of polymer micelles from poly(ethylene glycol)-poly(D, L-lactide) block copolymers as potential drug carrier, *J. Controlled Release*, **1999**, 62, 89–100.
11 Nakanishi, T., Fukushima, S., Okamoto, K., Suzuki, M., Matsumura, Y., Yokoyama, M., Okano, T., Sakurai, Y., Kataoka, K., Development of the polymer micelle

carrier system for doxorubicin, *J. Controlled Release*, **2001**, 74, 295–302.
12 ALLEN, T. M., CULLIS, P. R., Drug delivery systems: Entering the mainstream, *Science*, **2004**, 303, 1818–1822.
13 LÜBBE, A. S., ALEXIOU, C., BERGEMANN, C., Clinical applications of magnetic drug targeting, *J. Surg. Res.*, **2001**, 95, 200–206.
14 ULBRICH, K., SUBR, V., Polymeric anticancer drugs with pH-controlled activation, *Adv. Drug Deliv. Rev.*, **2004**, 56, 1023–1050.
15 KIM, C. K., LIM, S. J., Recent progress in drug delivery systems for anticancer agents, *Arch. Pharm. Res.*, **2002**, 25, 229–239.
16 KREUTER, J., Influence of the surface properties on nanoparticle-mediated transport of drugs to the brain, *J. Nanosci. Nanotechnol.*, **2004**, 4, 484–488.
17 JULYAN, P. J., SEYMOUR, L. W., FERRY, D. R., DARYANI, S., BOIVIN, C. M., DORAN, J., DAVID, M., ANDERSON, D., CHRISTODOULOU, C., YOUNG, A. M., HESSLEWOOD, S., KERR, D. J., Preliminary clinical study of the distribution of HPMA copolymers bearing doxorubicin and galactosamine, *J. Controlled Release*, **1999**, 57, 281–290.
18 WYMAN, T. B., NICOL, F., ZELPHATI, O., SCARIA, P. V., PLANK, C., SZOKA, F. C., Design, synthesis and characterization of a cationic peptide that binds to nucleic acids and permeabilize bilayers, *Biochemistry*, **1997**, 36, 3008–3017.
19 ETRYCH, T., JELINKOVA, M., LHOVA, B., ULBRICH, K., New HPMA copolymers containing doxorubicin bound via pH-sensitive linkage: Synthesis and preliminary in vitro and in vivo biological properties, *J. Controlled Release*, **2001**, 73, 89–102.
20 LEROUX, J. C., ROUX, E., LEGARREC, D., HONG, K., DRUMMOND, D. C., N-isopropylacrylamide copolymers for the preparation of pH-sensitive liposomes and polymeric micelles, *J. Controlled Release*, **2001**, 72, 71–84.
21 STRAUBINGER, R. M., pH-sensitive liposomes for delivery of macromolecules into cytoplasm of culture cells, *Methods Enzymol.*, **1993**, 221, 361–376.
22 PRASMICKAITE, L., HOGSET, A., BERG, K., Evaluation of different photosensitizers for use in photochemical gene transfection, *Photochem. Photobiol.*, **2001**, 73, 388–395.
23 BRIGGER, I., CHAMINADE, P., MARSAUD, V., APPEL, M., BESNARD, M., GURNY, R., RENOIR, M., COUVREUR, P., Tamoxifen encapsulation within polyethylene glycol-coated nanospheres. A new antiestrogen formulation, *Int. J. Pharm.*, **2001**, 214, 37–42.
24 BRANNON-PEPPAS, L., BLANCHETTE, J. O., Nanoparticle and targeted systems for cancer therapy, *Adv. Drug Deliv. Rev.*, **2004**, 56, 1649–1659.
25 OH, I., LEE, K., KWON, H. Y., LEE, Y. B., SHIN, S. C., CHO, C. S., KIM, C. K., Release of adriamycin from poly(gamma-benzyl-L-glutamate)/poly(ethylene oxide) nanoparticles, *Int. J. Pharm.*, **1999**, 181, 107–115.
26 CHEN, J. H., LING, R., YAO, Q., WANG, L., MA, Z., LI, Y., WANG, Z., XU, H., Enhanced antitumor efficacy on hepatoma-bearing rats with adriamycin-loaded nanoparticles administered into hepatic artery, *World J. Gastroenterol.*, **2004**, 10, 1989–1991.
27 SOMA, C. E., DUBERNET, C., BARRATT, G., BENITA, S., COUVREUR, P., Investigation of the role of macrophages on the cytotoxicity of doxorubicin and doxorubicin-loaded nanoparticles on M5076 cells in vitro, *J. Controlled Release*, **2000**, 68, 283–289.
28 GULYAEV, A. E., GELPERINA, S. E., SKIDAN, I. N., ANTROPOV, A. S., KIVMAN, G. Y., KREUTER, J., Significant transport of doxorubicin into the brain with polysorbate 80-coated nanoparticles, *Pharm. Res.*, **1999**, 16, 1564–1569.
29 OH, K. S., LEE, K. E., HAN, S. S., CHO, S. H., KIM, D., YUK, S. H., Formation of core/shell nanoparticles with a lipid core and their application as a drug delivery system, *Biomacromoles*, **2005**, 6, 1062–1067.
30 SERPE, L., CATALANO, M. G., CAVALLI,

R., Ugazio, E., Bosco, O., Canaparo, R., Muntoni, E., Frairia, R., Gasco, M. R., Eandi, M., Zara, G. P., Cytotoxicity of anticancer drugs incorporated in solid lipid nanoparticles on HT-29 colorectal cancer cell line, *Eur. J. Pharm. Biopharm.*, **2004**, 58, 673–680.

31 Barraud, L., Merle, P., Soma, E., Lefrancois, L., Guerret, S., Chevallier, M., Dubernet, C., Couvreur, P., Trepo, C., Vitvitski, L., Increase of doxorubicin sensitivity by doxorubicin-loading into nanoparticles for hepatocellular carcinoma cells in vitro and in vivo, *J. Hepatol.*, **2005**, 42, 736–743.

32 ten Tije, A. J., Verweij, J., Loos, W. J., Sparreboom, A., Pharmacological effects of formulation vehicles: Implications for cancer chemotherapy, *Clin. Pharmacokin.*, **2003**, 42, 665–685.

33 Wang, Y. M., Sato, H., Dachi, I., Horikoshi, I., Preparation and characterization of poly(lactic-co-glycolic acid) microspheres for targeted delivery of novel anticancer agent, taxol, *Chem. Pharm. Bull.*, **1996**, 44, 1935–1940.

34 Si-Shen, F., Guofeng, H., Effect of emulsifiers on the controlled release of paclitaxel/taxol) from nanospheres of biodegradable polymer, *J. Controlled Release*, **2001**, 71, 53–69.

35 Fonseca, CM, Simoes, S., Gaspar, R., Paclitaxel-loaded PLGA nanoparticles: Preparation, physicochemical characterization, and in vitro antitumoral activity, *J. Controlled Release*, **2002**, 83, 273–286.

36 Mu, L., Feng, S. S., A novel controlled release formulation for the anticancer drug paclitaxel (taxol):PLGA nanoparticles containing vitamin E TPGS, *J. Controlled Release*, **2003**, 86, 33–48.

37 Chen, D. B., Yang, T. Z., Lu, W. L., Zhang, Q., In vitro and in vivo study of two types of long-circulating solid lipid nanoparticles containing paclitaxel, *Chem. Pharm. Bull.*, **2001**, 49, 1444–1447.

38 Feng, S. S., Mu, L., Win, K. Y., Huang, G., Nanoparticles of biodegradable polymers for clinical administration of paclitaxel, *Curr. Med. Chem.*, **2004**, 11, 413–424.

39 Dong, Y., Feng, S. S., Methoxy poly(ethylene glycol)-poly(lactide) (MPEG-PLA) nanoparticles for controlled delivery of anticancer drugs, *Biomaterials*, **2004**, 25, 2843–2849.

40 Hamaguchi, T., Matsumura, Y., Suzuki, M., Shimizu, K., Goda, R., Nakamura, I., Nakatomi, I., Yokoyama, M., Kataoka, K., Kakizoe, T., NK105, a paclitaxel-incorporating micellar nanoparticle formulation, can extend in vivo antitumour activity and reduce the neurotoxicity of paclitaxel, *Br. J. Cancer*, **2005**, 92, 1240–1246.

41 Li, S., Jiang, W. Q., Wang, A. X., Guan, Z. Z., Pan, S. R., Studies on 5-FU/PEG-PBLG nano-micelles: Preparation, characteristics, and drug releasing in vivo, *Aizheng*, **2004**, 23, 381–385.

42 Shenoy, D. B., Amiji, M. M., Poly(ethylene oxide)-modified poly(epsilon-caprolactone) nanoparticles for targeted delivery of tamoxifen in breast cancer, *Int. J. Pharm.*, **2005**, 293, 261–270.

43 Maillard, S., Ameller, T., Gauduchon, J., Gougelet, A., Gouilleux, F., Legrand, P., Marsaud, V., Fattal, E., Sola, B., Renoir, J. M., Innovative drug delivery nanosystems improve the anti-tumor activity in vitro and in vivo of anti-estrogens in human breast cancer and multiple myeloma, *J. Steroid Biochem. Mol. Biol.*, **2005**, 94, 111–121.

44 Avgoustakis, K., Beletsi, A., Panagi, Z., Kleptsanis, P., Karydas, A. G., Ithakissios, D. S., PLGA-mPEG nanoparticles of cisplatin: In vitro nanoparticles degradation, in vitro drug release and in vivo drug residence in blood properties, *J. Controlled Release*, **2002**, 79, 123–135.

45 Lu, C., Perez-Soler, R., Piperdi, B., Walsh, G. L., Swisher, S. G., Smythe, W. R., Shin, H. J., Ro, J. Y., Feng, L., Truong, M., Yalamanchili, A., Lopez-Berestein, G., Hong, W. K., Khobar, A. R.,

SHIN, D. M., Phase II study of a liposome-entrapped cisplatin analog (L-NDDP) administered intrapleurally and pathologic response rates in patients with malignant pleural mesothelioma, *J. Clin. Oncol.*, **2005**, 23, 3495–3501.

46 MARR, A. K., KURZMAN, I. D., VAIL, D. M., Preclinical evaluation of a liposome-encapsulated formulation of cisplatin in clinically normal dogs, *Am. J. Vet. Res.*, **2004**, 65, 1474–1478.

47 HOVING, S., VAN TIEL, S. T., EGGERMONT, A. M., TEN HADEN, T. L., Effect of low-dose tumor necrosis factor-alpha in combination with STEALTH liposomal cisplatin (SPI-077) on soft-tissue- and osteosarcoma-bearing rats, *Anticancer Res.*, **2005**, 25, 743–750.

48 ZHANG, Y., ZHUO, R. X., Synthesis and drug release behavior of poly(trimethylene carbonate)-poly(ethylene glycol)-poly(trimethylene carbonate) nanoparticles, *Biomaterials*, **2005**, 26, 2089–2094.

49 KUKOWSKA-LATALLO, J. F., CANDIDO, K. A., CAO, Z., NIGAVEKAR, S. S., MAJOROS, I. J., THOMAS, T. P., BALOGH, L. P., KHAN, M. K., BAKER, J. R., Nanoparticles targeting of anticancer drug improves therapeutic response in animal model of human epithelial cancer, *Cancer Res.*, **2005**, 65, 5317–5324.

50 CEGNAR, M., KOS, J., KRISTL, J., Cystatin incorporated in poly(lactide-co-glycolide) nanoparticles: Development and fundamental studies on preservation of its activity, *Eur. J. Pharm. Sci.*, **2004**, 22, 357–364.

51 CEGNAR, M., PREMZL, A., ZAVASNIK-BERGANT, V., KRISTL, J., KOS, J., Poly(lactide-co-glycolide) nanoparticles as a carrier system for delivering cysteine protease inhibitor cystatin into tumor cells, *Exp. Cell Res.*, **2004**, 301, 223–231.

52 EL-SAYED, I. H., HUANG, X., EL-SAYED, M. A., Surface plasmon resonance scattering and absorption of anti-EGFR antibody conjugated gold nanoparticles in cancer diagnostics: Applications in oral cancer, *NanoLetters*, **2005**, 5, 829–834.

53 ITO, A., KUGA, Y., HONDA, H., KIKKAWA, H., HORIUCHI, A., WATANABE, Y., KOBAYASHI, T., Magnetite nanoparticle-loaded anti-HER2 immunoliposomes for combination of antibody therapy with hyperthermia, *Cancer Lett.*, **2004**, 212, 167–1675.

54 KOCH, A. M., REYNOLDS, F., MERKLE, H. P., WEISSLEDER, R., JOSEPHSON, L., Transport of surface-modified nanoparticles through cell monolayers, *Chembiochem*, **2005**, 6, 337–345.

55 LIM, S. J., LEE, M. K., KIM, C. K., Altered chemical and biological activities of all-trans retinoic acid incorporated in solid lipid nanoparticle powders, *J. Controlled Release*, **2004**, 100, 53–61.

56 YANG, K., WEN, Y., WANG, C., Clinical application of anticancer nanoparticles targeting metastasis foci of cervical lymph nodes in patients with oral carcinoma, *Hua Xi Kou Qiang Yi Xue Za Zhi*, **2003**, 21, 447–450.

57 ROY, I., OHULCHANSKYY, T. Y., PUDAVAR, H. E., BERGEY, E. J., OSEROFF, A. R., MORGAN, J., DOUGHERTY, T. J., PRASAD, P. N., Ceramic-based nanoparticles entrapping water-insoluble photosensitizing anticancer drugs: A novel drug-carrier system for photodynamic therapy, *J. Am. Chem. Soc.*, **2003**, 125, 7860–7865.

58 TANG, W., XU, H., KOPELMAN, R., PHILBERT, M. A., Photodynamic characterization and in vitro application of methylene blue-containing nanoparticle platforms, *Photochem. Photobiol.*, **2005**, 81, 242–249.

59 ALLEMAN, E., ROUSSEAU, J., BRASSEUR, N., KUDREVICH, S. V., LEWIS, K., VAN LIER, J. E., Photodynamic therapy of tumors with hexadecafluoro zinc phthalocyanine formulated in PEG-coated poly(lactic acid) nanoparticles, *Int. J. Cancer*, **1996**, 66, 821–824.

60 PACK, D. W., HOFFMAN, A. S., PUN, S., STAYTON, P. S., Design and development of polymers for gene

therapy, *Nat. Rev. Drug Discov.*, **2005**, 4, 581–593.
61 BRIGGER, I., DUBERNET, C., COUVREUR, P., Nanoparticles in cancer therapy and diagnosis, *Adv. Drug Deliv. Rev.*, **2002**, 54, 631–651.
62 KUMAR, R. M., HELLERMANN, G., LOCKEY, R. F., MOHAPATRA, S. S., Nanoparticle-mediated gene delivery: State of the art, *Expert Opin. Biol. Ther.*, **2004**, 4, 1213–1224.
63 CAO, Z. G., ZHOU, S. W., SUN, K., LU, X. B., LUO, G., LIU, J. H., Preparation and feasibility of superparamagnetic dextran iron oxide nanoparticles as gene carrier, *Aizheng*, **2004**, 23, 1105–1109.
64 LIU, T., TANG, A., ZHANG, G., CHEN, Y., ZHANG, J., PENG, S., CAI, Z., Calcium phosphate nanoparticles as a novel nonviral vector for efficient transfection of DNA in cancer gene therapy, *Cancer Biother. Radiopharm.*, **2005**, 20, 141–149.
65 KIANG, T., BRIGHT, C., CHEUNG, C. Y., STAYTON, P. S., HOFFMAN, A. S., LEONG, K. W., Formulation of chitosan-DNA nanoparticles with poly(propyl acrylic acid) enhances gene expression, *J. Biomater.*, **2004**, 15, 1405–1421.
66 CHIANNIKULCHAI, N., AMMOURY, N., CAILLOU, B., DEVISSAGUET, J. P., COUVREUR, P., Hepatic tissue distribution of doxorubicin-loaded particles after i. v. administration in reticulosarcoma M 5076 metastasis-bearing mice, *Cancer Chemother. Pharmacol.*, **1990**, 26, 122–126.
67 VERDUN, C., BRASSEUR, F., VRANCKS, H., COUVREUR, P., ROLAND, M., Tissue distribution of doxorubicin associated with polyhexylcyanoacrylate nanoparticles, *Cancer Chemother. Pharmacol.*, **1990**, 26, 13–18.
68 CHIANNIKULCHAI, N., DRIOUCH, Z., BENOIT, J. P., PARODI, A. L., COUVREUR, P., Doxorubicin-loaded nanoparticles: Increased efficiency in murine hepatic metastasis, *Sel. Cancer Ther.*, **1989**, 5, 1–11.
69 BRASSEUR, F., COUVREUR, P., KANTE, B., DECKERS-PASSAU, L., ROLAND, M., DECKERS, C., SPEISER, P., Actinomycin D adsorbed on polymethylcyanoacrylate nanoparticles: Increased efficiency against an experimental tumor, *Eur. J. Cancer*, **1980**, 10, 1441–1445.
70 COUVREUR, P., KANTE, B., LENAERTS, V., SCAILTEUR, V., ROLAND, M., SPEISER, P., Tissue distribution of antitumor drugs associated with polyalkylcyanoacrylate nanoparticles, *J. Pharm. Sci.*, **1980**, 69, 199–202.
71 TU, R. S., TIRRELL, M., Bottom-up design of biomimetic assemblies, *Adv. Drug Deliv. Rev.*, **2004**, 56, 1537–1563.
72 NAGASAKI, Y., KUTSUNA, T., IIJIMA, M., KATO, M., KATAOKA, K., KITANO, S., KADOMA, Y., Formyl-ended heterobifunctional poly(ethylene oxide) synthesis of poly(ethylene oxide) with a formyl group at one end and a hydroxyl group at the other end, *Bioconj. Chem.*, **1996**, 6, 231–233.
73 LEE, K. B., YOON, K. R., WOO, S. I., CHOI, I. S., Surface modification of poly(glycolic acid) (PGA) for biomedical applications, *J. Pharm. Sci.*, **2003**, 92, 933–937.
74 SHAO, H., ZHANG, Q., GOODNOW, R., CHEN, L., TAM, A new polymer-bound N-hydroxysuccinimidyl active ester linker, *Tetrahedron Lett.*, **2000**, 41, 4257–4260.
75 LU, Y., CHEN, S. C., Micro- and nano-fabrication of biodegradable polymers for drug delivery, *Adv. Drug Deliv. Rev.*, **2004**, 56, 1621–1333.
76 MARTINEZ BARBOSA, M. E., CAMMAS, S., APPEL, M., PONCHEL, G., Investigation of the degradation mechanisms of poly(malic acid) esters in vitro and their related cytotoxicities on J774 macrophages, *Biomacromolecules*, **2004**, 5, 137–143.
77 CAMMAS, S., BEAR, M. M., MOINE, L., ESCALUP, R., PONCHEL, G., KATAOKA, K., GUERIN, P., Polymers of malic acid and 3-alkylmalic acid as synthetic PHAs in the design of biocompatible hydrolyzable devices, *Int. J. Biol. Macromol.*, **1999**, 25, 273–282.
78 CAMMAS, S., NAGASAKI, Y., KATAOKA, K., Heterobifunctional poly(ethylene oxide): Synthesis of alpha-methoxy-omega-amino and alpha-hydroxy-omega-amino PEOs with the same

molecular weights, *Bioconj. Chem.*, **1995**, 6, 226–230.
79 Missirlis, D., Tirelli, N., Hubbell, J. A., Amphiphilic hydrogel nanoparticles. Preparation, characterization, and preliminary assessment as new colloidal drug carriers, *Langmuir*, **2005**, 21, 2605–2613.
80 Rossignol, H., Bousta, M., Vert, M., Synthetic poly(beta-hydroxyalkanoates) with carboxylic acid or primary amine pendent groups and their complexes, *Int. J. Biol. Macromol.*, **1999**, 25, 255–264.
81 Grazia Cascone, M., Zhu, Z., Borselli, F., Lazzeri, L., Poly(vinyl alcohol) hydrogels as hydrophilic matrices for the release of lipophilic drugs loaded in PLGA nanoparticles, *J. Mater. Sci. -Mater. Med.*, **2002**, 13, 29–32.
82 Solaro, R., Chiellini, F., Signori, F., Fiumi, C., Bizzarri, R., Chiellini, E., Nanoparticle systems for the targeted release of active principles of proteic nature, *J. Mater. Sci. -Mater. Med.*, **2003**, 14, 705–711.
83 Win, K. Y., Feng, S. S., Effects of particle size and surface coating on cellular uptake of polymeric nanoparticles for oral delivery of anticancer drugs, *Biomaterials*, **2005**, 26, 2713–2711.
84 Lemarchand, C., Gref, R., Couvreur, P., Polysaccharide-decorated nanoparticles, *Eur. J. Pharm. Biopharm.*, **2004**, 58, 327–341.
85 Vauthier, C., Dubernet, C., Chauvierre, C., Brigger, I., Couvreur, P., Drug delivery to resistant tumors: The potential of poly(alkyl cyanoacrylate) nanoparticles, *J. Controlled Release*, **2003**, 93, 151–160.
86 Xiong, X. Y., Tam, K. C., Gan, L. H., Release kinetics of hydrophobic and hydrophilic model drugs from pluronic F127/poly(lactic acid) nanoparticles, *J. Controlled Release*, **2005**, 103, 73–82.
87 Zeng, Y., Pitt, W. G., Poly(ethylene oxide)-β-poly(N-isopropylacrylamide) nanoparticles with cross-linked cores as drug carriers, *J. Biomater. Sci. Polymer Ed.*, **2005**, 16, 371–380.
88 Hilt, J. Z., Byrne, M. E., Configurational biomimesis in drug delivery: Molecular imprinting of biologically significant molecules, *Adv. Drug Deliv. Rev.*, **2004**, 56, 1599–1620.
89 Lo, C. L., Lin, K. M., Hsiue, G. H., Preparation and characterization of intelligent core-shell nanoparticles based on poly(d, l-lactide)-g-poly(N-isopropyl acrylamide-co-methacrylic acid), *J. Controlled Release*, **2005**, 104, 477–488.
90 Xia, X., Hu, Z., Marquez, M., Physically bonded nanoparticle networks: A novel drug delivery system, *J. Controlled Release*, **2005**, 103, 21–30.
91 Rehor, A., Hubbell, J. A., Tirelli, N., Oxidation-sensitive polymeric nanoparticles, *Langmuir*, **2005**, 21, 411–417.
92 Alexiou, C., Arnold, W., Klein, R. J., Parak, F. G., Hulin, P., Bergemann, C., Erhardt, W., Wagenpfeil, S., Lübbe, A. S., Locoregional cancer treatment with magnetic drug targeting, *Cancer Res.*, **2000**, 60, 6641–6648.
93 Rudge, S., Peterson, C., Vessely, C., Koda, J., Stevens, S., Catterall, L., Adsorption and desorption of chemotherapeutic drugs from a magnetically targeted carrier (MTC), *J. Controlled Release*, **2001**, 74, 335–340.
94 Jain, T. K., Morales, M. A., Sahoo, S. K., Leslie-Pelecki, D. L., Labhasetwar, V., Iron oxide nanoparticles for sustained delivery of anticancer agents, *Mol. Pharm.*, **2005**, 2, 194–205.
95 Gupta, A. K., Gupta, M., Synthesis and surface engineering of iron oxide nanoparticles for biomedical applications, *Biomaterials*, **2005**, 26, 3995–4021.
96 Tkachenko, A. G., Xie, H., Coleman, D., Glomm, W., Ryan, J., Anderson, M. F., Franzen, S., Feldheim, D. L., Multifunctional gold nanoparticle-peptide complexes for nuclear targeting, *J. Am. Chem. Soc.*, **2003**, 125, 4700–4701.
97 Lubbe, A. S., Bergemann, C., Huhnt, W., Fricke, T., Riess, H.,

Brock, J. W., Huhn, D., Preclinical experiences with magnetic drug targeting: Tolerance, and efficacy, *Cancer Res.*, **1996**, 56, 4694–4701.

98 Flynn, E. R., Bryant, H. C., A biomagnetic system for in vivo cancer imaging, *Phys. Med. Biol.*, **2005**, 50, 1273–1293.

99 Gong, L. S., Zhang, Y. D., Liu, S., Target distribution of magnetic albumin nanoparticles containing adriamycin in transplanted rat liver cancer model, *Hepatobil. Pancr. Dis. Int.*, **2004**, 3, 365–368.

100 Zhang, Y., Kohler, N., Zhang, M., Surface modification of superparamagnetic magnetite nanoparticles and their intracellular uptake, *Biomaterials*, **2002**, 23, 1553–1561.

101 Wilhelm, C., Billotey, C., Roger, J., Pons, J. N., Bacri, J. C., Gazeau, F., Intracellular uptake of anionic superparamagnetic nanoparticles as a function of their surface coating, *Biomaterials*, **2003**, 24, 1001–1011.

102 Panyam, J., Zhou, W. Z., Prabha, S., Sahoo, S. K., Labhasetwar, V., Rapid endo-lysosomal escape of poly(D, L-lactide-co-glycolide) nanoparticles: Implications for drug and gene delivery, *FASEB J.*, **2002**, 16, 1217–1226.

103 Raynal, I., Rrigent, P., Peyramaure, S., Najid, A., Rebuzzi, C., Corot, C., Macrophage endocytosis of superparamagnetic iron oxide nanoparticles: Mechanisms and comparison of ferrumoxides and ferrumoxtran-10, *Invest. Rad.*, **2004**, 39, 56–63.

104 Stella, B., Arpicco, S., Peracchia, M. T., Desmaele, D., Hoebeke, J., Renoir, M., D'Angelo, J., Cattel, L., Couvreur, P., Design of folic acid conjugated nanoparticles for drug targeting, *J. Pharm. Sci.*, **2000**, 89, 1452–1464.

105 Petri-Fink, A., Chastellain, M., Juillerat-Jeanneret, L., Ferrari, A., Hofmann, H., Development of functionalized superparamagnetic iron oxide nanoparticles for interaction with human cancer cells, *Biomaterials*, **2005**, 26, 2685–2694.

106 Sundstrom, J. B., Mao, H., Santoianni, R., Villinger, F., Little, D. M., Huynh, T. T., Mayne, A. E., Hao, E., Ansari, A. A., Magnetic resonance imaging of activated proliferating rhesus macaque T cells labeled with superparamagnetic monocrystalline iron oxide nanoparticles, *J. AIDS*, **2004**, 35, 9–21.

107 Berry, C. C., Charlses, S., Wells, S., Dalby, M. J., Curtis, A. S., The influence of transferrin stabilised magnetic nanoparticles on human dermal fibroblasts in culture, *Int. J. Pharm.*, **2004**, 269, 211–225.

108 Porter, C. J. H., Moghimi, S. M., Illum, L., Davis, S. S., The polyoxoethylene/polyoxopropylene block copolymer poloxamer-407 selectively redirects intravenously injected microspheres to sinusoidal endothelial cells of rabbit bone-marrow, *FEBS Lett.*, **1992**, 305, 62–66.

109 McLean, J. W., Fox, E. A., Baluk, P., Bolton, P. B., Haskell, A., Pearlman, R., Thurston, C., Unemoto, E. Y., McDonald, D. M., Organ-specific endothelial uptake of cationic liposome-DNA complexes in mice, *Am. J. Physiol.*, **1997**, 273, H387–H404.

110 Shenoy, V. S., Vijay, I. K., Murthy, R. S., Tumour targeting: Biological factors and formulation advances in injectable lipid nanoparticles, *J. Pharm. Pharmacol.*, **2005**, 57, 411–422.

111 Videira, M. A., Botelho, M. F., Santos, A. C., Gouveia, L. F., de Lima, J. J., Almeida, A. J., Lymphatic uptake of pulmonary delivered radiolabelled solid lipid nanoparticles, *J. Drug Target.*, **2002**, 10, 607–613.

112 Chen, Q. R., Zhang, L., Gasper, W., Mixson, A. J., Targeting tumor angiogenesis with gene therapy, *Mol. Gen. Metab.*, **2001**, 74, 120–127.

113 Reynolds, A. R., Moghimi, S. M., Hodivala-Dilke, K., Nanoparticle-mediated gene delivery to tumor vasculature, *Trends Mol. Med.*, **2003**, 9, 2–4.

114 Hood, J. D., Bednarski, M., Frausto, R., Guccione, S., Resifeld, R. A., Xiang, R., Cheresh, D. A., Tumor

regression by targeted gene delivery to the vasculature, *Science*, **2002**, 296, 2404–2407.

115 Suzawa, T., Nagamura, S., Saito, H., Ohta, S., Hanai, N., Kanazawa, J., Okabe, M., Yamasake, M., Enhanced tumor cell selectivity of adriamycin monoclonal antibody conjugate via a poly(ethylene glycol)based cleavable linker, *J. Controlled Release*, **2002**, 79, 229–242.

116 Moghimi, S. M., Hunter, A. C., Murray, J. C., Long-circulating and target-specific nanoparticles: Theory to practice, *Pharm. Rev.*, **2001**, 53, 283–318.

117 Zhang, Y., Zhang, J., Surface modification of monodisperse magnetite nanoparticles for improved intracellular uptake to breast cancer cells, *J. Colloid Interf. Sci.*, **2005**, 283, 352–357.

118 Zhang, L., Hou, S., Mao, S., Wei, D., Song, X., Lu, Y., Uptake of folate-conjugated albumin nanoparticles to the SKOV3 cells, *Int. J. Pharm.*, **2004**, 287, 155–162.

119 Park, J. W., Benz, C. C., Martin, F. J., Future directions of liposome- and immunoliposome-based cancer therapeutics, *Semin. Oncol.*, **2004**, 31(Suppl 13), 196–205.

120 Nielsen, U. B., Kirpotin, D. B., Pickering, E. M., Hong, K., Park, J. W., Shalaby, M. R., Shao, Y., Benz, C. C., Marks, J. D., Therapeutic efficacy af anti-ErbB2 immunoliposomes targeted by a phage antibody selected for cellular endocytosis, *Biochim. Biophys. Acta*, **2002**, 1591, 109–118.

121 Zhang, Z. R., Gong, Y., Huang, Y., He, Q., Conjugation of mitoxantrone-loaded nanospheres and anti-C-erbB-2 monoclonal antibodies, *Yao Hsueh Hsueh Pao – Acta Pharm. Sin.*, **2001**, 36, 151–154.

122 Loo, C., Lowery, A., Halas, N., West, J., Drezek, R., Immunotargeted nanoshells for integrated cancer imaging and therapy, *NanoLetters*, **2005**, 5, 709–711.

123 Kim, I. S., Kim, S. H., Development of a polymeric nanoparticulate drug delivery system. In vitro characterization of nanoparticles based on sugar-containing conjugates. *Int. J. Pham.*, **2002**, 245, 67–73.

124 Kim, I. S., Kim, S. H., Development of polymeric nanoparticulate drug delivery systems: Evaluation of nanoparticles based on biotinylated poly(ethylene glycol) with sugar moiety, *Int. J. Pharm.*, **2003**, 257, 195–203.

125 Iinuma, H., Maruyama, K., Okinaga, K., Sasaki, K., Sekine, T., Ishida, O., Ogiwara, N., Johkura, K., Yonemura, Y., Intracellular targeting therapy of cisplatin-encapsulated transferrin-polyethylene glycol liposome on peritoneal dissemination of gastric cancer, *Int. J. Cancer*, **2002**, 99, 130–137.

126 Dubowchik, G. M., Walker, M. A., Receptor-mediated and enzyme-dependent targeting of cytotoxic anticancer drugs, *Pharmacol. Ther.*, **1999**, 83, 67–123.

127 Savic, R., Luo, L., Eisenberg, A., Maysinger, D., Micellar nanocontainers distribute to defined cytoplasmic organelles, *Science*, **2003**, 300, 615–618.

128 Panyam, J., Labhasetwar, V., Sustained cytoplasmic delivery of drugs with intracellular receptors using biodegradable nanoparticles, *Mol. Pharm.*, **2004**, 1, 77–84.

129 Moghimi, S. M., Hunter, A. C., Murray, J. C., Szewczyk, A., Cellular distribution of non-ionic micelles, *Science*, **2004**, 303, 626–627.

130 Drummond, D. C., Zignani, M., Leroux, J. C., Current status of pH-sensitive liposomes in drug delivery, *Prog. Lipid Res.*, **2000**, 39, 409–460.

131 Haining, W. N., Erson, D. G., Little, S. R., vonBerwelt-Baildon, M. S., Cardoso, A. A., Alves, P., Kosmatopoulos, K., Nadler, L. M., Langer, R., Kohane, D. S., pH-sensitive microparticles for peptide vaccination, *J. Immunol.*, **2004**, 173, 2578–2585.

132 Ma, Z., Lim, L. Y., Uptake of chitosan and associated insulin in Caco-2 cell monolayers: A comparison between

chitosan molecules and chitosan nanoparticles, *Pharm. Res.*, **2003**, 20, 1812–1819.

133 Mo, Y., Lin, L. Y., Mechanistic study of the uptake of wheat germ agglutinin-conjugated nanoparticles by A549 cells, *J. Pharm. Sci.*, **2004**, 93, 20–28.

134 Panyam, J., Labhsetwar, V., Dynamics of endocytosis and exocytosis of poly(D, L-lactide-co-glycolide) nanoparticles in vascular smooth muscle cells, *Pharm. Res.*, **2003**, 20, 212–220.

135 Console, S., Marty, C., Garcia-Echeverria, C., Schwendener, R., Ballmer-Hofer, K., Antennapedia and HIV transactivator of transcription (TAT) "protein transduction domains" promote endocytosis of high molecular weight cargo upon binding to cell surface glycosaminoglycans, *J. Biol. Chem.*, **2003**, 278, 35 109–35 114.

136 Silhol, M., Tyagi, M., Giacca, M., Lebleu, B., Vives, E., Different mechanisms for cellular internalization of the HIV-1 Tat-derived cell penetrating peptide and recombinant proteins fused to Tat, *Eur. J. Biochem.*, **2002**, 269, 494–501.

137 Wadia, J. S., Stan, R. V., Dowdy, S. F., Transducible TAT-HA fusogenic peptide enhances escape of TAT-fusion proteins after lipid raft macropinocytosis, *Nat. Med.*, **2004**, 10, 310–315.

138 Torchilin, V. P., Recent advances with liposomes as pharmaceutical carriers, *Nat. Rev. Drug Discov.*, **2005**, 4, 145–160.

139 Colin de Verdière, A., Dubernet, C., Némati, F., Soma, M., Appel, M., Ferté, J., Bernard, S., Puisieux, F., Couvreur, P., Reversion of multidrug resistance with polyalkylcyano acrylate nanoparticles: Toward a mechanism of action, *Br. J. Cancer*, **1997**, 76, 198–205.

140 Soma, C. E., Dubernet, C., Bentolila, D., Benita, S., Couvreur, P., Reversion of multidrug resistance by co-encapsulation of doxorubicin and cyclosporin A in polyalylcyanoacrylate nanoparticles, *Biomaterials*, **2000**, 21, 1–7.

141 Müller, R. H., Lherm, C., Herbort, J., Couvreur, P., In vitro model for the degradation of alkylcyanoacrylate nanoparticles, *Biomaterials*, **1990**, 11, 590–595.

142 Treupel, L., Poupon, M. F., Couvreur, P., Puisieux, F., Vectorization of doxorubicin in nanospheres and reversion of pleiotropic resistance of tumor cells, *C. R. Acad. Sci.*, **1991**, 313, 1–174.

143 Cuvier, C., Roblot-Treupel, L., Millot, J. M., Lizard, G., Chevillard, S., Manfait, M., Couvreur, P., Poupon, M. F., Doxorubicin-loaded nanospheres bypass tumor cell multidrug resistance, *Biochem. Pharmacol.*, **1992**, 44, 509–517.

144 Nemati, F., Dubernet, C., Fessi, H., Colin de Verdière, A., Poupon, M. F., Puisieux, F., Couvreur, P., Reversion of multidrug resistance using nanoparticles in vitro: Influence of the nature of the polymer, *Int. J. Pharm.*, **1996**, 138, 237–246.

145 Batrakova, E. V., Li, S., Alakhov, V. Y., Miller, D. W., Kabanov, A. V., Optimal structure requirements for pluronic acid block copolymers in modifying P-glycoprotein drug effelux transporter activity in bovine brain microvessels endothelial cells, *J. Pharm. Exp. Ther.*, **2003**, 304, 845–854.

146 Kabanov, A. V., Batrakova, E. V., Alakhov, V. Y., An essesntial relationship between ATP depletion and chemosensitizing activity of pluronic acid block copolymers, *J. Controlled Release*, **2003**, 91, 75–83.

147 Colvin, V. L., The potential environmental impact of engineered nanomaterials, *Nat. Biotechnol.*, **2003**, 21, 1166–1170.

148 Hunter, A. C., Moghimi, S. M., Therapeutic synthetic polymers: A game of Russian roulette? *Drug Discov. Today*, **2002**, 7, 998–1001.

149 Nishiyama, N., Kiozumi, F., Okazaki, S., Matsumura, Y., Nishio, K., Kataoka, K., Differential gene expression profile between PC14 cells treated with free cisplatin and

cisplatin-incorporated polymeric micelles, *Bioconj. Chem.*, **2003**, 14, 449–457.
150 LAM, K. H., SCHAKENRAAD, J. M., ESSELBRUGGE, H., FEIJEN, J., NIEUWEHUIS, P., The effect of phagocytosis of poly(L-lactic acid) fragments on cellular morphology and viability, *J. Biomed. Mater. Res.*, **1993**, 27, 1569–1577.
151 MURRAY, J. C., MOGHIMI, S. M., HUNTER, A. C., SYMONDS, P., DEBSKA, G., SZEWCZYK, A., Lymphocytic death by cationic polymers: A role for mitochondrion and implications in gene therapy, *Br. J. Cancer*, **2004**, 91, S75.
152 SZEBENI, J., Complement activation-related pseudoallergy caused by liposomes, micellar carriers of intravenous drugs, and radiocontrast agents, *Crit. Rev. Ther. Drug Carr. Syst.*, **2001**, 18, 587–606.
153 DUNCAN, R., Polymer conjugates for tumour targeting and intracytoplasmic delivery. The EPR effect as a common gateway? *PSTT*, **1999**, 2, 441–449.
154 LU, Z. R., SHIAH, J. G., SAKUMA, S., KOPECHOVA, P., KOPECHEK, J., Design of novel bioconjugates for targeted drug delivery, *J. Controlled Release*, **2002**, 78, 165–173.
155 REENTS, R., JEYARAJ, D. A., WALDMANN, H., Enzymatically cleavable linker groups in polymer-supported synthesis, *Drug Discov. Today*, **2002**, 7, 71–76.
156 BRAUN, K., PESCHKLE, P., PIPKORN, R., LAMPEL, S., WACHAMUTH, M., WALDECK, W., FRIEDRICH, E., DUBUS, J., A biological transporter for the delivery of peptide nucleic acids (PNAs) to the nuclear compartment of living cells, *J. Mol. Biol.*, **2002**, 318, 237–243.
157 LU, Y., MILLER, M. J., Syntheses and studies of multiwarhead siderophore-5-fluorouridine conjugates, *Bioorg. Med. Chem.*, **1999**, 7, 3025–3038.
158 MARSCHÜTZ, M. K., VERONESE, F. M., BERNKOP-SCHNURCH, A., Influence of the spacer on the inhibitory effect of different polycarbophil-protease inhibitor conjugates, *Eur. J. Pharm. Biopharm.*, **2001**, 52, 137–144.
159 SCHOENMAKERS, R. G., VAN DE WETERING, P., ELBERT, D. L., HUBBELL, J. A., The effect of the linker on the hydrolysis rate of drug-linked ester bonds, *J. Controlled Release*, **2004**, 95, 291–300.
160 BHAVANE, R., KARATHANASIS, E., ANNAPRAGADA, A. V., Agglomerated vesicle technology: A new class of particles for controlled and modulated pulmonary drug delivery, *J. Controlled Release*, **2003**, 93, 15–28.
161 GANGULY, S., DASH, A. K., A novel in situ gel for sustained drug delivery and targeting, *Int. J. Pharm.*, **2004**, 276, 83–92.
162 BÖHM, G., DOWDENM, J., RICE, D. C., BURGESS, I., PILARD, J. F., GUILBERT, B., HAXTON, A., HUNTER, R. C., TURNER, N. J., FLITSCH, S. L., A novel linker for the attachment of alcohols to solid supports, *Tetrahedron Lett.*, **1998**, 39, 3819–3822.
163 MEHVAR, R., Dextrans for targeted and sustained delivery of therapeutic and imaging agents, *J. Controlled Release*, **2000**, 69, 1–25.
164 LAZNY, R., NODZEWSKA, A., KLOSOWSKI, P., A new strategy for synthesis of polymeric supports with triazene linkers, *Tetrahedron*, **2004**, 60, 121–130.
165 BRÄSE, S., KÖBBERLING, J., ENDERS, D, LAZNY, R., WANG, M., Triazenes as robust and simple linkers for amines in solid-phase organic synthesis, *Tetrahedron Lett.*, **1999**, 40, 2105–2108.
166 KRCHNAK, V., SLOUGH, G. A., Dual linker with a reference cleavage site for information rich analysis of polymer-supported transformations, *Tetrahedron Lett.*, **2004**, 45, 5237–5241.

8
Nanoparticles for Thermotherapy

Andreas Jordan, Klaus Maier-Hauff, Peter Wust, and Manfred Johannsen

8.1
Introduction

The biological effectiveness of heat in treating cancer has been known for decades and many of the corresponding molecular mechanisms are understood. Elevation of tissue temperature to above 40–41 °C is termed hyperthermia. It alters the function of many structural and enzymatic proteins within cells as a function of time and temperature, which in turn alters cell growth and differentiation and can induce apoptosis [1, 2]. Modest temperature rises are particularly effective against cells that tend to be resistant to radiation, cells in the S phase of the cell cycle and nutrient-deprived, low pH hypoxic cells [3]. Hyperthermia inhibits repair of sublethal radiation damage and also induces increasing radiosensitivity due to tumor reoxygenation [4, 5]. Treatments with tissue temperatures above 46 °C are termed thermoablation and have direct cytotoxic effects.

Thermotherapy is a physical therapy with fewer limitations than chemotherapy or radiotherapy and is typically used in combination with both of these therapies [6–8]. This allows a greater number of repeated treatments without accumulation of toxic side effects.

Although successful clinical trials have been conducted, thermotherapy is not yet established in clinical routine. This discrepancy probably derives from technical limitations in achieving effective temperature distributions in the depth of the human body rather than from a general lack of biological effectiveness [9].

Common thermotherapy techniques use different energy sources for generating heat in body tissue: Electromagnetic waves radiated by antennas (radiofrequency- or microwave-hyperthermia) [10, 11], ultrasonic sound [12, 13], magnetically excited thermoseeds as well as tubes with hot water [14, 15]. For reviews see Refs. [16, 17].

The major problem with all present conventional thermotherapy systems is achieving a homogenous heat distribution and deep regional therapeutic temperatures in the treated tumor tissue. According to this hypothesis, treatment failure results from an insufficient temperature rise in parts of the tumor, enabling tumor regrowth. However, excessive intratumoral temperatures might on the other hand induce damage of adjacent structures.

The heating of tissues using magnetic nanoparticles as a new treatment modality for interstitial thermotherapy has the potential to overcome these shortcomings.

When describing the properties of magnetic nanoparticles for thermotherapy, one has to consider that the magnetic properties are influenced to a large extent by the size and shape of the particles. Although there has been much progress regarding nanoparticle synthesis, the control of particle size and shape on the nanoscale level remains a synthetic challenge. Nevertheless, nanoscaled particles have several advantages over larger particles that favor their application in biomedicine. One advantage arises because particles change their magnetic properties when entering the size regime below approximately 20 nm, giving rise to distinct mechanisms of heating in alternating magnetic fields.

The use of magnetic nanoparticles for heating purposes was investigated comprehensively by Jordan et al. in 1993 [18]. Different studies have revealed that the specific absorption rate (SAR) of magnetic nanoparticles depends strongly on the diameter of the particle core [19–22]. In general, the heat production of magnetic nanoparticles in alternating magnetic fields can be attributed to hysteresis and relaxation losses, the latter being more pronounced in particles below 20 nm. Hysteresis losses play a role in larger particles, which consist of more than one magnetic domain and magnetization reversal occurs due to motions of the domain boundaries. In this case, the energy produced is proportional to the coercivity of the particles [19]. Magnetic fluids are dispersions of ultrafine magnetic particles, which are mostly superparamagnetic, meaning that each particle has only a single magnetic domain. For these so-called subdomain particles (SDP), coercivity is zero and magnetization reversal is only possible by relaxation processes. In alternating magnetic fields heat can be produced by two distinct relaxation mechanisms, Brownian and Néel relaxation. The first can be attributed to the rotation of the single-domain particle, which is related to the Brownian motion. The second corresponds to rotation of the magnetization vector in the crystal, if the particle is considered to be immobile [18].

To date, all magnetic nanoparticles intended to be used *in vivo* are composed of magnetic iron oxides. The main reason for this is their low toxicity and the known pathways of metabolism. The only two clinically relevant materials used are magnetite (Fe_3O_4) and maghemite (γ-Fe_2O_3). The crystal structures of both oxides are based on a cubic dense packing of oxide atoms, but they differ in the distribution of Fe ions in the crystal lattice. Magnetite is the most common ferrite and has an inverse spinel structure, $Fe(II)Fe(III)_2O_4$.

Pure iron-oxide nanoparticles have a high tendency to agglomerate and thus to build larger structures even in the absence of a magnetic field, which has a strong influence on the biomedical and magnetic properties of the particles. To prepare magnetic fluids for biomedical applications, the iron-oxide particles have to be coated with a protecting shell that prevents agglomeration. Furthermore, the protecting shell is responsible for the interaction of the particles with its surrounding (e.g., tumor tissue) and can provide functional domains that are useful for the coupling of biomolecules or drugs to the surface. Common shell materials are polymers such as dextran, starch or poly(ethylene glycol).

Most techniques established so far concerning the use of magnetic particles for thermotherapy are basically based on direct instillation of magnetic fluids into the tumor tissue followed by exposure to an externally applied alternating magnetic field.

Ferromagnetic embolization hyperthermia is another technique for the local application of nanoparticles, in which the arterial supply of certain body regions is used as a pathway to carry magnetic particles directly into the tumor area.

8.2
Thermotherapy following Intratumoral Administration of Magnetic Nanoparticles

Magnetic particles suspended in a carrier fluid can be injected directly into tumor tissue and subsequently be heated in an alternating magnetic field.

The history of magnetic particles for selective heating of tumors had already begun in 1957, when Gilchrist et al. used magnetic particles a few micrometers in size for inductive heating of lymph nodes in dogs [23].

In 1979, Gordon et al., for the first time, used magnetic fluids of dextran-coated magnetite-particles (often referred to as "dextran-magnetites") for hyperthermia of implanted mammary tumors of rats [24]. After systemic application of the nanoparticles with a core size of up to 6 nm and exposition to an alternating magnetic field, the authors demonstrated histological evidence of tumor necrosis after intratumoral temperature increases of 8 °C.

Direct injection of microscaled ferromagnetic particles into renal carcinomas of rabbits and subsequent hysteresis heating in an alternating magnetic field was reported by Rand et al. in 1981 [25]. They measured tumor surface temperatures of 55 °C, leading to destruction of most of the treated tumor tissue within 3 days.

Direct injection of "truly nanoscaled" magnetic particles into tumors was first reported by our group in 1997 [26]. We studied single high dose thermotherapy on intramuscularly implanted mammary carcinoma of mice. Dextran-coated magnetite particles with a core size of approximately 3 nm were injected intratumorally 20–30 min before excitation and trapped by magnetic targeting (50 mT), which yielded a 2.5-fold enhancement of the intratumoral iron concentration. An intratumoral steady state temperature of 47 °C was maintained for 30 min with whole-body alternating magnetic fields of 6–12.5 kA m^{-1} at 520 kHz. The study demonstrated that a single high thermal dose was able to induce local tumor control in many of the treated animals within 30 days after therapy. Histological examinations of tumor tissue after thermotherapy showed widespread tumor necrosis. Tumor growth after thermotherapy was heterogeneous; some tumors did not show evidence for regrowth after 50 days whereas others had grown rapidly. This was most probably due to inhomogeneities of the intratumoral particle distribution, which was also confirmed qualitatively by magnetic resonance imaging (MRI).

In 2002 Hilger et al. injected colloidal suspensions of coated nanoparticles with average total particle sizes of 10 nm and 200 nm into human breast adenocarcinoma implanted into immunodeficient mice (4–18 mg magnetite per 100 mg tumor tissue) [27]. During exposure to an alternating magnetic field temperature increases between 12 and 73 °C in circumscribed areas within the tumor could be induced (termed "magnetic thermal ablation"). Histological examination revealed the presence of early stages of coagulation necrosis in the treated tumor cells.

Ohno et al. have investigated a new type of magnetite for interstitial hyperthermia [28]. They inserted stick-type carboxymethylcellulose-magnetite containing nanoparticles (Fe_3O_4, average particle size approximately 10 nm) stereotactically into gliomas of rats and exposed the animals to an alternating magnetic field. This investigation revealed a characteristic spreading of the magnetite particles through the tumor after three heat treatments and an approximately three-fold prolongation of survival time.

Groups at the University of Nagoya in Japan have developed "magnetic cationic liposomes" (MCLs) with improved adsorption and accumulation properties and demonstrated the efficacy of this approach in several animal tumor models:

Induction of antitumoral immunity to rat glioma by intratumoral hyperthermia using these particles was investigated by Yanase et al. in 1998 [29]. For intratumoral hyperthermia, the cationic liposomes containing magnetite (Fe_3O_4 core with approximately 10 nm diameter) were directly injected into glioma tissue transplanted subcutaneously into the thigh of rats. Three subsequent treatments in an alternating magnetic field led to killing of the tumor cells and to the induction of a host immune response, which could be demonstrated by suppressed tumor growth after rechallenge with glioma cells 3 months later. In similar experiments using the same tumor model and the same particles, published one year later [30], complete tumor regression was observed in about 90% of the animals.

Le et al. have described successful tumor control (over two weeks) after hyperthermia of 43 °C using immuno-targeted magnetoliposomes [31].

In 2003, Ito et al. investigated the effect of intratumoral hyperthermia in combination with an immunotherapy of melanoma in mice [32]. After direct injection of MCLs into the tumors, the animals were exposed to an alternating magnetic field that raised the tumor temperature up to 43 °C; 24 h later, in a second treatment, interleukin-2 (IL-2) or granulocyte macrophage-colony stimulating factor (GM-CSF) was injected directly into the melanomas. In 75% (receiving IL-2) and 40% (receiving GM-CSF) of the animals treated by both therapies complete regression of the tumors was observed, while no tumor regression was observed in mice receiving only one of the treatments.

In 2005, Tanaka et al. investigated the therapeutic effects of tumor-specific hyperthermia using MCLs combined with an immunotherapy (intratumorally injected immature dendritic cells) on mouse melanoma [33]. Complete regression of tumors was observed in 60% of the mice and the cured animals rejected a second challenge of melanoma cells.

Kawai et al. have used the same liposomes to heat prostate cancer cells injected subcutaneously into the flank of rats [34]. They reported on tumor regression after generating intratumoral temperatures of 45 °C in an alternating magnetic field.

Carcinoma of the prostate represents an attractive target for minimally invasive, interstitial treatments due to its relatively easy accessibility via the transrectal or transperineal route.

To explore the potential of magnetic nanoparticle thermotherapy in an orthotopic prostate carcinoma, we have carried out experiments using the Dunning R3327 tumor model (MatMyLu-subtype), an established rat model developed to study prostate cancer progression [35]. This model resembles the human prostate with respect to physiological blood supply and local growth and may thus allow a more accurate analysis of the effects of a local hyperthermia treatment than a heterotopic model. In a preliminary *in vivo* evaluation, the feasibility and good overall tolerability of this technique in prostate cancer could be demonstrated [36].

Thermotherapy was carried out using an alternating magnetic field applicator system for small animals, operating at a frequency of 100 kHz and variable field strength of 0–18 kA m^{-1}.

Thermoablative temperatures of 50 °C were achieved at a field strength of 15 kA m^{-1}. Iron measurements were carried out to determine the percentage of nanoparticles present in the prostate tumors at different time points following intratumoral application. Without thermotherapy, 7 days after injection of magnetic fluid into the rat prostates, 53% of the injected amount of iron oxide was still present in the prostates. At 4 and 13 days after application 79% and 64%, respectively, of the injected amount of iron was still retained in the prostates after two sequential thermotherapy treatments. This difference in iron content between untreated and treated animals was attributed to the so-called "thermal bystander effect", which describes a hyperthermia-induced spread of coated nanoparticles in the target tissue, where they seem to escape rapid removal by macrophages [26]. This effect could also be demonstrated by means of histopathology. After magnetic fluid application alone, a loose distribution of nanoparticles within the tumor tissue with deposits in capillaries could be seen in sections of rat prostate tumors, whereas homogeneous distribution and co-localization of nanoparticles with necrotic areas could be demonstrated in animals that had undergone thermotherapy.

In a further systematic *in vivo* analysis using a more concentrated magnetic fluid containing approximately 112 mg-iron mL^{-1}, as well as a slightly modified aminosilane-based coating, intraprostatic temperatures of 70 °C could be achieved at a maximum magnetic field strength of 18 kA m^{-1} in the same tumor model [37]. Up to 0.5 mL of magnetic fluid per mL of tumor volume was injected into the prostate tumors in these experiments. The animals received two thermotherapy treatments carried out 48 h apart to avoid heat shock protein-induced thermoresistance. At a constant field strength of 12.6 kA m^{-1}, mean minimal intratumoral temperatures during the first and second thermotherapy sessions were 41.2 and 41.4 °C, respectively, whereas mean maximal temperatures were 54.8 and 54.2 °C (averaged over 12 animals). In this study, animals were sacrificed on day 20 after tumor cell inoculation (10 days following the first thermotherapy treatment) to compare tumor weights in the treatment and control groups and for iron measurements. A significant inhibition of prostate cancer growth of 44–51% over controls was demonstrated in the thermotherapy group. Mean iron content of the prostates of treated and untreated control animals was 81.6% and 83.7%, respectively, whereas 7.6% and 2.3% of the injected dose of iron was found in the liver. No detectable amount of iron was found in the blood stream. While thermal treatment itself had a significant impact on intratumoral distribution and stable deposition of nanoparticles in the previous pilot study, no such effect was observed using the modified magnetic fluid. In fact, intratumoral deposition of the MFL AS particles was the same with or without thermotherapy treatment. Histological analysis revealed a co-localization of nanoparticles with areas of necrosis, but a rather inconsistent pattern of intratumoral distribution, which appeared less homogeneous than with the previously used magnetic fluid preparation.

Further experimental studies have investigated the effect of combined magnetic nanoparticle thermotherapy and external irradiation in the Dunning tumor model [38].

Mean maximal and minimal intratumoral temperatures obtained in these experiments were 59 (centrally) and 43 °C (peripherally) during the first thermotherapy and 55 and 42 °C, respectively, during the second of two treatment sessions. Combined thermotherapy and radiation with 20 Gy was significantly more effective than radiation with 20 Gy alone and reduced tumor growth by 88% versus controls, achieving an equal tumor growth inhibition as irradiation alone with 60 Gy. As in the previous experiments, sequential heat treatments were possible without repeated injection of magnetic fluid. The mean iron content in the prostates on day 20 was 88% of the injected dose of ferrites, whereas only 2.5% was found in the liver.

In conclusion, thermotherapy using magnetic nanoparticles achieved significant growth inhibition, but not tumor control or eradication in the orthotopic Dunning tumor model of the rat. The results of preclinical evaluation of magnetic nanoparticle thermotherapy suggest that this technique may, in principle, be suitable for both local hyperthermia and thermoablation of prostate carcinoma, since the desired treatment temperatures can be freely selected by modulating the magnetic field strength [39].

However, for complete thermal ablation of the prostate, homogeneous infiltration of the whole prostatic tissue with magnetic nanoparticles would be required and high magnetic field strengths would have to be applied. In view of the synergistic effects of hyperthermia and radiotherapy, a combination of magnetic nanoparticle thermotherapy with irradiation may also be an attractive option for the treatment of prostate cancer.

In a rat model of glioblastoma multiforme we could also demonstrate the high efficacy of our new technique, leading to an up to 4.5-fold prolongation of survival [40]. The animals received two thermotherapy treatments within 48 h after a single stereotactic injection of the magnetic fluid into the tumor. Intratumoral temperatures between 43 and 47 °C correlated well with prolonged survival. As in our prostate tumor model, the application of aminosilane-coated nanoparticles led to the formation of stable deposits in the brain, thus allowing for repeated magnetic field treatments without repeated applications of the particles.

8.3
Ferromagnetic Embolization Hyperthermia

Another technique of using magnetic nanoparticles for thermotherapy is ferromagnetic embolization hyperthermia, in which a feeding artery is used to carry ferromagnetic particles into a tumor.

The technique seems to be well suited for the treatment of hepatic malignancies due to the differences in blood supply of hepatic tumor cells and normal liver parenchyma. Liver tumors mainly derive their blood supply from the hepatic arterial system, while the normal liver parenchyma receives most of its blood supply from the portal venous system [41]. Hence any substance infused into the arterial system will have the potential to preferentially target liver tumors [42].

Several preclinical studies have reported on arterial embolization hyperthermia of liver cancer.

Moroz et al. have demonstrated successful ferromagnetic embolization hyperthermia of rabbit hepatic carcinomas after arterial infusion of magnetic nanoparticles suspended in lipiodol [43, 44].

In 1994, Mitsumori et al. tested the thermotherapeutical properties of dextran-coated magnetite particles suspended in lipiodol or in degradable starch microspheres [45]. The study reported an increase of over 12 °C after 10 min of heating after embolization of the renal artery and selective heating of the embolized kidney by exposure to an alternating magnetic field. The authors followed up this work with a study in 1996 [46], in which hepatic carcinoma-bearing rabbits were arterially infused with the same sub-domain sized particles (75 nm in diameter with a 7.4 nm magnetite core). Slight tumor temperature increases were recorded *in vivo*.

In a study performed by Minamimura et al. a novel preparation of microspheres incorporating a dextran magnetite complex was used for arterial embolization and inductive hyperthermia of liver tumors in rats [47]. Tumor temperatures of around 43 °C were maintained for 30–40 min and three days after treatment the increase of tumor volume in the treated animals differed significantly from that of the control groups.

Jones et al. have achieved positive temperature differences between tumor and normal liver and consequent therapeutic responses in experimental rabbit liver tumors after arterially infused ferromagnetic microspheres [48]. Heating to over 42 °C for 20 min by exposure to an alternating magnetic field resulted in large areas of tumor necrosis 14 days after treatment.

After arterial infusion of iron-oxide particles suspended in lipiodol into rabbit hepatic carcinoma, Moroz et al. have found that an iron concentration of 2–3 mg per gram of tumor was necessary to produce heating rates up to 11.5× greater than those in adjacent normal liver parenchyma [43]. The authors reported a mean tumor to normal liver iron concentration ratio of 5.3.

Further work by this group employing the same tumor model and infusion regimen has shown that large hepatic tumors are more amenable to thermotherapy after arterial embolization than small tumors [44]. The authors concluded from their data that, for a given tumor iron concentration, larger tumors heat at a greater rate than small tumors, owing to poorer tissue cooling and better heat conduction in the necrotic regions of large tumors.

In a later study the authors investigated the clearance of ferromagnetic particles from the liver [49]. The normal liver of pigs was arterially embolized with γ-Fe_2O_3 particles (150 nm) suspended in lipiodol and polymer microspheres (32 µm) containing ferromagnetic particles suspended in 1% Tween-water. Both types of particles were extensively phagocytosed in the liver and there was no significant reduction in the hepatic iron concentration in either treatment group 28 days after infusion. The suspension of 150 nm ferromagnetic particles in lipiodol was too vaso-occlusive for use in hepatic tissue, while the suspension of 32 µm spheres was safe and well tolerated.

Although few preclinical studies have been described, arterial embolization hyperthermia has demonstrated encouraging results. The technique seems to be particularly well suited for the treatment of hepatic malignancies. Potential advantages are a very selective local heat deposition and a more homogeneous tissue temperature distribution than other hyperthermia techniques.

8.4
First Clinical Experiences with Thermotherapy using Magnetic Nanoparticles: MagForce Nanotherapy

The pathways developed so far for thermotherapy using magnetic nanoparticles have in common that they deliver the magnetic material directly into or adjacent to the tumor tissue. To facilitate cellular uptake by tumor cells, the particle coatings often were modified with targeting ligands such as antibodies or peptides.

Magnetic fluids can be instilled into tumor tissue percutaneously under CT, ultrasound or fluoroscopy guidance. Owing to inductive excitation of the nanoparticles this technique allows very precise heating of almost every part of the body.

A major advantage of direct intratumoral application of the particles is the good control of deposition, leading to high concentrations of the magnetic material within the tumor while healthy tissue can be spared. The magnetic fluid can be distributed in very small portions and, therefore, almost continuously within the targeted area. Therefore, the requirement of maximal heat deposition within the target area while sparing neighboring healthy tissue can be met.

MagForce Nanotherapy, formerly designated magnetic fluid hyperthermia, has been developed at Charité – University Medicine Berlin over the last 15 years. Our preclinical studies have demonstrated that this innovative approach has the potential to improve heating capabilities with certain cancer types and thus overcome existing prejudices against routine application of heat to treat cancer.

It is the first nanotechnology-based local thermotherapy to enter clinical trials.

Beginning in 2003, we started four different clinical studies to investigate the feasibility of our new thermotherapy approach on different tumor entities.

The magnetic fluid MFL AS used in these studies is manufactured by MagForce Nanotechnologies, Berlin, and consists of superparamagnetic iron-oxide nanopar-

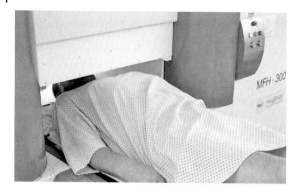

Fig. 8.1. Thermotherapy treatment of a glioblastoma multiforme patient in the magnetic field applicator MFH® 300F (MagForce Nanotechnologies, AG, Berlin, Germany).

ticles in aqueous solution with an iron concentration of 2 mol L^{-1}. The iron oxide core is covered by an aminosilane-type shell and has a diameter of approximately 15 nm. The particles generate heat in an alternating magnetic field by Brownian and Néel relaxation processes.

Thermotherapy is performed in a magnetic field applicator (MFH® 300F, MagForce Nanotechnologies, AG, Berlin, Germany), generating an alternating magnetic field (100 kHz) and a variable field strength of 0–18 kA m^{-1} (Fig. 8.1). This applicator meets the safety and practicability criteria for medical use imposed by the respective European authorities. Owing to its universal design, it can be used for treatment of malignancies in every location of the human body. Fixation or anesthesia of the patients during treatment is not necessary.

Minimally invasive measurement of treatment temperatures is carried out by fiber-optic thermometry probes as part of the therapy system.

8.4.1
Feasibility Study on Thermotherapy using Magnetic Nanoparticles in Recurrent Glioblastoma Multiforme

From March 2003 to June 2004 we performed the world's first phase I trial on tumor thermotherapy using magnetic nanoparticle, with 14 glioblastoma multiforme patients.

Malignant gliomas (anaplastic astrocytoma and glioblastoma multiforme, WHO grades III and IV, respectively) have an incidence of approximately 5 in 100 000 per year and represent approximately 40% of primary brain tumors in adults, with 50% of them belonging to the most malignant phenotype, glioblastoma multiforme. Clinically, they are highly problematic due to their widely invasive nature, which makes complete resection almost impossible. Median overall survival after first-line therapy does not exceed 12–15 months [50, 51] and no significant increase has been achieved over the last decade, despite modern diagnostics and treatments with surgery, radio- and chemotherapy.

All patients of our trial received stereotactic injection of the magnetic fluid into the tumor area. Before starting thermotherapy, the position of the instilled nanoparticles was determined by computed tomography (CT). These data were matched to presurgical MR images using specially designed software (MagForce NanoPlan®, not commercially available), thus allowing calculation of the expected heat distribution within the treatment area in relation to the magnetic field strength [52].

The preoperatively planned and neuro-navigationally guided procedure led to an almost atraumatic instillation of the magnetic fluid and rise of intracranial pressure could be avoided by slow injection of the magnetic fluid.

Patients received 4–10 thermotherapy sessions where intratumoral temperatures of 42–49 °C could be measured; body temperatures increased by 1.0–1.5 °C.

All patients tolerated the instillation of the nanoparticles without complications and side effects such as headache, nausea, vomitus or allergic reactions were not observed. Neurological deficits or infections did not occur.

As documented by CT-scans and thermometry measurements, a high stability of the nanoparticle deposits can be assumed. In fact, reproducible intratumoral temperatures could be achieved during the therapy, even several weeks after administration.

Data of the trial concerning overall survival and time to progression suggest that intracranial thermotherapy using magnetic nanoparticles can be safely applied with therapeutic temperatures and without side effects [53]. A phase II study is in progress to evaluate the efficacy of this treatment approach.

8.4.2
Feasibility Study on Thermotherapy using Magnetic Nanoparticles in Recurrent and Residual Tumors

Another phase I trial started in February 2004, which enrolled 21 patients suffering from non-resectable and pre-treated local relapses of different tumor entities (e.g., rectum, ovarian, prostate, and cervix carcinoma as well as soft tissue sarcoma). All patients received thermotherapy in combination with radio- or chemotherapy [54].

The objectives of the study were treatment planning, application and subsequent control of magnetic fluid distributions in circumscribed lesions (<4–5 cm) and, furthermore, assessment of safety, quality and feasibility of the heating patterns achieved *in situ*.

Patients with metallic implants <30 cm from the treatment area had to be excluded. Teeth amalgam fillings or gold crowns had to be replaced by ceramics if involved in the treatment field (head and neck, upper thorax). Metallic clips or seeds several millimeters long and <1 mm in diameter were permitted due to their very low power absorption capacities.

Two principally different techniques for the instillation of the magnetic fluid were evaluated.

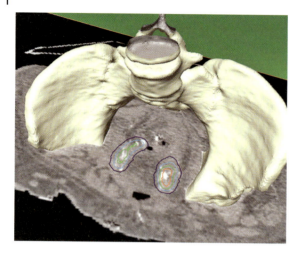

Fig. 8.2. Thermotherapy using magnetic nanoparticles of a patient suffering from metastases of a cervical carcinoma. Three-dimensional reconstruction of a pelvis sliced by a CT scan: The calculated temperature distributions (isotherms) during thermotherapy decrease from orange to blue. White spots within the colored lines show the magnetic nanoparticle deposits.

The first strategy implied a prospective planning of the "ideal" nanofluid distribution on the basis of a three-dimensional CT dataset performed in treatment position. This dataset is then transferred via the radiological information system into a software platform to calculate the temperature distribution during thermotherapy within the treatment area (Fig. 8.2).

The second implantation technique consisted of intraoperative injection under visual control. In this case prospective planning is not possible, because after tumor debulking an individual (non-resectable) tumor rest or an area at risk (R1/2-situation) has to be infiltrated. Cervical cancer recurrences at the pelvic wall (after primary treatment) are typical clinical examples.

Nanofluid concentration in the target area was claimed to be as high as possible. In these patients a retrospective planning based on a post-operative CT dataset was performed.

Direct temperature measurements via implanted catheters were suitable to estimate a mean perfusion and to calculate the temperature distributions in and around the target volume.

Tolerated H-fields were limited by local discomfort to 3–6 kA m^{-1} in the pelvic region and up to 7.5 kA m^{-1} in the upper thorax.

From our experience with brain tumor patients we know that magnetic fields of 10–14 kA m^{-1} are technically achievable and tolerated well. The H-field limitations are related to the periphery of the body where the accompanying electric fields and the current densities are highest. The larger the cross section, the lower the tolerated H field strength (comparing pelvis with head). However, treatment-limiting hot spots arise from skin folds, where the current path narrows and the current density, therefore, increases.

We also observed relative temperature maxima at tissue boundaries such as the anal sphincter and pelvic floor. These locations were, however, never treatment-limiting during our study, but are a typical problem during RF hyperthermia, where the E-fields are considerably higher.

One particular advantage of MagForce Nanotherapy is the option to plan the magnetic fluid distribution prospectively and to control the heat distribution thereafter to a high degree of reliability, thus increasing the safety of the application significantly.

Another major advantage of the new treatment is that, owing to the stability of the nanoparticle deposits, it can be repeated over weeks without additional injection of the magnetic fluid. This enables multimodal treatment concepts.

The heat treatments were tolerated well with only moderate side effects. The follow-up showed encouraging results for severe oncological diseases. Several strategies are available to further improve the effectiveness of treatment, e.g., to elevate the H-field (after modification of the applicator) or to increase the amount of magnetic fluid in the target.

8.4.3
Feasibility Study on Thermotherapy using Magnetic Nanoparticles in Recurrent Prostate Carcinoma

Despite much progress in the development of hyperthermia application and treatment planning systems for prostate cancer, clinical hyperthermia of this deep-seated organ is still a challenging problem [55].

Reported mean intraprostatic temperatures obtained with conventional heating methods have been between 40 and 42 °C, rarely reaching or exceeding the critical temperature of 43.0 °C, where a measurable cytotoxic effect is documented for various cell types [56, 57].

The first published report on thermotherapy treatment of human cancer using magnetic nanoparticles describes the preliminary experience with this technique in patients with locally recurrent, radioresistant prostate cancer [58]. In this investigation, treatment planning based on computerized tomography of the prostate preceded intraprostatic application of magnetic fluid. The number and position of magnetic fluid depots required for sufficient heat deposition was calculated according to the anatomy of the prostate and the estimated SAR of magnetic fluids in prostatic tissue. Nanoparticle suspensions were injected transperineally into the prostate under transrectal ultrasound and fluoroscopy guidance and six weekly treatments were delivered in an alternating magnetic field. Temperatures of up to 48.5 °C with a minimum of 40.0 °C were achieved in the prostate during the 1st treatment at a field strength of 4–5 kA m^{-1}, whereas during the 6th treatment, intraprostatic temperatures between 42.5 and 39.4 °C were recorded by means of invasive thermometry. Maximum and minimal temperatures, respectively, measured in the urethra were 42.4 and 38.0 °C whereas in the rectum 42.1 and 37.9 °C were recorded. Treatment was well tolerated without anesthesia. Since the injected magnetic fluid depots appear as signal-intense areas on the CT-scan, it could be documented that nanoparticles are retained in the prostate well beyond the treatment interval of 6 weeks. Thus, further applications of magnetic fluid are unnecessary for effective thermal treatments and at least hyperthermic temperatures can be achieved many weeks following intratumoral injection of nanoparticles.

Further refinement of the technique appears necessary regarding several important aspects. Firstly, the intraprostatic application and homogeneous distribution of nanoparticles appeared suboptimal due to fibrotic changes induced by the previous radiotherapy and may be improved. Most importantly, however, only one-third of the available power was used in these first clinical treatments, since an increase of magnetic field strength beyond 5 kA m^{-1} led to substantial patient discomfort in the perineum and the groin. This may be due to accompanying electric fields and current densities, which are highest in skin folds and at areas with the largest cross-sectional volume such as the pelvis. Since SAR increases quadratically with the magnetic field strength, significantly higher treatment temperatures can be achieved in the prostate by applying higher magnetic field strengths. Interposition of devices for focusing of the magnetic field may improve tolerability of treatment in the future. Furthermore, non-invasive temperature calculations based on evaluation of CT data are another area of ongoing research.

A phase I study evaluating feasibility, toxicity and quality of life during hyperthermia using magnetic nanoparticles, involving ten patients with radio-recurrent prostate carcinoma, is nearing completion. Preliminary clinical results suggest that thermotherapy using magnetic nanoparticles may be suitable for thermoablation of the prostate, since intratumoral temperatures well above 44 °C were achieved at low magnetic field strengths. However, further technical developments regarding the magnetic field applicator may be necessary to fully exploit the potential of this technique as a monotherapy. In addition, this treatment modality may also be combined with interstitial or external irradiation in patients with localized prostate cancer.

References

1 SELLINS KS, COHEN JJ, Hyperthermia induces apoptosis in thymocytes, *Radiation Res.*, **1991**, 126, 88–95.
2 CHRISTOPHI C, WINKWORTH A, MURALIHDARAN V, EVANS P, The treatment of malignancy by hyperthermia, *Surg. Oncol.*, **1999**, 7, 83–90.
3 DEWEY WC, FREEMAN ML, RAAPHORST GP, CLARK EP, WONG RSL, HIGHFIELD DP, SPIRO IJ, TOMASOVIC SP, DENMAN DL, COSS RA, Cell biology of hyperthermia and radiation, in MEYN RE, WITHERS HR (eds.), *Radiation Biology in Cancer Research*, Raven Press, New York, **1980**.
4 OLESON JR, Eugene Robertson special lecture. Hyperthermia from the clinic to the laboratory: A hypothesis, *Int. J. Hyperthermia*, **1995**, 11, 315–322.
5 OVERGAARD J, HORSMAN MR, Modification of hypoxia-induced radioresistance in tumors by the use of oxygen and sensitizers, in TEPPER JE (ed.) *Seminars in Radiation Oncology*, W. B. Saunders Company, Philadelphia, **1996**, pp. 10–19.

6 Overgaard J, The current and potential role of hyperthermia in radiotherapy, *Int. J. Radiat. Oncol. Biol. Phys.*, **1989**, 16, 535–549.

7 Anderson RL, Kapp DS, Hyperthermia in cancer therapy: Current status, *Med. J. Aust.*, **1990**, 152, 310–315.

8 Overgaard J, Gonzalez D, Hulshof MC, Arcangeli G, Dahl O, Mella O, Bentzen SM, Randomised trial of hyperthermia as adjuvant to radiotherapy for recurrent or metastatic malignant melanoma. European Society for Hyperthermic Oncology, *Lancet*, **1995**, 345, 540–543.

9 Wust P, Hildebrandt B, Sreenivasa G, Rau B, Gellermann J, Riess H, Felix R, Schlag PM, Hyperthermia in combined treatment of cancer, *Lancet Oncol.*, **2002**, 3, 487–497.

10 Stahl H, Wust P, Maier-Hauff K, Seebass M, Mischel M, Gremmler M, Golde G, Loffel J, Felix R, The use of an early postoperative interstitial-hyperthermia combination therapy in malignant gliomas, *Strahlenther Onkol.*, **1995**, 171, 510–524.

11 Sneed PK, Stauffer PR, McDermott MW, Diederich CJ, Lamborn KR, Prados MD, Chang S, Weaver KA, Spry L, Malec MK, Lamb SA, Voss B, Davis RL, Wara WM, Larson DA, Phillips TL, Gutin PH, Survival benefit of hyperthermia in a prospective randomized trial of brachytherapy boost +/− hyperthermia for glioblastoma multiforme, *Int. J. Radiat. Oncol. Biol. Phys.*, **1998**, 40, 287–295.

12 Clement GT, Hynynen K, A non-invasive method for focusing ultrasound through the human skull, *Phys. Med. Biol.*, **2002**, 47, 1219–1236.

13 Mitsumori M, Hiraoka M, Okuno Y, Nishimura Y, Li YP, Fujishiro S, Nagata Y, Abe M, Koishi M, Sano T, Marume T, Takayama N, A phase I and II clinical trial of a newly developed ultrasound hyperthermia system with an improved planar transducer, *Int. J. Radiat. Oncol. Biol. Phys.*, **1996**, 36, 1169–1175.

14 Schreier K, Budihna M, Lesnicar H, Handl-Zeller L, Hand JW, Prior MV, Clegg ST, Brezovich IA, Preliminary studies of interstitial hyperthermia using hot water, *Int. J. Hyperthermia*, **1990**, 6, 431–444.

15 Wust P, Seebass M, Nadobny J, Felix R, Electromagnetic deep heating technology, in Seegenschmidt MH, Fessenden P, Vernon CC (eds.), *Medical Radiology, Principles and Practice of Thermoradiotherapy and Thermochemotherapy*, Springer Verlag, Berlin, **1995**, pp. 219–251.

16 Wust P, Rau B, Gremmler M, Schlag P, Jordan A, Löffel J, Riess H, Felix R, Radio-thermotherapy in multimodal surgical treatment concepts, *Onkologie*, **1995**, 18, 110–121.

17 van der Zee J, Heating the patient: A promising approach? *Ann. Oncol.*, **2002**, 13, 1173–1184.

18 Jordan A, Wust P, Fahling H, John W, Hinz A, Felix R, Inductive heating of ferrimagnetic particles and magnetic fluids: Physical evaluation of their potential for hyperthermia, *Int. J. Hyperthermia*, **1993**, 9, 51–68.

19 Ma M, Wu Y, Zhou H, Sun YK, Zhang Y, Gu N, Size dependence of specific power absorption of Fe_3O_4 particles in AC magnetic field, *J. Magn. Magn. Mater.*, **2004**, 268, 33–39.

20 Jordan A, Wust P, Scholz R, Faehling H, Krause J, Felix R, Magnetic fluid hyperthermia (MFH), in Haefeli U, Schuett W, Teller J, Zborowsky M (eds.), *Scientific and Clinical Applications of Magnetic Carriers*, Plenum Press, New York, **1997**, pp. 569–595.

21 Jordan A, Rheinlander T, Waldofner N, Scholz R, Increase of the specific absorption rate (SAR) by magnetic fractionation of magnetic fluids, *J. Nanoparticle Res.*, **2003**, 5, 597–600.

22 Hergt R, Hiergeist R, Zeisberger M, Glockl G, Weitschies W, Ramirez P, Hilger I, Kaiser WA, Enhancement of AC-losses of magnetic nanoparticles for heating applications, *J. Magn. Magn. Mater.*, **2004**, 280, 358–368.

23 Gilchrist RK, Shorey WD, Hanselman RC, Parrott JC, Taylor CB, Medal R, Selective inductive heating of lymph nodes, *Ann. Surg.* **1957**, 146, 596–606.

24 Gordon RT, Hines JR, Gordon D, Intracellular hyperthermia. A biophysical approach to cancer treatment via intracellular temperature and biophysical alterations, *Med. Hypotheses*, **1979**, 5, 83–102.

25 Rand RW, Snow HD, Brown WJ, Thermomagnetic surgery for cancer, *J. Surg. Res.*, **1981**, 33, 177–183.

26 Jordan A, Scholz R, Wust P, Fahling H, Krause J, Wlodarczyk W, Sander B, Vogl T, Felix R, Effects of magnetic fluid hyperthermia (MFH) on C3H mammary carcinoma in vivo, *Int. J. Hyperthermia*, **1997**, 13, 587–605.

27 Hilger I, Hiergeist R, Hergt R, Winnefeld K, Schubert H, Kaiser WA, Thermal ablation of tumors using magnetic nanoparticles: An in vivo feasibility study, *Invest. Radiol.*, **2002**, 37, 580–586.

28 Ohno T, Wakabayashi T, Takemura A, Yoshida J, Ito A, Shinkai M, Honda H, Kobayashi T, Effective solitary hyperthermia treatment of malignant glioma using stick type CMC-magnetite. In vivo study, *J. Neurooncol.*, **2002**, 56, 233–239.

29 Yanase M, Shinkai M, Honda H, Wakabayashi T, Yoshida J, Kobayashi T, Antitumor immunity induction by intracellular hyperthermia using magnetite cationic liposomes, *Jpn. J. Cancer Res.*, **1998**, 89, 775–782.

30 Shinkai M, Yanase M, Suzuki M, Honda H, Wakabayashi T, Yoshida J, Kobayashi T, Intracellular hyperthermia for cancer using magnetite cationic liposomes, *J. Magn. Magn. Mater.*, **1999**, 194, 176–184.

31 Le B, Shinkai M, Kitade T, Honda H, Yoshida J, Wakabayashi T, Kobayashi T, Preparation of tumor-specific magnetoliposomes and their application for hyperthermia, *J. Chem. Eng. Jpn.*, **2001**, 34, 66–72.

32 Ito A, Tanaka K, Kondo K, Shinkai M, Honda H, Matsumoto K, Saida T, Kobayashi T, Tumor regression by combined immunotherapy and hyperthermia using magnetic nanoparticles in an experimental subcutaneous murine melanoma, *Cancer Sci.*, **2003**, 94, 308–313.

33 Tanaka K, Ito A, Kobayashi T, Kawamura T, Shimada S, Matsumoto K, Saida T, Honda H, Intratumoral injection of immature dendritic cells enhances antitumor effect of hyperthermia using magnetic nanoparticles, *Int. J. Cancer*, **2005**, 116, 624–633.

34 Kawai N, Ito A, Nakahara Y, Futakuchi M, Shirai T, Honda H, Kobayashi T, Kohri K, Anticancer effect of hyperthermia on prostate cancer mediated by magnetite cationic liposomes and immune-response induction in transplanted syngeneic rats, *Prostate*, **2005**, 64, 373–381.

35 Lucia MS, Bostwick DG, Bosland M, Cockett AT, Knapp DW, Leav I, Pollard M, Rinker-Schaeffer C, Shirai T, Watkins BA, Workgroup I: Rodent models of prostate cancer, *Prostate*, **1998**, 36, 49–55.

36 Johannsen M, Jordan A, Scholz R, Koch M, Lein M, Deger S, Roigas J, Jung K, Loening S, Evaluation of magnetic fluid hyperthermia in a standard rat model of prostate cancer, *J. Endourol.*, **2004**, 18, 495–500.

37 Johannsen M, Thiesen B, Jordan A, Taymoorian K, Gneveckow U, Waldofner N, Scholz R, Koch M, Lein M, Jung K, Loening SA, Magnetic fluid hyperthermia (MFH) reduces prostate cancer growth in the orthotopic Dunning R3327 rat model, *Prostate*, **2005**, 64, 283–292.

38 Johannsen M, Thiesen B, Gneveckow U, Taymoorian K, Waldöfner N, Scholz R, Deger S, Jung K, Loening S, Jordan A, Thermotherapy using magnetic nanoparticles combined with external radiation in an orthotopic Dunning rat model of prostate cancer, *Prostate*, **2005**, 66, 97–104.

39 Deger S, Taymoorian K, Boehmer D, Schink T, Roigas J, Wille AH,

Budach V, Wernecke KD, Loening SA, Thermoradiotherapy using interstitial self-regulating thermoseeds: An intermediate analysis of a phase II trial, *Eur. Urol.*, **2004**, 45, 574–579; discussion 580.

40 Jordan A, Scholz R, Maier-Hauff K, van Landeghem F, Waldoefner N, Teichgraeber U, Pinkernelle J, Bruhn H, Neumann F, Thiesen B, von Deimling A, Felix R, The effect of thermotherapy using magnetic nanoparticles on rat malignant glioma, submitted for publication, **2005**.

41 Archer SG, Gray BN, Vascularization of small liver metastases, *Br. J. Surg.*, **1989**, 76, 545–548.

42 Archer SG, Gray BN, Comparison of portal vein chemotherapy with hepatic artery chemotherapy in the treatment of liver micrometastases, *Am. J. Surg.*, **1990**, 159, 325–329.

43 Moroz P, Jones SK, Winter J, Gray BN, Targeting liver tumors with hyperthermia: Ferromagnetic embolization in a rabbit liver tumor model, *J. Surg. Oncol.*, **2001**, 78, 22–29; discussion 30–31.

44 Moroz P, Jones SK, Gray BN, The effect of tumour size on ferromagnetic embolization hyperthermia in a rabbit liver tumour model, *Int. J. Hyperthermia*, **2002**, 18, 129–140.

45 Mitsumori M, Hiraoka M, Shibata T, Okuno Y, Masunaga S, Koishi M, Okajima K, Nagata Y, Nishimura Y, Abe M, et al., Development of intra-arterial hyperthermia using a dextran-magnetite complex, *Int. J. Hyperthermia*, **1994**, 10, 785–793.

46 Mitsumori M, Hiraoka M, Shibata T, Okuno Y, Nagata Y, Nishimura Y, Abe M, Hasegawa M, Nagae H, Ebisawa Y, Targeted hyperthermia using dextran magnetite complex: A new treatment modality for liver tumors, *Hepatogastroenterology*, **1996**, 43, 1431–1437.

47 Minamimura T, Sato H, Kasaoka S, Saito T, Ishizawa S, Takemori S, Tazawa K, Tsukada K, Tumor regression by inductive hyperthermia combined with hepatic embolization using dextran magnetite-incorporated microspheres in rats, *Int. J. Oncol.*, **2000**, 16, 1153–1158.

48 Jones SK, Winter JG, Gray BN, Treatment of experimental rabbit liver tumours by selectively targeted hyperthermia, *Int. J. Hyperthermia*, **2002**, 18, 117–128.

49 Moroz P, Jones SK, Metcalf C, Gray BN, Hepatic clearance of arterially infused ferromagnetic particles, *Int. J. Hyperthermia*, **2003**, 19, 23–34.

50 Fine HA, Dear KB, Loeffler JS, Black PM, Canellos GP, Meta-analysis of radiation therapy with and without adjuvant chemotherapy for malignant gliomas in adults, *Cancer*, **1993**, 71, 2585–2597.

51 Stupp R, Mason WP, van den Bent MJ, Weller M, Fisher B, Taphoorn MJ, Belanger K, Brandes AA, Marosi C, Bogdahn U, Curschmann J, Janzer RC, Ludwin SK, Gorlia T, Allgeier A, Lacombe D, Cairncross JG, Eisenhauer E, Mirimanoff RO, Radiotherapy plus concomitant and adjuvant temozolomide for glioblastoma, *N. Engl. J. Med.*, **2005**, 352, 987–996.

52 Gneveckow U, Jordan A, Scholz R, Bruess V, Waldoefner N, Ricke J, Feussner A, Hildebrandt B, Rau B, Wust P, Description and characterization of the novel hyperthermia- and thermoablation-system MFH 300F for clinical magnetic fluid hyperthermia, *Med. Phys.*, **2004**, 31, 1444–1451.

53 Maier-Hauff K, Rothe R, Scholz R, Gneveckow W, Wust P, Thiesen B, Feussner A, von Deimling A, Waldoefner N, Felix R, Jordan A, Deep regional thermotherapy using magnetic nanoparticles: Results of a feasibility study with 14 glioblastoma multiforme patients, submitted for publication, **2006**.

54 Wust P, Gneveckow U, Ricke J, Feussner A, Henkel T, Kahmann F, Johannsen M, Kümmel S, Sehouli J, Felix R, Jordan A, Magnetic nanoparticles for interstitial thermotherapy – feasibility, tolerance, achieved temperatures, submitted for publication, **2006**.

55 Tilly W, Gellermann J, Graf R, Hildebrandt B, Weissbach L, Budach V, Felix R, Wust P, Regional hyperthermia in conjunction with definitive radiotherapy against recurrent or locally advanced prostate cancer T3 pN0 M0, *Strahlenther Onkol.*, **2005**, 181, 35–41.

56 Algan O, Fosmire H, Hynynen K, Dalkin B, Cui H, Drach G, Stea B, Cassady JR, External beam radiotherapy and hyperthermia in the treatment of patients with locally advanced prostate carcinoma, *Cancer*, **2000**, 89, 399–403.

57 Hurwitz MD, Kaplan ID, Hansen JL, Prokopios-Davos S, Topulos GP, Wishnow K, Manola J, Bornstein BA, Hynynen K, Association of rectal toxicity with thermal dose parameters in treatment of locally advanced prostate cancer with radiation and hyperthermia, *Int. J. Radiat. Oncol. Biol. Phys.*, **2002**, 53, 913–918.

58 Johannsen M, Gneveckow U, Eckelt L, Feussner A, Waldoefner N, Scholz R, Deger S, Wust P, Loening S, Jordan A, Clinical hyperthermia of prostate cancer using magnetic nanoparticles: Presentation of a new interstitial technique, *Int. J. Hyperthermia* E-pub, 27th June **2005**.

9
Ferromagnetic Filled Carbon Nanotubes as Novel and Potential Containers for Anticancer Treatment Strategies

Ingolf Moench, Axel Meye, and Albrecht Leonhardt

9.1
Introduction

Especially during recent years, the discovery of various species and modifications of carbon nanotubes (CNTs) have stimulated research on their applications, including in human medicine. The success of these applications depends significantly on the physical, chemical and biological properties of the CNTs and their modifications. For application in the human body, CNTs must be pure and biologically inert.

The development and testing of novel and alternative therapeutic concepts against cancer are needed, especially for advanced tumor types were no curative conservative treatment option is established. Attractive novel and potential tools for anticancer treatment strategies are nanoparticles (e.g., for hyperthermia) and nano-scaled tubes (e.g., for targeted drug release).

Here, we propose novel types of functionalized and ferromagnetic filled multi-walled CNTs (fff-MWCNTs) with various advantages for application in human medicine, especially in anti-tumor therapeutic concepts. These structures represent multifunctional nano-scaled containers for possible use in different medical treatments, including (a) magnetically guided hyperthermia, (b) heat-inducible drug delivery/carrier and stepwise drug release and enhancement systems, (c) internal tracer/drug carrier systems for the detection and/or (d) combinational applications with conservative treatment modalities. Such fff-MWCNT containers are schematically illustrated in Fig. 9.1. These structures combine the advantages of ferromagnetic containers (for hyperthermia or other mechanisms of heat transfer) with those from a broad spectrum for encapsulations and modifications within the carbon tube as well as outside.

Presently, human prostate cancer (PCa) – the most abundant tumor in men – with the well-known limitations of conventional therapies serves as an attractive model object for developing and optimizing treatment concepts based on and/or associated with fff-MWCNTs. First, a short overview and a clinical introduction about this tumor entity are given, followed by the different methods and techniques of synthesis and characterization of the MWCNTs, including novel techniques for their modification and functionalization. The treatment concept pro-

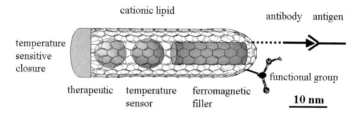

Fig. 9.1. Schematic of a multifunctional nanocontainer. (Modified from Ref. [232].)

posed will be realized by principles of nanotechnology; therefore, the magnetic properties of nano-scaled objects and materials are introduced. Furthermore, we discuss the relevant parameters for and mechanisms of heat transfer as well as of the specific absorption rate.

Recent knowledge of modified MWCNT will be summarized, including our own results for synthesis strategies for ffMWCNT, (bio)functionalization for the *in vitro* and *in vivo* transfer into different cellular systems and the accumulation in target cells and tissues. Future steps in the development of these nanocontainers are also discussed. These containers must fulfill numerous criteria, especially basic safety and efficacy requirements, before being used in humans.

For a potential future application *in vivo*, fundamental issues that need to be resolved include the homogeneity (uniformity of the carbon multi-wall sidewall, grade and composition of the ferromagnetic and additional filling materials) and purity of the MWCNT at the various stages of intrinsic and extrinsic functionalization.

This chapter describes the feasibility of different applications using MWCNTs, e.g., as heat mediators for hyperthermia of solid tumors. The extraordinary potential of the proposed MWCNTs is shown by their functionality as multiple containers. These therapies are based on heat dissipation in the ferromagnetic filling material of the MWCNTs. Therefore, the requirements for the heat transfer of the nanocontainer are related to those of the hyperthermia. Thus, the material requirements were derived from the knowledge of hyperthermia but are not limited to this application field. For this reason it is difficult to present a complete and comprehensive overview of all modifications and variants of functionalized MWCNTs, and so for technical and medical reasons the focus here will be on novel types of fff-MWCNTs and the model system of prostate cancer.

9.2
Prostate Cancer

9.2.1
Incidence, Risk Factors and Diagnostic Criteria

Prostate cancer (PCa) is the most commonly diagnosed cancer in men of western countries. However, except for the risk factors of age, race, geographic dependence, and family history of PCa, the etiology of this tumor entity remains poorly understood. PCa is primarily a tumor of older men. The incidence rate for men over the age of 65 is about 20-fold greater than that for men between 50 and 54. Less than 1% of PCa are diagnosed in patients in their first three decades of life. About 2–3% of all male deaths are attributable to PCa.

A familial history of PCa, *per definitionem* an affected father or brother with an association with at least a two-fold increase of disease, remains a consistent and important risk factor. About 40% of PCa cases are estimated to have a genetic component. However, PCa is a outstanding example of a tumor entity characterized by tumorbiological and genetic heterogeneity (reviewed in Ref. [1]). The prostate of a man diagnosed with PCa contains an average of five apparently independent lesions [2], and this multifocality is independent of familial history.

In the general population of the U.S., the risk of PCa is highest in African-American men, being approximately double that of their Caucasian counterparts. Moreover, Caucasians and African-American men in the U.S. have a PCa incidence that is 5–50× greater than that of Japanese men residing in Japan, and the incidence of PCa on Japanese immigrants to the U.S. is four times that of their native Japanese counterparts. This marked racial and cultural disparity indicates that dietary factors might affect PCa onset and growth.

As a serum tumor marker, prostate specific antigen (PSA), a serine protease belonging to the human glandular kallikrein family, has revolutionized the detection and management of PCa like no other marker in the history of oncology. With early identification of PCa using serum PSA, less than 10% of PCa with distant metastases are now diagnosed, a stage with five-year survival of only about one-third.

A patient's serum level of PSA, clinical tumor stage and Gleason grade provide valuable information to clinicians. The Tumor-Node-Metastases (TNM) classification system is used for the staging of PCa, whereas the Gleason score system is used for histologic grading characterizing the aggressiveness of this malignancy. Classification of PCa is based on size, invasion of the prostate capsule, and clinical stage. PCa develops in two different regions of the gland, with about 80% being found in the periphery, whereas the remaining cancers are found in the periurethral region, termed transitional zone (Fig. 9.2).

Primary PCa growth is characterized by an extremely heterogeneous and multifocal pattern. Initially, the tumor growth is an androgen-dependent and mediated by the androgen receptor. Despite an initial sensitivity, PCa become more or less quickly androgen-independent. Then the PCa becomes metastatic and, finally, hormone refractory [3].

The entity of PCa represents a significant public health problem and underscores the need for the development of improved diagnostic markers [4] as well as of treatment modalities.

9.2.2
Treatment Options, Outcome and Limits

The serum marker PSA improves the detection of clinically important tumors without significantly increasing the detection of unimportant tumors; most PSA-

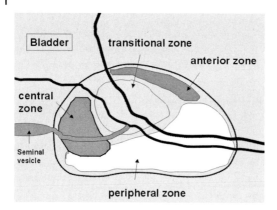

Fig. 9.2. Staging modality section of the prostate. The prostate gland with a branched structure can be divided in four distinct regions that reveal a different incidence of PCa: the anterior zone is purely fibro-muscular, the central zone contains the ejaculatory ducts (8% of cancers), the transitional zone (25% of cancers) is characterized by two lateral lobes, together with the periurethral glands, and the peripheral zone gives rise to 67% of cancers.

detected tumors are curable using current techniques, but there is no cure for metastatic disease. Treatment options, mostly radical prostatectomy, external beam radiation and brachytherapy, have become increasingly used to manage localized disease.

Contemporary methods of radical prostatectomy for patients with clinically localized disease are generally associated with excellent outcomes (reviewed in Ref. [5]). The cancer specific survival 10 years after radical perineal prostatectomy for patients with organ confined disease is approximately 90%. Crude survival rates at 10 and 15 years after surgery were similar to those of age-matched men from the general population without PCa. However, about 30–50% of patients thought to have organ-confined PCa at the time of surgery are later found to have disease beyond the prostate based on a careful review of the surgical specimen [5].

Neoadjuvant therapy with hormonal ablation has been evaluated in several studies and resulted in reduced positive surgical margins, and decreased volume of the PCa, equivalent to a pathological down-staging [6].

Virtually all recurrent clinical and metastatic diseases are preceded by a rising PSA, and only a few sporadic cases of recurrence have been reported in the absence of a detectable serum PSA. Biochemical failure is defined as either the persistence of detectable PSA after surgery or the development of detectable PSA in those with previously undetectable postoperative level.

For advanced disease, ablating androgenic hormones is the mainstay of therapy (reviewed in Ref. [7]), and will result in tumor response in about 80% of men.

For men considered to be at high-risk of treatment failure after local therapy alone, multimodal treatment strategies, for instance the combination of radiotherapy followed by hormonal therapy, may result in improved cancer-control outcomes. The benefit of neoadjuvant or adjuvant hormonal and/or chemotherapy followed by radical prostatectomy in high-risk patients is unclear but is the subject of ongoing or planned Phase III clinical trials. These studies will help to examine the role of multimodal treatment strategies in these high-risk patients.

In summary, alternative therapeutic strategies are needed for the treatment, especially, of locally advanced and metastatic PCa.

9.2.3
MWCNT Model

Figure 9.1 depicts a schematic diagram of a multifunctional nanocontainer that is possibly applicable for future anticancer therapy. The main component is a multi-wall carbon nanotube, filled with a ferromagnetic material, with an external diameter of 20–60 nm and a length of approx. 100–10 000 nm, as required. The case is made up of a pre-defined number of graphene layers (2–20), which guarantee an extremely high chemical stability. The external case is already filled during production. Iron [8–12], nickel [11], cobalt [13, 11], or FeCo [11, 14] may be used as ferromagnetic materials. The deposits can grow without a substrate in powder form [8, 9] or on substrates [10, 12], which is of particular technological advantage. Of these materials, we favor iron as a filling material for future application in antitumor therapy, owing to the possible toxic side effects of Ni, Co or FeCo alloys. Even at this early stage, the ff-MWCNT can be put to practical use as part of a therapy. One particularly notable consideration is magnetic particle hyperthermia. The basic effectiveness of this therapy and the biological compatibility of the material system have been confirmed in animal testing (mice). With regard to superparamagnetic particles (SPM, see Chapter 8) also discussed for these applications – using magnetite and maghemite – the different mechanism for heat generation must be born in mind.

The primary field of application is the single "magnetic particle hyperthermia" (MPH) or in combinations with other treatment modalities. Therapeutic efficacy and the biocompatibility of ff-MWCNTs have been confirmed in cell culture and animal experiments in recent years by our group.

For these applications and in comparison with other nanoparticles, especially superparamagnetic particles containing magnetite or maghemite (see Chapter 8), the function of ff-MWCNTs is based on a completely different mechanism of heat induction.

Apart from the basic functionality (for hyperthermia), the container system proposed allows a broad spectrum of further applications. Since the ferromagnetic filling is often partial and the ff-MWCNTs can be opened after synthesis, a secondary filling with an additional agent and a subsequent defined closing is applicable. The agents could be (a) enhancer substances for hyperthermia, (b) chemotherapeutics with better efficiency at higher temperatures (e.g., cisplatin). Furthermore, minimalized sensors can be integrated for contact-free measurement of local temperatures. The inclusion of rare elements for the localization of ff-MWCNT within the body has been suggested.

The opened ff-MWCNT is re-closed by the chemical reconstruction of the carbon tube or by addition of foreign materials such as heat-inducible and biocompatible polymers. Moreover, specific functionalization increases the accumulation within the target tissue and can stimulate the uptake by the target cells [15]. The proposed nanocontainers are suitable (a) alone for hyperthermia, (b) as adjuvant therapeutics or (c) alone or in combination with other nanoparticles (e.g., γ-Fe_2O_3 as SPM) for heat-inducible drug-releasing containers.

9.3
Carbon Nanotubes

9.3.1
General Remarks

After detailed systematic studies of very thin carbon filaments in a high-resolution transmission microscope Iijima reported the existence of SWCNT and MWCNT in 1991 [16, 17]. Since then there have been intensive investigations into the synthesis, structure and properties of these new nanostructured materials. Their unique electrical, thermal, optical and mechanical properties have already generated great interest for applications in many different fields. They have shown increasing potential for use as field emission devices, nanoscale transistors, tips for scanning microscopy or components for composite materials [18]. Moreover, CNTs filled with ferromagnetic metals such as Fe, Ni, or Co represent a fascinating new material. Owing to their size and enhanced magnetic coercivities these carbon-covered ferromagnetic nanowires have significant possibilities in different areas too. They can be used in magnetic recording media or, as individually filled ff-CNTs, as sensors for magnetic force microscopy [19]. In addition to pure CNTs, derivative nanotubes with attached chemical or biochemical groups have been prepared that should find applications in biomedicine [20].

Two types of CNTs can be distinguished, single-walled (SWCNTs) and multi-walled (MWCNTs). They can be described as cylindrically shaped molecules made of rolled up single or multilayer sheets of graphitic planes. Typical diameters are approximately 1 nm for a SWCNT and 10–100 nm for a MWCNT and a length reaching up to the centimeter-range (Fig. 9.3). Besides catalytic chemical vapor deposition (cCVD) [21, 22], electric discharge [17, 23] and laser ablation methods [24], various types of plasma-enhanced chemical vapor deposition methods [25–27] are practicable and controllable methods for the preparation of CNTs.

The electrical and electronic properties of carbon nanotubes depend on their geometrical structure and, for SWCNTs, on the so-called chirality. SWCNTs can be either metallic or semiconducting. CNTs are among the strongest and most resilient materials in nature.

The properties of MWCNTs strongly depend on the structure of their shells. Especially, the electrical conductivity varies between nearly metallic and insulator. The quality of shells depends on the synthesis method used.

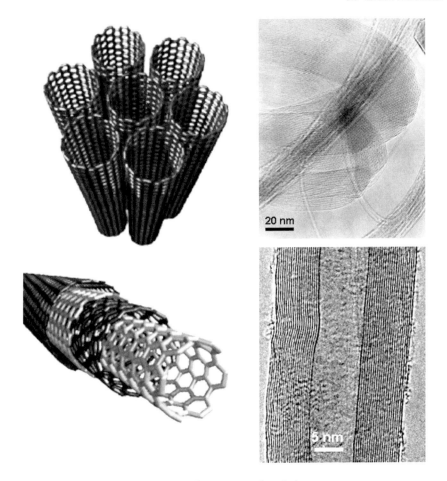

Fig. 9.3. Schematic and TEM images of SWCNT (single-walled carbon nanotubes) bundles and MWCNT (multi-walled CNT carbon nanotubes). (Modified from Ref. [233].)

Both SWCNTs and MWCNTs consist of two regions: the first is the sidewall of the tube and the second is the cap of the tube. By using different chemical and physical methods, the end caps can be opened and caps and sidewalls can be derivatized with different functional groups, radicals or molecules. Such manipulable nanotubes are well suited for application in biology and/or medicine. With a special material as filling (e.g., a ferromagnetic metal) CNTs can play an important role as a multifunctional nanocontainer in the diagnosis and therapeutic treatment of different diseases.

The next section describes the synthesis of f-MWCNTs and shows the unusual magnetic properties and possibilities of functionalization and manipulation (open, fill, close).

9.3.2
Preparation and Structure of Filled Multi-walled Carbon Nanotubes

9.3.2.1 Synthesis of Ferromagnetic Filled Multi-walled Carbon Nanotubes

f-MWCNTs can be synthesized by various methods. We classify two general versions: the first is an *"in situ* method", in which the formation of the CNTs and their filling with different elements or compounds take place simultaneously. The other method is a "step by step process": synthesis of empty CNTs – opening the tubes by chemical treatment – depositing and diffusion of filling material using different methods, mostly by wet chemical techniques – and, finally, closing the filled nanotubes by redeposition with a polymer or other carbon-containing phases (post-filling method).

In the following we only deal with *"in situ* techniques" for which there are physical and chemical methods. A physical, promising method is the "arc discharge". This method delivers long, relatively continuous nanowires of many different elements, e.g., transition metals (Cr, Ni, Re, Au) rare earth metals (Sm, Gd, Dy, Yb) and other elements (S, Ge, Se, Sb) [28, 29]. The alternative, well-established method for synthesizing CNTs filled with metals as Fe, Co or Ni is catalytic chemical vapor deposition (cCVD). Different groups world-wide have used this method in a similar way.

With the suitable precursors, cCVD can deliver well-defined MWCNTs with a relatively high grade of filling and on a large scale, especially with elements of the iron triad. Compounds of the organometallic complex family "metallocenes" are such precursors. They have a "sandwich structure" of two parallel cyclopentadienyl rings and a metal in the center between these rings [$M(C_5H_5)_2$ with M = Fe, Co, Ni]; they are solid at room temperature, readily applicable, soluble in different organic solvents but also show suitable decomposition behavior in the temperature range 600–1150 °C. Pyrolysis of these metallocenes (Fe, Co, Ni) has been analyzed by Dyagileva et al. [30]. They showed that the decomposition is a complex process of homogeneous–heterogeneous nature of a series of consecutive, parallel and catalytic reactions.

Thus, the synthesis process of ff-MWCNTs, based on the decomposition of metallocenes, will also be catalytically determined and, consequently, a support (substrate) covered (completely or partially) with an active catalyst material will promote the formation of filled nanotubes.

In principle, all known methods for manufacturing (Fe,Co,Ni)-filled CNTs have the same basic concept. Figure 9.4 shows a typical synthesis (two-band) reactor, as used by different groups of researchers.

Sen et al. reported the first experimental results for the synthesis of f-MWCNTs [31]. The pyrolysis of metallocenes (ferrocene, cobaltocene, nickelocene) was carried out in a quartz tube located in a two-stage furnace system. In a typical experiment a defined quantity of a metallocene was placed in a quartz boat, placed inside a furnace and a mixture of argon and hydrogen passed through the quartz tube (in some experiments acetylene is used as additional gas-phase component). The furnace temperature was increased to 200 °C, leading to the generation of the met-

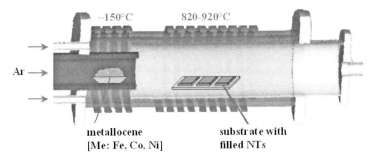

Fig. 9.4. Typical two-band furnace for the synthesis of filled carbon nanotubes, based on the decomposition of metallocenes. (Modified from Ref. [12].)

allocene vapor, which was carried by the Ar/H$_2$-stream into a quartz tube maintained at 900 °C in a second furnace. Depending on the deposition conditions, partially filled CNTs were unoriented or aligned deposited on the quartz tube walls, and no substrate was positioned in the reactor. Sen et al. have recognized that this method is a real alternative to previously known methods (arc discharge) [31].

Grobert et al. at Cambridge University (Kroto's group) has synthesized iron and Invar (iron/nickel alloy) f-CNTs in an aligned structure by using a mixture of ferrocene and fullerene (C$_{60}$) in the temperature range 900–1050 °C [9]. For the first time, the Cambridge group could also show the enhanced magnetic properties in such iron nanowires inside the CNTs. Later, Grobert used the aerosol-technique for the decomposition of ferrocene [32]. Here, a benzene solution containing ferrocene is nebulized by an aerosol generator in an argon flow and is injected in the reactor through a nozzle directly (Fig. 9.5).

In a further study the same group have prepared aligned, partly filled MWNTs (30–130 µm long) in high yield by using a compressed gas Ar driven atomizer

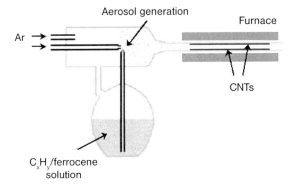

Fig. 9.5. Aerosol-technique for the injection of ferrocene in the deposition zone. (Modified from Ref. [32].)

[33]. They demonstrated that the yield of filling is proportional to the ferrocene concentration, but complete filling could be not achieved.

In 2002, Satishkumar et al. from Rao's group observed, besides increased coercivity, Barkhausen jumps with 5 emu g^{-1} steps in Fe-filled CNTs [34]. The ff-MWCNTs were prepared by pyrolysis of ferrocene along with acetylene. Selected area diffraction patterns of the nanowires show spots due to (011) planes of an α-phase, the ferromagnetic modification of iron. Additional X-ray investigations revealed γ-Fe and, as the minor phase, cementite (Fe$_3$C).

In our laboratory we have synthesized ff-MWCNTs on precoated substrates [10, 12]. On oxidized Si-wafers, thin layers (few nm) of the iron triad group (the catalyst material) have been deposited by using the sputter technique. Before beginning the deposition process, the pre-coated substrates were annealed at 900–1000 °C, directly in the deposition apparatus for a short time (few minutes). The annealing yielded some nm-sized islands on the substrate surface, which act as a source for nanotube growth. Hence, both iron from the surface in a solid or liquid-like state and the iron from the gas phase (ferrocene pyrolysis) deliver the material for the filling, causing a higher yield of filling than obtained by the groups of Kroto and Rao. With the higher filling yield the magnetic properties were also improved. The aligned-grown nanotubes show a pronounced magnetic anisotropy, with the easy axis perpendicular to the substrate plane and parallel to the axis of the aligned MWCNTs. Coercivity for the magnetic field direction perpendicular to the substrate amounts to 44.56 kA m^{-1} ($\mu_0 H = 56$ mT) compared with 19.89 kA m^{-1} (25 mT) for the direction parallel to the substrate. For comparison, the coercivity of bulk iron amounts to 0.072 kA m^{-1} (0.09 mT).

X-Ray diffraction analysis indicated the presence of bcc-Fe (α-Fe), fcc-Fe (γ-Fe) and, as minor phase, Fe$_3$C with a relatively strong $\langle 011 \rangle$ texture. An annealing process below the transition temperature of γ-Fe to α-Fe led to an increase of the ferromagnetic phase (α) and an enhancement of magnetization saturation [35]. Similar results could be realized in the same group for Co- and Co/Fe-filled MWCNTs. In the latter, the filled material enhanced coercivities to about 103.5 kA m^{-1} (130 mT) [14, 36].

Recently, our group has explored various means of improving the deposition procedure in regard to increasing the grade of filling and the abundance of ff-MWCNTs. We have also investigated aligned and relatively small, short f-CNTs that are suitable for special applications.

Figure 9.6 illustrates a so-called two-stage reactor. It shows a first conveyor belt system for a constant and reproducible transport of the ferrocene precursor. The ferrocene is dissolved in cyclopentane and dropped on the moving band continuously. In the first part the solvent is vaporized and the ferrocene is transported alone in the reactor at a defined temperature and constant transport velocity. In the deposition reactor a second moving band, populated with Fe-precoated substrates is positioned. In principle, such an apparatus can work continuously [37]. By using this equipment, very strong aligned f-MWCNTs can be synthesized with extremely high magnetic anisotropies and coercivities. The tube lengths are linearly dependent on the concentration of ferrocene in the solvent cyclopentane and

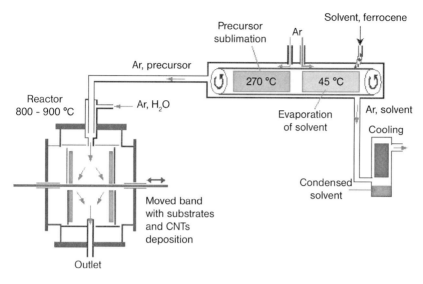

Fig. 9.6. A two-stage reactor for continuous deposition of filled carbon nanotubes. (Modified from Ref. [37].)

range between a few μm to 40 μm. In contrast, the diameter of filled nanotubes is independent of ferrocene concentration and constantly in the range < 10–50 nm [inner (Fe-wire) diameter] and 20–180 nm (outer tube diameter).

9.3.2.2 Crystallographic Structure of Core Material in Filled Multi-walled Carbon Nanotubes

Some research groups have determined the structure of the core material in f-MWCNTs. Detailed results exist about Fe-filled tubes, which we only report on here. Figure 9.7 shows typical TEM images of CNTs with a partial and a complete

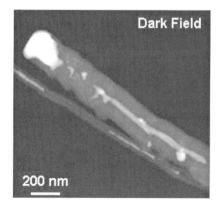 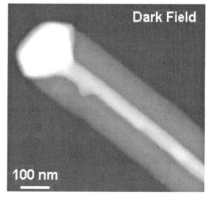

Fig. 9.7. Partly and completely filled multi-walled carbon nanotubes (MWCNT).

iron filling. The core material seems to be homogeneous. But which crystallographic phases does this material consist of? Several studies have investigated the X-ray-diffraction patterns of Fe-filled carbon nanotubes synthesized by cCVD in the temperature range 830–900 °C [33, 38, 39]. In most cases, the basic-centered cubic phase of iron (α-Fe) is the main component of the core material and is $\langle 100 \rangle$-textured along the tube axis.

Only Che et al. have reported that the filling material is amorphous if the cooling rate is very high after the synthesis process [40]. A post-annealing leads to crystalline α-iron. In addition, Pichot et al. have shown that inside their aligned grown MWCNTs carpets faced-centered cubic iron (γ-Fe) is the dominant phase, with a $\langle 110 \rangle$ texture parallel to the tube axis [41]. Finally, if the argon transport gas contains small amounts of oxygen (~1%) it is also possible that the filling consists of iron oxides such as hematite or magnetite [42]. These partly contrary results deserved detailed study. Recently, Kim and Sigmund have examined the crystallographic structures and orientations of iron nanowires inside carbon nanotubes with and without heat treatments at various temperatures [43]. The core material was found to consist of both bcc- and fcc-iron, in general, with their ratio dependent on the post-annealing temperature (without post-annealing and at room temperature the ratio γ/α is <1; after post-annealing at 1000 °C the ratio is near 1.0 and at 1400 °C approx. 2.3). Interestingly, the high temperature γ-Fe remains stabile in a large quantity at room temperature. The reason why is explained below. The bcc-structure of iron is oriented in the $\langle 001 \rangle$ or $\langle 111 \rangle$ directions, while the fcc-structure is aligned only in the $\langle 110 \rangle$ directions, along the axis of CNTs.

Besides the two modifications of iron a third phase, the thermodynamically unstable iron carbide phase Fe_3C, is often detected as a core material component. This phase plays a decisive role in the growth of empty and filled carbon nanotubes [44, 45]. Both Kim [43] and Shaper [45] have found that the iron carbide is enriched in the tips of the aligned CNT, and Shaper concluded that this iron carbide is an important intermediate in the catalyst-mediated growth of the tubes.

The application of iron-filled CNTs as nanomagnets requires complete filling with ferromagnetic iron, namely the bcc-phase α-Fe.

As reported, at a synthesis temperature of 850–920 °C the core material consists of α- and γ-Fe in various ratios, depending on the temperature, and a small amount of Fe_3C. Comparison with the Fe–C phase diagram shows that we do not have a thermodynamically stable state. The γ-Fe and Fe_3C cores are in a "frozen" state. Therefore, for the paramagnetic γ-phase to be transformed into the ferromagnetic α-phase, a post-annealing process at a temperature below the eutectic line at 723 °C is needed. However, the $\gamma \rightarrow \alpha$ transformation is connected with a 9% volume dilatation and if the γ-Fe is in a tight contact with the carbon shells it cannot expand and transform into α-Fe because of the high elastic modulus of CNTs. Nevertheless, such a post-annealing has been realized at 645 °C for 20 h by our laboratory [46]. After this treatment the fcc-phase was completely transformed into the bcc-phase, as proved by X-ray diffraction.

Investigations of the crystallographic structures of core material, especially filled with iron, have led to differing views on the growth mechanism of empty and filled MWCNTs.

9.3.2.3 Growth Mechanism of Multi-walled Carbon Nanotubes

The growth mechanism of MWCNTs, especially of filled MWCNTs is still under discussion. However, some experimental facts are certain. First, growth is always carried out by a catalytically determined process, which means that a catalyst material is absolutely necessary. Secondly, tubular MWCNTs grow by a vapor–liquid–solid (VLS) mechanism with the catalyst particle in a liquid-like constitution, in contrast to the bamboo-like structures produced with a solid catalyst particle (VSS growth mechanism, see Fig. 9.8) [47, 48].

When a CNT is grown on a substrate, two types of growth modes are observed: so-called tip growth and the base or root growth mode (Fig. 9.9). The actual mode depends on the contact forces between the particle and the substrate. Strong adhesion promotes the base or root growth mode – it is often observed on plain substrates.

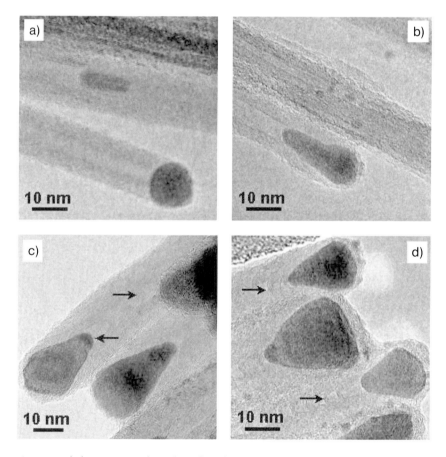

Fig. 9.8. Tubular MWCNT with a spherical catalyst (a, b) and a bamboo-like nanotube with a conic-shaped catalyst (c, d). (Modified from Ref. [230].)

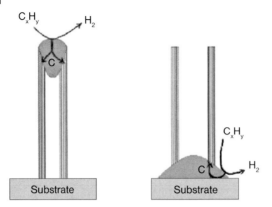

Fig. 9.9. Tip and root growth modes of a CNT (carbon nanotube) on a substrate [231].

The diameter of the growing nanotubes depends on the size of catalyst particle on the substrate and on the size of a stable catalyst particle on the Young contact angle between substrate and particle. Figure 9.10 shows an example.

Catalytic acting metals are mainly the iron triad group, i.e., iron, cobalt and nickel. Thus, metallocenes of these elements are used as precursors for the synthesis of CNTs. An effective, well-known method for the preparation of tubular MWCNTs is the injection method [49, 50] using ferrocene dissolved in a hydrocarbon (e.g., xylene or benzene). This mixture is injected in the hot zone of the reactor chamber. By spontaneous decomposition of both ferrocene and xylene a rapid and aligned growth of MWCNT is performed on the reactor walls or on a used substrate.

Dyagileva et al. have investigated the decomposition behavior of (Fe, Co, Ni)-metallocenes and ascertained that their thermal stability decreases in the order $Cp_2Fe > Cp_2Co > Cp_2Ni$ [30]. Furthermore, ferrocene decomposes mainly into H_2 and CH_4 above 500 °C, in contrast to the Co- and Ni-metallocenes, which mainly decompose to give cyclopentadiene (C_5H_6) and less H_2 and CH_4. This decomposition behavior of ferrocene is advantageous for growth because by hydrogen formation the Cp-complex is completely cleaved and carbon is released for tube shell growth. The hydrogen produced additionally reduces fresh ferrocene in the

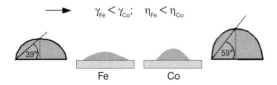

Fig. 9.10. Calculated Young contact angles for Fe and Co on oxidized silicon. Iron has a lower surface energy and a lower viscosity than cobalt. (Modified from Ref. [54].)

gas flux and inhibits oxidation of the iron and carbon. In fact, Dormans has observed the reduction of ferrocene at between 673 and 1173 K in a hydrogen atmosphere, whereas no decomposition was detected up to 1173 K in helium [51].

Consequently, the first step of the CNT growth mechanism is the homogeneous gas-phase decomposition of ferrocene:

$$Fe(C_5H_5)_2 \rightarrow Fe + 2C_5H_5 \rightarrow H_2 + CH_4 + C_5H_6 + \cdots \tag{1}$$

As additional carbon feedstock the solvent xylene or benzene is used in the injection method. For the hydrocarbon xylene, a derivative of benzene, Endo et al. have considered two gas-phase and four surface reactions and found very good agreement between experimentally determined and calculated production rates of MWNTs allowing for these six elementary equations [52].

Xylene reacts with the hydrogen formed by ferrocene decomposition to afford toluene which then yields benzene:

$$C_8H_{10} + H_2 \rightarrow C_7H_8 + CH_4 \tag{2}$$

$$C_7H_8 + H_2 \rightarrow C_6H_6 + CH_4 \tag{3}$$

These four hydrocarbons (xylene, toluene, benzene, methane) will decompose on the catalyst particle surface, and Eqs. (2) and (3) are named as the second step in the growth mechanism of MWNTs.

Promotion of reactions according to Eqs. (1–3) is always observed on the walls of the reactor chamber or on an inert substrate, positioned in the hot zone of the oven, because a heterogeneous reaction, two-dimensional nuclei formation, is always energetically favored. At relatively low temperatures (820–850 °C) such heterogeneous reactions are exclusively dominant.

It is assumed that the iron particles impinge on the substrate and solve the carbon. The solved carbon diffuses through the particles and precipitates again as tube shells. As mentioned above, a tip or root growth mode can be observed, depending on the adhesion forces on the exclusive substrate material. Wafers composed of oxidized silicon are often as substrates. The tip growth mode is often realized on such materials, where the catalyst material is deposited by a sputter technique on the substrate before nanotube growth is started. Nanometer-sized droplets are subsequently formed by a thermal pre-treatment of these very thin sputtered layers (a few nm thick) in either an argon or hydrogen atmosphere [12, 52]. The subsequent CNT growth is affected by the introduction of a catalyst-free hydrocarbon in the reaction chamber. In contrast, using spray pyrolysis of ferrocene dissolved in a hydrocarbon, spontaneous deposition on the reactor walls leads to base growth [53].

Normally, a grown MWCNT has only one catalyst particle on either the base or tip. The core of tube is hollow and free of inclusions.

By using the ferrocene pyrolysis technique with and without an additional hydrocarbon the synthesized carbon nanotubes show a partial filling with catalyst mate-

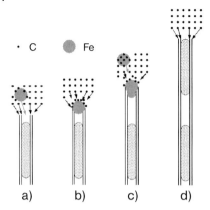

Fig. 9.11. Continued growth of partially filled carbon nanotubes. (Modified from Ref. [234].)

rial; the tubes exhibit multiple particles or wires along the middle of the tube (Fig. 9.11).

Investigations aiming to explain this phenomena have shown that the root or base growth mode is a suitable model for describing the growth of such partially ff-MWCNTs [50, 52, 53]. Figure 9.12 shows the initial state schematically. Short catalyst-containing tubes deposit on the reactor wall or on the substrate. The tubes always show an open tube end (root growth). With continued growth the gas-phase delivered carbon and iron and iron particles can fall into these open tube ends. Simultaneously, carbon dissolves in the particle and graphite precipitation occurs into the wall of the CNT, thereby increasing the tube length. The particle is deformed due to the squeeze of the tip. As the distance between the open tube end and the catalyst particle increases, it becomes harder for carbon to reach this particle. After a defined time, a new iron particle falls on the open end and growth is continued. The model explains not only the growth of partially filled tubes but also the growth of very long tubes.

Is it is possible to influence the yield of filling by deposition conditions? Cer-

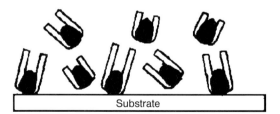

Fig. 9.12. Initial state of growth of partially filled carbon nanotubes when using the spray technique with continuous catalyst addition. (Modified from Ref. [53].)

tainly, the catalyst addition rate plays an important role. This addition rate can be increased by a higher Fe/carbon ratio in the gas phase (ferrocene decomposition without hydrocarbon) or by first and additional catalyst particles on the substrate. Using latter method we have reached a filling yield of nearly 50 vol% [46]. In doing so, we could prove that both the catalyst material from the substrate and from the gas phase participates in the filling of nanotubes. Furthermore, the deposition of filled tubes using cobaltocene as catalyst and a carbon feedstock was realized on an iron pre-coated substrate. The result was a carpet of aligned Fe/Co filled MWCNTs. The iron could be detected in the tips of tubes, meaning that it diffused into the Co filling material that was delivered from the gas phase.

As mentioned above, the open tube end model can explain the rapid growth of super long MWCNTs. Nevertheless, with longer deposition times the growth rate decreases and more Fe particles and no nanotubes are formed [11, 54]. This means that although new catalyst material is supplied again and again the catalytic reaction at a tube material will be ever more deactivated. This effect of catalyst deactivation is well known and is indicated by an overcoating of the iron particles with an amorphous carbon layer. It is caused by a momentary supersaturation of carbon on the surface of the catalyst particle due to too high a carbon supply from the gas phase.

9.3.3
Post-treatment: Opening, Filling and Closing of MWCNTs

The application of ff-MWCNTs as multifunctional nanocontainers (for instance in the medicine) requires the ends of these tubes to be opened, further filled with drugs or agents and, finally, re-closed (cf. Fig. 9.1).

The opening process of both SWCNTs and MWCNTs is relatively well controlled. This procedure is often connected with a necessary purification of the synthesized nanotube material. CNTs are effectively purified and opened by an oxidation process, either by a treatment in an oxidizing atmosphere at increased temperature (O_2 annealing or treatment in a O_2-containing Ar-plasma [55–57]) or by a wet-chemical post-treatment in an oxygen acid (HNO_3) [58, 59].

Tsang et al. first published the method of wet-chemical opening with HNO_3 in 1993 [60, 61] and today this method is used by many research groups. However, all these procedures, which are primarily developed for the purification of CNTs, i.e., as means of eliminating of amorphous carbon and metal particles, are very aggressive. They not only remove the ends of the carbon nanotubes, they also damage parts of the tube walls. Such defects in the tube walls reduce the mechanical properties and limit the application fields. Recently, Raymondo-Pinero have reported on the simultaneous purification and opening of MWCNTs by using of solid NaOH [62]. By such a solid–solid reaction between NaOH and as-grown carbon nanotubes the MWCNTs will be purified and opened without defects in the tube shells (Fig. 9.13).

The next necessary step for the construction of a nanocontainer is to fill the opened carbon with the desired agents, drugs or sensor materials. In principle,

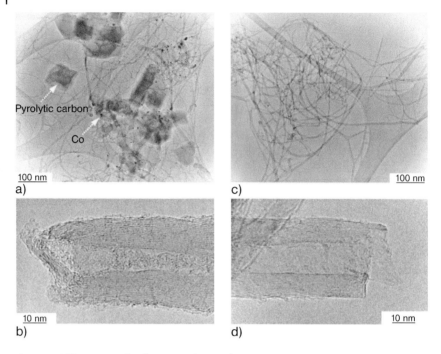

Fig. 9.13. TEM images of carbon nanotubes: (a, b) as grown and (c, d) after reaction with NaOH at 800 °C for 1 h. (Modified from Ref. [62].)

this is generally possible because the hollow interior can serve as a nanometer-sized capillary.

Ajayan and Iijima could already show in 1993 that CNTs can be filled with molten material through such a capillary action (Pb and Bi melts) [63]. By using a simple wet chemical method (suspending the CNTs in a nitric acid solution containing nickel nitrate), Tsang et al. have filled the CNTs with Ag [61]. In 1996, Chu et al. described a two-step method for filling CNTs with silver and gold metal with a high grade of filling [64]. Ugarte et al. have studied in detail the wetting and capillarity by metal salts by a similar two-step method [65]. First, nanotubes were opened by oxidation in the air and, subsequently, nanotubes were immersed a in molten salt (e.g., $AgNO_3$).

Wu et al. have successfully prepared Fe-Ni alloy nanoparticles inside carbon nanotubes by wet chemistry [66a]. Using nitrates of iron and nickel, filling lengths of up to 50 nm were reached. Gao and Bando have developed a special nanothermometer by filling carbon nanotubes with liquid gallium [66]. Gallium has one of the greatest liquid ranges of any metal and has a low vapor pressure and a high thermal expansion coefficient. Such a thermometer can be used over a wide temperature range.

Besides filling with metals or salts, CNTs can be also filled with oxides [67, 68] and halides [69, 70]. Recently, CNTs have been filled controllably with fluorescent particles by immersion in liquid ethylene glycol, by the complementary action of capillary forces and evaporation of the liquid [71]. In addition, Korneva et al. have described a simple capillary action technique to fill CNTs with paramagnetic iron oxide particles using commercial ferrofluids (particle size $\sim$ 10 nm) [72].

Both of the latter results show important progress and that filling a CNT with solid particles (up to 10 nm diameter!) is generally possible. This is one of the most important requirements for filling CNTs with drugs or agents and thus for possible application as a nanocontainer to transport these drugs to specific locations in the human body.

After the post-filling process it is necessary to re-close the nanotubes. This can be achieved by reaction with different reagents such as ethylene glycol or by treatment with benzene or another hydrocarbon vapor in a reducing atmosphere of argon and hydrogen at increased temperatures [73].

Another possibility is an additional thermal treatment with a biocompatible monomer that will polymerize at higher temperatures.

9.4
Magnetism in Nano-sized Materials

9.4.1
General Remarks

The principles suggested here for putting the multifunctional nanocontainer to use can only really become effective when it is produced on a nm scale. Understanding the difference between bulk and nm materials and the changes in properties they cause is the main condition for appreciating how magnetic materials can be used in biomedical applications [74, 75]. To better understand the special uses of ferromagnetic materials, especially those with geometric forms at the nm scale, some fundamental magnetic parameters, and dimension-related changes where necessary, must be briefly elaborated. These relations are, however, highly complex and extensive, meaning that only some selected challenges can be briefly described at this point. For complementary, continuing deliberations, reference should be made to the relevant standard works on magnetism [76–80], and to studies that deal specifically with physical characteristics specific to nanoscale arrays of ferromagnetic materials [81–86]. For an immediate understanding of the problem, we explain briefly some of the parameters necessary for the suggested therapy plan using a nanocontainer to be effective. In particular, these include those that affect heat development by AC heating. Magnetic reorientation is responsible for losses in ferro- or ferrimagnetic materials such ff-MWCNT. The reorientation depends on the type of demagnetization process. These processes are determined by intrinsic magnetic properties such as magneto crystalline anisotropy and magnetization, on the one hand, and extrinsic properties such as shape, particle size and aspect ratio on the other hand. The size and shape dependence of H_c are well known.

H_c will be maximized when the size reaches a critical low size (single domain particle). In the same way, it can be enhanced for particles having large aspect ratios [75]. To study these dependences in detail, the challenge of producing uniform nanomagnets must be solved. To prepare nanomagnets, complex technologies are often used, not only based on conventional thin film or CVD technology but also on a combination of wet chemical methods and special technology. One technique that has proved particularly valuably is the electrolytic deposition of ferromagnetic materials in the pores of suitable substrates (e.g., Al_2O_3) [82–85, 87, 88].

Another point that needs to be explained is that the preferred ferromagnetic metals (iron, cobalt and nickel) are, for example, susceptible to oxidation [89–92] and that the oxides may have different magnetic characteristics than the pure metal. If the characteristics of nanoparticles with diameters of <100 nm are to be determined and used, the volume/surface ratio detrimental to oxidization must be considered here. The improvement of biocompatibility is often the reason why nanoparticles are coated with paramagnetic materials. For this purpose, Al or Al_2O_3 [93–95], Au [96], Si or SiO_2 [97–99], and other materials are discussed. Several groups suggest the use carbon [92, 100–104]. In the case of the ff-MWCNT, protection from oxidation is guaranteed by the carbon [9]. Even a small number of graphene layers provide excellent protection against chemical changes. The magnetic characteristics do not change over a long time, even in unfavorable conditions [9]. For example, Fe-filled MWCNTs can be stored in acids and brines [105] with no demonstrable reaction and thus no demonstrable change in their magnetic characteristics. This avoidance of oxidation is acquired at the cost of possible limited biodegradability, however. Nonetheless, these external cases/coatings can also indirectly have an (sometimes positive) effect on the magnetic qualities, e.g., by mechanical stresses [106], even if these case materials are paramagnetic by nature. As well as the applications described here, magnets in the desirable single-domain state generally demonstrate a wide range of potential uses in biomedicine [107], if the challenges of production can be solved.

9.4.2
Magnetization in Nano-sized Materials

Ferromagnetism is the result of an ordered alignment of the atomic magnetic moments. Fe, Co, Ni and their alloys are the most important substances of this material class. Ferromagnetic materials are materials that can remain magnetized after application of an external magnetic field. This external field is typically applied by another permanent magnet, or by an electromagnet. If the temperature of a ferromagnetic material is raised above a certain point, called the Curie temperature (T_c), the ferromagnet loses its long-range magnetic order that establishes the spontaneous magnetization and becomes simply paramagnetic [108].

The magnetic behavior of a ferromagnetic material can be illustrated in a plot of the magnetization (M) versus the applied field (H). For ferromagnets, which are the focus of this work, the "hysteresis loop" [109] is the most evident distinction. The remnant magnetization (M_r), the saturation magnetization (M_{sat}) and the co-

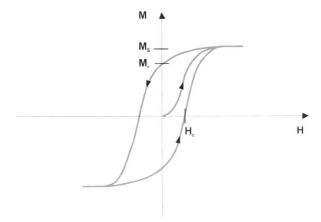

Fig. 9.14. Typical hysteresis loop (M vs. H) for ferromagnetic materials.

ercive field (H_c) are important magnetic properties (Fig. 9.14). M_r is the magnetization of the sample in the absence of an external applied field and is often used in relation with M_{sat} (i.e., M_r/M_{sat}). H_c is the applied field in the direction opposite to the current magnetization direction, which is necessary to bring the global magnetization to zero [79]. This does not imply that $M = 0$, but the magnetization breaks up into domains that are magnetically oriented such that the integration over the entire sample disappears. Table 9.1 shows the values of some magnetic characteristics for bulk Fe, Co and Ni.

In contrast to bulk material, enhanced values of the coercivity have been obtained for ferromagnetic nanoparticles [110, 111] or nanowires encapsulated inside a carbon envelope, where the coercivity reaches tens of millitesla [9]. Generally, the origin of this phenomenon is related to the small size and the single domain nature of the encapsulated metal crystals [9].

9.4.3
Influence of the Dimensions on the Magnetization Distribution

Without a magnetic field applied, a piece of a ferromagnet often assumes a state with global magnetization vanishing. Such behavior indicates the presence of domain closures inside the material. The overall magnetization $M_{overall}$ of the magnetic object is given by Eq. (4).

$$M_{overall} = \frac{1}{V_{overall}} \sum_i M_i V_i \qquad (4)$$

where M_i and V_i are the magnetization and the volume of the domain, respectively, and $V_{overall}$ is the volume of the ferromagnet.

Tab. 9.1. Some magnetic characteristics of the ferromagnetic elements Fe, Co and Ni (bulk and nano-sized materials).

		Fe (bcc)	Co (hcp)	Ni (fcc)
Bulk material				
Curie-temperature	T_c (°C)	770	1131	355
Saturation magnetization	M_s (emu cm^{-3})	1710	1422	484
First anisotropy constant	K_1 (erg cm^{-3})	4.6×10^5	45×10^5	-0.5×10^5
Exchange constant	A (erg cm^{-1})	2.5×10^{-6}	1.3×10^{-6}	0.86×10^{-6}
Coercive field (bulk)	H_c (A m^{-1})	71.6	795.8	159.2
Nano-sized material				
Coercive field (nano-sized)	H_c (A m^{-1})	31 830 [96] 127 337 [54] 99 470 [11] 21 200 (par.) [136] 37 900 (perp.) [136] 6446 [110] 49 800 [111] 9700 [166] 24 020 [166] 270 572 [171]	55 900 [111] 47 100 [36]	1190 [172] 1800 [111]
Critical diameter	D_{cr} (nm)	32 ($l/d \approx 3$) [84] 45 ($l/d \approx 2$) [86] 85 ($l/d \approx 5$) [86] 140 ($l/d = 10$) [86]	28 ($l/d \approx 3$) [84] 140 ($l/d = 10$) [86]	64 ($l/d \approx 3$) 600 ($l/d = 10$) [86]

Figure 9.15(a) depicts a ferromagnet consisting of many domains with an overall magnetic moment ($M_{overall}$) equal to zero. At zero applied fields, a domain circuit or closure is created to approach the lowest energy state and the minimum demagnetizing field [78]. In the following, we consider the changes in domain state that depend on the diameter of the ferromagnetic material (D_p) relative to the critical diameter D_{cr} for a single domain state.

When D_p of a magnetic nanowire is larger than D_{cr}, the creation of more magnetic domains or a strong non-uniform magnetization ($\nabla M \neq 0$) allows a more stable lower energy state. When D_p is less than the critical single domain diameter, the magnetization of the nanowire is almost uniform ($\nabla M \approx 0$) and spreads along the wire axis (Fig. 9.15b). Micromagnetic calculations for magnetic nanowires with an aspect ratio (length/diameter) of ≈ 3 have shown the following dependence:

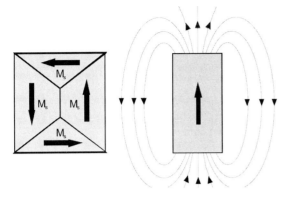

a) Multi-domain particle   b) Single-domain particle

Fig. 9.15. Magnetization in magnetic elements: (a) multi- and (b) single-domain particle. (Modified from Ref. [84].)

$$D_{cr} \approx 3.5 \frac{\sqrt{A}}{M_s} \qquad (5)$$

where A is the exchange constant (Table 9.1). Estimated D_{cr} for Fe, Co and Ni are 32, 28 and 64 nm, respectively. The critical diameter should also increase with increasing aspect ratio (length/diameter). These dependences for Fe, Co, and Ni have been determined by Sun et al. using micromagnetic calculations [86]. They reported critical diameters (Fe) of ≈ 45 nm (aspect ratio = 2), ≈ 85 nm (aspect ratio 5) and 140 nm (aspect ratio 10).

Figure 9.16 shows several magnetization configurations and their respective hysteresis curves, which depend on the nanowire diameter. Magnetic nanowires with a diameter significantly larger than D_{cr} break into domains of uniform magnetization to approach the lowest energy state (Fig. 9.16a). The overall remnant magnetization of this multidomain state is negligible. When the nanowire diameter decreases down to D_{cr}, the formation of a domain wall will not be energetically favored and no domains are formed. In this case, the magnetization is non-uniform. Such a magnetization configuration is called a "flower" [112] (Fig. 9.16b).

Another configuration observed in this type of magnetic material is the "vortex" structure. There is a flower–vortex transition that depends on the particle size and the aspect ratio. Micromagnetic calculations for Ni nanowires (aspect ratio < 3) made by Ross et al. point to the existence of a "flower-state" magnetization for small diameters [113]. As the diameter increases, a gradual transition to a "vortex" state occurs.

For ff-MWCNTs with filler diameters of 10–70 nm and aspect ratios of 1 → 100 (Fig. 9.7), from the point of view of magnetism we are in the transitional area from

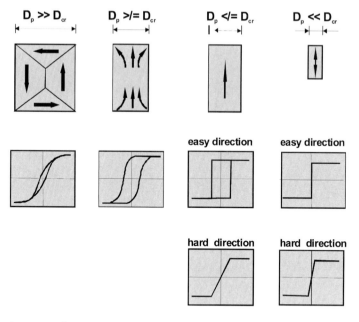

Fig. 9.16. Schematic magnetization configuration and hysteresis loops of ferromagnetic materials of different sizes. (Modified from Ref. [84].)

single domain to "flower/vortex" structures. For specific applications it seems of advantage, in the long run, to aim for a single domain structure. However, to carry out alternating field heating with this single domain material, taking into consideration the limiting values for magnetic fields in medical applications, a specific alteration is necessary; seen from the point of view of the maximum values attained on the material we synthesized, it is a *reduction* of $H_c \leq 15$ kA m^{-1} (18.8 mT).

When the nanowire diameter is $<D_{cr}$, the magnetization is almost uniform and is oriented along the wire axis (Fig. 9.16c). In this case, the hysteresis loop for the direction along the nanowire axis has the form of a rectangle. As the nanowire size continues to decrease within the single domain range, another critical threshold is reached, at which the remnant magnetization and the coercivity go to zero. When this happens, the material becomes superparamagnetic (Fig. 9.16d). Superparamagnetism occurs when the wire becomes so tiny that random thermal vibrations at room temperature cause them to lose their ability to hold their magnetic orientation, resulting in a spontaneous reversal of the magnetization. The geometric dimensions at which this transition from the ferromagnetic to the superparamagnetic state takes place are known as the superparamagnetic limit (SPL). For

iron [114, 115] and carbon-wrapped Fe wires [116], arranged in an array, these dimensions have been estimated at <10 nm.

9.4.4
Anisotropy and Interaction

Magnetic anisotropy is another characteristic evident from the hysteresis. This is described as the tendency of a ferromagnetic material to have a preferential direction, "easy axis", of magnetization [80]. For Fe, which has a bcc structure, the (100) axis is the easy axis, whereas the (110) and (111) axes correspond to the medium and hard axis, respectively. The anisotropy of magnetic nanowires is usually determined by the form of the ferromagnetic material and its crystal structure. In our experience, the crystal structure of the filler is a given fact. As described above, ff-MWCNT filler consists mainly of monocrystal α- or γ-Fe. The γ-Fe phase can be transformed into ferromagnetic α-Fe(110) by subsequent heat treatment.

In addition to the magnetocrystalline anisotropy, the preferential direction of magnetization can be determined by the shape of the material (i.e., a wire prefers the easy axis to be along its large dimensions) – shape anisotropy. The essential anisotropy contribution in thin metal nanowires is the shape anisotropy with the easy axis of magnetization parallel to the wire axis [117]. This anisotropic contribution can be decisively influenced by production technology, by means of the filler diameter and length. The shape anisotropy of a ferromagnetic nanowire is determined by the stray field energy of each nanomagnet, the decisive factor being the stray field vector H_D, which determines the easy magnetic direction. According to Eq. (6) the local stray field energy density E_D is lowest when M points in the direction of the stray field determined by the nanomagnet.

$$E_D = -\tfrac{1}{2} |\overrightarrow{H_D}| \cdot |\overrightarrow{M}| \cdot \cos \beta \tag{6}$$

To estimate the stray field energy, the conventional model of Stoner and Wohlfarth is often drawn upon [118]. This model assumes that the cylinder (ff-MWCNT filler) can be approximately represented by a homogeneously magnetized rotationally symmetric ellipsoid. The aspect ratio of a Stoner–Wohlfarth particle, $a = l/D_p$, is decisive for its shape anisotropy energy. With further simplifications, such as presuming an aspect ratio of $\gg 1$ and the orientation of the magnetic moment along the axis, the effect of the stray field can be described in a simplified manner as an additional anisotropy field H_D of the shape anisotropy. H_D is given by Eq. (7).

$$H_D = 2\pi M_{sat} \{1 + 3/a^2 [1 - \ln(2a)]\} \tag{7}$$

This means it is possible to determine the magnetic field necessary to rotate the magnetization in the axis 180°. In the borderline case of infinitely long nanowires, H_D approaches the value given in Eq. (8).

$$H_{D,\infty} = 2\pi M_{sat} \tag{8}$$

The model of Stoner and Wohlfarth drawn upon here can only describe a simple estimation, but clearly shows the principal influencing variables.

For H_D we examined a single nanomagnet. When a collection of several nanowires is involved, they affect each other interactively.

In terms of material production, a distinction must be made here between the conditions of the production of ff-MWCNTs on a substrate and the conditions of ff-MWCNTs in tumor tissues. Whereas magnetic nanowires on a substrate are very closely packed in great numbers, once they are detached and transferred into the tissue they spread out. The magnetostatic interaction of the nanomagnets with the external field is described by the *Zeeman* therm.

$$E_{zee} = -HM_{sat} V \cos \Phi \tag{9}$$

In Eq. (9), Φ is the angle between the magnetic field and the magnetization. When the magnetic field is realized along the easy axis of the nanomagnet, the hysteresis loop is rectangular. The coercivity H_c is equal to the anisotropy field (H_A, Eq. 10).

$$H_A = \frac{2K_u}{M_{sat}} \tag{10}$$

In evaluating the magnetic properties of nanomagnets like ff-MWCNTs we have to differentiate between the properties of structures arranged on a substrate and disordered in a tissue. Here it is meaningful to appropriate an ensemble of non-interacting and randomly oriented Stoner–Wohlfarth particles. Averaging over all possible spatial direction results in a hysteresis loop characterized by $M_r = 0.5 M_{sat}$ and $H_c = 0.48 H_{sat}$. Hysteresis losses are reduced by a factor of ≈ 0.25 in comparison with the aligned situation [119, 120]. These differences mean that the hysteresis loops of a single material can be markedly different on the substrate (after production) and after being transferred to biological systems. For further examination of this problem, reference should be made to the relevant technical literature.

9.4.5
Magnetic Reversal

The mechanisms of magnetic reversal are also important for future therapy options using ff-MWCNTs. What is of significance here is the relative size of the nanomagnets expressed by the D_P/D_{cr} ratio (as in Fig. 9.16). To turn the magnetization in a nanomagnet around 180°, a reversal field, H_{SW}, is necessary. This field depends mainly on the stray fields and the anisotropic contributions. For the magnetic reversal of particles with the dimensions examined here, a homogeneous rotation is mostly discussed [86, 118] or, by Curling's rotation [86, 121], an inhomogeneous rotation. The magnetic bipolar moments remain in parallel alignment in the case of homogeneous rotation during magnetic reversal. They rotate to the new orientation together. This mechanism is often adopted for mesoscopic, single-domain ferromagnets with a uniaxial anisotropy. The accompanying magnetization curve is rectangular when the external magnetic field H_{ex} is parallel to the axis of the nanomagnets; it is a linear function of the external field in the case of a perpendicular field direction. Curling's inhomogeneous rotation is discussed for $D_P \geq D_{cr}$, i.e., for flower structures, for example. At the ends of the ferromagnetic cylinder, vortex structures are expected to form and then spread out across the cylinder as the opposing field increases. In practice, especially when a very high number of nanomagnets are spatially randomly dispersed and interact with one another, magnetic reversal can not be satisfactorily described using a model. It must be assumed that all energy contributions according to Eq. (11) are moving towards a common minimum; E_K is the magnetocrystalline anisotropic density, E_A is the exchange energy density, E_{zee} is the magnetostatic energy density (*Zeeman* energy), and E_D the stray field energy density.

$$E_{ges} = E_K + E_A + E_{zee} + E_D \tag{11}$$

9.4.6
Magnetic Properties of Filled Multi-walled Carbon Nanotubes

The effects described above are crucial for ff-MWCNTs as a material system, which is the central element of the proposed multifunctional nanocontainer. In evaluating the magnetic properties, we must take into account the following aspects:

1. The fillers are mainly produced in a monocrystalline form (Fig. 9.17). They may exist not only as α-Fe but in some cases also as γ-Fe or Fe_3C. In the present case, α-Fe is the desired ferromagnetic phase. A transformation of $\gamma \rightarrow \alpha$ is partly possible using temperature treatment. The higher volume of the α-phase may in some cases change the mechanical stress state. By this means the shape of the magnetization curve and H_c could sometimes be influenced.

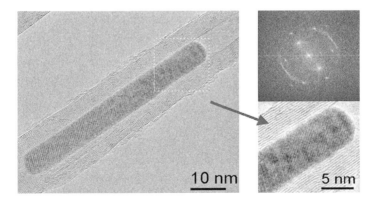

Fig. 9.17. TEM image and SAED pattern of a partly filled Fe-MWCNT.

2. Consideration must be given to variation in the level of filling in general, the filler diameter and the length of the ferromagnetic nanoparticles (aspect ratio $1 \rightarrow 100$, see Fig. 9.7). Because of these, both single domain and flower/vortex states may be achieved in nanotubes from one batch.
3. During therapy, the ff-MWCNTs are no longer arranged parallel to one another in the form of up to 10^9 tubes per cm^2, but face in all directions in a statistically even manner. This affects the way the ff-MWCNTs interact magnetically.
4. The number of carbon shells can vary considerably, which is another reason why the influence of mechanical stresses on magnetic properties can not be ruled out.

These few points suffice to make it clear that many different variables must be considered when evaluating the magnetic properties of ff-MWCNTs. Much fundamental research is still necessary for a complete understanding of these relations. Rising to this fascinating challenge is important not only for the biomedical applications discussed but also for a whole array of other interesting applications.

CNTs filled with most different materials have numerous applications in different areas. These unique nanostructured materials with a nanowire as material inside and the carbon shell around are receiving increasing attention. The closed CNT saves the sensitive core material by their relatively chemically inert carbon shells against oxidation or reduction. Therefore, the material will open up new applications, e.g., in nanoelectronics. ff-MWCNTs have attracted increasing interest. "Saved magnetic wires" are fascinating objects in itself, revealing unusual magnetic properties [122, 123], e.g., they will find application as magnetic electrodes in future molecular spintronic devices or as new material for high-density magnetic recording media [124]. As already mentioned, one very interesting application area is their utilization as magnetic nanoparticles in medicine, as material for hyperthermic therapy of cancer tissues.

The major advantage of filled CNTs is that their catalyst materials, the metals of the iron triad group (Fe, Co, Ni), are ferromagnetic over a wide temperature range. Thus, the catalyst and the necessary ferromagnetic material are identical (Section 9.3.2.1).

All three metals are suitable for hyperthermic application at temperatures up to 45 °C as they are all ferromagnetic in this temperature range.

Nevertheless, the ferromagnetic properties of the filling material are determined by the presence of the ferromagnetic Fe-phase (α-Fe). Therefore, the synthesis temperature is one of the most important parameters. Conforming with the phase diagram, unsurprisingly, at a synthesis temperature of 850 °C the Fe filling consists of both para- and ferromagnetic iron and, particularly in the presence of carbon, iron carbide too. Complete transformation into the ferromagnetic phase is not possible, because the transformation of paramagnetic (fcc lattice) into the ferromagnetic phase (bcc lattice) is associated with a 10% volume dilatation. Iron tightly embedded in the nanotube would cause a permanent strain on the carbon shells; however, that is not possible due to the high elastic modulus of CNTs. The alternative is, necessarily, destruction of the nanotubes.

Several groups have investigated the magnetic properties of ff-MWCNTs. Figure

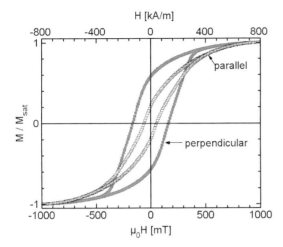

Fig. 9.18. Typical hysteresis loops of Fe-MWCNTs grown on a silicon substrate (parallel = magnetic field parallel to the substrate on which the Fe-MWCNTs are in perpendicular positions).

9.18 shows a typical hysteresis loop of an ff-CNT carpet on a silicon substrate. The measurements always reveal a uniaxial magnetic anisotropy with the easy axis along the tube axis. The magnetic characteristics saturation magnetization (M_{sat}) and coercivity (H_C) depend not only on the synthesis parameters and the γ/α-Fe ratio, as mentioned above, but also on the alignment level, the diameter and the length of the CNTs on the substrate (Fig. 9.19). For instance, Satishkumar et al. have found that these magnetic properties are significantly dependent on the used hydrocarbon [34] (Table 9.2). This is because of the different diameters, lengths and alignment levels produced as a result of the significantly different growth rates obtained with different hydrocarbons. The higher the growth rate the stronger the alignment. With elongation of the aligned CNTs the magnetic anisotropy increased. This means that the aspect ratio determined the magnetic anisotropy too (so-called shape anisotropy). This phenomenon causes the extremely high coercivities observed in such CNT carpets.

Table 9.3 summarizes experimentally determined coercivities parallel to the tube axis. The values were determined using superconducting quantum interference devices (SQUID), an alternating gradient magnetometer (AGM) and a vibrating-sample magnetometer (VSM). As can be seen, the single values differ greatly, but are always higher at low temperature. In addition, the coercivities are far higher than in bulk material [$H_{C(Fe-bulk)} = 0.072$ kA m^{-1} (0.09 mT), $H_{C(Ni-bulk)} = 0.159$ kA m^{-1} (0.2 mT) and $H_{C(Co-bulk)} = 0.796$ kA m^{-1} (1 mT)]. This is due to the shape anisotropy and to the aligned arrangement of the CNTs on Si substrates.

Figure 9.19 illustrates clearly that at a strong alignment the anisotropy and the H_C values are very high in contrast to the case of sub-optimal alignment, where the shape anisotropy is weak and the H_C values are identical. This very interesting behavior is typical for arrangement on a substrate. After removal from the sub-

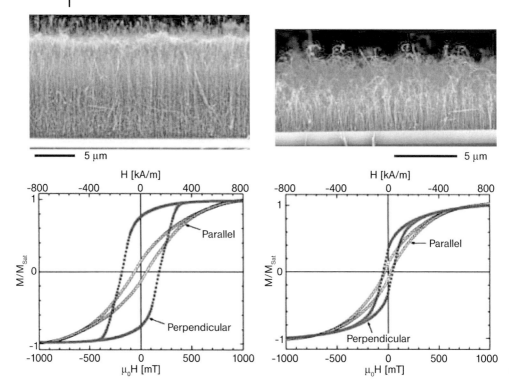

Fig. 9.19. Comparison of hysteresis loops of strong and sub-optimal aligned fCNT on Si-substrates. Left: strong, optimal alignment; right: sub-optimal alignment.

strate the ff-MWCNs lose their alignment. To study this effect, the removed nanotubes were transferred into human cancer cells [15] on a sample holder substrate. Figure 9.20 shows a typical hysteresis loop for this situation. The measured curves for magnetic fields in the x, y, and z directions are almost identical. This is most likely due to the statistical random orientation of the ff-MWCNTs. The H_cs for the individual directions in kA m^{-1} are H_{cx} = 18.35 (23 mT),

Tab. 9.2. Magnetic properties of Fe-filled carbon nanotubes obtained from the pyrolysis of ferrocene–hydrocarbon mixtures.

Hydrocarbon	Ar flow (sccm)	M_S (emu g^{-1})	H_C (kA m^{-1})
Methane (50 sccm)	950	20	40.9
Acetylene (50 sccm)	950	29	45.4
Butane (50 sccm)	950	48	46.2
Acetylene (100 sccm)	900	90	45.4

Tab. 9.3. Coercivities of filled carbon nanotubes.

Filling material	H_C (5 K) (kA m^{-1})	H_C (300 K) (kA m^{-1})	Ref.
Fe	85.2	34.2	9
Fe	74.8	27.7	199
Fe	111.4	63.7	39
Fe		≤79.6	200
Fe	52.5	24.4	38
Fe	46.2[a]	6.4[a]	136
Fe	87.6	31.8	19
Fe		52.5	34
Fe		42.2[a]	11
Fe		44.6[a]	46
Fe		127.3[a]	54
Fe/Co		100.3[a]	14
Co		47.0[a]	36
Co		26.6	201
Ni		14.6	202
Ni	3.2		203

[a] Aligned ff-MWCNTs on Si-substrates, measured along the tube axis.

$H_{cy} = 18.43$ (23.1 mT) and $H_{cz} = 18.83$ (23.6 mT). These values give the relation $H_{c\_perpendicular}/H_{c\_parallel} \approx 1.02$, which is significantly smaller than for parallel aligned nanotubes on a Si substrate. H_{c_par}/H_{c_perp} for ff-MWCNTs grown on a wafer lies in the range 1.8–6. The magnetic characteristics determined for ferromagnetic filled nanotubes in cells and tissues are much more representative for

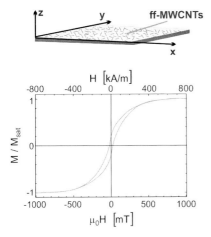

Fig. 9.20. Hysteresis loop of Fe-MWCNTs in cancer cells on a sample holder substrate.

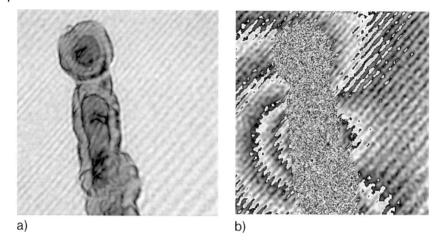

Fig. 9.21. Electron holography images of a single Fe-filled MWCNT. (a) Reconstructed intensity distribution and (b) reconstructed phase distribution outside the structure. The phase shift reflects the enclosed magnetic flux line pattern. (Modified from Ref. [122].)

applications in therapy than those obtained for ff-MWCNTs on a substrate. Our data – based on hysteresis loops, holographic experiments (Fig. 9.21) and MFM measurements – strongly suggest that *single* nanotubes can display switching behaviors as single-domain element. Because of the large number of tubes in an issuing batch it is possible to find different filling diameters and different aspect ratios. In the present state of technology, you can produce particles with an aspect ratio in the range $1 \rightarrow 100$ in one batch (Fig. 9.7). The fraction of single domain and flower/vortex structures is not well-defined. The present state of technological knowledge is insufficient to predetermine the fraction of single domain and flower/vortex structures in one batch. Remarkably, concerning Fe-MWCNTs there is a gap in the evidence regarding the general transferability of results, because either 1 or $>10^4$ tubes have been evaluated. It is a challenge to transfer the results from one area to another. Study of the intermediate area should be expanded experimentally. Detailed investigations would be important in addressing these issues.

9.5
Heat Generation

9.5.1
General Remarks

When exposing materials in alternating magnetic fields, apart from the qualities of the materials, the geometric conditions and the frequency applied, we may in some cases also have to take into account several heating principles, which differ due to their individual dependency on frequency (f) and modulation ($\hat{B}$) [120, 125–127]. The possible heating principles do not generally all occur at the same time in the different applications. What are principally seen are relaxation losses, dielectric losses, hysteresis losses and eddy current losses. In the technical literature, different formulae are sometimes used for the dependencies that apply, according to whether the work is of a magnetic or biomedical nature. Here, we mainly use the formulae that have become established in the biomedical literature. For hyperthermia using superparamagnetic (very small) particles, Rosenszweig has refined a model that was originally developed to describe dielectric dispersion in polar liquids [128].

For small field amplitudes, and assuming minimal interactions between the particles, the response of the magnetization to an AC field can be described in terms of the its complex susceptibility $\chi = \chi' + i\chi''$. Both parts, χ' and χ'', are frequency dependent. The out-of-phase χ'' components result in heat generation, given by Eq. (12) [128].

$$P_{spm} = \mu_0 \pi \chi'' H^2 \qquad (12)$$

This can be interpreted physically as meaning that if M lags H there is a positive conversion of magnetic energy into internal energy. This simple theory compares favorably with experimental results, e.g., in predicting a square dependence of P_{spm} on H, and the dependence of χ'' on the driving frequency.

As this heating principle does not play a decisive part in the ferromagnetic particles we favor, for further considerations and a detailed description of the dependencies reference should be made to Chapter 8 (Jordan et al.) of this book. Another principle considered is that of dielectric losses. These result in Eq. (13).

$$P_d \approx c_1 f^3 \hat{B} \qquad (13)$$

For a concrete example of the application of ff-MWCNTs in alternating magnetic fields <20 kA m^{-1} (<25.1 mT) with frequencies of up to 250 kHz (the operating range we prefer for ferromagnetic filled nanocontainers) the dielectric losses and the energy transfer in the tissue are also not the determining factor. They are only of importance for far higher frequencies [129–134]. In the application suggested here, the heating of ferromagnetic materials (e.g., ff-MWCNTs) with dimensions above the superparamagnetic limit, the determining heating principle is assumed to be that of heating by hysteresis losses. The general view is that the such losses are determined according to Eq. (14).

$$P_{FM} = \mu_0 f \int H \, dM \qquad (14)$$

Apart from the frequency used, the area under the magnetization curve is also an especially decisive factor in heat transfer. Equation (14) is sometimes also used in the form of Eq. (15).

$$P_{FM} \approx c_2 f H_c B_r \tag{15}$$

From this it can be seen that the hysteresis losses of different materials can be compared using the material parameters H_c and B_r. To achieve the high heat transfer desirable in the application of hyperthermia, it appears expedient to choose a material with a high H_c and a high B_r. However, in this case, notably, the application of very high magnetic fields is not possible for medical reasons (meaning hard magnetic materials can be considered only to a limited extent) and, physiologically, not all possible materials can be used. For example, the elements cobalt and nickel and their alloys, which are often used for magnetic materials, must be considered toxic and therefore questionable [135]. For this reason we prefer to discuss the use of iron as a possible material. The magnetic parameters that characterize iron have been investigated in many studies and determined with great precision. The coercivity of bulk Fe is considered verified at 0.072 kA m^{-1} (0.09 mT). As described above, the dimensionality of the ferromagnetic materials has a significant influence on the parameters examined [9, 19, 136]. In Fe-CNTs synthesized by our group, as the filler mainly consisted of single-domain particles, and because of the shape anisotropy, we were able to verify values from 103.45 kA m^{-1} (130 mT) [11]. This rise by a factor of >1300 compared with the H_c of the bulk material has a direct influence, according to Eq. (15), on the energy transfer that can be achieved and is therefore of fundamental importance. Pure iron can only be considered as a possible material for hyperthermia when there is a rise of H_c compared with the bulk material values and in particular when the *shape* of the magnetization curve is specifically modified. Iron with any other magnetic configuration (multi-domain particles) does not allow sufficient energy transfer [137]. Furthermore, the demands on the magnetizing field concerning the field strength needed also depend on the H_c and B_r of the material to be re-magnetized. For an isotropic dispersal of magnetically uniaxial particles, $B_r \approx J_s/2$, where J_s is the saturation polarization. The magnetic field amplitude should fulfill the condition of Eq. (16) in order for the material to be nearly entirely re-magnetized (see also Ref. [120]).

$$H_{ext} > H_c + \tfrac{1}{3}\mu_0 J_s \quad \text{with } \mu_0 = 4\pi \times 10^{-7} \text{ Vs m}^{-2} \tag{16}$$

The magnetic reversal losses of various hard magnetic materials have been studied [138]. At a diameter of 30 μm, the particle size was such that eddy current losses are no longer negligible. Nonetheless, it was still possible to show that the conditions of Eq. (16) must be fulfilled to achieve high-performance yields. In Ref. [120] this is also confirmed for the use of magnetite.

According to Eq. (15), as well as the aspect that can be technologically influenced (shape of magnetization curve), the energy transfer is linearly dependant on the frequency of the alternating field applied.

As well as the alternating magnetic field losses discussed above, one last substantial contribution in particular must be considered, namely energy transfer due to eddy current losses that occur. This heating, known technically as inductive heating, occurs when conductive materials – even non-magnetic ones – are subjected to an alternating field. Several important parameters, including conductivity, essentially determine the possible energy transfer. As the human body also possesses an appreciable degree of electrical conductivity it too can be heated according to this principle. Considering these losses is extremely important as they directly determine the boundary conditions of the alternating field heating actually desired. For inductive heating, both the geometric dimensions and electrical conductivity are determining factors. There are great differences between the factors to be considered. While ff-MWCNTs are geometrically very small (10^1 nm) and exhibit very high electrical conductivity, biological tissue displays only low electrical conductivity, but the dimensions of a real patient are significant (10^8 nm). For this reason, both systems reach the limiting values when considered as a whole and both materials connect differently. It therefore seems advisable to examine these cases separately, taking that of ff-MWCNTs first. In the usual technical discussions on inductive heating, the losses that occur are generally described by Eq. (17).

$$P_{\text{eddy}} \approx c_3 f^2 \hat{B}^2 d^2 \sigma_e \tag{17}$$

In Eq. (17), f is the frequency, $\hat{B}$ the working induction, σ is the penetration depth, and c is a parameter that depends on the materials and geometry; d represents the work-piece diameter.

An essential difference between the heating methods discussed here is the frequency dependency of the energy transfer. Whereas alternating field losses increase linearly with frequency during magnetization reversal processes, the square of the frequency is a determinant of eddy current losses. Figure 9.22 presents these

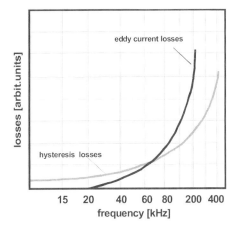

Fig. 9.22. Variation of both hysteresis losses and eddy current losses with frequency. (Modified from Ref. [204].)

Tab. 9.4. Dependence of the degree of efficiency on the ratio d/σ ($d/\sigma = 0.4$–8). (Modified from [204].)

d/σ	8	6	4	2	1	0.6	0.4	0.0001 (Fe-MWCNTs)
Degree of efficiency (%)	95	85	65	30	10	4	1	≪1*

* ≪1 is a postulation, but the value could not be determined experimentally.

ratios for iron with large geometric dimensions. Clearly, above approx. 70 kHz – i.e., in the area of interest for a therapy – the eddy current losses very quickly come to dominate in the case of electrical conductivity *and* dimensions on a mm level. A special feature of inductive heating is that energy transfer tends to occur more in the outer areas of the work piece. Because of the skin effect, eddy current flow is mostly on the surface of the material. The depth at which the energy density drops to $1/e = 0.368$ is known as the current penetration depth. Some 86% of the energy is transformed between the edge of the work piece and σ, and the rest heats lower-lying layers.

The penetration depth itself depends on the electrical conductivity of the material to be heated. For energy transfer to be very effective, the geometric conditions must be optimally adjusted. The most important (influencable) parameters here in particular are the work-piece diameter d and the construction of the coil system. Once the boundary conditions, which can be technically altered, have been optimized, the ratio work-piece diameter d to penetration depth σ is of decisive importance for the energy transfer that can be achieved. Table 9.4 illustrates the effect of this parameter on energy transfer. Whereas energy transfer still reaches 10% at a d/σ ratio of 1, at a ratio of 0.4 it already drops to 1%. When Fe-containing ff-MWCNTs are used a d/σ ratio of $\approx 10^{-4}$ is achieved. Figure 9.23 illustrates these dependencies. The figure 9.23(a) shows a screw with a diameter of 10 mm 10 s after start of AC-heating. The induced increase in temperature was >500 K. Moreover, the so-called "skin" effect is clearly seen. The large temperature difference was mediated by the high electric conductivity and the nearly optimal geometry. Figure 9.23(b) shows a substrate placed at the same site (10×10 mm) with aligned Fe-MWCNTs. In this case and under the same conditions no change in temperature was found. The difference between the two cases is based on the geometry because both consist of the same material (Fe).

Gilchrist et al. have already dealt with this challenge in their fundamental publication describing the initial application of magnetic particle hyperthermia [130]. They were able to show that, in metallic powders, the eddy current loss becomes negligible at a level of <5 μm. With the ff-MWCNTs the diameter of the magnetic particles is not 5 μm but only 0.015 μm. This seems to verify the assumption that eddy current heating does not dominate in nanoparticles of this dimension; an as-

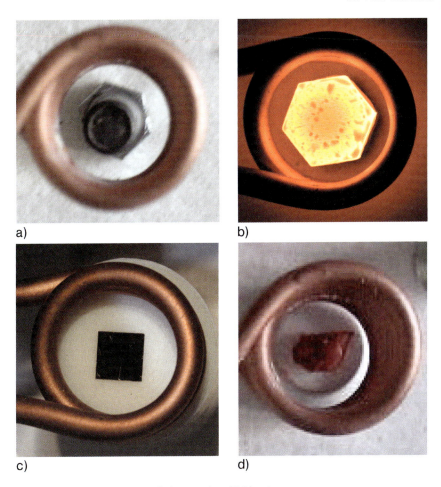

Fig. 9.23. Schematic images and photographs of AC-heating experiments using iron in the cm- (a) or nm-range (b).

sumption shared by other authors [120, 125, 128, 139]. Taking these results into consideration, we conclude that for hyperthermia using Fe-containing MWCNTs, eddy current losses are negligible and magnetic alternating field losses can be assumed to dominate.

The other borderline case with eddy current heating is the direct heating of human tissue. In the biomedical literature, the modification of the Eq. (17) written as Eq. (18) has become established [131, 133, 140], where σ_T is the conductivity of tissue and r is the distance from the central axis of the body.

$$P_{\text{tissue}} = \frac{\pi^2 \mu^2 \mu_0^2 \sigma_T r^2 (f^2 H_0^2)}{2} \tag{18}$$

Equation (18) is valid only for $H_0 = $ const and $\sqrt{(\omega \sigma \mu_0 r_{max})} \ll 1$ [135]. Biological tissue fulfills this condition because of its low electrical conductivity.

In principle, this heating method can also be used to treat patients. It has not become established as defined localized heating, e.g., in lower-lying tissue, is only possible to a very limited extent. Instead, the objective is to concentrate the heat precisely in the tumor tissue. For thermoseed treatment, Stauffer et al. called for a performance ratio of at least 10:1 [131].

When therapy focuses on magnetization reversal heating it is, therefore, wise to find ways to minimize eddy current heating. According to Eq. (18) r, f and H are the alterable parameters. As the square of the radius is a determinant of performance rate, an undesirably high level of heating is most likely to occur if the body is surrounded by the coil system at its widest point (the chest cavity). The problem is then lessened if treatment can be carried out in the extremities with far smaller radii. Theoretical calculations and experiments have shown that an $H \times f$ product of $\leq 4 \times 10^8$ A m^{-1} s^{-1} can be safely applied when treating tumors in the chest cavity. Higher values can be achieved if the radius can be reduced. Hilger et al., for example, specify an $H \times f$ product of 1.5×10^9 A m^{-1} s^{-1} when treating breast cancer (coil diameter 150 mm) [141]. For this very important boundary value to be determined with certainty, however, in our opinion too few experimental data are as yet available.

9.5.2
Requirements for the Development of Materials for Hyperthermia and Magnetism

The use of magnetic nanoparticles in hyperthermia goes back to Gilchrist et al., who treated different tissue samples with particles of γ-Fe$_2$O$_3$ (20–100 nm) in a magnetic field of up to 20 kA m^{-1} (25.1 mT) and a frequency of 1200 kHz [130]. Since then, numerous different materials and technical boundary conditions have been proposed.

The concept of locally and contact-free heating in the tumor tissue itself by raising the temperature up to >42 °C for 30 min to influence apoptosis and/or to destroy the tumor is fascinating, but requires reproducible conditions for particle production as well as reproducible heat dispersal in the tumor. At the same time, the therapeutic effects on humans depends on the actual conditions in the appropriate target tissue and organ. In particular, heat transport in different kinds of tissue and the differing blood supply in the organs mean that only limited theoretical predictions can be made. What is more, the temperature rise required for long-term therapy success has been estimated at varying levels. Hilger et al., for example, have suggested implementing temperatures around 55 °C, the thermoablative level, when treating breast cancer, as when the temperature is limited to 44 °C approx. 50% of the tumors recur [142].

Since the pioneering work of Gilchrist et al. [130] various materials have been suggested for hyperthermia using magnetic nanoparticles. These can be classified conveniently according to their magnetic condition at the temperature of application (41–55 °C). There is a differentiation between:

1. Particles that are in a superparamagnetic state under the application conditions due to their geometric dimensions or their physical properties. These include Fe_3O_4 (magnetite) or γ-Fe_2O_3 particles (maghemite) with *small* dimensions

and

2. Particles that are in a ferri- or ferromagnetic state under the application conditions due to their geometric dimensions or their physical properties, e.g., γ-Fe_2O_3 particles (maghemite) of *large* dimensions, multidomain ferrite or ferromagnetic materials.

Several interesting studies have been carried out using materials of these two groups [120, 130, 137, 140–156]. This summary of different studies can not be complete. It reflects the different goals concerning the material and the heating mechanism. Many different materials – such as alloy-based systems – have been used by other groups. Our focus is directed towards pure iron or some iron compounds.

Another interesting alternative option in terms of heat generation, although with geometric dimensions at the μm or mm scale, is:

3. "Self-regulating thermoseeds" with a Curie temperature of >42 to approx. 57 °C [157–160], i.e., above the Curie temperature these materials change their properties and the energy transfer is greatly reduced. This strategy shows that self-regulation mechanisms can be achieved by optimizing material properties – here the Curie temperature – in a targeted manner. If the AC losses of these types of alloys can successfully be raised considerably in moderate fields, then this principle of temperature limiting could also be of interest for magnetic particle hyperthermia.

Materials can only be sorted into categories (1) and (2) above as suggested because the materials of one composition can be placed in either group solely on the basis of their geometric dimensions. For maghemite and magnetite particles, it is not always possible to state unequivocally that the material is in a superparamagnetic state. The division does, however, take into account the fact that the mechanism for heat transfer is fundamentally different [120, 161], meaning that different steps must be taken to selectively develop the material. Based on this division, we now confine this review to particles with ferromagnetic properties. With regard to superparamagnetic nanoparticles, which are very important for hyperthermia, the reader is referred to Chapter 8 by Jordan et al., which describes this point in detail.

As the first, simplest application of ff-MWCNTs – note that this principle only applies here – we begin from a therapeutical point of view by focusing on the treatment of urological carcinomas, including PCa. These are diseases that are, statistically, more likely to occur in later life. It must be taken into account that quite a substantial number of patients may already be fitted with metal implants, e.g., hip prostheses. This means that the use of high frequencies is only possible and/or practical in a limited way due to the occurrence of eddy current heating [129, 132, 133, 157]. For this reason, within the limits of the admissible $H \times f$ product [133, 141], we focus our attention on higher fields and not the commonly-implemented higher frequencies. According to the conditions described in Section 9.5.1, high frequencies produce eddy current losses on a mm scale, and thus possibly very high temperatures (Fig. 9.23) that can result in serious health problems.

For the successful therapeutic application of magnetic particle hyperthermia, the key data contained in Table 9.5 must be taken as a starting point. In the last few years, superparamagnetic material (SPM) particles have increasingly been discussed, as they are attributed with better heat generation. This judgment is based on the magnetic values of the materials in a multidomain state [137]. The production, for instance, of Fe nanoparticles that are, magnetically, in the desirable single domain or flower condition constitutes a technological challenge. For example, Fe

Tab. 9.5. SAR values of different materials.

Material	Diameter (nm)	SAR (W g^{-1})	Frequency (kHz)	H (kA m^{-1})	Ref.
Magnetite (Fe$_3$O$_4$)					
Fe$_3$O$_4$		7.5	400	6.5	143
Fe$_3$O$_4$		950	410	10	
Fe$_3$O$_4$		50–200	410	6.5	161
Fe$_3$O$_4$		<0.1–45	300	6.5	161
Fe$_3$O$_4$	10	≤45	300	14	120
Fe$_3$O$_4$	350	≤75	300	14	120
Maghemite (γ-Fe$_2$O$_3$)					
γ-Fe$_2$O$_3$		210	880	7.2	176
γ-Fe$_2$O$_3$	12	240	880	9.3	177
	240				
γ-Fe$_2$O$_3$		250–370	1100	6.84	205
γ-Fe$_2$O$_3$	100–150	42	880	7.2	175
γ-Fe$_2$O$_3$	6–12	12–240	880	7.2	175
γ-Fe$_2$O$_3$	20–160	≤400	410	11	206
γ-Fe$_2$O$_3$	3.3	120	520	13.2	137
	13.1	146			
Pure metals					
Fe	1000–2500	21	880	7.2	175
Fe	20–50	100[a,b]	230	20	188
Fe	20–50	720[a,c]	230	34	188

[a] SAR [W g(α-Fe)$^{-1}$].
[b] Quality: high grade.
[c] Quality: best grade.

as a favored material is susceptible to oxidation [89] and at this particle size a substantial proportion of the volume has been transformed into Fe oxides. Furthermore, iron is pyrophoric [162], meaning that conventional production techniques such as milling, which is often used in industry, can only be used to a limited extent and/or after further technological measures have been taken. Only the much-used wet-chemical, dry-chemical methods or thin film technologies, sometimes with the implementation of extra protective coatings where applicable, have led to any obvious progress [92, 102, 111, 163–171]. Materials with unusual magnetic properties are described in these studies.

These materials have are also been rendered interesting for this kind of application by alternative nanotechnology processes. Natural nanostructures, such as magnetotactic bacteria, are not the focus of our interest. Using these particles, a defined influencing of the magnetic properties is hardly possible. Useful magnetic nanostructures can also be obtained by the production of ff-MWCNTs using CVD, affording $> 1000\times$ higher H_cs, magnetization properties that are more favorable to this kind of application compared with bulk material, and permanent oxidation protection. However, notably, the characteristics of the nanoparticles are always partly determined by structure and by production-caused contamination, mechanical stresses, etc.

Several parameters play an important role in optimizing a material; some parameters affect, negatively or positively, optimization of the magnetic properties. For example, the nickel nanoparticles encapsulated in graphite by Hwang et al. showed distinctly lower H_cs than in bulk material [172]. Pirota et al., however, showed from the example of magnetic-phase multilayer microwires at the µm scale that by adding another coating to the microwires the hysteresis loop could be clearly improved towards single-domain characteristics [106]. Thus, production technology is of decisive importance, and the hysteresis loops can and must be adjusted. On the basis of these methods of influencing magnetic parameters and their effect on heat generation, it must again be underlined that the therapeutic method we suggest using the multifunctional nanocontainer is only made possible when the entire system, or the decisive parameters, is optimized (not maximized).

Taking as a basis an $H \times f$ product of 1.5×10^9 A m^{-1}s^{-1} [141], the maximum that can be applied during therapy, a magnetic field that can usefully be applied is at most approx. 20 kA m^{-1} (25.1 mT). As the hysteresis loop must be modulated to exceed H_c for heat to develop, seen from this angle it appears pointless to try to achieve H_cs higher than 100 kA m^{-1} (125.3 mT), even though this is possible with the ff-MWCNTs (Table 9.3). The point is much rather to influence the shape of the magnetization curve, as in Fig. 9.24, with an H_c of approximately 15 kA m^{-1} (18.8 mT). To do this, it is advantageous to provide a material in a single domain state. This, again, means that the aim should be a Fe filler diameter of between approx. 10 nm (<10 nm SPL) and approx. 32 nm [84] or approximately 45 nm [86] (no single-domain particles). As mentioned above, the higher values depend on the aspect ratio. The given values are representative for $d/l \leq 3$. For favorable shape anisotropy, the particles should have a length/diameter ratio of ≥ 2, requiring a

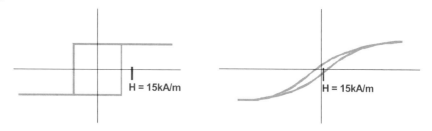

Fig. 9.24. Schematic of well- and ill-conditioned hysteresis loops for medical applications (e.g., hyperthermia).

nanoparticle length of at least 20–60 nm. Furthermore, as high a magnetization as possible should be guaranteed. Based on these benchmark figures, high heat dispersion is possible in the case of single-domain-ff-MWCNT. However, this remains a distant goal.

Our data strongly suggests that *single* ff-MWCNTs can display switching behaviors as a single-domain element. Because of the large number of tubes in an issuing batch the fraction of single domain and flower/vortex structures is not well-defined. The present state of knowledge is insufficient to determine the optimal conditions (single domain, flower/vortex or mixing) for heat generation. To answer this question the conditions of the therapy must be considered.

Single-domain elements can generate a large amount of heat only in magnetic fields $H_{ext} > (H_c + \Delta H)$. From a medical viewpoint, the therapy might need be performed in a low magnetic field. Under this condition the application of another type of ff-MWCNTs could be advantageous. Detailed investigations are important in further answering these questions.

The ff-MWCNTs described at present are, most likely, not yet entirely in a single-domain state. Because of the magnetization behavior and the varying diameters, we must assume that significant proportions of the materials are in a flower/vortex condition. Nonetheless, considerable heat dispersion could be proved.

9.5.3
Specific Absorption Rate (SAR)

A characteristic parameter is necessary to evaluate the effectiveness of a material system. For hyperthermia two processes must be considered. In principle, we have to evaluate the process of heat generation. It is determined by the loss power or specific heating power of the material. We also have to consider the process of heat absorption by a biological tissue. The specific absorption rate (SAR), the mass-normalized rate of energy absorption by a biological body, in W g^{-1}, has become the most important parameter in characterizing the effectiveness of a material system in hyperthermia. It is determined according to Eq. (19) [140, 173], where c is the specific heat capacity.

$$\text{SAR} = c\,dT/dt \tag{19}$$

For water, $c_{\text{water}} = 4.118$ J g^{-1} K^{-1}. For better comparability with different magnetic materials, some groups have used a particular SAR that is related to the active magnetic component [174–177]. This SAR of iron is given by Eq. (20), with m_{tot} the total mass of the specimen, and m_{Fe} the mass of the iron content.

$$\text{SAR}_{\text{Fe}} = \text{SAR}(m_{\text{tot}}/m_{\text{Fe}}) \tag{20}$$

This means the specific absorption rate SAR_{Fe} allows comparison of the efficiencies of different types of magnetic particles. In the description of our own material, we use this type of evaluation too.

With ff-MWCNTs, magnetic reversal losses are assumed to dominate for the mechanism of heat generation. On account of the opinion sometimes raised in the literature that sufficient heat transfer is not possible by this means with compatible magnetic fields [137, 139], it seems expedient to start with a calculated estimate. According to Eq. (21), the amount of heat generation per unit volume can be obtained by the frequency multiplied by the area of the hysteresis loop [75, 139].

$$P_{\text{FM}} = \mu_0 f \int H\,dM \tag{21}$$

Here, the substantiated assumption is made that the magnetization curve in the area examined is not dependent on frequency.

Under these conditions, P_{FM} can be readily determined from quasi-static measurements of the hysteresis loop [120, 139, 150] using an AGM, VSM or SQUID. To make these estimates, Fe-MWCNTs were transferred into tumor cells [15] and washed, and 50 000 cells were separated using fluorescence-activated cell sorting (see below). A defined volume of this solution was transferred onto a suitable substrate material. The magnetic properties of samples of this kind were studied by recording hysteresis loops in an alternating gradient magnetometer (AGM, 2900 MicroMag). As well as measuring complete hysteresis loops (major loops) the minor loops, which are decisive for the conditions of application, were also determined. For this purpose, when recording individual minor loops the maximum magnetic fields were limited to $H_{\max}$ (kA m^{-1}) = 0.796, 1.591, 9.549, 12.732, 15.915, 39.788, 59.683, 79.577, 119.366, 159.155, 198.943, 397.887, 238.732, 318.310, 477.464, 557.042, 636.620, 716.197, and 795.775 (1 mT–1 T). Typical magnetization curves are presented in Fig. 9.25. Notably, the curves shown here are not those of Fe-containing ff-MWCNTs in parallel alignment on substrates, but of Fe-containing ff-MWCNTs at a random spatial orientation in tumor cells. Where a parallel magnetic field orientation is mentioned this means, in this case, parallel to the substrate and no longer parallel to the tube axis. Because of the random ori-

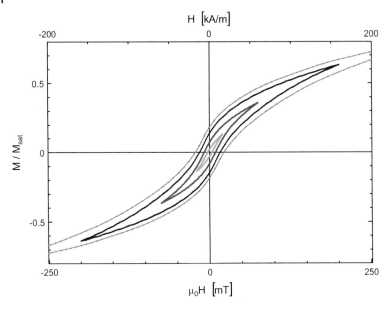

Fig. 9.25. Hysteresis loops of Fe-MWCNTs in cancer cells. Outermost line: major loop; inner lines: minor loops.

entation, to some extent an averaging-out of all the spatial axes takes place. This is reflected in the directional dependence of the magnetization curves, which is now only weak. By determining the areas of the magnetization curves and multiplying them with an assumed frequency, the energy dissipation calculated for each magnetic field was estimated at 0.796–796 kA m^{-1} (1 mT–1 T). For the calculated frequency, we drew upon the limiting frequencies of 50 and 1200 kHz generally accepted in the literature as well as our favored frequency of 250 kHz. Figure 9.26 shows the results obtained. Based upon quasi-stationary measurements, sufficiently high energy losses can already be achieved with magnetic fields of 10 kA m^{-1} (12.5 mT). Energy losses > 250 W (g-α-Fe)$^{-1}$ are, however, only to be expected for this material from above 30 kA m^{-1} (37.7 mT). Although fields of 20 kA m^{-1} (25.1 mT) at 1200 kHz [130] or 45 kA m^{-1} (56.4 mT) at 53 kHz [153] have been used for hyperthermia treatments, lower magnetic fields are often preferred, in the range of the recommended $f \times H$ product of 4.85×10^8 [133] to 1.5×10^9 A (m s)$^{-1}$ [141]. The range of possible combinations of frequency and magnetic field that result from this fit in with the range defined as acceptable for clinical applications by Pankhurst et al. [139], $f = 50$–1200 kHz and $H = 0$–15 kA m^{-1} (0–18.8 mT). Based on these first fundamental assertions on the possible achievement of high energy losses by magnetic reversal, in a further step the SAR was also determined experimentally.

Determination by measurement often takes place using a calorimetric (time-resolved) method, with certain boundary conditions concerning technology and/or

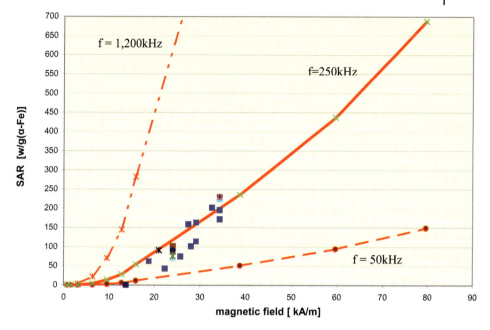

Fig. 9.26. Variation of specific absorption rate (SAR_{Fe}) of Fe-MWCNTs with magnetic field (lines: calculated from hysteresis loops; symbols: experimental measurements).

measuring techniques due to the high-frequency alternating fields. As illustrated in Fig. 9.27, the set-up we used consists of a high-frequency generator with an impedance matching network and the magnetic coil system. Water-cooled copper tubes are wound into a coil system (e.g., 4–10 turns, diameter of bore 20–100 mm, $l = 60–120$ mm) in which the sample to be examined is placed. The sample must have good heat insulation and the coils must be sufficiently cooled. The sample in the coil system is heated by applying an alternating magnetic field. The temperature change per time unit is determined using an appropriate device. Fiber-optic systems such as the FLUOROPTIK® Thermometer have proven effective for high-frequency magnetic fields: these metal-free systems are not affected by the alternating magnetic field and eddy current effects do not occur. With these systems, the temperature can be measured *in situ*. The output resolution and the accuracy also meet the demands posed, at 0.001 °C and ±0.2 °C respectively.

Another notable point is that the magnetic field amplitudes are often determined by measuring the coil current and then calculating H. This is possible with sufficient precision in the case of coil geometries, which are generally simple. For the simplest case of a single circular conductor loop with radius R in the x–y plane (if the origin is the center of the circle) we obtain Eq. (22) for the field components in the z direction, with I the current passing through the conductor.

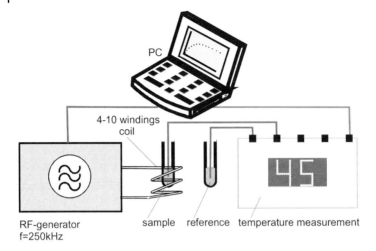

Fig. 9.27. Experimental setup for time-dependent calorimetric measurements.

$$H_z = \frac{1}{2} IR^2 \frac{1}{(R^2 + z^2)^{3/2}} \tag{22}$$

Clearly, for $z \gg R$ the field decreases at about $1/z^3$, and therefore outside the coil there is only slight stress on the tissue during treatment. For therapy the field is of importance within a long coil. It applies to Eq. (23), where n is the number of turns and l is the length of the coil.

$$H = \frac{nI}{l} \tag{23}$$

Initial attempts to determine the SAR experimentally used water, physiological salt solution and agarose gel blocks, to each of which was added a defined quantity of ff-MWCNTs. Heat insulation of the sample is extremely important, as mentioned above. When simple glass vessels were used we observed a distinct time dependency of the determined SAR. As shown in Fig. 9.28, the values vary from about 750 W (g-α-Fe)$^{-1}$ (30 s measuring time) up to 150 W (g-α-Fe)$^{-1}$ (900 s measuring time). We ascribe these differences to an unfavorable ratio of sample volume/sample surface at the time and, in particular, to heat conduction from the sample into the sample vessel. After the geometry of the materials used was optimized and heat conduction was minimized, this problem was reduced. It was possible to increase the accuracy and to realize the well-known t–T characteristics. The initial slope was used to determine the SAR.

Figure 9.26 shows the SAR obtained by experiment under these altered conditions at a frequency of 250 kHz, depending on the magnetic field strength for the mid-quality material used for the quasi-static measurements. The figure also shows the SAR curves calculated for this material. The material used was from

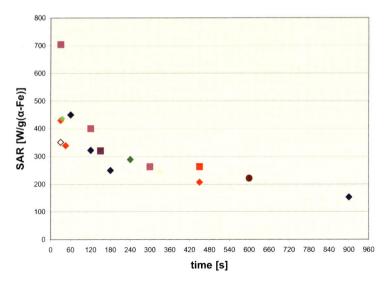

Fig. 9.28. Variation of specific absorption rate (SAR$_{Fe}$) of Fe-MWCNTs with time ($H = 30$ kA m^{-1}). Different colors represent data points from independent experimental series.

one batch but was not that used for the magnetic measurements. Clearly, the values obtained by experiment in a magnetic field area of up to about 28 kA m^{-1} (35.1 mT) tend to be lower than expected from the quasi-static measurements. What is more, there is distinct scattering of the measured results in all magnetic fields. We ascribe this scattering to instrument-related problems in controlling the HF current, as when the HF transmitter was at full drive, values were determined that could easily be reproduced. The SARs determined by experiment at 34 kA m^{-1} (42.6 mT) came out at 286, 280, 286, 292, 286, and 292 W (g-α-Fe)$^{-1}$ in a series of measurements, for example. At the level of magnetic fields <12 kA m^{-1} (<15 mT) it was not yet possible to obtain any sufficiently firm measured values. One reason could be the switching behavior of single-domain particles. To throw light on the real cause, further equipment-related improvements and basic studies of the switching characteristics of *individual* ff-MWCNTs must be carried out. The results already gathered by experiment, however, tend to confirm the expected course. In the range that can be applied in therapy, up to approximately 20 kA m^{-1} (25.1 mT), we were able to verify SARs of ≤ 100 W (g-α-Fe)$^{-1}$ for this not yet optimized material. However, clearly, far higher values could be obtained in the range >30 kA m^{-1} (37.6 mT). From this, central questions emerge concerning:

1. The maximum $H \times f$ product that is possible to use;
2. how these values appear in relation to the superparamagnetic particles;
3. whether the present situation achieved is sufficient for treating urological tumors.

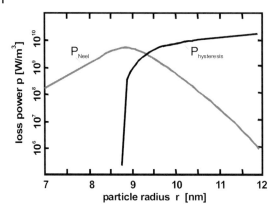

Fig. 9.29. Grain size dependence of the loss power density for small ellipsoidal particles of magnetite ($H = 6.5$ kA m^{-1}). (Modified from Ref. [120].)

Table 9.5 lists typical SARs for the most important material systems, magnetite and maghemite, as well as for iron. The differing values for the frequencies and magnetic fields must be taken into account when comparing the results. Even after numerical correction of the SARs to "standard frequencies" and "standard magnetic fields" in accordance with the known dependences given by Eqs. (12) or (15), a great deal of scattering still clearly takes place within each material system. One reason for this can be found in the dependence of the energy loss on particle diameter and working frequency. As shown by Hergt et al. (Figs. 9.29 and 9.30), when using superparamagnetic nanoparticles, a distinct reduction in energy transfer achieved can be expected, even from a deviation of around ±1 nm from the op-

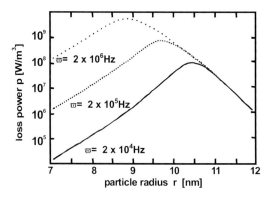

Fig. 9.30. Dependence of magnetic loss power density on particle size for magnetite ($H = 6.5$ kA m^{-1}). (Modified from Ref. [120].)

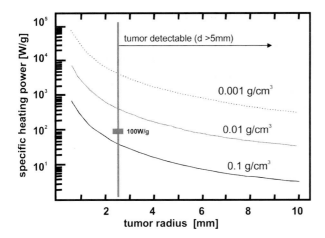

Fig. 9.31. Power requirements of magnetic nanoparticles. (Modified from Ref. [120]; 3 curves.)

timal diameter [120]. With ferromagnetic particles up to the superparamagnetic limit, however, only a relatively minor change occurs depending on the diameter. Here too, the marked influence of technology on the results that can be obtained is apparent. As has already been established, controlling heat conduction in the tumor and the surrounding tissue is of decisive importance in connection with the SAR achieved. The group of Andrä, Hergt and Hilger [141, 143, 178] have stated more precisely results [143] for the range of thermoablation ($T = 51–55$ °C). The calculations imply that the specific heating power needed to achieve the required rise in temperature in the tumor goes up sharply as the diameter of the tumor falls. With thermoablation, the target is a temperature difference of 15 K.

Figure 9.31 shows the specific heating power required for this for three different concentrations of magnetite, depending on the tumor diameter. The low and high concentrations of 0.001 and 0.1 g cm^{-3} are approximately the limiting values that are usefully applicable in practice. From the graphic representation, the required specific heating power clearly varies by several orders of magnitude, depending on the concentration of particles achieved. According to this, for an average particle concentration of 0.01 g cm^{-3} and 100 W g^{-1} (Table 9.5), tumors from a diameter of about 1 cm can be treated. By analogy, a higher particle concentration results in a size of 4 mm. These observations apply to the range of thermoablation. For hyperthermia, the proportions are slightly more favorable due to the lower temperature difference required. As only slight differences can be made to the heat conductivity conditions in the tumor tissue, one urgent task for materials development is to raise the performance of the material. The heat condition problem, on the one hand, and the amount of heat generated by a few magnetic nanoparticles inside a biological cell, on the other hand, are the background for the discussion: Is hyperthermia on a cellular level possible [137, 141, 179]?

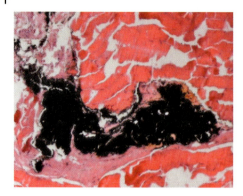 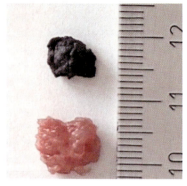

Fig. 9.32. AC-heating using model tissue: (a) Muscle tissue containing Fe-MWCNTs (black), (b) muscle tissue: before and after AC-heating.

Taking as a basis the SARs for Fe-containing ff-MWCNTs at the *level used in therapy* of approx. 100 W (g-α-Fe)$^{-1}$, the minimum SARs postulated in the literature to achieve the necessary temperature rise in the tissue, it can be concluded that this ferromagnetic material in a single-domain state should be suited for the treatment of a diagnosed PCa with an extension of at least several millimeters. Assuming this to be the case, experiments have been made in heating model systems. Beef was used as a model substance for the first heating experiments (specimen 7 × 7 × 7 mm). Fe-MWCNTs were injected into this tissue (see Fig. 9.32a) and the sample was heated (at $H = 20$ kA m^{-1} and $f = 250$ kHz). Figure 9.33 shows the temperature curve within the muscle tissue (43 °C after 7.5 min AC-heating). By compari-

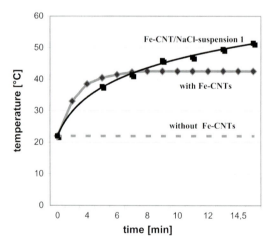

Fig. 9.33. Temperature development during AC-heating.

son, without an injection of Fe-MWCNTs no temperature difference was measured for at least 15 min after the start of magnetic induction. Figure 9.32(b) shows typical photographs of model specimens/tissues before and after AC-heating. Based on these results, further experiments were carried out on cell cultures, tissue and organ samples and animals.

9.6
Study Results for *In Vitro* and *In Vivo* Applications of ff-MWCNTs

9.6.1
Efficient Endocytosis *In Vitro*, Lipid-mediated Could Enhance the Internalization Rate and Efficiency

Using different types of ff-MWCNTs we have studied the uptake by different urologic cancer cell lines. In cell culture experiments an efficient delivery into human EJ28 bladder cancer cells after complex formation of the ff-MWCNTs with cationic lipids was detected [15]. Our original conclusion was that the lipid is necessary for efficient uptake of ff-MWCNTs by cancer cells in our settings. However, systematic investigations of several cancer cell lines indicated that the ff-MWCNTs with different sizes (1–10 µm long) were also internalized without lipid addition. Therefore, we hypothesized that lipid addition *in vitro* can stimulate the internalization, probably via endocytosis. Internalization by cancer cells under optimal growth conditions *in vitro* guarantees an efficient transfer in the cytoplasm of most of the cells. Our data implicate a potential direct association between the proliferating capacity of the individual cell type and the rate of internalization, which is in accordance with an active uptake mechanism. Further analyses should clarify this point in more detail because of its potential importance for the development of future therapeutic concepts.

Using TEM we have detected clusters of ff-MWCNTs mainly in the cytoplasm of the PCa model cell line PC-3. This uptake had no influence on the cell viability quantified by WST-1-assay in the first three days after short-term (4 h) treatment with ff-MWCNTs. These data confirm several reports for unfilled SWCNT [180–186] and also indicate that ff-MWCNTs appear non-toxic once internalized into mammalian cells of malignant (PCa and bladder cancer cells) and non-malignant origin, and without adverse effects to cell viability.

Beside the intracellular detection of ff-MWCNTs in the form of clusters or bundles the association with cells was further evaluated by flow cytometry. The results indicate that clusters of ff-MWCNTs are detectable in solution by FSC-SSC scatter flow cytometric analyses (FACS) (Fig. 9.34). They can be discriminated from cancer cells because of their higher granularity and smaller size (cf. Fig. 9.34b). After addition of ff-MWCNTs in serum-free culture medium to the cancer cells for 2 h, FACS analysis revealed a high percentage of cells characterized by a larger granularity, indicating an association between cells and ff-MWCNTs. Cells with internalized ff-MWCNTs and those with ff-MWCNTs at their surface can not be differenti-

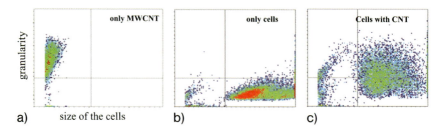

Fig. 9.34. Representative results of flow cytometric analyses of human PC3 cells without MWCNT transfection (b). In comparison, 4 h after incubation of cells with Fe-MWCNTs (c) the granularity and size of the cells have increased. Remarkably, bundles of Fe-MWCNTs are also detectable in solution (a).

ated by this technique. However, the detection of a broad spectrum of increasing granularity confirms the TEM studies, showing that cancer cells have internalized different amounts of ff-MWCNTs. Moreover, on repeating the flow cytometry two days later, the same percentages of cells associated with ff-MWCNTs were found. Summarizing these experiments, clearly, by addition to the culture medium an intracellular accumulation of ff-MWCNTs takes place within a few hours, especially as cytoplasmic localized aggregates (Fig. 9.35). In addition, cell-associated and internalized ff-MWCNTs have the same magnetic properties as ff-MWCNTs in solution without cells.

A pilot animal study has checked the distribution of Fe-MWCNTs in mice [187, 188]. Briefly, male mice at 8 weeks of age were narcotized, and different doses of ff-MWCNTs were injected once intraperitoneally (i.p.) or intravenously (i.v.; via the tail vein). The animals were sacrificed 20 h after treatment. Tissue samples of different organs were conserved for TEM and histological analyses. Mice without treatment were used as controls. The remaining but treated animals were observed over a period of 150 days. For several mice of this treatment group, the injections were repeated. All animals survived and showed no abnormalities in their behavior or food consumptions. Remarkably, one mouse with a five-fold injection over a period of 3 months was treated with a total of >1 g ff-MWCNTs per kg of body weight.

In sacrificed animals, large agglomerates of ff-MWCNTs were detected at various organs in the retroperitoneum for the i.p. treatment but not in the i.v. treatment group. This indicates that most of the intraperitoneally administrated ff-MWCNTs is retained within the retroperitoneum over several weeks. Interestingly, the macroscopically visible agglomerates were attached in most cases to the surface of organs and penetrated to the peripheral zone of different organs, including kidney and liver. So, remarkably, magnetic force was transferred to the organs (Fig. 9.36).

The amount of magnetic material required to produce the appropriate temperature for hyperthermia or thermoablation depends to a large extent on the method of *in vivo* administration. Intra- or peritumoral injection or deposit allows for sub-

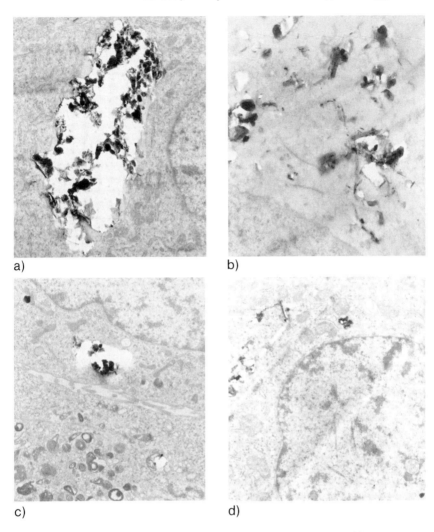

Fig. 9.35. Representative TEM photographs showing the internalization of ff-MWCNT by the PCa cell line PC-3. Black structures represent an accumulation of Fe-MWCNTs. (a) One large complex with many single CNTs was detected within the cytoplasm (magnification ×7000); (b) intracytoplasmatically localized CNTs with different grades of ferromagnetic filling (dark) (magnification ×12 100); (c) different types of complex formation within the cytoplasm of PC-3 cells containing many CNTs (4 h after incubation, magnification each ×7000).

stantially greater amounts of magnetic material to be concentrated in a tumor than do other methods employing intravascular, intradermal or intraperitoneal administration and/or targeting. A general assumption is the concentration of at least 5 mg of the magnetic material within 1 cm^3 of target (tumor) tissue.

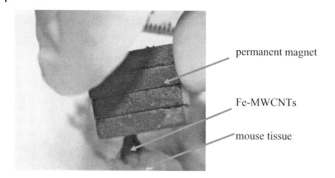

Fig. 9.36. Transfer of magnetic force to tissue.

9.6.2
Production of Two Types of ff-MWCNTs for *In Vivo* Application

ff-MWCNTs can be synthesized by two different types of CVD. The first, solid source CVD (SSCVD), uses a simple quartz-tube gas flow reactor inside a two-zone furnace system [12]. Ferrocene, $Fe(C_5H_5)_2$, is sublimated in the first zone at about 150 °C. Decomposition is then realized in the second, the hot, zone at about 830 °C in an Ar flow on oxidized Si-wafers.

The second CVD system (so-called liquid source CVD, LSCVD) consists of a band evaporator with continuous action and a hot wall reactor with a tape on which the precoated substrates are positioned. This tape can be moved through the hot zone of the reactor. Here the ferrocene is solved in cyclopentane and drops continuously on the moving band evaporator. At 45 °C the solvent evaporates, at 270 °C the ferrocene is completely sublimated and, finally, the vapor is transported into the CVD reactor. At 900 °C the precursor decomposes and ff-MWCNTs are deposited on the precoated substrates directly. The CVD system has been described in detail previously [189].

Therefore, this method has the important advantage of producing filled nanotubes continuously and is, thus, favorable for a large-scale application.

The nanotube material was investigated by scanning (SEM, XL 30, Philips) and transmission electron microscopy (TEM; TECNAI F30 with GiF 200). Thermal gravimetric analysis (TGA, RUBOTHERM) in an Ar/O_2 atmosphere up to 450 °C was used to determine the thermal stability and the filling grade of the nanotubes (oxidation of nanotubes up to the remaining Fe_2O_3). Alternating gradient magnetometry (AGM) revealed the ferromagnetic behaviors of the filled nanotube-ensembles on the Si-substrates.

Depending on the deposition conditions, the length, structure, filling grade and the magnetic properties of the CNTs varied. Here, two different batches of ff-MWCNTs were used. One batch was synthesized by SSCVD, affording ff-MWCNTs about 10 μm long (type 1), with a filling yield of 25–30 wt.%. The ff-MWCNTs of

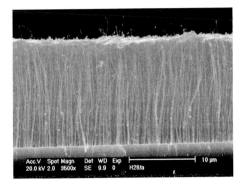

Fig. 9.37. SEM image of Type 1 ff-MWCNTs used for administration studies in mice.

the other batch (type 2), produced by LSCVD, were about 15 μm long (Fig. 9.37), with filling yields of 30–35 wt.%. Both types had the same diameter distribution, i.e., 10–30 nm (inner) and 20–130 nm (outer). The number of carbon shells ranged between 20 and 100.

9.6.3
Outlook/Next Steps in Evaluation of these fff-MWCNTs

The future goal is the design and optimization of biofunctionalized ff-MWCNTs for different anticancer applications (Table 9.6). These multifunctional nanocontainers are finely regulated regarding the site of action, the temperature induced by external magnetic field application, and the step-by-step release of different drugs. In addition to therapeutic applications, nanocontainers could also be functional units for diagnostic and therapeutic monitoring purposes. They fulfill all the criteria for nanoscaled mediators of anticancer treatment with an optional intrinsic sensoric unit. The latter can function both for the detection of small tumor lesions as well as for the control of therapeutic effects. Note that either all of the individual characteristics can be incorporated in a single MWCNT species or a combination of different species with various behaviors can be synthesized, produced on a large scale, and applied under standardized conditions.

In our view, a variable experimental model is needed to test these possibilities. Therefore, we suggest a chamber model for the administration and characterization of *in vitro* and *in vivo* effects using a standardized device (Fig. 9.38).

Potential applications (Tables 9.6 and 9.7) of improved functionalization of ff-MWCNTs include:

1. Selective tumor-targeting by coupling of specific antibodies to the sidewall or the ends of ff-MWCNTs.
2. The formation of complexes with suitable lipids for better internalization of ff-MWCNTs inside target cells and tissues.

Purposes for the experiments with the model chamber:
- Objects of analyses: solution with different species of fff-MWCNT, adherente growing normal and cancer cell lines, slices of explanted model tissues (normal & malignant prostate from mouse or rat) including xenografts, organ bathes, or whole animals (mouse, rat)
- Study of the instrinsic magnetic properties of ff-MWCNT and fff-MWCNT in aqueous solution & within tissue slices
- Targeting, selective labeling & detection of tumor cells
- Study and optimization of different treatment modalities (microscopy, measurements within the chamber, external temperature control): intracellular intratumoral hyperthermia, drug release, combinational therapies

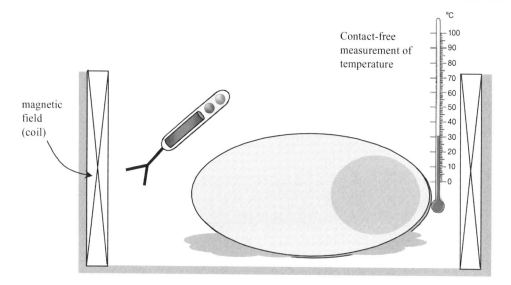

Fig. 9.38. Principle of the chamber used to analyze the interaction between ff-MWCNTs and tissues or organs.

3. Conjugation with other nanomaterials (nanoparticles, nanotubes, nanowires) characterized by other intrinsic biological, chemical or physical properties for an enhanced effect in diagnosis and treatment of malignant tumors.
4. Biofunctionalization and/or labeling of the ends of CNTs.
5. Potential as dual container (intrinsic chambers) and cargo system (functionalization, e.g., conjugation at the sidewall and at both ends of ff-MWCNTs).

Kam and coworkers have described recently the overall potential of functionalized CNT for anticancer therapies in general because of the intrinsic advanced physical and biological properties of these nanomaterials [185]. They reported that short (≈ 150 nm long and ≈ 1.2 nm in diameter) and unfilled HIPCO-SWCNT solubilized in the aqueous phase by noncovalently and strongly adsorbing either fluorescently labeled 15-mer oligonucleotides (ODN) or phospholipid-PEG-grafted folic

Tab. 9.6. Advantages of CNTs and particles.

Parameter/behavior	CNT	Particles (and SWCNT?)
Functionality	Filling with diagnostics and therapeutics in defined chambers	
	Possibility of different locations of the anti-tumor substances (as intrinsic filling) and the tumor/tissue targeting system (as extrinsic bio-functional units)	
	Minimum of ferromagnetic filling (for necessary magnetic properties and energy transfer)	
	High energy transfer within a target cell and tissue	
	Low(er) immunogenecity and/or toxicity?	
	Relatively inert tissue distribution of deposed long CNT (>1 μm long)	
Morphology	Length–width ratio	Optimal surface area–diameter ratio
	Intrinsic depot function	
	Compared to the filling volume, a relatively large and single surface area/sidewall for functionalization or coupling to other structures and molecules; f.e. MWCNT offer a higher available surface for interaction with DNA (compared with SWCNT and round beads)	
	Possibility of defined (and temperature-dependent) release of filling substances in two directions	
	Relative flexibility of the tubular structure	
	Inert covering by the carbon multi-wall	
	Relatively large surface area improves functionality as shuttling system/vector (cf. Singh et al. 2005 [196])	

Tab. 9.7. Reported examples of studies with functionalized CNTs.

Type of functionalization	Methods and results	Ref.
Drug/vaccine delivery Delivery of nucleic acids and proteins	Crossing the cell membrane and cytoplasmic accumulation enhance the immune (ab) response against peptides with no cross-reactivity to CNT, SWCNT as molecular cargoes into cells by noncovalently and non-specifically bound proteins to the nanotube sidewall and intracellular delivery by endocytosis, delivery of plasmid DNA by binding and condensation onto ammonium-functionalized SWCNT and MWCNT and onto lysine-functionalized SWCNT, selective targeting and destruction of tumor cells	180, 182–186, 190–193, 207–212
Usage as new types of biosensors based on DNA or proteins	Examples: CNT functionalized with PNA bind DNA containing a complementary sequence, nanoassembly by DNA–DNA interaction	213–220
Complement activation and protein adsorption	Description only for SWCNT and double-wall CNT	221
Imaging of cells, use in atomic force microscopy	By covalent linkage to visible-wavelength fluorophores or as Pluronic SWCNT fluorophores through near-infrared microscopy	180, 181, 192, 222, 223
Enzyme immobilization, usage as ion channel blockers		224, 225
Molecular electronics	Formation of nanoassemblies useful as molecular switches	226–229

acid were internalized inside folate-overexpressing HeLa cancer cells by simple incubation in cell culture medium at 37 °C (Fig. 9.39). This was in contrast to the results of the same experiment at 4 °C, which indicated that uptake is based on receptor-mediated and energy-dependent endocytosis. The SWCNT were heated by external near-infrared laser pulses, resulting in an intracytoplasmic release of ODN

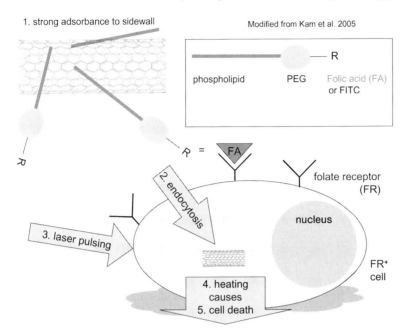

Fig. 9.39. Selective targeting and killing of tumor cells using functionalized SWNT. After internalization of modified SWCNT via folate receptor (FR) binding on FR expressing tumor cells laser radiation induces selective tumor cell death. (Modified from Ref. [191].)

from the SWCNT and translocation into the nucleus. Note that radiation of the aqueous SWCNT solution (25 μg mL^{-1}, without cells) by an 808 nm laser (1.4 W cm^{-2}) continuously for 2 min caused heating to ≈ 70 °C. The authors suggested that the SWCNTs can act as tiny "heaters" in living cells overexpressing the receptor molecule (folate receptor) targeted by the functionalized SWCNTs. Selective targeting and induction of necrosis of HeLa cells was observed after the same laser pulse treatment, without harming receptor-free normal cells. Dying cells released the ODN cargoes and were mixed with cell debris in black aggregates.

The same properties should be transferred and adapted to our fff-MWCNT model because of the intrinsic heating function of the ferromagnetic filling. In general, new PCa-specific antigens recognized by PCa-specific receptors can be applied for a selective fff-MWCNT-based hyperthermia in the histologically remarkably heterogeneous PCa tissue containing also receptor-negative stromal and normal prostatic cells.

A major advantage of using ff-MWCNTs would be the external control available for localization and intrinsic heating. Note that the finely-regulated, slow and controlled induction of heating should also improve the internalization of fff-

MWCNTs by target cells that did not primarily take them up; this implies a possible multiplicity of anticancer therapeutic efficacy once or in multiple treatments: stepwise (a) heating, (b) improvement of uptake within the target tissue, (c) release of drug from the container at a defined temperature, (d) repeating the hyperthermia as secondary local treatment and release of other drug classes by realization of higher temperatures than at the time of primary hyperthermia.

Biocompatibility is a major concern when introducing a therapeutic agent inside the human body. For different, mainly short and pure, species of unfilled SWCNT no adverse or toxic effects have been reported by various groups on testing the cellular viability and the proliferation [180–182, 184–186, 190].

Studies with SWCNTs indicate that simple mixing of oxidized SWCNT with solutions of various types of proteins resulted in covalent and non-covalent bonds to the sidewall [191]. The sidewall was functionalized with oxygen-containing groups was by refluxing and sonication of SWCNTs in nitric acid. Interestingly, the adsorption of proteins with a molecular weight of ≤ 80 kDa (bovine serum albumine and cytochrome c) on SWCNTs imparts hydrophilicity to the CNTs. The spontaneously formed nanotube–protein conjugates were internalized by different cell types of non-malignant as well as malignant origin. While single proteins were unable to cross the cell membranes by themselves, SWCNTs efficiently traffick protein cargos inside the cells. An energy-dependent endocytosis mechanism for the uptake of the conjugates was proposed [191].

However, other mechanisms as the dominant internalization processes *in vitro* have also been discussed recently in the literature, including uptake of the CNTs by insertion and diffusion through cellular membrane phagocytosis [181, 182, 184, 190, 192]. The detailed uptake mechanisms and efficacy for the different target and non-target cells should be evaluated in various animal models.

Kam et al. have further described the release of internalized conjugates with SWCNTs and proteins from endosomes by adding the membrane-passing base chloroquine to the culture medium [191]. The functionality of adsorbed protein was shown for cytochrome-c because, after release of this protein from the endosomes, cells died by active apoptosis (mediated by the delivered cytochrome c).

Recently, derivatization of unfilled MWCNTs with N-protected amino acids based on a cycloaddition reaction to the external sidewall was described by Georgakilas et al. [193]. The modified MWCNTs had a lower solubilization than functionalized SWCNTs. However, 12 mg mL^{-1} of functionalized MWCNTs gave a clear solution in water. The authors conclude that this result represents the basis towards the synthesis of peptide-based CNTs.

Other reports have shown the interactions between SWCNT (HiPco tubes) and nucleic acids. Refs. [194, 195] describe the extremely strong binding of short single stranded DNA (oligonucleotide) by π-stacking. The binding of the oligonucleotide was realized by helical wrapping with right- and left-hand turns or simply by adsorption at the sidewall with linearly extended structures. This DNA coating of bundled SWCNTs was effectively dispersed in water by sonication. Moreover, subsequent functionalization was made by labeling one terminus of the oligonucleotides with biotin.

Ammonium-functionalized SWCNTs and MWCNTs facilitate the delivery of plasmid DNA in different murine [180, 196]. Complexes of DNA and unfilled, ammonium-functionalized MWCNTs formed aggregates >4 µm and possessed a planar lattice structure, whereas the double-stranded plasmid molecules formed a planar structure with MWCNTs buried within. Compared with SWCNTs, the sidewall of MWCNTs had a higher charge density and an increased surface, resulting in a closer DNA association with the MWCNTs. The authors found >96% of plasmid DNA is condensed by ammonium-functionalized MWCNTs at a charge ratio of 1:1.

Future developments in this fascinating application spectrum will depend on whether the specific synthesis methods and, afterwards, the defined functionalization for CNTs of a desired structure can be realized (Tables 11.6 and 11.7). According to the planned therapeutic application of fff-MWCNTs, the focus should be on the stepwise shortening of the length of the sidewall, the definition of a feasible amount of sidewall. In addition, the dependence of chemical reactivity on the detailed MWCNT structure is particularly interesting.

As well as investigation of the general toxicities of ff-MWCNTs, the principal biodistribution has been studied in a mouse model for the i.v. and i.p. administration route. For biological TEM analyses, different organs were fixed immediately after removal from sacrificed mice by fixating them in glutaraldehyde. Ultrathin slices were prepared as described recently by us [187]. Additionally, other samples of the same tissues were also collected, fixed in formalin and embedded in paraffin for routine H&E staining and subsequent histological analyses.

The major goal of these studies was to investigate the short-term accumulation and distribution of the containers within the body. Furthermore, by histopathological investigations potential toxic effects after single or multiple *in vivo* administrations were assessed.

Recent results of both TEM and histological light microscopy studies indicate that aggregated to form clusters, independently of the localization site *in vivo* [187]. The aggregation can be observed even macroscopically after sonication and during injection. Therefore, the ff-MWCNTs can reach different tissues by diffusing through capillaries. The cluster sizes, detected by TEM, ranged from 0.5 to several microns. The structure of these clusters was identical to those found in cell culture incubation experiments. From the calculated sizes each cluster, mainly detected intracellular, may consist of several to hundreds or thousands of individual ff-MWCNTs. Remarkably, the tendency of *in vivo* formation of such large clusters did not influence the health conditions of the animals, and did not reduce survival. All animals ($n = 14$) treated with different dosages of MWCNTs of both subtypes (once or several times) survived 150 days.

In several animals injected once with one of both types of ff-MWCNTs, the CNTs were detected within different cell types of different organs, including lung, heart, liver, colon or accumulated within the tissues (Fig. 9.40). Only in the minority of TEM positive samples were individual ff-MWCNTs found.

Moreover, after i.p.-injection low amounts of bundled ff-MWCNTs were also detected in feces. This observation needs further detailed investigation.

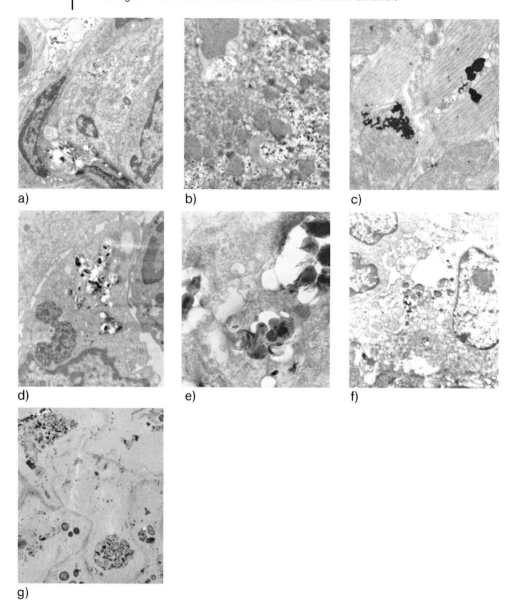

Fig. 9.40. Representative results of ff-MWCNTs detected by TEM analysis in different mouse organs 20 h after injections (i.v. or i.p.). (a) Lung section collected 20 h after i.v. injection of type 2 (mouse No 8, 9 weeks of age); magnification ×7000. (b) Liver section collected 20 h after i.v. injection of type 1 (mouse No 4, 9 weeks old); magnification ×7000. (c) Diaphragm section derived from the same mouse as (b); magnification ×7000. (d) Colon section collected 20 h after i.p. injection of type 1 (mouse No 6, 9 weeks old); magnification ×7000. (e) Same as (d) but at magnification ×32 000. (f) Heart section collected 20 h after i.v. injection of type 1 (mouse No 3, 9 weeks old); magnification ×4700. (g) Faeces collected 20 h after i.p. injection of type 1 (mouse No 15, 9 weeks old).

Other animals injected once or several times with doses of approximately 100–400 µg ff-MWCNTs per injection survived more than 150 days without any indication of toxic or other adverse events. Analysis of the body weight indicated no influence of the administration of ff-MWCNTs on health condition.

Moreover, for example, several animals were sacrificed 6 weeks after first injections and analyzed by an experienced pathologist for potential alterations caused by the accumulation of ff-MWCNTs. In general, 6 weeks after the first injection ff-MWCNTs were detected cell-associated in most cases. Several microphotographs revealed an association of ff-MWCNTs accumulation with macrophages.

No formation of giant cells or granuloma was detected, indicating the absence of inflammation signs. The unrestricted localization of bundles and clusters of ff-MWCNTs was not associated with encapsulation of the accumulated tubes within the different tissues and organs. Especially in lung tissue specimens, the clusters were found near the alvealoe and beside microvessels (Fig. 9.41). Interestingly, after i.p.-administration, in some cases tubes were also detected in the lung but in significantly lower amounts and number than after the i.v. administration.

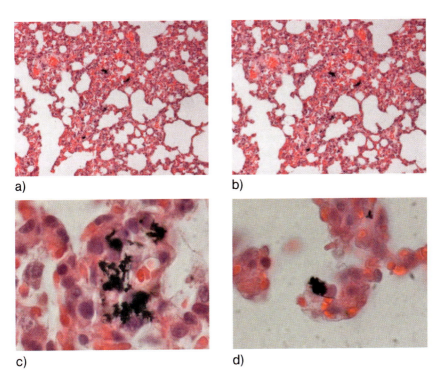

Fig. 9.41. Detection of Fe-MWCNTs by H&E-stained histological slices of different mouse tissue 6-weeks after i.v. or i.p. administration. Black structures represent an accumulation of Fe-MWCNTs. Lung section at a magnification of (a) ×100, (b) ×1000, (c) ×10000, and (d) with accumulation of ff-MWCNTs in the alveole at ×1000.

Tab. 9.8. Temperature developed during AC-heating in mouse tissue containing Fe-MWCNTs.

Time (min)	Mouse muscle (T/°C)	Kidney from mouse (T/°C)
0	28.0	29.2
5	40.5	40.0
10	42.6	40.7

The determined long-term accumulation of ff-MWCNTs in different organs and tissue indicated the biocompatibility of this compound over several weeks without dramatic toxicity. The comparable patterns of accumulation between short- and long-term treated animals showed the injected ff-MWCNTs can accumulate within hours and persist in these depot tissues without encapsulation.

Partially needle-like formations of bundles were detected. In the liver, low numbers of clusters of ff-MWCNTs were localized at the sinus.

In an additional experiment series the principle of ff-MWCNTs-mediated hyperthermia was tested using freshly explanted mouse tissue specimens. At seven months of age, one untreated mouse was sacrificed, and one kidney and muscle from the leg ($5 \times 3 \times 3$ mm) was collected for heating analyses. After injection of approximately 10–20 µL of freshly sonicated 0.9% saline containing type 1 ff-MWCNTs (20 µg µL^{-1}) the heating potential was tested immediately using the device described in Section 9.5.3. A magnetic field of 20 kA m^{-1} was realized at a frequency of 230 kHz. The explanted tissue specimens were incubated in the magnetic field. The temperature was measured in the centre of the treated specimens at the beginning, after 5 and 10 min (Table 9.8). Remarkably, for both explanted tissues an increase of >10 K was detectable after 10 min of treatment with <400 µg of ff-MWCNTs. In comparison with control experiments (applying the same magnetic field for equivalent tissue specimens without injections of ff-MWCNTs) the heat injection clearly confirmed our working hypothesis that ff-MWCNTs are useful for hyperthermia purposes. These pilot experiments *ex vivo* clearly showed that doses of lower than the mg range of ff-MWCNTs can mediate an intrinsic temperature shift to >40 °C after a couple of minutes of treatment.

The traditional approaches to hyperthermia, including radiofrequency, microwave and ultrasound, showed only limited efficacy in clinical studies because of (a) mediating of inhomogeneous temperature conditions, (b) difficulties in focusing in the target tissue, and/or (c) relatively high absorption of the heat by the intratumoral capillaries and connective tissues. The prostate as the target organ for local treatment primary PCa represents an extraordinary example because of the anatomy of the pelvis and the high perfusion of this gland.

Jordan et al. (Magforce GmbH, Berlin) have developed amino-silane-coated iron-oxide superparamagnetic nanoparticles with a mean diameter of 15 nm.

Aqueous ferromagnetic fluid (nanoparticles at a concentration of 120 mg mL^{-1}) were injected directly in the target tissue and subsequently heated by intracellular hyperthermia. The particles were taken up efficiently by endocytosis. The so-called magnetic fluid thermotherapy (MFH) was proposed in principal as suitable for both local hyperthermia and thermoablation of PCa, since the intratumoral temperature conditions can be selected over a broad range by modulating the magnetic field. Moreover, in contrast to E-field dominant systems used for regional hyperthermia, the boundaries of differently conductive tissues do not interfere with power absorption in the MFH treatment.

In a subsequent animal study using orthotopically implanted MatLyLu-cells (Dunning R3327 model), magnetic fluid (200–400 µL) was injected into the prostate tumors (0.5 mL per cm^3 of tumor). Immediately after injection, an AC field was applied (100 kHz, 0–18 kA m^{-1}; gradually increased field strength from 45% to 70% over a period of 17 min, heat treatment at 70% field strength for 30 min) using the MFH 12-TS (MagForce Nanotechnologies GmbH, Berlin) [197, 198]. By intratumoral temperature control, the authors described mean maximal temperatures of up to 58 °C. The observation that >85% of the nanoparticles were retained in the prostate 10 days after injection indicates that repeated thermal treatment is possible. Afterwards, external radiation of 20 Gy was applied. Combined thermoradiotherapy resulted in a higher inhibition of tumor growth than radiation with the same dose alone. The combinational treatment was as effective in tumor growth control as radiation therapy alone with a three-times higher dose.

Recent data from a phase I clinical study has been used to evaluate the toxicity and tolerability of MFH using a novel hyperthermia- and thermoablation-system (MFH 300F) [174] in PCa patients [147]. To our knowledge, this is the first report of the clinical application of interstitial hyperthermia using magnetic nanoparticles in the treatment on human carcinomas. A patient with locally progressive growing PCa following a high dose brachytherapy received a total of 12.5 mL of magnetic fluid in 24 depots (each of 0.5 mL) in the prostate (intraperineal injection under rectal ultrasound control and fluoroscopy guidance) with an approximate volume of 35 mL. Thermotherapy was applied using an initial magnetic field strength of 2.5 kA m^{-1}, which was gradually increased to 5 kA m^{-1}, and was kept constant for one hour. Five subsequent MFH applications took place at weekly intervals. The maximum intra-prostatic temperatures determined (48.5 and 42.2 °C during the first and second treatment, respectively) achieved were in the thermoablative range.

It was assumed that for sufficient heating of the prostate gland, and ideally a homogenous distribution of the nanoparticles field strength of 10 kA m^{-1} toxicity, 0.2 mL of the magnetic fluids per mL of tissue volume is necessary to guarantee an SAR$_{tissue}$ of 300 W kg^{-1}. However, the measured particle distribution within the tumor tissue was sub-optimal and must be improved.

As side effects, the patients experienced local pain several times (especially when the field strength was >5 kA m^{-1}) and moderate bladder spasms occurred after the first treatment cycle.

Moreover, in this report the authors also described preliminary experiences in glioblastoma and recurrent soft-tissue sarcomas of the ongoing phase I study. Interestingly, the particles are retained in the PCa tissue for at least several weeks.

The authors conclude that this observation represents an important pre-requisite for the clinical application of MPH in locally advanced PCa.

In comparison, an aqueous solution of ff-MWCNTs *per se* represents a homogenous population of particles based on the principles of their synthesis. This could be a major advantage over superparamagnetic particles for a cost-effective clinical use as thermal transmitters.

This development of nanoscaled containers, including the proposed fff-MWCNTs, opens a new chapter in the field of anticancer therapeutic concepts based on nanotechnology. Further steps of evaluation and optimization must undergo very different and detailed tests to ensure their therapeutic efficacy and mechanical stability and safety after local implantation or systemic injection. Progress in medical device technology is strongly linked to progress in material science technology, and new combinations of different materials that have been developed for different application in human medicine have influenced the design, and also the structural, chemical and biological properties of CNTs.

Acknowledgments

The authors thank Brigitte Hamann for TEM and Dr. M. Haase (both from the Institute of Pathology, Technical University Dresden) for histopathological studies, and Dietmar Meiler (IFW Dresden) for expert technical assistance. We thank Dr. Gerd Hammermann (HAMSTEIN-Consult, Dresden, Germany) for helpful cooperation and Systenanix GmbH (Dresden, Germany) for support. Furthermore, we are thankful to Kai Krämer, and to Dr. Vladimir Novotny for participation and support in animal experiments. The ff-MWCNTs were synthesized and characterized by Dr. Manfred Ritschel, Radinka Kozhuharova-Koseva, Dr. Silke Hampel and Christian Müller. The magnetic measurements were performed by Dr. Dieter Elefant. We thank Kai Krämer and Vicky Rutschmann-Will for critical language editing. The project was generously supported by Prof. Dr. M.P. Wirth (Head of the Department of Urology, Technical University Dresden) and by Prof. Dr. B. Büchner (Head of the Institute for Solid State Research; IFW Dresden).

Abbreviations

AC	Alternating current
CNT	Carbon nanotube
CVD	Chemical vapor deposition
FACS	Fluorescence-activated cell sorting
fff-MWCNT	Functionalized and ferromagnetic filled MWCNT
ff-MWCNT	Ferromagnetic filled MWCNT
f-MWCNT	Filled MWCNT

H_c	Coercivity
MFH	Magnetic fluid hyperthermia
MPH	Magnetic particle hyperthermia
M_r	Remnant magnetization
M_{sat}	Saturation magnetization
MWCNT	Multi-walled carbon nanotube
PCa	Prostate cancer
P_{FM}	Power loss of ferromagnetic material
PSA	Prostate specific antigen
P_{SPM}	Power loss of superparamagnetic material
SAR_{Fe}	Specific absorption rate of iron
SAR	Specific absorption rate
SEM	Scanning electron microscope
SWCNT	Single-walled carbon nanotube
TEM	Transmission electron microscope
T	Temperature
t	Time

References

1 GONZALGO ML, ISAACS WB, Molecular pathways to prostate cancer, *J. Urol.*, **2003**, 170, 2444–2452.
2 BASTACKY SI, WOJNO KJ, WALSH PC, CARMICHAEL MJ, EPSTEIN JI, Pathological features of hereditary prostate cancer, *J. Urol.*, **1995**, 153, 987–992.
3 FELDMAN BJ, FELDMAN D, The development of androgen-independent prostate cancer, *Nat. Rev. Cancer*, **2001**, 1, 34–45.
4 TRICOLI JV, SCHOENFELDT M, CONLEY BA, Detection of prostate cancer and predicting progression: Current and future diagnostic markers, *Clin. Cancer Res.*, **2004**, 10, 3943–3953.
5 DOWNS TM, KANE GD, GROSSFELD GD, MENG MV, CAROLL PR, Surgery for prostatate cancer: Rational, techniques and outcome, *Cancer Metastasis Rev.*, **2002**, 21, 29–44.
6 GOMELLA LG, ZELTSER I, VALICENTI RK, Use of neoadjuvant and adjuvant therapy to prevent or delay recurrence of prostate cancer in patients undergoing surgical treatment for prostate cancer, *Urology*, **2003**, 62(Suppl 1), 46–54.
7 MIYAMOTO H, MESSING EM, CHANG C, Androgen deprivation therapy for prostate cancer: Current status and future prospects, *Prostate*, **2004**, 61, 332–353.
8 RAO CNR, SEN R, SATISKUMAR BC, GOVINDARAJ A, Large aligned-nanotube bundles from ferrocene pyrolysis, *Chem. Commun.*, **1998**, 1525.
9 GROBERT N, HSU WK, ZHU YQ, HARE JP, KROTO HW, WALTON DRM, TERRONES M, TERRONES H, REDLICH P, RUHLE M, ESCUDERO R, MORALES F, Enhanced magnetic coercivities in Fe nanowires, *Appl. Phys. Lett.*, **1999**, 75, 3363–3365.
10 KOZHUHAROVA R, RITSCHEL M, ELEFANT D, GRAFF A, LEONHARDT A, MONCH I, MUHL T, SCHNEIDER CM, Synthesis and characterization of aligned Fe-filled carbon nanotubes on silicon substrates, *J. Mater. Sci.-Mater. Electron.*, **2003**, 14, 789–791.
11 KOZHUHAROVA R, PhD-thesis: "Preparation and characterization of ferromagnetic filled carbon nano-tubes", in *Fakultät Maschinenwesen*, Technische Universität, Dresden, **2006**.

12 Leonhardt A, Ritschel A, Kozhuharova R, Graff A, Muhl T, Huhle R, Monch I, Elefant D, Schneider CM, Synthesis and properties of filled carbon nanotubes, *Diamond Relat. Mater.*, **2003**, 12, 790–793.

13 Kozhuharova R, Ritschel M, Monch I, Muhl T, Leonhardt A, Graff A, Schneider CM, Selective growth of aligned Co-filled carbon nanotubes on silicon substrates, *Fullerenes Nanotubes Carbon Nanostruct.*, **2005**, 13, 347–353.

14 Kozhuharova R, Ritschel M, Elefant D, Graff A, Monch I, Muhl T, Schneider CM, Leonhardt A, (FexCo1-x)-alloy filled vertically aligned carbon nanotubes grown by thermal chemical vapor deposition, *J. Magn. Magn. Mater.*, **2005**, 290–291, 250–253.

15 Monch I, Meye A, Leonhardt A, Kramer K, Kozhuharova R, Gemming T, Wirth MP, Büchner B, Ferromagnetic filled carbon nanotubes and nanoparticles: Synthesis and lipid-mediated delivery into human tumor cells, *J. Magn. Magn. Mater.*, **2005**, 290–291, 276–278.

16 Iijima S, Helical microtubules of graphitic carbon, *Nature* **1991**, 354, 56–58.

17 Iijima S, Ichihashi T, Single-shell carbon nanotubes of 1-nm diameter, *Nature*, **1993**, 363, 603–605.

18 Dresselhaus MS, Dresselhaus G, Avouris P, Carbon nanotubes, synthesis, structure, properties, applications, *Top. Appl. Phys.*, **2001**, 80.

19 Prados C, Crespo P, Gonzalez JM, Hernando A, Marco JF, Gancedo R, Grobert N, Terrones M, Walton RM, Kroto HW, Hysteresis shift in Fe-filled carbon nanotubes due to gamma-Fe, *Phys. Rev. B (Condensed Matter Mater. Phys.)*, **2002**, 65, 113 404–113 405.

20 Wong Shi Kam N, O'Connell M, Wisdom JA, Dai H, *Proc. Natl. Acad. Sci. U.S.A.*, **2005**, 102, 11 600.

21 Cheng HM, Li F, Sun X, Brown SDM, Pimenta MA, Marucci A, Dresselhaus G, Dresselhaus MS, Bulk morphology and diameter distribution of single-walled carbon nanotubes synthesized by catalytic decomposition of hydrocarbons, *Chem. Phys. Lett.*, **1998**, 289, 602–610.

22 Colomer J-F, Stephan C, Lefrant S, Van Tendeloo G, Willems I, Konya Z, Fonseca A, Laurent C, Nagy BJ, Large-scale synthesis of single-wall carbon nanotubes by catalytic chemical vapor deposition (CCVD) method, *Chem. Phys. Lett.*, **2000**, 317, 83–89.

23 Joumet C, Maser WK, Bernier P, *Nature*, **1997**, 388, 756.

24 Guo T, Nikolaev P, Thess A, Colbert DT, Smalley RE, Catalytic growth of single-walled nanotubes by laser vaporization, *Chem. Phys. Lett.*, **1995**, 243, 49–54.

25 Täschner C, Pacal F, Leonhardt A, Spatenka P, Bartsch K, Graff A, Kaltofen R, *Surf. Coatings Technol.*, **2003**, 81.

26 Chhowalla M, Teo KBK, Ducati C, Rupesinghe NL, Amaratunga GAJ, Ferrari AC, Roy D, Robertson J, Milne WI, Growth process conditions of vertically aligned carbon nanotubes using plasma enhanced chemical vapor deposition, *J. Appl. Phys.*, **2001**, 90, 5308–5317.

27 Ren ZF, Huang ZP, Xu JW, Wang JH, Bush P, Siegal MP, Provencio PN, Synthesis of large arrays of well-aligned carbon nanotubes on glass, *Science*, **1998**, 282.

28 Loiseau A, Pascard H, Synthesis of long carbon nanotubes filled with Se, S, Sb and Ge by the arc method, *Chem. Phys. Lett.*, **1996**, 256, 246–252.

29 Loiseau A, Willaime F, Filled and mixed nanotubes: From TEM studies to the growth mechanism within a phase-diagram approach, *Appl. Surf. Sci.*, **2000**, 164, 227–240.

30 Dyagileva LM, Mar'in VP, Tsyganova EI, Razuvaev GA, Reactivity of the first transition row metallocenes in thermal decomposition reaction, *J. Organomet. Chem.*, **1979**, 175, 63–72.

31 Sen R, Govindaraj A, Rao CNR,

Carbon nanotubes by the metallocene route, *Chem. Phys. Lett.*, **1997**, 267, 276–280.

32 GROBERT N, MAYNE M, TERRONES M, SLOAN J, DUNIN–BORKOWSKI RE, KAMALAKARAN R, SEEGER T, TERRONES H, RÜHLE M, WALTON DRM, KROTO HW, HUTCHISON JL, *Chem. Commun.*, **2001**, 471.

33 MAYNE M, GROBERT N, TERRONES M, KAMALAKARAN R, RUHLE M, KROTO HW, WALTON DRM, Pyrolytic production of aligned carbon nanotubes from homogeneously dispersed benzene-based aerosols, *Chem. Phys. Lett.*, **2001**, 338, 101–107.

34 SATISHKUMAR BC, GOVINDARAJ A, VANITHA PV, RAYCHAUDHURI AK, RAO CNR, Barkhausen jumps and related magnetic properties of iron nanowires encapsulated in aligned carbon nanotube bundles, *Chem. Phys. Lett.*, **2002**, 362, 301–306.

35 LEONHARDT A, RITSCHEL A, ELEFANT D, MATTERN N, BIEDERMANN K, HAMPEL S, MÜLLER C, GEMMING T, BÜCHNER B, *J. Appl. Phys.*, **2005**, 98, 074315-1-5.

36 KOZHUHAROVA R, RITSCHEL M, ELEFANT D, GRAFF A, LEONHARDT A, MONCH I, MUHL T, GROUDEVA-ZOTOVA S, SCHNEIDER CM, Well-aligned Co-filled carbon nanotubes: Preparation and magnetic properties, *Appl. Surf. Sci.*, **2004**, 238, 355–359.

37 HAMPEL S, LEONHARDT A, SELBMANN D, BIEDERMANN K, ELEFANT D, BÜCHNER B, Growth and characterization of filled carbon nanotubes with ferromagnetic properties, *Carbon*, accepted for publication, **2006**.

38 KARMAKAR S, SHARMA SM, MUKADAM MD, YUSUF SM, SOOD AK, Magnetic behavior of iron-filled multiwalled carbon nanotubes, *J. Appl. Phys.*, **2005**, 97, 54 305–54 306.

39 MUNOZ-SANDOVAL E, LOPEZ-URIAS F, DIAZ-ORTIZ A, TERRONES M, REYES-REYES M, MORAN-LOPEZ JL, Magnetic and transport properties of Fe nanowires encapsulated in carbon nanotubes, *J. Magn. Magn. Mater.*, **2004**, 272–276, E1255–E1257.

40 CHE R, PENG L-M, CHEN Q, DUAN XF, ZOU BS, GU ZN, Controlled synthesis and phase transformation of ferrous nanowires inside carbon nanotubes, *Chem. Phys. Lett.*, **2003**, 375, 59–64.

41 PICHOT V, LAUNOIS P, PINAULT M, MAYNE-L'HERMITE M, REYNAUD C, Evidence of strong nanotube alignment and for iron preferential growth axis in multiwalled carbon nanotube carpets, *Appl. Phys. Lett.*, **2004**, 85, 473–475.

42 SCHNITZLER MC, OLIVEIRA MM, UGARTE D, ZARBIN AJG, One-step route to iron oxide-filled carbon nanotubes and bucky-onions based on the pyrolysis of organometallic precursors, *Chem. Phys. Lett.*, **2003**, 381, 541–548.

43 KIM H, SIGMUND W, Iron nanoparticles in carbon nanotubes at various temperatures, *J. Crystal Growth*, **2005**, 276, 594–605.

44 KIM H, SIGMUND W, Iron particles in carbon nanotubes, *Carbon*, **2005**, 43, 1743–1748.

45 SHAPER AK, HOU H, GREINER A, PHILLIPP F, *J. Catal.*, **2004**, 222, 250.

46 LEONHARDT A, MÜLLER C, HAMPEL S, SELBMANN D, RITSCHEL A, GEMMING T, BIEDERMANN K, ELEFANT D, BÜCHNER B, *Proc. Electrochem. Soc. Inc., Euro CVD*, **2005**, 15, 372.

47 BARTSCH K, LEONHARDT A, Growth and morphology of aligned carbon nanotube layers, *Thin Solid Films*, **2004**, 469–470, 115–119.

48 BARTSCH K, BIEDERMANN K, GEMMING T, LEONHARDT A, On the diffusion-controlled growth of multiwalled carbon nanotubes, *J. Appl. Phys.*, **2005**, 97, 114 301–114 307.

49 ANDREWS R, JACQUES D, RAO AM, DERBYSHIRE F, QIAN D, FAN X, DICKEY EC, CHEN J, Continuous production of aligned carbon nanotubes: A step closer to commercial realization, *Chem. Phys. Lett.*, **1999**, 303, 467–474.

50 ZHANG X, CAO A, WEI B, LI Y, WEI J, XU C, WU D, Rapid growth of well-aligned carbon nanotube arrays, *Chem. Phys. Lett.*, **2002**, 362, 285–290.

51 DORMANS GJM, OMCVD of transition

52 Endo H, Kuwana K, Saito K, Qian D, Andrews R, Grulke EA, CFD prediction of carbon nanotube production rate in a CVD reactor, *Chem. Phys. Lett.*, **2004**, 387, 307–311.

53 Deck CP, Vecchio K, Prediction of carbon nanotube growth success by the analysis of carbon-catalyst binary phase diagrams, *Carbon*, **2006**, 43, 2608–2617.

54 Müller C, Hampel S, Elefant D, Biedermann K, Leonhardt A, Ritschel A, Büchner B, Iron filled carbon nanotubes grown on substrates with thin metal layers and their magnetic properties, *Carbon*, accepted for publication, **2006**.

55 Erlanger BF, Chen BX, Zhu M, Bruns L, *Nano Lett.*, **2001**, 1, 465.

56 Yu X, Chattopadhyay D, Galeska I, Papadimitrakopoulos F, Rusling JF, Peroxidase activity of enzymes bound to the ends of single-wall carbon nanotube forest electrodes, *Electrochem. Commun.*, **2003**, 5, 408–411.

57 Zhang L, Zhao, GC, Wie XW, Yang, ZS, *Chem. Lett.* **2004**, 33, 86.

58 Dai H, Petil A, Vaia RA, in *Molecular Structures, XVII. International Winterschool/Euroconference on Electronic Properties of Novel Materials; AIP Conference Proceedings*, ed. Kuzmany H, Mehring M, Roth S, Fink J, Melville, New York, **2003**.

59 Ahn KS, Kim JS, Kim CO, Hong JP, Non-reactive rf treatment of multiwall carbon nanotube with inert argon plasma for enhanced field emission, *Carbon*, **2003**, 41, 2481–2485.

60 Tsang SC, Harris PJF, Green M, *Nature*, **1993**, 362, 520.

61 Tsang SC, Chen YK, Harris PJF, Green M, *Nature*, **1994**, 327, 159.

62 Raymondo-Pinero E, Cacciaguerra T, Simon P, Beguin F, A single step process for the simultaneous purification and opening of multiwalled carbon nanotubes, *Chem. Phys. Lett.*, **2005**, 412, 184–189.

63 Ajayan PM, Iijima S, *Nature*, **1993**, 361, 333.

64 Chu A, Cook J, Heesom RJR, Hutchison JL, Green MLH, Sloan J, *Chem. Mater.*, **1996**, 8, 2751.

65 Ugarte D, Stäckli T, Bonard JM, Châtelain A, de Heer WA, Filling carbon nanotubes, *Appl. Phys. A, Mater. Sci. & Process.*, **1998**, 67, 101–105.

66 Gao Y, Bando Y, *Nature*, **2002**, 415, 599.

66a Wu H, Wei X, Shao M, Gu J, Qu M, Preparation of Fe-Ni-alloy nanoparticles inside carbon nanotubes via wet chemistry, *J. Mater. Chem.*, **2002**, 12, 1919–1921.

67 Corio P, Santos AP, Santos PS, Temperini MLA, Brar VW, Pimenta MA, Dresselhaus MS, Characterization of single wall carbon nanotubes filled with silver and with chromium compounds, *Chem. Phys. Lett.*, **2004**, 383, 475–480.

68 Mittal J, Monthioux M, Allouche H, Stephan O, Room temperature filling of single-wall carbon nanotubes with chromium oxide in open air, *Chem. Phys. Lett.*, **2001**, 339, 311–318.

69 Sloan J, Cook J, Chu A, Zwiefka-Sibley M, Green MLH, Hutchinson J, *J. Solid State Chem.*, **1998**, 140, 83.

70 Hsu WK, Li WZ, Zhu YQ, Grobert N, Terrones M, Terrones H, Yao N, Zhang JP, Firth S, Clark RJH, KCl crystallization within the space between carbon nanotube walls, *Chem. Phys. Lett.*, **2000**, 317, 77–82.

71 Kim BM, Quian S, Bau HH, *Nano Lett.*, **2005**, 5, 873.

72 Korneva G, Ye H, Gogotsi Y, Halverson D, Friedman G, Bradley J-C, Kornev KG, *Nano Lett.*, **2005**, 5, 879.

73 Satishkumar BC, Govindaraj A, Mofokeng J, Subbanna GN, Rao CNR, *J. Phys. B, At. Mol. Opt. Phys.*, **1996**, 29, 4925.

74 Kodama RH, Magnetic nanoparticles, *J. Magn. Magn. Mater.*, **1999**, 200, 359–372.

75 Bahadur D, Giri J, Biomaterials and magnetism, *Sadhana*, **2003**, 28, 639–656.

76 CHIKAZUMI S, *Physics of Magnetism*, John Wiley & Sons, Inc., New York, London, Sydney, **1964**.

77 AHARONI A, *Introduction to the Theory of Ferromagnetism*, Oxford University Press, Oxford, UK, **1996**.

78 CULLITY BD, *Introduction to Magnetic Materials*, Addison-Wesley, Boston, MA, U.S.A., **1972**.

79 YOSHIDA K, *Theory of Magnetism*, Springer, New York, **1996**.

80 ZIJLSTRA H, *Ferromagnetic Materials 3*, North-Holland Publishing Company, Amsterdam, **1982**.

81 SELLMYER DJ, ZHENG M, SKOMSKI R, Magnetism of Fe, Co and Ni nanowires in self-assembled arrays, *J. Phys., Condensed Matter*, **2001**, R433–R460.

82 LI F, REN L, NIU Z, WANG H, WANG T, Magnetic moment orientations in α-Fe nanowire arrays embedded in anodic aluminum oxide templates, *J. Phys., Condensed Matter*, **2002**, 6875–6882.

83 NIELSCH K, WEHRSPOHN RB, BARTHEL J, KIRSCHNER J, GOSELE U, FISCHER SF, KRONMULLER H, Hexagonally ordered 100 nm period nickel nanowire arrays, *Appl. Phys. Lett.*, **2001**, 79, 1360–1362.

84 NIELSCH K, PhD-thesis, Hochgeordnete ferromagnetische nanostabensembles: Elektrochemische herstellung und magnetische charakterisierung, in *Mathematisch-Naturwissenschaftlich-Technische fakultät*, Halle (Saale), Martin-Luther-Universität Halle-Wittenberg, **2002**.

85 ROSS CA, HWANG M, SHIMA M, CHENG JY, FARHOUD M, SAVAS TA, SMITH HI, SCHWARZACHER W, ROSS FM, REDJDAL M, HUMPHREY FB, Micromagnetic behaviour of electrodeposited cylinder arrays, *Phys. Rev. B*, **2002**, 65, 144 411–144 418.

86 SUN L, HAO Y, CHIEN C-L, SEARSON PC, Tuning the properties of magnetic nanowires, *IBM J. Res. & Dev.*, **2005**, 49, 79–102.

87 CHENG GS, ZHANG LD, ZHU YQ, FREI GT, LI L, MO CM, MAO YQ, *Appl. Phys. Lett.* **1999**, 75, 2455.

88 WANG XF, ZHANG LD, SHI HZ, PENG XS, ZHENG MJ, FANG J, CHEN JL, GAO BJ, *J. Phys. D, Appl. Phys.*, **2001**, 34, 418.

89 MORAS K, SCHAARSCHUCH R, RIEHEMANN W, MORAS K, SCHAARSCHUCH R, RIEHEMANN W, Production of nanoscale fesi and fecr powders for magnetofluide application using laser evaporation. Magnetohydrodynamics, **2003**, 39, 35–39.

90 BAO Y, KRISHNAN KM, Preparation of functionalized and gold-coated cobalt nanocrystals for biomedical applications, *J. Magn. Magn. Mater.*, **2005**, 293, 15–19.

91 BUTTER K, PHILIPSE AP, VROEGE GJ, Synthesis and properties of iron ferrofluids, *J. Magn. Magn. Mater.*, **2002**, 252, 1–3.

92 SAITO Y, YOSHIKAWA T, OKUDA M, FUJIMOTO A, YAMAMURO S, WAKOH K, SUMIYAMA K, SUZUKI K, KASUYA A, NISHINA Y, Cobalt particles wrapped in graphitic carbon prepared by an arc discharge method, *J. Appl. Phys.*, **1994**, 75, 134–137.

93 BÖNNEMANN H, BRIJOUX W, BRINKMANN R, MATOUSSEVITCH N, WALDÖFNER N, *Magnetohydrodynamics*, **2003**, 39, 29.

94 BÖNNEMANN H, BRAND RA, BRIJOUX W, HOFSTADT H-W, FRERICHS M, KEMPTER V, MAUS-FRIEDRICHS W, MATOUSSEVITCH N, NAGABHUSHANA KS, VOIGTS F, CAPS V, Air stable Fe and Fe–Co magnetic fluids – synthesis and characterization, *Appl. Organomet. Chem.*, **2005**, 19, 790–796.

95 GIRI J, RAY A, DASGUPTA S, DATTA D, BAHADUR D, Investigations on Tc tuned nano particles of magnetic oxides for hyperthermia applications, *Bio-Med. Mater. Eng.*, **2003**, 13, 387–399.

96 CARPENTER EE, Iron nanoparticles as potential magnetic carriers, *J. Magn. Magn. Mater.*, **2001**, 225, 17–20.

97 CONNOLLY J, PIERRE TGS, RUTNAKRNPITUK M, RIFFLE JS, Silica coating of cobalt nanoparticles increase their magnetic and chemical stability for biomedical applications, *Eur. Cells Mater.*, **2002**, 3, 106–109.

98 HÜLSER T, Strukturelle und

magnetische eigenschaften von eisenkarbid- und eisenoxid- nanopartikeln, in *Fakultät für Naturwissenschften der Universität Duisburg-Essen*, Duisburg-Essen, **2003**.

99 HE YP, WANG SQ, LI CR, MIAO YM, WU ZY, ZOU BS, Synthesis and characterization of functionalized silica-coated Fe_3O_4 superparamagnetic nanocrystals for biological applications, *J. Phys. D, Appl. Phys.*, **2005**, 1342–1350.

100 DRAVID VP, HORST JJ, TENG MH, ELLIOTT B, HWANG L, JOHNSON DL, MASON TO, WEERTMAN JR, Controlled-size nanocapsules, *Nature*, **1995**, 374, 602–602.

101 FU L, DRAVID VP, KLUG K, LIU X, MIRKIN CA, Synthesis and patterning of magnetic nanostructures, *Eur. Cells Mater.*, **2002**, 3, 156–157.

102 SAITO Y, YOSHIKAWA T, OKUDA M, FUJIMOTO N, SUMIYAMA K, SUZUKI K, KASUYA A, NISHINA Y, Carbon nanocapsules encaging metals and carbides, *J. Phys. Chem. Solids*, **1993**, 54, 1849–1860.

103 BRUNSMAN EM, SUTTON R, BORTZ E, KIRKPATRICK S, MIDELFORT K, WILLIAMS J, SMITH P, MCHENRY ME, MAJETICH SA, ARTMAN JO, DE GRAEF M, STALEY SW, Magnetic properties of carbon-coated, ferromagnetic nanoparticles produced by a carbon-arc method, in *38th Annual Conference on Magnetism and Magnetic Materials*, AIP, Minneapolis, Minnesota, **1994**, vol. 75, pp 5882–5884.

104 MCHENRY ME, MAJETICH SA, ARTMAN JO, DEGRAEF M, STALEY SW, Superparamagnetism in carbon-coated Co particles produced by Kratschmer carbon arc process, *Phys. Rev. B*, **1994**, 49, 11 358–11 363.

105 IVANTCHENKO A, Syntheses, investigations and applications on carbon materials from reactions of graphite oxide, acetylene, metal acetylides, and ferrocene through transition metal catalysts, in *Swiss Federal Institute of Technology Zürich*, Zürich, **2004**.

106 PIROTA KR, PROVENCIO M, GARCIA KL, ESCOBAR-GALINDO R, MENDOZA ZELIS P, HERNANDEZ-VELEZ M, VAZQUEZ M, Bi-magnetic microwires: A novel family of materials with controlled magnetic behavior, *J. Magn. Mater.*, **2005**, 290–291, 68–73.

107 BARBIC M, Single domain magnets in bio-medical applications, *Eur. Cells Mater.*, **2002**, 3, 132–134.

108 HALLIDAY, R, *Phys. Patr II*, **1968**, 820.

109 HUANG S, DAI L, MAU AWH, *J. Phys. Chem. B*, **1999**, 103, 4223.

110 HIHARA T, ONODERA H, SUMIYAMA K, SUZUKI K, KASUYA A, NISHINA Y, SAITO Y, YOSHIKAWA T, OKUDA M, Magnetic properties of iron in nanocapsules, *Jpn. J. Appl. Phys.*, **1994**, 33, L24–L25.

111 SUN X, GUTIERREZ A, YACKMANN MJ, DONG X, JIN S, Investigations on magnetic properties and structure for carbon encapsulated nanoparticles of Fe, Co, Ni, *Mater. Sci. Eng. A*, **2000**, 286, 157–160.

112 SEBERINO C, BERTRAM HN, *IEEE Trans. Magn.*, **1997**, MAG-33, 3055–3057.

113 ROSS CA, HWANG M, SHIMA M, CHENG JY, FARHOUD M, SAVAS TA, SMITH HI, SCHWARZACHER W, ROSS FM, REDJDAL M, HUMPHREY FB, Micromagnetic behavior of electro-deposited cylinder arrays, *Phys. Rev. B (Condensed Matter Mater. Phys.)*, **2002**, 65, 144 417–144 418.

114 GANGOPADHYAY S, HADJIPANAYIS GC, DALE B, SORENSEN CM, KLABUNDE KJ, PAPAEFTHYMIOU V, KOSTIKAS A, Magnetic properties of ultrafine iron particles, *Phys. Rev. B*, **1992**, 45, 9778–9787.

115 WANG SX, TARATORIN AM, Magnetic information storage technology, in *Magnetic Information Storage Technology*, Academic Press, San Diego, **1999**.

116 WIRTH S, VON MOLNAR S, Hall cross size scaling and its application to measurements on nanometer-size iron particle arrays, *Appl. Phys. Lett.*, **2000**, 76, 3283–3285.

117 SKOMSKI R, ZENG H, SELLMYER DJ, Incoherent magnetization reversal in nanowires, *J. Magn. Magn. Mater.*, **2002**, 249, 175–180.

118 STONER EC, WOHLFARTH EP, A mechanism of magnetic hysteresis in heterogeneous alloys, *Phil. Trans. Royal Soc. London; Ser. A Math. Phys. Sci.*, **1948**, 240, 599–642.

119 KNELLER E, Theory of the magnetization curve of small crystals, in *Encyclopedia of Physics, vol. XVIII/2, Ferromagnetism*, ed. WIJN HPJ, Springer Verlag, New York, **1966**.

120 HERGT R, ANDRA W, D'AMBLY CG, HILGER I, KAISER W, RICHTER U, SCHMIDT H-G, Physical limits of hyperthermia using magnetite fine particles, *IEEE Trans. Magn.*, **1998**, 34, 3745–3754.

121 SEBERINO C, BERTRAM HN, Numerical study of hysteresis and morphology in elongated tape particles, *J. Appl. Phys.*, **1999**, 85(8), 5543–5545.

122 ZHAO B, KOZHUHAROVA R, MÜHL T, MÖNCH I, VINZELBERG H, RITSCHEL M, GRAFF A, HUHLE M, LICHTE H, SCHNEIDER CM, Magnetic systems with carbon nanotubes, in *Structural and Electronic Properties of Molecular Nanostructures*, ed. KUZMANY H, American Institute of Physics, Springer, Heidelberg, **2002**, pp 583–587.

123 SCHNEIDER CM, ZHAO B, KOZHUHAROVA R, GROUDEVA-ZOTOVA S, MUHL T, RITSCHEL M, MONCH I, VINZELBERG H, ELEFANT D, GRAFF A, Towards molecular spintronics: Magnetotransport and magnetism in carbon nanotube-based systems, *Diamond Relat. Mater.*, **2004**, 13, 215–220.

124 CHOU SY, WEI MS, KRAUSS PR, FISCHER PB, Single-domain magnetic pillar array of 35 nm diameter and 65 Gbits/in.[sup 2] density for ultrahigh density quantum magnetic storage, *J. Appl. Phys.*, **1994**, 76, 6673–6675.

125 DREIKORN J, DREYER L, MICHALOWSKY L, ROSSEL J, SICKER U, *J. Phys. IV (Proc. Soft Magn. Mater. 13)*, **1998**, Pr2-457–460.

126 JONES SK, GRAY BN, BURTON MA, CODDE JP, STREET R, Evaluation of ferromagnetic materials for low-frequency hysteresis heating of tumours, *Phys. Med. Biol.*, **1992**, 37, 293–299.

127 ROTH S, Polymer mit eingelagerten Magnetteilchen, IFW-Dresden, personal communication.

128 ROSENSWEIG RE, Heating magnetic fluid with alternating magnetic field, *J. Magn. Magn. Mater.*, **2002**, 252, 370–374.

129 OLESON JR, A review of magnetic induction methods for hyperthermia treatement of cancer, *IEEE Trans. Biomed. Eng.*, **1984**, BME-31, 91–97.

130 GILCHRIST RK, MEDAL R, SHOREY WD, HANSELMAN RC, PARROTT JC, TAYLOR CB, Selective inductive heating of lymph nodes, *Ann. Surg.*, **1957**, 146, 596–606.

131 STAUFFER PR, CETAS TC, FLETCHER AM, DEYOUNG DW, DEWHIRST MW, OLESON JR, ROEMER RB, Observations on the use of ferromagnetic implants for inducing hyperthermia, *IEEE Trans. Biomed. Eng.*, **1984**, BME-31, 76–90.

132 STAUFFER PR, CETAS TC, JONES RC, Magnetic induction heating of ferromagnetic implants for inducing localized hyperthermia in deep-seated tumors, *IEEE Trans. Biomed. Eng.*, **1984**, BME-31, 235–251.

133 ATKINSON WJ, BREZOVICH IA, CHAKRABORTY DP, Usable frequencies in hyperthermia with thermal seeds, *IEEE Trans. Biomed. Eng.*, **1984**, 31, 70–75.

134 GORDON RT, HINES JR, GORDON D, Intracellular hyperthermia – A biophysical approach to cancer treatement via intracellular temperature and biophysical alterations, *Med. Hypotheses*, **1979**, 5, 83–102.

135 TARTAJ P, MORALES MP, GONZALEZ-CARRENO T, VEINTEMILLAS-VERDAGUER S, SERNA CJ, Advances in magnetic nanoparticles for biotechnology applications, *J. Magn. Magn. Mater.*, **2005**, 290–291, 28–34.

136 MUHL T, ELEFANT D, GRAFF A, KOZHUHAROVA R, LEONHARDT A, MONCH I, RITSCHEL M, SIMON P, GROUDEVA-ZOTOVA S, SCHNEIDER CM, Magnetic properties of aligned Fe-filled carbon nanotubes, *J. Appl. Phys.*, **2003**, 93, 7894–7896.

137 JORDAN A, SCHOLZ R, WUST P,

Fahling H, Roland F, Magnetic fluid hyperthermia (MFH): Cancer treatment with AC magnetic field induced excitation of biocompatible superparamagnetic nanoparticles, *J. Magn. Magn. Mater.*, **1999**, 201, 413–419.

138 Jones SK, Gray BN, Burton MA, Codde JP, Street R, *Phys. Med., Biol.*, **1992**, 37, 293–299.

139 Pankhurst QA, Connolly J, Jones SK, Dobson J, Applications of magnetic nanoparticles in biomedicine, *J. Phys. D, Appl. Phys.*, **2003**, R167–R181.

140 Jordan A, Wust P, Fähling H, John W, Hinz A, Felix R, Inductive heating of ferrimagnetic particles and magnetic fluids: Physical evaluation of their potential for hyperthermia, *Int. J. Hyperthermia*, **1993**, 9, 51–68.

141 Hilger I, Hergt R, Kaiser W, Use of magnetic nanoparticle heating in the treatement of breast cancer, *IEE Proc.-Nanobiotechnol.*, **2005**, 152, 33–39.

142 Hilger I, Andra W, Hergt R, Hiergeist R, Schubert H, Kaiser W, Electromagnetic heating of breast tumors in interventional radiology: In vitro and in vivo studies in human cadavers and mice, *Radiology*, **2001**, 218, 570–575.

143 Andra W, d'Ambly CG, Hergt R, Hilger I, Kaiser WA, Temperature distribution as function of time around a small spherical heat source of local magnetic hyperthermia, *J. Magn. Magn. Mater.*, **1999**, 194, 197–203.

144 Jordan A, Scholz R, Maier-Hauff K, Johannsen M, Wust P, Nadobny J, Schirra H, Schmidt H, Deger S, Loening S, Presentation of a new magnetic field therapy system for the treatment of human solid tumors with magnetic fluid hyperthermia, *J. Magn. Magn. Mater.*, **2001**, 225, 118–126.

145 Maehara T, Konishi K, Kamimori T, Aono H, Naohara T, Kikkawa H, Watanabe Y, Kawachi K, Heating of ferrite powder by an AC magnetic field for local hyperthermia, *Jpn. J. Appl. Phys.*, **2002**, 41, 1620–1621.

146 Babincova M, Leszczynska D, Sourivong P, Cicmanec P, Babinec P, Superparamagnetic gel as a novel material for electromagnetically induced hyperthermia, *J. Magn. Magn. Mater.*, **2001**, 225, 109–112.

147 Johannsen M, Gneveckow U, Eckelt L, Feussner A, Waldöfner N, Scholz R, Deger S, Wust P, Loening SA, Jordan A, Clinical hyperthermia of prostate cancer using magnetic nanoparticles: Presentation of a new interstitial technique, *Int. J. Hyperthermia*, preview article, **2005**, 1–11.

148 Hilger I, Andra W, Bahring R, Daum A, Hergt R, Kaiser WA, Evaluation of temperature increase with different amounts of magnetite in liver tissue samples, *Invest. Radiol.*, **1997**, 32, 705–712.

149 Hilger I, Kießling A, Romanus E, Hiergeist R, Hergt R, Andra W, Roskos M, Linss W, Weber P, Weitschies W, Kaiser W, Magnetic nanoparticles for selective heating of magnetically labelled cells in culture: Preliminary investigation, *Nanotechnology*, **2004**, 15, 1027–1032.

150 Jordan A, Wust P, Scholz R, Tesche B, Fähling H, Mitrovics T, Vogl T, Cervos-Navarro J, Felix R, Cellular uptake of magnetic fluid particles and their effects on human adenocarcinoma cells exposed to AC magnetic fields in vitro, *Int. J. Hyperthermia*, **1996**, 12, 705–722.

151 Jordan A, Scholz R, Wust P, Schirra H, Thomas S, Schmidt H, Felix R, Endocytosis of dextran and silan-coated magnetite nanoparticles and the effect of intracellular hyperthermia on human mammary carcinoma cells in vitro, *J. Magn. Magn. Mater.*, **1999**, 194, 185–196.

152 Li GC, Mivechi NF, Weitzel G, Heat shock proteins, thermotolerance, and their relevance to clinical hyperthermia, *Int. J. Hyperthermia*, **1995**, 11, 459–488.

153 Moroz P, Jones SK, Gray BN, Tumor response to arterial embolization hyperthermia and direct injection hyperthermia in a rabbit liver tumor model, *J. Surg. Oncol.*, **2002**, 80, 148–156.

154 MOROZ P, JONES SK, GRAY BN, Status of hyperthermia in the treatment of advanced liver cancer, *J. Surg. Oncol.*, **2001**, 77, 259–269.

155 MOROZ P, JONES SK, WINTER J, GRAY BN, Targeting liver tumors with hyperthermia: Ferromagnetic embolization in a rabbit liver tumor model, *J. Surg. Oncol.*, **2001**, 78, 22–29.

156 HILGER I, HERGT R, KAISER WA, Effects of magnetic thermoablation in muscle tissue using iron oxide particles: An in vitro study, *Investig. Radiol.*, **2000**, 35, 170–179.

157 BREZOVICH IA, MEREDITH RF, Practical aspects of ferromagnetic thermoseed hyperthermia, *Radiol. Clin. North Am.*, **1989**, 1, 589–602.

158 KUZNETSOV AA, SHLYAKHTIN OA, BRUSENTSOV NA, KUZNETSOV NA, "Smart" mediators for self-controlled inductive heating, *Eur. Cells Mater.*, **2002**, 3, 75–77.

159 CETAS TC, GROSS EJ, CONTRACTOR Y, *IEEE Trans. Biomed. Eng.*, **1998**, 45, 68.

160 HÄFELI UO, Magnetically modulated therapeutic systems, *Int. J. Pharmaceut.*, **2004**, 277, 19–24.

161 HIERGEIST R, ANDRA W, BUSKE N, HERGT R, HILGER I, RICHTER U, KAISER W, Application of magnetite ferrofluids for hyperthermia, *J. Magn. Magn. Mater.*, **1999**, 201, 420–422.

162 HUBER DL, Synthesis, properties, and applications of iron nanoparticles, *Small*, **2005**, 1, 482–501.

163 KIM DK, ZHANG Y, VOIT W, RAO KV, MUHAMMED M, Synthesis and characterization of surfactant-coated superparamagnetic monodispersed iron oxide nanoparticles, *J. Magn. Magn. Mater.*, **2001**, 225, 30–36.

164 KIM DK, MIKHAYLOVA M, ZHANG Y, MUHAMMED M, Protective coating of superparamagnetic iron oxide nanoparticles, *Chem. Mater.*, **2003**, 15, 1617–1627.

165 JAIN KT, MORALES MA, SAHOO SK, LESLIE-PELECKY DL, Labhasetwar, Iron oxide nanoparticles for sustained delivery of anticancer agents, *Mol. Pharmaceut.*, **2005**, 2, 194–205.

166 BÖNNEMANN H, BRANDT RA, BRIJOUX W, HOFSTADT H-W, FRERICHS M, KEPTER V, MAUS-FRIEDRICHS W, MATOUSSEVITCH N, NAGABHUSHANA KS, VOIGTS F, CAPS V, Air stable Fe and Fe-Co magnetic fluids – synthesis and characterization, *Appl. Organomet. Chem.*, **2005**, 19, 790–796.

167 BÖNNEMANN H, A review – How nanoparticles emerged from organometallic chemistry, *Appl. Organomet. Chem.*, **2004**, 18, 566–572.

168 CANNON WR, DANFORTH SC, FLINT JH, HAGGERTY JS, MARRA RA, *J. Am. Ceram. Soc.*, **1982**, 65, 324.

169 KUZNETSOV AA, FILIPPOV VI, KUZNETSOV OA, GERLIVANOV VG, DOBRINSKY EK, MALASHIN SI, New ferro-carbon adsorbents for magnetically guided transport of anticancer drugs, *J. Magn. Magn. Mater.*, **1999**, 194, 22–30.

170 RACKA K, GICH M, SLAWSKA-WANIEWSKA A, ROIG A, MOLINS E, Magnetic properties of Fe nanoparticle systems, *J. Magn. Magn. Mater.*, **2005**, 290–291, 127–130.

171 MOHADDES-ARDABILI L, ZHENG H, OGALE SB, HANNOYER B, TIAN W, WANG J, LOFLAND SE, SHINDE SR, ZHAO T, JIA Y, SALAMANCA-RIBA L, SCHLOM DG, WUTTIG M, RAMESH R, Self-assembled single-crystal ferromagnetic iron nanowires formed by decomposition, *Nat. Mater.*, **2004**, 3, 533–538.

172 HWANG J-h, DRAVID VP, TENG MH, HOST JJ, ELLIOTT BR, JOHNSON DL, MASON TO, Magnetic properties of graphitically encapsulated nickel nanocrystals, *J. Mater. Res.*, **1997**, 12, 1076–1082.

173 CHOU C-K, Use of heating rate and specific absorption rate in the hyperthermia clinic, *Int. J. Hyperthermia*, **1990**, 6, 367–370.

174 GNEVECKOW U, JORDAN A, SCHOLZ R, BRÜß V, WALDÖFNER N, RICKE J, FEUSSNER A, HILDEBRANDT B, RAU B, WUST P, Description and characterization of the novel hyperthermia- and thermoablation-system MFH 300F for clinical magnetic fluid hyperthermia, *Med. Phys.*, **2004**, 31, 1444–1451.

175 BRUSENTSOV NA, GOGOSOV VV,

Brusentsova TN, Sergeev AV, Jurchenko NY, Kuznetsov AA, Kuznetsov OA, Shumakov LI, Evaluation of ferromagnetic fluids and suspensions for the site-specific radiofrequency-induced hyperthermia of MX11 sarcoma cells in vitro, *J. Magn. Magn. Mater.*, **2001**, 225, 113–117.

176 Brusentsov NA, Nikitin LV, Brusentsova TN, Kuznetsov AA, Bayburtskiy FS, Shumakov LI, Jurchenko NY, Magnetic fluid hyperthermia of the mouse experimental tumor, *J. Magn. Magn. Mater.*, **2002**, 252, 378–380.

177 Brusentsov NA, Komissarova LK, Ksnetzov AA, et al., Evaluation of ferrifluids containing photosensitizer, *Eur. Cells Mater.*, **2002**, 3, 70–73.

178 Hergt R, Hiergeist R, Hilger I, Kaiser W, Magnetic nanoparticles for thermoablation, in *Recent Research Developments in Materials Science*, ed. Pandalai SG, Research Signpost, Trivandrum, India, **2002**, vol. 3, pp. 723–742.

179 Rabin Y, Is intracellular hyperthermia superior to extracellular hyperthermia in the thermal sense? *Int. J. Hyperthermia*, **2002**, 18, 194–202.

180 Pantarotto D, Shingh R, McCarthy D, Erhardt M, Briand J-P, Prato M, Kostarelos K, Bianco A, Functionalized carbon nanotubes for plasmid DNA gene delivery, *Angew. Chem.*, **2004**, 116, 5354–5358.

181 Cherukuri P, Bachilo SM, Litovsky SH, Weisman RB, Near-infrared fluorescence microscopy of single-walled carbon nanotubes in phagocytic cells, *J. Am. Chem. Soc.*, **2004**, 126, 15 638–15 639.

182 Lu X, Tian F, Xu X, Wang N, Zhang Q, A theoretical exploration of the 1,3-dipolar cycloadditions onto the sidewalls of (n,n) armchair single-wall carbon nanotubes, *J. Am. Chem. Soc.*, **2003**, 125, 10 459–10 464.

183 Bianco A, Kostarelos K, Partidos CD, Prato M, Biomedical applications of functionalised carbon nanotubes, *Chem. Commun. (Camb)*, **2005**, 571–577.

184 Bianco A, Kostarelos K, Prato M, Applications of carbon nanotubes in drug delivery, *Curr. Opin. Chem. Biol.*, **2005**.

185 Kam NWS, Dai H, Carbon nanotubes as intracellular protein transporters, generality and biological functionality, *J. Am. Chem. Soc.*, **2005**, 127, 6021–6026.

186 Kam NWS, Jessop TC, Wender PA, Dai H, Nanotube molecular transporters: Internalization of carbon nanotube-protein conjugates into mammalian cells, *J. Am. Chem. Soc.*, **2004**, 126, 6850–6851.

187 Meye A, et al., submitted for publication, **2005**.

188 Mönch I, Meye A, Leonhardt A, Elefant D, Hampel S, Müller C, Wirth MP, Büchner B, et al. submitted for publication, **2005**.

189 Selbmann D, Krellmann M, Leonhardt A, Eickemeyer J, *J. Phys. IV*, **2000**, Pr2-27.

190 Bianco A, Hoebeke J, Godefroy S, Chaloin O, Pantarotto D, Briand JP, Muller S, Prato M, Partidos CD, Cationic carbon nanotubes bind to CpG oligodeoxynucleotides and enhance their immunostimulatory properties, *J. Am. Chem. Soc.*, **2005**, 127, 58–59.

191 Kam NW, Liu Z, Dai H, Functionalization of carbon nanotubes via cleavable disulfide bonds for efficient intracellular delivery of siRNA and potent gene silencing, *J. Am. Chem. Soc.*, **2005**, 127, 12 492–12 493.

192 Pantarotto D, Briand JP, Prato M, Bianco A, Translocation of bioactive peptides across cell membranes by carbon nanotubes, *Chem. Commun. (Camb)*, **2004**, 16–17.

193 Georgakilas V, Tagmatarchis N, Pantarotto D, Bianco A, Briand JP, Prato M, Amino acid functionalisation of water soluble carbon nanotubes, *Chem. Commun. (Camb)*, **2002**, 3050–3051.

194 Zheng M, Jagota A, Strano MS, Santos AP, Barone P, Chou SG, Diner BA, Dresselhaus MS, McLean RS, Onoa GB, Samsonidze GG, Semke ED, Usrey M, Walls DJ,

Structure-based carbon nanotube sorting by sequence-dependent DNA assembly, *Science*, **2003**, 302, 1545–1548.

195 ZHENG M, JAGOTA A, SEMKE ED, DINER BA, MCLEAN RS, LUSTIG SR, RICHARDSON RE, TASSI NG, DNA-assisted dispersion and separation of carbon nanotubes, *Nat. Mater.*, **2003**, 2, 338–342.

196 SINGH R, PANTAROTTO D, MCCARTHY D, CHALOIN O, HOEBEKE J, PARTIDOS CD, BRIAND JP, PRATO M, BIANCO A, KOSTARELOS K, Binding and condensation of plasmid DNA onto functionalized carbon nanotubes: Toward the construction of nanotube-based gene delivery vectors, *J. Am. Chem. Soc.*, **2005**, 127, 4388–4396.

197 JOHANNSEN M, THIESEN B, GNEVECKOW U, TAYMOORIAN K, WALDOFNER N, SCHOLZ R, DEGER S, JUNG K, LOENING SA, JORDAN A, Thermotherapy using magnetic nanoparticles combined with external radiation in an orthotopic rat model of prostate cancer, *Prostate*, **2005**.

198 JOHANNSEN M, THIESEN B, JORDAN A, TAYMOORIAN K, GNEVECKOW U, WALDOFNER N, SCHOLZ R, KOCH M, LEIN M, JUNG K, LOENING SA, Magnetic fluid hyperthermia (MFH) reduces prostate cancer growth in the orthotopic Dunning R3327 rat model, *Prostate*, **2005**, 64, 283–292.

199 LIU B, WEI L, DING Q, YAO J, Synthesis and magnetic study for Fe-doped carbon nanotubes (CNTs), *J. Crystal Growth*, **2005**, 277, 293–297.

200 FUJIWARA Y, TAKEGAWA H, SATO H, MAEDA K, SAITO Y, KOBAYASHI T, SHIOMI S, Magnetic properties of carbon nanotubes grown on Fe catalyst layer by microwave plasma enhanced chemical vapor deposition, *J. Appl. Phys.*, **2004**, 95, 11, 7118–7120.

201 BAO J, TIE C, XU Z, SUO Z, ZHOU Q, HONG J, A facile method for creating an array of metal-filled carbon nanotubes, *Adv. Mater.*, **2002**, 14, 1483–1486.

202 BAO J, ZHOU Q, HONG J, XU Z, Synthesis and magnetic behavior of an array of nickel-filled carbon nanotubes, *Appl. Phys. Lett.*, **2002**, 81, 4592–4594.

203 TYAGI PK, SINGH MK, MISRA A, PALNITKAR U, MISRA DS, TITUS E, ALI N, CABRAL G, GRACIO J, ROY M, KULSHRESHTHA SK, Preparation of Ni-filled carbon nanotubes for key potential applications in nanotechnology, *Thin Solid Films*, **2004**, 469–470, 127–130.

204 BENKOWSKY G, *Induktionserwärmung*, Verlag Technik, Berlin, **1977**.

205 CHAN DCF, KIRPOTIN DB, BRUNN JR. PA, Synthesis and evaluation of colloidal magnetic iron oxides for the site-specific radiofrequency-induced hyperthermia of cancer, *J. Magn. Magn. Mater.*, **1993**, 122, 374–378.

206 HERGT R, HIERGEIST R, ZEISBERGER M, GLÖCKL G, WEITSCHIES W, RAMIREZ LP, HILGER I, KAISER WA, Enhancement of AC-losses of magnetic nanoparticles for heating applications, *J. Magn. Magn. Mater.*, **2004**, 280, 358–368.

207 GEORGAKILAS V, KORDATOS K, PRATO M, GULDI DM, HOLZINGER M, HIRSCH A, Organic functionalization of carbon nanotubes, *J. Am. Chem. Soc.*, **2002**, 124, 760–761.

208 KAM NW, LIU Z, DAI H, Functionalization of carbon nanotubes via cleavable disulfide bonds for efficient intracellular delivery of siRNA and potent gene silencing, *J. Am. Chem. Soc.*, **2005**, 127, 12 492–12 493.

209 PANTAROTTO D, PARTIDOS CD, HOEBEKE J, BROWN F, KRAMER E, BRIAND JP, MULLER S, PRATO M, BIANCO A, Immunization with peptide-functionalized carbon nanotubes enhances virus-specific neutralizing antibody responses, *Chem. Biol.*, **2003**, 10, 961–966.

210 PANTAROTTO D, PARTIDOS CD, GRAFF R, HOEBEKE J, BRIAND JP, PRATO M, BIANCO A, Synthesis, structural characterization, and immunological properties of carbon nanotubes functionalized with peptides, *J. Am. Chem. Soc.*, **2003**, 125, 6160–6164.

211 KAM NW, DAI H, Carbon nanotubes as intracellular protein transporters: Generality and biological functionality,

J. Am. Chem. Soc., **2005**, 127, 6021–6026.

212 SINGH R, PATAROTTO D, MCCARTHY D, CHALOIN O, HOEBEKE J, PARTIDOS CD, BRIAND J-P, PRATO M, BIANCO A, KOSTARELOS K, Binding and condensation of plasmid DNA onto functionalized carbon nanotubes: Toward the construction of nanotube-based gene delivery vectors, *J. Am. Chem. Soc.*, **2005**, 127, 4388–4396.

213 CAI Y, JIANG G, LIU J, ZHOU Q, Multiwalled carbon nanotubes as a solid-phase extraction adsorbent for the determination of bisphenol A, 4-n-nonylphenol, and 4-tert-octylphenol, *Anal. Chem.*, **2003**, 75, 2517–2521.

214 CAI H, CAO X, JIANG Y, HE P, FANG Y, Carbon nanotube-enhanced electrochemical DNA biosensor for DNA hybridization detection, *Anal. Bioanal., Chem.*, **2003**, 375, 287–293.

215 GOODING JJ, WIBOWO R, LIU J, YANG W, LOSIC D, ORBONS S, MEARNS FJ, SHAPTER JG, HIBBERT DB, Protein electrochemistry using aligned carbon nanotube arrays, *J. Am. Chem. Soc.*, **2003**, 125, 9006–9007.

216 WANG SG, WANG R, SELLIN PJ, ZHANG Q, DNA biosensors based on self-assembled carbon nanotubes, *Biochem. Biophys. Res. Commun.*, **2004**, 325, 1433–1437.

217 WANG J, KAWDE AN, JAN MR, Carbon-nanotube-modified electrodes for amplified enzyme-based electrical detection of DNA hybridization, *Biosens. Bioelectron.*, **2004**, 20, 995–1000.

218 WANG Y, LI Q, HU S, A multiwall carbon nanotubes film-modified carbon fiber ultramicroelectrode for the determination of nitric oxide radical in liver mitochondria, *Bioelectrochemistry*, **2005**, 65, 135–142.

219 CHEN RS, HUANG WH, TONG H, WANG ZL, CHENG JK, Carbon fiber nanoelectrodes modified by single-walled carbon nanotubes, *Anal. Chem.*, **2003**, 75, 6341–6345.

220 CHEN RJ, BANGSARUNTIP S, DROUVALAKIS KA, KAM NW, SHIM M, LI Y, KIM W, UTZ PJ, DAI H, Noncovalent functionalization of carbon nanotubes for highly specific electronic biosensors, *Proc. Natl. Acad. Sci. U.S.A.*, **2003**, 100, 4984–4989.

221 SALVADOR-MORALES C, FLAHAUT E, SIM E, SLOAN J, GREEN MLH, SIM RB, Complement activation and protein adsorption by carbon nanotubes, *Mol. Immunol.*, **2006**, 43, 193–201.

222 SCHNITZLER GR, CHEUNG CL, HAFNER JH, SAURIN AJ, KINGSTON RE, LIEBER CM, Direct imaging of human SWI/SNF-remodeled mono- and polynucleosomes by atomic force microscopy employing carbon nanotube tips, *Mol. Cell. Biol.*, **2001**, 21, 8504–8511.

223 HAFNER JH, CHEUNG CL, WOOLLEY AT, LIEBER CM, Structural and functional imaging with carbon nanotube AFM probes, *Prog. Biophys. Mol. Biol.*, **2001**, 77, 73–110.

224 MOGHIMI SM, SZEBENI J, Stealth liposomes and long circulating nanoparticles: Critical issues in pharmacokinetics, opsonization and protein-binding properties, *Prog. Lipid Res.*, **2003**, 42, 463–478.

225 PARK J, PASUPATHY AN, GOLDSMITH JI, CHANG C, YAISH Y, PETTA JR, RINKOSKI M, SETHNA JP, ABRUNA HD, MCEUEN PL, RALPH DC, Coulomb blockade and the Kondo effect in single-atom transistors, *Nature*, **2002**, 417, 722–725.

226 WEISMAN RB, Carbon nanotubes: Four degrees of separation, *Nat. Mater.*, **2003**, 2, 569–570.

227 SERVICE RF, Molecular electronics. Nanodevices make fresh strides toward reality, *Science*, **2003**, 302, 1310.

228 SERVICE RF, Nanotechnology. Sorting technique may boost nanotube research, *Science*, **2003**, 300, 2018.

229 JAVEY A, GUO J, WANG Q, LUNDSTROM M, DAI H, Ballistic carbon nanotube field-effect transistors, *Nature*, **2003**, 424, 654–657.

230 BARTSCH K, LEONHARDT A, An approach to the structural diversity of aligned grown multi-walled carbon nanotubes on catalyst layers, *Carbon*, **2004**, 42, 1731–1736.

231 Dupuis AC, *Progr. Mater. Sci.*, **2005**, 50, 929–961.
232 Mönch I, Leonhardt A, Büchner B, Medical application of filled nanotubes, spin-off: *SYSTENANIX. Annual Report 2004*, Leibniz-Institute for Solid State and Materials Research Dresden; http://www.ifw-dresden.de/forsch/jb2004/Techtrans.pdf, **2004**, 46.
233 Vostrowsky O, Hirsch A, Forscher gucken durch die Nano-Röhre – nanotubes aus Kohlenstoff abayfor-Zukunft im Brennpunkt, **2003**; http://www.abayfor.de/abayfor/_media/pdf/ZIB2/zib2_vostrowsky.pdf, **2003**, 85–88.
234 Zhang X, Cao A, Li Y, Xu C, Liang J, Wu D, Wei B, Self-organized arrays of carbon nanotube ropes, *Chem. Phys. Lett.*, **2002**, 351, 183–188.

10
Liposomes, Dendrimers and other Polymeric Nanoparticles for Targeted Delivery of Anticancer Agents – A Comparative Study

Yong Zhang and Dev K. Chatterjee

10.1
Introduction

Cancer management through chemotherapy, surgery and radiotherapy often fails because of high toxicity and poor target selectivity. Nanotechnology, through the synthesis and modification of nanoparticles, has the potential to overcome these barriers by providing drug molecule stability in circulation, reduced toxicity and better targeting of tumors.

Several excellent reviews deal with the anti-tumor activities of nanoparticles. A look at recent ones in this fast changing field shows that most have tackled the wider issue of nanotechnology as a tool against cancer. In an excellent review, Ferrari has dealt with the whole field of cancer nanotechnology, including *in vitro* diagnostics as well as *in vivo* targeting [1]. In a very recent review Jain (of Jain PharmaBiotech in Switzerland) has focused on the field of "nanobiotechnologies" and its effect on drug delivery in cancer [2]. The same author has also reviewed nanotechnology in the setting of the diagnostic clinical laboratory [3]. Another recent review, by McNeil, deals more widely with the ramifications of nanotechnology for the biologist [4]. A similar review by Fortina et al. details the uses and promises of nanobiotechnology in the whole field of molecular recognition, mainly for enhanced *in vitro* molecular diagnostics [5]. As these and other examples show, most reviews deal either with the wider aspects of nanotechnology or with different subsets.

This chapter focuses specifically on the targeting of drug0-loaded nanoparticles to tumor sites. This is perhaps the most important aspect of anticancer therapy with nanoparticles. However, even with a wish to limit ourselves to detailing the means of targeting nanoparticles and discussing and comparing current reports about the three major types of nanoparticles, we find it imperative to include short introductions, wherever applicable, so that even recent entrants to the field can peruse this chapter.

Nanoparticles that are currently under consideration as drug delivery vehicles can be considered to be of three major types: liposomes, dendrimers and other polymeric nanoparticles. When modified with certain chemicals, mainly poly(ethylene glycol), they enjoy long circulation times and selectively accumulate at tumor sites due to an enhanced permeation and retention (EPR) effect. When a targeting ligand or antibody is attached this selectivity is enhanced. Suitable targets include antigens of the neo-angiogenic process associated with tumor growth, altered antigens specific for tumors (tumor specific antigens) and growth factor receptors such as transferrin and folic acid receptors. Drugs delivered to the site include conventional anti-neoplastic drugs such as doxorubicin and paclitaxel, and also other biological response modifiers such as cytokines and antibodies. Corporate interest in the future of this technology has resulted in a few start ups that have already marketed a few anti-tumor nanoparticles formulations; hope for a better future is appreciably high.

Section 10.2 deals with the effectiveness of chemotherapy and other conventional therapies and discusses their limitations in halting tumor growth. We introduce the concept of targeted drug delivery in Section 10.3 and elaborate on targeting ligands in Section 10.4. Section 10.5 deals with the achievements of each of the three major types of nanoparticles. Section 10.6 wraps up the review with a look at the overall picture that is emerging and ends on an optimistic note.

10.2
Cancer Chemotherapy: so Far, but not so Good

The fundamental differences between cancer cells and normal human cells are still not clear. None of the empirically developed anti-neoplastic drugs in conventional use appear to involve a mechanism or target that is completely unique to the cancer cell. They appear to achieve some degree of selectivity in their action by acting on certain characteristics that are found altered in cancerous tissues as compared to normal ones. These include a rapid rate of cell division, differences in the rate of uptake of or the sensitivity to the drug in different types of cells, and occasional presence in the cancer cells of hormonal responses characteristic of the original cells from which they developed.

All cells undergo divisions in a cyclical manner. This cell cycle has been divided into several phases. Very broadly, cells can undergo DNA synthesis (S phase) and mitotic and cytokinetic activity (M phase) while dividing; or can be in a dormant non-dividing state (G_0 phase). Anti-neoplastic drugs can be broadly classified according to whether they are cycle sensitive or insensitive. Cycle sensitive agents act on dividing cells, during the M or S phase, and have first-order kinetics and exponential dose–response curves. They are highly active on rapidly dividing tumor cells, but fail to kill cells that are dormant or dividing slowly.

The selection of drugs for a particular cancer depends on several factors: the type of cancer, the stage and grade of the tumor, the condition of the patient, and the ability to afford the therapy. The selected regimen may aim for palliation, i.e., reduction in the severity or extent of the disease, or remission, i.e., complete absence of cancerous growth clinically and with laboratory tests. Current cancer therapy suffers from the serious disadvantage of high toxicity to the patient's normal body tissues. The effectiveness of therapeutic regimens has to be balanced against their side effects. Poor selectivity for cancer cells by the anticancer drugs is the underlying cause.

The major reason for poor selectivity is the relative paucity of marker molecules that will unequivocally distinguish neoplastic from non-neoplastic cells. Most cancer chemotherapeutic drugs interfere with the mechanism of cell division. Some cancers are intrinsically highly drug-sensitive, such as childhood acute lymphoblastic leukemia (ALL), Hodgkin's disease, some non-Hodgkin's lymphomas, and testicular cancer. For these, relatively lower doses of chemotherapy may be effective. However, for most other cancers, promiscuous interaction of significant amounts of potent drug molecules with normal cells cannot fail to have serious consequences. The immune system, hematopoietic system and the internal lining of the gastrointestinal tract are the major sufferers as they have the highest cell turnovers. Side effects of therapy can be acute or delayed. Acute complications include nausea, vomiting, diarrhea, anorexia, allergic reactions and anaphylaxis. Delayed toxicities include myelosuppression, which can cause neutropenia and repeated infections, anemia from multiple causes, and hemorrhagic manifestations from thrombocytopenia. As a consequence of all these, and perhaps other unknown causes, the patient may develop a severe form of cachexia that is almost exclusively found in advanced cancers.

To avoid or reduce toxicities, several approaches have been adopted. Supportive therapy in cancer aims to reduce the specific toxicities of the drugs. Potent anti-emetics for gastrointestinal toxicities, diuretics for nephrotoxic agents such as cis-platin, anti-histaminics for allergic reactions, and the more recently introduced recombinant therapies such as erythropoietin and GM-CSF to treat myelosuppression are a few examples. Another is the use of several drugs at lower doses for a single cancer, rather than a single drug in high doses. Multidrug regimens, or protocols, have become the norm in cancer treatment. This is advantageous because the anticancer effects of the drugs will be additive when the drugs have different mechanisms of action, while the side effects will be distributive if their toxic effects are on different cells. Combination of a cycle-specific and non-specific drug may prevent tumor resistance by killing actively dividing as well as dormant cells. Combination chemotherapy also suppresses drug resistance, which is another major drawback of conventional chemotherapy.

Drug resistance, or the lack of responsiveness to the chemotherapeutic agent, can be due to several causes. The drug may not be reaching the tumor site at effective concentrations, either because of poor blood flow, or even after reaching the site may not achieve high enough concentration in the cell due to poor absorption, metabolic degradation, or rapid excretion from the cell (pharmacokinetic resistance). Failure to achieve total cell kills result in cytokinetic resistance: a population of survivors serve as source for repeated proliferation of the tumor. Tumor cells can evolve under these conditions to develop biochemical resistance by altering uptake and target molecules, developing or upgrading drug metabolism and excretion pathways, increasing the intracellular concentration of the drug-affected molecule, or incorporating protective genetic changes.

When multidrug regimens are not effective on their own, they are supplemented with radiotherapy and surgery. Unfortunately, radiotherapy also has the same depressive effect on the immune system and dividing cells, and results in nausea, vomiting, diarrhea and anemia. Surgery in early stages in some cancers can be curative, with only limited associated morbidity and very low fatality. However, most deep-seated cancers in the body are detected after they have metastasized, while some cancers, notably those of intracranial origin, are not easily approachable surgically, and carry high mortality and morbidity risks. Various ingenious combinations of the above, such as adjuvant (and neo-adjuvant) chemotherapy and radiotherapy with surgery to reduce tumor size or treat disseminated tumor cell nests, have been used to maximize the benefit to the patient.

In recent years, in recognition of the limited success of "classical" chemotherapy to treat cancers, several new methods have been investigated. Most important among them are gene therapy for correcting the altered genetic profile in cancer cells; and the biologic response modifiers that aim to enhance the innate antitumor immunity of the human body. Nonspecific immune modulators like BCG (for bladder cancer) and cimetidine (for melanoma) have shown limited efficacy. Better results have been obtained with the introduction of lymphokines and cytokines (IL-1, IL-2, TNF and more recently CCL21). These are natural human molecules produced in minute quantities for the purpose of signaling among immune cells. However, nonspecific use of these cytokines, especially IL-2 and TNF, has shown unacceptable levels of toxicity. This is primarily because of the short half-lives of these small molecules. To achieve acceptable levels inside the tumor microenvironment, very high doses need to be introduced: a common theme in cancer therapy regimens. A way to increase the stability of these molecules in circulation, along with the means to target them to selective neoplastic tissue, has become the current need.

10.3
Nanoparticles and Drug Delivery in Cancer: a new Road

10.3.1
Importance of Nanoparticles in Cancer Therapy

Nanotechnology, especially nanoparticulate drug delivery systems, may provide the solution to the problems facing current cancer therapy. Nanoparticles can be defined as spherical (or spherical-like) particles with at least one dimension less than 100 nm [6]. The history and technology involved in nanoparticle synthesis can be found elsewhere (e.g. [6]). Nanoparticles, along with liposomes and block copolymer micelles, form the group of submicron sized colloidal systems used as targeted drug delivery vehicles (Fig. 10.1). Their small size allows intravenous administration without the risk of embolization, and passage through capillary vessels [7] and mucosae [8], while affording special properties of large surface

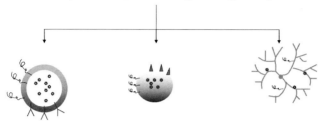

Fig. 10.1. Types of nanoparticulate drug delivery systems and their common targeting methods. Left to right: Liposomes with enclosed drug molecules; other polymeric nanoparticles, and dendrimers.

area, significant surface properties, greater solubility (especially for oil-based drugs) and better tissue adhesion [9].

Nanoparticles can have several roles in cancer therapy. They can be useful in targeting the tumor and achieving local therapeutic concentration of the toxic drug while keeping circulating levels of free drug fairly low, thus reducing systemic toxicity. They can stabilize lipophilic drugs in circulation, and help them enter previously inaccessible regions of the body like the central nervous system by crossing the blood–brain barrier, and treat previously untreatable tumors and reach "tumor sanctuaries". They can increase the circulatory period of drugs by controlled release, thus overcoming the toxicity associated with initial high concentrations in periodic doses. They can also, conceivably, overcome multidrug resistance by a combination of these effects [6]. Apart from their therapeutic benefits, nanoparticles – as fluorescent nanoparticles and quantum dots – can also help in the early detection of tumors.

Therapeutic effects of nanoparticles can be achieved by "physical", "chemical" or "biological" means. Physical means include the recently developed methods of hyperthermia and magnetic therapy. Chemical means include the delivery of more conventional chemotherapeutic drugs to the tumor sites. Biological response modifiers like immunotherapeutic agents are also gaining favor.

It is debatable whether the term "nanoparticles" applies only to the more recently developed polymeric nanoparticles or if, applying the definition in a broader sense, it should encompass block copolymer micelles and liposomes. Taking the emerging field of targeted drug delivery to cancer as the leit motif of this chapter, and acknowledging the significant roles played by both these types of "nanoparticles" in the development of this important branch of cancer therapeutics, we have included short subsections on both. The reader will thus get a better overview of the current state of knowledge and can peruse the reference section to further his/her interest. However, henceforth, we use the term "nanoparticles" in the narrower sense of polymeric spherical coiled particles, whether modified or unmodified.

10.3.2
An Overview of Targeting Methods

In this chapter we concentrate on the methods and materials by which nanoparticles are targeted to neoplastic tissue. Targeting can be loosely defined in this context as any means that increases the specificity of localization of nanoparticles in tumor cell masses. Targeting does not intrinsically imply improved sensitivity but, as we shall see, the different methods employed to increase the specificity allow administration of higher doses of the drugs, thus also favorably increasing sensitivity. Also, as mentioned earlier, the ability of nanoparticles to cross the blood–brain barrier and other impediments to conventional therapy increases its volume of distribution. This also results in increased sensitivity.

Targeting can be divided into two major types – passive and active. Passive targeting involves modifications of nanoparticles that increase the circulation time without addition of any component/involvement of any method that is specific to the tumor; increased circulation time helps in accumulation of the particles in the tumor by an EPR effect described below. Active targeting of tumors can be divided into "physical" and "chemical/biological" targeting. Physical targeting involves directing magnetic nanoparticles to tumor sites under the influence of an external magnetic field. Chemical/biological targeting involves the modification of nanoparticles' surfaces with tumor-specific ligands.

Notably, any of these methods can used in conjunction with others. For example, common modifications for passive targeting, such as PEGylation, are frequently used with more "active" modifications like antibodies. The methods are almost independent of each other, and can be judiciously combined to increase the effectiveness of the drugs.

This chapter does not deal extensively with the magnetic therapy of cancer and induced hyperthermic killing of tumor cells (see Chapters 5, 8 and 9). However, in keeping with the idea of overviewing the whole field of targeting, we have incorporated a short discussion of physical targeting by magnetically directing particles to tumor tissues.

10.4
Means to the End: Methods for Targeting

10.4.1
Passive Targeting

Here we must make a clearer distinction between what can be described as "active" and "passive" targeting. In general terms, a targeted drug can be defined as one modified in a fashion that allows its preferential uptake in the desired cells/tissues, in this case cancer. Evidence has been provided that many nanoparticles, especially long circulating ones, will show a preferential distribution to cancer sites over healthy tissues, even without any specific targeting molecule. This was demon-

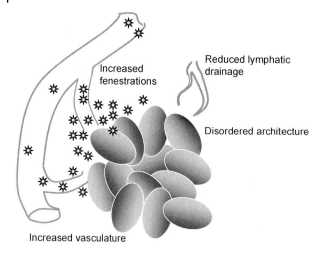

Fig. 10.2. The enhanced permeation and retention (EPR) effect accumulates nanoparticles to tumor sites without active targeting.

strated for the first time my Matsumara and Maeda [10]. This is likely due to the increased vasculature of these regions, larger fenestrations in the capillary walls for rapid delivery of nutrients, general disordered architecture that is symbolic of the neoplastic process, and the reduced lymphatic drainage in these regions. All these factors lead to the enhanced permeation and retention effect (EPR) [11, 12] (Fig. 10.2). This fortuitous distribution has been taken advantage of by numerous researchers in designing targeted drug delivery systems. As an example, in a recent study tamoxifen was encapsulated in poly(ethylene oxide)-modified poly(ε-caprolactone) (PEO-PCL) nanoparticles and administered to a murine model of breast cancer [13]. The poly(ethylene oxide) (PEO) coating made it a "stealth nanoparticle", i.e., able to avoid detection by the body's MPS system for a considerable time. The PEO surface modified nanoparticles showed significantly increased levels of accumulation within tumor with time than did the native drug or surface unmodified nanoparticles. An earlier effort by one of the authors to incorporate tamoxifen in nanoparticles for preferential uptake by estrogen receptor (ER) positive breast cancer cells had been shown to be successful *in vitro* [14]. However, to translate that success *in vivo*, a long-circulating nanoparticle was necessary. Another current example is the incorporation of cisplatin in liposomal formulations [15] with poly(ethylene glycol) (PEG) coating for gastric tumors: in preclinical and clinical trials, this formulation has been demonstrated to have longer half-life in circulation without the attendant side effects.

This targeted distribution of nanoparticles, together with the controlled release of anticancer drugs, can bring about a type of "passive" but effective targeted delivery. The other type of targeting is by specific ligand–receptor interaction. Possibly, better specificity and sensitivity can only be achieved by this more "active" means. However, the introduction of a biological entity on the surface of the nanoparticle can create several problems: the size and complexity will increase, with attendant complexities of synthesis and characterization; the biomolecule may react adversely with the immune system or any of the other myriad proteins in the blood and tissues. This may give rise to unpredictable adverse reactions.

However, potential advantages include the ability to deliver larger amounts of drug per target biorecognition event; increase selectivity by including more than one type of targeting molecule, better avoidance of barriers by using different avoidance methods and the ability to administer "localized integrated combination therapy" by including multiple drugs in the same nanoparticle [1].

A method for enhancing this passive delivery to the cancer using nanoparticles is by the use of pH- and heat-sensitive nanoparticles. The fact that the pH in the region of cancerous growth is lower than in the rest of body, while the temperature is raised locally because of enhanced metabolism and ongoing inflammation, has been used to design pH-responsive nanoparticles: a lower pH causes enhanced drug release. Wei et al. have given a recent example of this [16]. The temperature- and pH-sensitive amphiphilic polymer poly(N-isopropylacrylamide-co-acrylic acid-co-cholesteryl acrylate) [P(NIPAAm-co-AA-co-CHA)] was synthesized and used to encapsulate paclitaxel in core–shell nanoparticles fabricated by a membrane dialysis method. The spherical nanoparticles are below 200 nm and have a pH-dependent lower critical solution temperature (LCST). *In vitro* release of paclitaxel from the nanoparticles was responsive to external pH changes, and was faster in a lower pH environment.

10.4.2
Magnetic Targeting of Nanoparticles

Magnetic drug targeting has been defined as "the specific delivery of chemotherapeutic agents to their desired targets, e.g., tumors, by using magnetic nanoparticles (ferrofluids) bound to these agents and an external magnetic field which is focused on the tumor" [17]. This type of tumor targeting is aimed at concentrating the toxic drug at the cancer, hence enhancing its efficacy and reducing systemic side effects from high doses.

Gilchrist et al. described the first magnetic nanoparticles in 1957 [18]. These are generally 10–200 nm in diameter, magnetic or superparamagnetic and usually composed of iron oxide, magnetite, nickel, cobalt or neodymium; of these, magnetite (Fe_3O_4) and maghemite (γ-Fe_2O_3) are more biocompatible and preferred. Iron oxide is degradable in the body and has been shown to be safe for *in vivo* applications [6]. However, to be effective, magnetic nanoparticles should demonstrate high magnetization for external magnetic field control and should exceed linear blood flow rates of 10 cm s^{-1} in arteries and 0.05 cm s^{-1} in capillaries. For this purpose, particles composed of, for example, 20% magnetite require a field of 0.8 Tesla. Tissue depth is a limiting factor in active targeting, with deeply localized tumors being hard to access.

Lubbe et al. have utilized a complex of 4-epirubicin bound to ferrofluids in the treatment of squamous carcinoma of the breast or head and neck [19]. They used a field of 0.5–0.8 T at the tumor site while infusing the particles. Results showed accumulation in tumors in 6 of 14 patients with reduced toxicity, a transient rise of serum iron, but no demonstrable damage to kidney or hepatic function. Alexiou et al. have used mitoxantrone complexed to 100 nm ferromagnetic particles against VX2 squamous cell carcinoma among rabbits to demonstrate the superiority of intra-arterial infusion to other routes of delivery, especially the intravenous route where the particles are removed from circulation by the reticulo-endothelial system [20]. They have also demonstrated that circulating mitoxantrone becomes concentrated at the tumor tissue when complexed to ferrofluids; a higher concentration of the complexed drug (compared to the free drug) is achieved at the tumor site using only a half to one-fifth dose [17]. The drug mainly resides in the intraluminal space, the tumor interstitium and peritumoral area (region surrounding the tumor ≤ 1 cm). Other researchers have investigated ferrous nanoparticles in glioblastoma (a tumor of the brain), and demonstrated that these particles effectively cross the blood–brain barrier [21].

The clinical applications of magnetic drug targeting have been reviewed in more detail [22].

10.4.3
Ligands for Active Targeting

Cancer cells arise from normal cells through a complex series of genetic events. Unlike infectious agents like bacteria, they largely share the same proteins as normal cells. However, some proteins are found in much larger numbers on cancer cells than in normal cells [23]. These overexpressed antigens are called tumor-associated antigens (TAA). Some proteins derived from normally silent genes or mutated forms of normal proteins are found exclusively on cancer cells. These are known as tumor-specific antigens (TSA). The obvious targets for targeted cancer therapy are TSAs. However, TSAs are often difficult to characterize for a particular tumor. When found, they are usually not extensive, i.e., they are not found in all patients affected by the tumor, nor are they found in all the cells in a particular tumor in the same patient. The aberrantly expressed tumor specific antigens are produced by aberrant glycosylation in glycolipids, glycoproteins, proteoglycans, and mucin [24]. For example, the glycosylation pattern of MUC1 membrane mucin of breast cancer epithelial cells differs from normal breast epithelial cells, possibly as a result of changes in expression of glycosyltransferases [25]. TAG-72 is a mucin-like tumor-associated glycoprotein [26] that is found in some epithelial tumors. Melanoma cells aberrantly express GM3 ganglioside on their surface [27]. Gastrointestinal cancer cells abnormally express LeX antigens [28].

Tumor-associated antigens are often growth factor receptors on the tumor that are overexpressed to meet the rapidly dividing neoplastic cells' demands. A classic example is folic acid and congeners. These are low molecular weight pterin-based

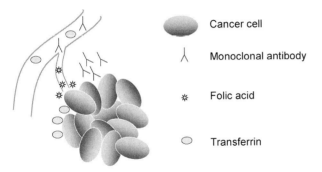

Fig. 10.3. Cancer cell antigens are targeted with monoclonal antibodies; ligands such as folic acid and transferrin also show specificity.

vitamins essential for carbon metabolism and *de novo* nucleotide synthesis. The presence of elevated levels of folate receptors has been demonstrated from epithelial tumors of various organs such as the colon, lungs, prostate, ovaries, mammary glands, and brain [29–40]. These present attractive targets for drug delivery. Another promising tumor receptor target is Her2_neu, also known as c-erbB-2. This is a transmembrane protein with an epidermal growth factor receptor that possesses intrinsic tyrosine kinase activity [41–43]. The normal human Her-2_neu proto-oncogene is frequently found to be overexpressed in breast and ovarian cancers among others. Its level may correlate with the metastatic potential of the cancer cells [44, 45].

Another such membrane-associated tumor antigen is the transferrin receptor. This is overexpressed in different types of cancers [46]. The levels may also correlate with the malignant potential of these cells [47] (Fig. 10.3).

The presence of various other tumor antigens has been demonstrated: membrane-associated Carcinoembryonic antigen; CD10 or CALLA in leukemias, melanomas and myelomas [48, 49]; CD20 in B cell malignancies [50]; etc. Many others are being recognized routinely. They all represent potential goals for targeted drug delivery.

10.4.3.1 Monoclonal Antibodies against Tumor-specific Antigens

Specific targeting became a possibility with the discovery of monoclonal antibodies in 1975 [51]. As the name suggests, these are antibodies derived from a single clone of cells. They can be mass produced in the laboratory from a single clone and recognize only one antigen. Monoclonal antibodies are usually made by fusing a short-lived, antibody-producing B cell to a fast-growing cell, such as a cancer cell. The resulting hybrid cell, or hybridoma, multiplies rapidly, creating a clone that produces large quantities of the antibody. It was hoped that monoclonal antibodies would be able to target specific antigens on the surface of cancer cells, and initially raised high hopes for targeted cancer therapy. Animal-origin monoclonal antibodies such as murine antibodies were strongly immunogenic; to avoid this, chimeric antibodies [52] and humanized antibodies [53, 54] were produced. More recently, completely human monoclonal antibodies and single-chain human antibody fragments [55] have been introduced for immunotherapy and targeted drug delivery.

Antigen targets for monoclonal antibody therapy (or for any other targeted anticancer therapies) must have the property of specificity, i.e., present only on the cancer cell surface and be absent from the surfaces of all normal cells; they should be extensive, i.e., present on all the cells in a given tumor, preferably on all such tumors in every patient; there should not be any mutation or structural variation of the antigen; and they should be preferably involved with a critical function of the cell, so that adaptive response of the cancer to therapy by losing the antigen can be prevented [56]. However, naturally, such a perfect antigen is rarely, if ever, found. Several good candidates have been discovered, and monoclonal antibodies against these have been marketed. Examples include anti-CD20 antibody for relapsed/refractory CD20 positive B-cell non-Hodgkin's lymphoma and low-grade or follicular-type lymphoma, anti-CD53 antibody for relapsed/refractory AML, and anti-CD52 antibody for B-cell chronic lymphocytic leukemia [56, 57]. Recently, Zhou et al. have utilized BDI-1 monoclonal antibodies to target highly toxic to bladder cancer cells *in vitro* [58]. To improve therapeutic efficacy and decrease toxicity, we prepared arsenic trioxide-loaded aluminates immuno-nanospheres [As_2O_3-(HAS-NS)-BDI-1] targeted with monoclonal antibody (McAb) BDI-1 and tested its specific killing effect against bladder cancer cells.

Identifying Monoclonal Antibodies One strategy to identify ideal targets for monoclonal antibodies is to select the internalizing antibodies from phage libraries; accordingly, two antibodies to the breast cancer cell line SK-BR-3 which bind to ErbB2 have been identified [59]. Nielsen et al. have demonstrated the use of single-chain Fv (scFv) antibodies for internalization of nanoparticles in target cells [60]. The antibodies were recovered from a non-immune phage library and the scFv specific for ErbB2 (F5) was re-engineered and attached to liposomes. Doxorubicin loaded in these immunoliposomes was selectively active on ErbB2 positive breast cancer cells and showed targeted cancer cell cytotoxicity.

Liu et al. have described a method to map the "epitope space" of a tumor using monoclonal antibody libraries [61]. The expressed epitopes on the cell surface, which constitute the "epitope space", is essential to targeted drug delivery. This is "highly complex, composed of proteins, carbohydrates, and other membrane-associated determinants including post-translational modification products, which are difficult to probe by approaches based on gene expression" [61]. The authors used monoclonal antibody libraries against prostrate cancer cells and identified over 90 antibodies, which bind to the cancer cells selectively, with little or no binding to normal human cells. This approach does not attempt to identify the tumor-specific antigens, but rather takes a functional approach to the problem of tumor-target identification.

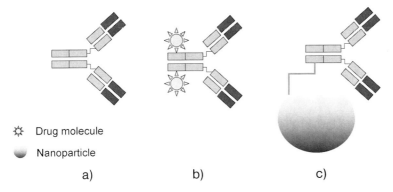

Fig. 10.4. Monoclonal antibodies against cancer were originally used either by themselves (A) or attached to drug molecules (B). However, attachment to nanoparticles (C) provides longer circulation times and lower toxicity.

Monoclonal Antibodies and Nanoparticles Monoclonal antibodies were initially hailed as therapy for cancers in their own right (Fig. 10.4). To introduce more cancer cell killing ability, they can also be conjugated with anticancer drugs. This complex can also be formulated as a prodrug, which is cleaved to release the anticancer agent at the cancer site. Usually, the link between the molecules would be peptidase cleavable, in order to separate inside the cancer cell. However, such formulations are not very stable *in vivo*, and may undergo early cleavage. The absence or reduction of cleavage at the site will, clearly, lower the potency of the anticancer agent.

The greatest difficulty in using monoclonal antibodies for cancer therapy is their rapid clearance from the bloodstream by the immune mechanisms of the body. Li et al. have estimated that <0.01% of the administered dose of antibodies reaches the target sites [62]. This has two serious drawbacks: firstly, monoclonal antibodies are relatively costly, and a large dose of the therapy becomes prohibitively expensive; secondly, the excess antibody can have serious toxic effects on rest of the patient's body. As we discussed in the introductory section on chemotherapy, this problem, in various proportions, dogs all current therapies of cancer. Hence, monoclonal antibodies by themselves provide no advantage over general anticancer drugs. Attachment of the molecules to nanoparticles circumvents some of these problems.

10.4.3.2 Targeting the Angiogenic Process

Judah Folkman discussed one of the prime candidates for targeted cancer therapy, angiogenic factors, in 1989 [63]. Cancer cells require a significant amount of nutrition to keep growing and reproducing. To obtain this they promote the ingrowth of capillary vessels into the tumor site. This is known as angiogenesis. Tumor cells secrete chemicals that promote angiogenesis. Since active promotion of angiogenesis is unnecessary in most body cells, the associated antigens are of interest in targeted therapies. The anti-VEGF (vascular endothelial growth factor) drug Avastin® has shown moderate effectiveness in several solid tumors. This has led to a gene-therapeutic approach to anti-angiogenesis using liposomes [64] and cationic nanoparticles [65]. In the latter, an integrin-targeting ligand was used for specific delivery of a mutant Raf gene, and selectivity was demonstrated.

Hallahan et al. have described novel targets that depend on targeting radiation-induced neoantigens in the cancer microvasculature [66]. They demonstrated the presence of an RGD-containing amino acid sequence in phage-displayed peptides obtained from irradiated tumors. This peptide binds to integrins. Immunohistochemical examination confirmed the presence of the fibrinogen receptor integrin in irradiated tumors. Targeting this receptor by fibrinogen-containing nanoparticles showed enhanced anti-tumor effects.

The Intradigm Corporation (MD, USA) has utilized the RGD sequence attached to PEG to target their siRNA-containing nanoparticles to the integrins present in the tumor microvasculature [67]. The siRNA provide potent selective gene inhibition with high specificity. They delivered siRNA-inhibiting vascular endothelial growth factor receptor-2 (VEGF R2) expression and, thereby, tumor angiogenesis to a mouse tumor model using PEI-PEG nanoparticles. The siRNA was found to be active in a highly cell-selective manner. They demonstrated that selectivity could be incorporated in the drug itself, along with the ligand–receptor interaction.

Nucleic acid conjugates have been favored as targeting ligands for their exquisite specificity for the targeted molecule. Farokhzad et al. have utilized nucleic acid ligands (aptamers) to target cells expressing prostate-specific membrane antigen, a tumor marker for prostate cancer acinar epithelial cells [68]. They created PEG-containing PLA nanoparticles that bind to aptamers by surface negative charge. These aptamer conjugated nanoparticles showed 77× greater uptake by the cancer cells than unmodified nanoparticles.

Other efforts [69] have demonstrated nanoparticle targeting to angiogenic epithelium using the $\alpha_v\beta_3$-integrin, which is found on endothelial cells. Anti-angiogenic effects were demonstrated in mouse models of melanoma and adenocarcinoma.

10.4.3.3 Folic Acid and Cancer Targeting

Perhaps the most interesting molecule to attract attention for its targeting abilities in recent times is the humble folic acid. Most normal human cells take up this molecule in its reduced form (Fig. 10.5). One of the notable exceptions is the

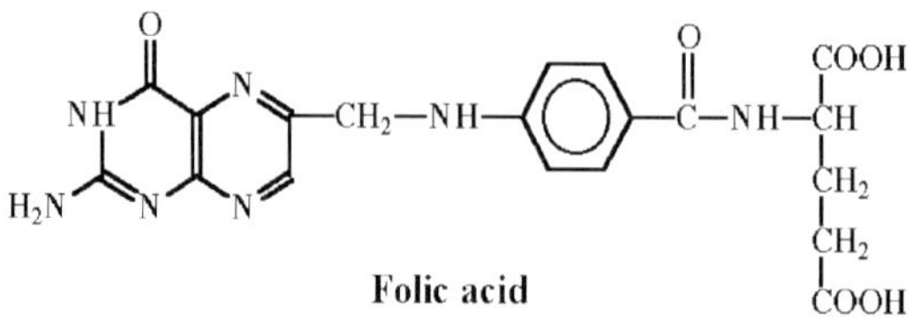

Folic acid

Fig. 10.5. Molecular structure of folic acid.

luminal sides of the gut endothelial layer, which take up the folic acid present in food, convert it into folate, and release it into our bloodstream. Many types of cancer cells, however, take up folic acid in its oxidized form. Since intravenously administered nanoparticles will only come into contact with the abluminal side of the gut epithelium, selectivity for tumor cells is very high. As mentioned above, several endothelial tumors (e.g., derived from the ovaries, mammary glands, colon, lungs, prostate and brain) possess elevated levels of folic acid receptors on their surface [29, 30, 32, 33, 36–40]. Already, a large body of research has accumulated regarding the use of folic acid receptors as cancer targets. Since folic acid, the natural choice for ligand to the folic acid receptor, can be quite easily coupled to the nanoparticle surface, it has been used for targeting these to cancer cells.

Hattori and Maitani have reported *in vitro* studies of folate-linked nanoparticles in human cancer cells [31]. They synthesized DC cholesterol–Tween 80 nanoparticles with incorporated folate-PEG conjugates for steric stability and targeting. The nanoparticles were complexed with plasmid DNA. They showed enhanced uptake in human oral cancer cells by a folate-dependent route. Human prostate cancer cells also showed high uptake of the nanoparticle–DNA complexes. However, based on their results, the authors suggest that the route of uptake may be different for the two cell types.

Oyewumi and Mumper have reported the cellular uptake, distribution and tumor retention of folate-coated and PEG-coated gadolinium nanoparticles in mice models of human nasopharyngeal carcinomas [34]. While this was more for imaging than therapeutic purposes, their results also indicate the efficacy of using folate as a targeting agent. Baker et al. at the Center for Biologic Nanotechnology, University of Michigan Medical School have focused on the use of dendrimers as targeted delivery vehicles (described in detail later). They used folic acid as targeting element to deliver a triplex-forming growth-inhibitory oligonucleotide to breast, ovarian and prostate cell lines [35]. A very recent example of folic acid receptor targeting in squamous cell carcinoma has been demonstrated by Santra et al. [70], who employed a novel technique that uses fluorescent silica nanoparticles (FSNPs) to detect overexpressed folate receptors.

Folic acid targeting to cancer cells has also been demonstrated in our own Cellular and Molecular Bioengineering Laboratory in the National University of Singapore. Although the use of folate as a targeting agent has been extensively reported, little has been done to continuously track the intracellular delivery of nanoparticles grafted with folate using imaging techniques such as confocal microscope. This is possibly due to the short lifetime of most biological fluorescent labels. The problem has been tackled using quantum dots (QDs).

QDs have several advantages over conventional fluorescent dyes and proteins like the Green Fluorescent Protein. They exhibit a strong fluorescence emission, a broad absorption spectra and a narrow, symmetric emission spectrum, and are photochemically stable. The most exciting finding is that QDs also exhibit a wide range of colors, which is exquisitely controlled by their size; a broad absorption spectra means that a series of different-colored dots can be activated using a single laser. These properties of QDs raise the possibility of using them to tag biomolecules with an optical coding technology. This can, for example, create QD bar codes, and the use of six colors and ten intensity levels can theoretically encode one million biomolecules. Techniques have been developed to incorporate QDs into polymer beads, to solve problems relating to their surface chemistry. In general, quantum dots exhibit water insolubility, poor biocompatibility, and low chemical stability in physiological media; encapsulation in polymer nanoparticles can reduce these problems. In their work, Zhang et al. have incorporated luminescent CdSe/ZnS QDs into polystyrene (PS) nanoparticles grafted with carboxyl groups using an emulsion polymerization method. Nanoscale QD-incorporated PS nanoparticles (30 nm diameter) were separated by centrifugation at high speed in viscous solution.

Finally, the nanoparticles were modified with folic acid on their surface and their intracellular uptake into NIH-3T3 (an immortalized epithelial cell line) and HT-29 (colon cancer epithelial cell line) cells was investigated using a confocal laser scanning microscope.

For nanoparticles without a folate cover, fluorescence was observed in the cytoplasm of the cells but not in the nuclei of most cells, indicating that the nanoparticles have difficulty in reaching the nuclei on their own. After surface modification of the nanoparticles with folic acid, fluorescence appeared in the nuclei of some cells (Fig. 10.6). This suggested that intranuclear uptake of the nanoparticles may be affected by the folic acid attached to the surface.

It took three hours of incubation for the PS-encapsulated QDs with folate-modified surfaces to spread throughout the cytoplasm of cells. Folic acid modified nanoparticles first attached to the folate receptors expressed on the cell membrane after 0.5 h incubation. After 1 h, more fluorescence was found on the cells' membranes and some fluorescence spots were already seen in the cytoplasm, indicating early migration of the nanoparticles into cells. After 1.5 h, more fluorescence was observed inside the cytoplasms. Throughout the observation period, fluorescence was seen on the cell membranes. This can be explained by the fact that internalized nanoparticles were rapidly replaced by nanoparticles from the surrounding media. This seems to argue in favor of a receptor-mediated endocytosis process. In contrast, slight cytoplasmic fluorescence was observed with unmodified nano-

Fig. 10.6. (top) Confocal images of NIH-3T3 cells after culture with (a) unmodified and (b) folic acid modified nanoparticles and (c, d) their corresponding bright field images.

Fig. 10.7. (bottom) Transferrin and its receptor. The levels of these receptors are elevated in several tumors. (Source: The protein data bank; http://www.rcsb.org/pdb/molecules/pdb35_2.html)

10.4 Means to the End: Methods for Targeting | 353

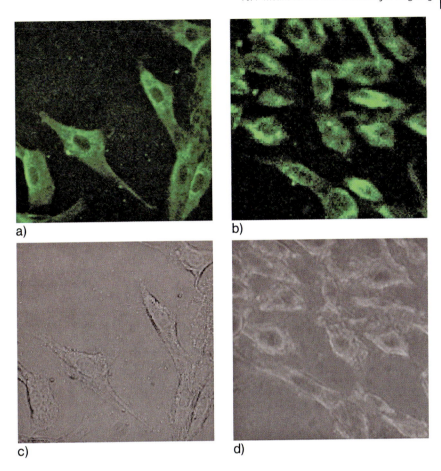

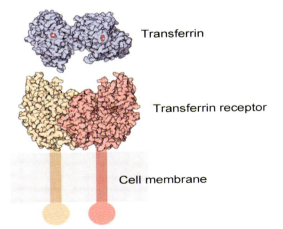

particles, and did not change throughout the observation period. This uptake was possibly by non-specific endocytosis of the PS beads by the cells.

10.4.3.4 Transferrin as a Targeting Ligand

Approximately 3 mg or 0.1% of body iron in the adult male circulates in the plasma as an exchangeable store. Essentially all circulating plasma iron normally is bound to an 80 kDa glycoprotein called transferrin. Plasma transferrin is synthesized by the liver and secreted into plasma. It has two homologous N-terminal and C-terminal iron-binding domains [71] (Fig. 10.7). This iron–protein chelate serves three functions: it makes iron soluble under physiologic conditions, it prevents iron-mediated free radical toxicity, and it facilitates transport of iron into cells. Transferrin is the most important physiological source of iron for red cells [72]. The liver synthesizes transferrin and secretes it into the plasma. Transferrin is a suitable ligand for tumor targeting because of the upregulation of its receptors in many cancers [46] in which their levels correlate with the malignant potential [47].

Insert Therapeutics (California, USA) have reported transferrin based targeting. They demonstrated tumor targeting in mice by transferrin-modified nanoparticles containing DNAzymes (short catalytic single-stranded DNA molecules) [73]. They synthesized β-cyclodextrin polymer–DNAzyme nanoparticles. Steric stabilization was obtained by adamantane-PEG conjugates complexed to the surface by adamantane-cyclodextrin interaction. Transferrin was included for targeting transferrin receptor expressing tumors. While the uptake of unmodified nanoparticles was demonstrated *in vivo* in nude mice, transfer modification was demonstrated to be necessary for intracellular delivery of the DNAzyme.

Previously, the same group demonstrated [74] the uptake of transferrin-modified cyclodextrin-adamantane inclusion complex nanoparticles to leukemia cells, which was competitively inhibited by excess free transferrin.

10.4.3.5 Other Targeting Ligands

Another targeting molecule used is biotin (vitamin H). Na et al. have incorporated this molecule into a hydrophobically modified polysaccharide, pullulan acetate, loaded with adriamycin [75]. The biotinylated nanoparticles exhibited very strong adsorption to the HepG2 cells, while the unmodified nanoparticles showed no significant interaction. Uptake was confirmed by confocal microscopy. While the degree of interaction increased with increasing vitamin H content, drug uptake by the nanoparticle was inversely related to the vitamin H content.

Various other ligands have been studied as probable candidates for targeted drug delivery in tumors. These include antibodies against tumor-associated antigens [76], lectins [77], and glycoproteins [78]. Recently, Yu et al. from Japan have detailed the use of the surface antigen (sAg) of the hepatitis B virus to accomplish liver-specific delivery of genes and drugs. They describe the association of approximately 110 molecules of this protein to form a polymeric capsule (which they call a "BioNanoCapsule" or BNC) with an average diameter of 130 nm. The nanoparticles that were produced delivered emerald green fluorescent protein (EGFP) to human hepatoma cells [79].

10.5
Targeting with Different Types of Nanoparticles

10.5.1
Liposomes in Cancer Targeting

Currently, the most exciting event in targeted drug delivery systems is undoubtedly the ongoing clinical trial of the world's first tumor-targeted gene delivery vector, Rexin-G.™, marketed by Epeius Biotechnologies Corporation. This nanoparticle formulation combines a proprietary vector system with a proprietary anticancer gene controlling cell cycle. The company website claims that "when given by intravenous infusions, Rexin-G™ has been shown to eradicate remote metastatic cancers in mice and to arrest cancer growth with shrinkage and necrosis of solid tumors in humans – determined by CT Scan and MRI – without eliciting systemic side effects." The first clinical trials were started in 2003 in Manila, Philippines; and following FDA approval shortly thereafter, clinical trials were started in New York from April, 2005 for metastatic pancreatic and colon cancer. Initial reports are favorable and results of ongoing studies are eagerly awaited [3, 80, 81].

In this context, a few words regarding immunoliposomes are in order. In our initial classification of colloid based systems, we separated nanoparticles from liposomes and block copolymer micelles. Many authors consider them part and parcel of the same package [1, 82]. Hence, in the interest of topicality and completeness, we consider both in some measure.

Liposomes (and, in some instances, micelles) have long been used for targeted delivery of drugs. Of all the proposed nanoparticulate drug delivery systems, liposome-based agents, particularly liposomal anthracyclines, have had the greatest impact on oncology to date [82]. These lipid based formulations have several advantages as drug carriers: they are amphiphilic, hence allowing the carriage of lipophilic drugs in plasma; they fuse with cell membranes and transfer their load inside cells; and they can be embedded with targeting molecules like antibodies on their outer surface. They have been likened to the Trojan horse in Greek mythology: the cancer cells fuse to the liposomal membrane and takes up the contents of the liposome, which contains the anticancer formulations (the company that is conducting the liposomal targeted-gene delivery trials has been appropriately named after Epeius, the maker of the original Trojan horse).

Several problems have dogged liposomal drug delivery systems from its conception and implementation. These include a short half-life in circulation, instability *in vivo*, and lack of target selectivity. Liposomes are removed rapidly from the circulation by the reticulo-endothelial system (RES) [83]. Stability and circulation times were increased, as in nanoparticles, by the introduction of a steric stabilizer on the surface, usually PEG [84]. Like other nanoparticles, enhanced passive uptake of liposomes has been demonstrated with longer circulation times. PEGylation has increased therapeutic efficacy of the liposomal formulations [85, 86]. These innovations, affording longer and stable circulation times, have resulted in approval of some liposomal formulations for clinical use [87]. The passive increase of liposomes in cancer sites was recognized as a means of delivering toxic, poorly soluble drugs. Liposome-encapsulated doxorubicin was approved for use in Kaposi's sarcoma more than a decade ago. Formulations are available now for breast and ovarian cancers. The most recent instance deals with liposomes encapsulating cisplatin; this has been demonstrated in preclinical and phase 1 clinical trials in humans to have almost negligible nephrotoxicity, ototoxicity and neurotoxicity, which are the major side-effects of free cisplatin in circulation [15].

Improved target selectivity has been demonstrated using ligands and antibodies. Active targeting of liposomes using ligands for tumor-associated receptors, and antibodies against tumor-associated antigens, has been investigated as a means of improving selectivity. Liposome–antibody complexes (immunoliposomes) have received most attention [88]. The problem of rapid removal by the RES is compounded by the attachment of antibodies [89, 90]. However, PEGylation has created a generation of "stealth" immunoliposomes that can effectively hide from the RES for long periods. In this context, immunoliposomes can be classified into two groups based on the relation of the antibody to the PEG group on the liposomal surface: antibody coupled to lipid head group 76 and antibody coupled to the distal end of PEG [91]. The latter composition has been considered to be better for efficient antibody–target coupling because of the absence of interference from PEG chains [92]. At optimal compositions of both types of formulations, antibodies do not interfere with PEG function and long circulation times have been demonstrated; conversely, PEG does not interfere with antibody function, and active targeting is unimpaired. Immunoliposomes complexed with PEG have been successfully demonstrated for *in vivo* targeting of brain [93] and lung [91] tumors.

Immunoliposomes with cancer targeting antibodies have been shown to be cytotoxic to cancer cells *in vitro* [94]. However, this did not translate to *in vivo* success. This was because not all antibodies used to target the cancer cells result in uptake and internalization of the liposome. The use of more specific internalizing antibodies aided in uptake and cytotoxic effects *in vitro*. As an example, the anticancer antibody anti-ErbB2 did not increase cytotoxicity of the liposome [95]; but when the anti-ErbB2 Fab fragment was used, the liposome was internalized and had high efficacy of cancer kill [96, 97]. More recently, scFV (single-chain Fv) antibodies against ErbB2 have been utilized for cancer targeting with doxorubicin-loaded liposomal formulations. The antibodies are selected from a phage display library, and demonstrate higher internalization efficacy [98–100]. Ongoing research by Park et al. at UCSF comprehensive cancer centre [82, 101] aims to develop immunoliposomes with antibodies targeted to HER 2, overexpressed in breast cancer, and EGFR, overexpressed in several cancers such as lung, pancreas and prostate. The targeted liposomes have been loaded with anticancer drugs like doxorubicin and vinorelbine. The loaded immunoliposomes have shown increased effectiveness and reduced toxicity compared with conventional chemotherapy.

The issue of internalization can be circumvented by a two-step method to release the anticancer drugs from the liposome near the tumor. This antibody-directed enzyme pro-drug therapy (ADEPT) involves first administering an enzyme–antibody conjugate that preferentially targets cancer cells, followed by administration of a non-toxic prodrug that is activated by the bound enzyme and releases the toxic product close to the cancer cells [102]. An improvement of the ADEPT method using immunoliposomes to carry the enzyme shows greater stability in the body and has been tested pre-clinically [103].

10.5.1.1 Beyond Immunoliposomes

Similar to the development of nanoparticles, scientists have looked beyond monoclonal antibodies as a means of targeting liposome to cancer. This was to avoid the intrinsic immunogenic nature of antibodies, their high cost, and the ease of procurement of ligands for overexpressed growth receptors on the tumor surface. Folic acid coupled liposomes have been developed and tested against various cancer cells both *in vitro* and *in vivo* [85, 104, 105]. Transferrin has also been evaluated as a targeting ligand for anticancer drug carrying liposomes [106].

Targeting with transferrin and folic acid has also been used to deliver anticancer genes to tumor cells. Cationic liposome–DNA complexes (lipoplexes) that are formed spontaneously by charge interaction exhibit the best gene-transfer efficiency. However, cationic liposomes were found to be unsuitable for active cancer cell targeting. Hence, a new anionic liposome–polyplex (anionic liposome-entrapped polycation-condensed DNA, LPD-II) was introduced for the gene therapy of cancer. Cationic lipoplexes containing liposomes and DNA were better internalized and transfected when transferrin was conjugated [107] and demonstrated selectivity for myeloblast cells [108], adenocarcinomas and squamous cell carcinomas [109]. Folic acid has also been used for tumor targeting the LPD-II polyplexes [110]; and systemic administration of anticancer gene therapy for selective uptake in squamous cell carcinomas of the head and neck, and breast cancer xenografts have been demonstrated [109].

10.5.2
Dendrimers

Block copolymers or dendrimers were described as far back as 1985. Initial studies elucidated the physical and chemical properties of these new materials. However, their entry into the field of targeted cancer therapy is relatively recent.

Dendrimers are globular macromolecules that have a "starburst"-like shape, with multiple branches radiating from a central core. The stepwise synthesis of dendrimers makes it possible to fine-tune its highly regular branched structure with defined peripheral groups and adjust physical properties such as the high molecular weight and low polydispersity index. Dendrimers have several advantages over polymeric nanoparticles derived from linear chains. Dendrimers are multivalent: several different drug molecules, targeting agents and other groups can be attached in a predefined fashion. Dendrimers have low polydispersity: all the synthesized molecules have molecular weights within a narrow range; this makes their behavior highly reproducible. Dendrimers have a globular rather than random coil structure, which could lead to better biologic properties [111]. Various types of

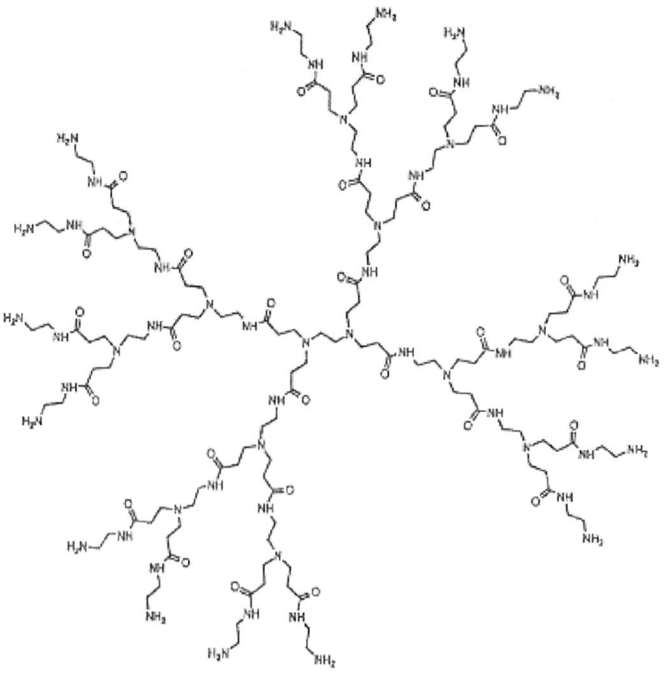

PAMAM dendrimers

Fig. 10.8. PAMAM dendrimers are the most common dendrimer nanoparticles currently used in drug delivery. (Source: http://www.dendritech.com/pamam.html)

dendrimers have been synthesized. The most common is the polyamidoamine (PAMAM) dendrimers developed by Tomalia and coworkers [112] (Fig. 10.8), which uses a divergent growth approach. Other types include polypropyleneimine dendrimers [113], polyaryl ether dendrimers [114] and peptide-based dendrimers, such as polylysine-based ones [115].

Passive targeting can be achieved by dendrimers having long circulation times. Unlike polymeric micelles, dendritic unimolecular micelles maintain their structure at all concentrations because of the covalent linkage [111]. This increased stability has the disadvantage that drug release from the central core cannot be controlled. This can be circumvented by attaching the drug molecule covalently to the periphery of the radiating arms of the dendrimer molecule. The amount of drug conjugated can be controlled by the number of generations (which increases the length of the radiating arms) and the release can be ensured by making the drug–dendrimer linkage easily degradable. Attachment of cisplatin to PAMAM dendrimers has been demonstrated by Duncan et al. [116]. Another of type of passive targeting that utilizes the lowered pH of the intracellular endosomal and lysosomal microenvironment for drug release has been demonstrated [117]. Doxorubicin was conjugated to hybrid dendrimers using hydrazone linkages; these are stable at normal physiologic pHs, but break down after uptake in the mildly acidic endosomes and lysosomes. They demonstrate longer circulation times and toxicity to several tumors [118].

Active targeting of dendrimers using folic acid conjugates has been demonstrated by Frechet and coworkers [119] using dendrimer folate and dendrimer methotrexate conjugates. Quintana et al. have created analogous PAMAM–methotrexate conjugates with degradable amide and ester linkages [35]. Folic acid conjugation enhanced cellular uptake of these compounds, resulting in increased cancer cell cytotoxicity and efficacy of the anticancer drug as compared with the free drug *in vivo*.

Dendrimers have enormous potential in targeted drug delivery systems because of their reproducibility, controlled synthesis, and multivalent structure, which ensures incorporation of different functions like targeting, imaging and multidrug delivery. However, relatively less literature is available on these exciting molecules and they suffer from lower circulation times and drug release rates than those of more established nanoparticles and liposomes, and their biodistribution is still being investigated. With time and experience, these technical obstacles will, hopefully, be solved and they will have an even wider impact on targeted drug delivery.

10.5.3
Other Polymeric Nanoparticles

Historically, nanoparticles were first introduced by Birrenbach and Speiser [120]. They were later developed to become drug delivery systems as a substitute for liposomes. Initial formulations were made by emulsion polymerizations [120]. However, later methods were developed (like phase separation, controlled gelation, etc.) that made use of preformed polymers with already characterized physicochemical properties. This allowed better control over the properties of the nanoparticles.

The most common polymers used as core material for polymeric nanoparticles are poly(alkyl cyanoacrylate), poly(lactide) derivatives and chitosan. However, several other materials have also been investigated, as noted in other volumes of this series. Recently, Gilbert et al. have developed fluorescent virus-like nanoparticles (fVLPs) from fusion protein VP2 and GFP that may serve as a model for the development of vehicles that can be designed for the delivery of large biomolecules to cancer cells [121]. These are taken up with high efficiency and transferred to the nucleus using the microtubular network. In another very recent example, Mo et al. have developed a novel lectin-conjugated isopropyl myristate (IPM)-incorporated PLGA nanoparticle for the local delivery of paclitaxel to the lungs [122].

Nanoparticles can be nanospheres or nanocapsules, the basic difference being that nanospheres have a matrix structure, while the capsules have a core-and-shell structure. While a wide range of sizes are available and have been tried, generally, the particles should be less than 100 nm in diameter [123]. The greater surface area of the nanoparticles allows attachment of targeting molecules and anticancer agents; alternatively, the drug may form the core of the nanocapsule with a polymeric shell.

Nanoparticles as vehicles of targeted drug delivery enjoy several advantages. Primarily, the attachment of different molecules to the same platform confers bi- and multi-functionality. This is important because conjugating targeting molecules directly to the anticancer drug may cause decreased functionality of either or both. Nanoparticles can be modified to incorporate not only two but more functions. One of could report the presence of the tumor using, for example, quantum dots.

Secondly, attachment to the nanoparticle surface increases the stability of molecules. This is particularly important with peptides, nucleic acids (like anti-HA-*ras* [124]) and anti-Ewing sarcoma [125] and small proteins (like antibodies), which are easily removed from the circulation in the free form. It is also important for carrying poorly soluble drugs (e.g., muramyl tripeptide cholesterol [126]) in significant quantities without the usual side effects. Controlled release can be achieved to ensure a long-term, high level of the drug without encountering the loading dose problem. This has been shown in the use of doxorubicin-loaded nanoparticles. Doxorubicin can cause acute and chronic cardiomyopathy at high levels. By attachment to nanoparticles, high levels of doxorubicin were achieved in the circulation, with reduced cardiac levels and, consequently, less toxic effects and increased effectiveness against the metastatic cancer.

Chemotherapy fails *in vivo* even when the cancer is shown to be highly sensitive to the drug *in vitro* because of cellular "sanctuaries". For example, the recurrence of acute lymphoblastic leukemia has been attributed to cell nests in the CNS. Intravenous drug formulations fail to cross the blood–brain barrier. Nanoparticles, however, can accomplish this feat, perhaps by trans-cellular movement after endocytosis by the endothelial cells. Some examples of this include the use of polysorbate 80-coated nanospheres to deliver Kytorphin, turbocurarine NMDA antagonist and doxorubicin to the brain. Polysorbate 80 adsorbs apolipoprotein E from the circulation and becomes attached to the low-density lipoprotein receptors on the endothelial surface.

The major drawback of nanoparticles is their propensity to be taken up by the macrophages lodged in the organs of the mononuclear phagocytic system (MPS). After intravenous administration, most of the nanoparticles accumulate in the spleen and the liver, and to a lesser extent in the bone marrow. Steric stabilization has been devised to avoid this. The attachment of PEG to the nanoparticle surface allows them to "hide" from the MPS and stay longer in circulation, eventually finding their way to the targeted regions. Poloxamine has been proposed as an alternative to PEG for producing steric stabilization.

In fact, PEG-coated poly(cyanoacrylate) (pCA) nanoparticles – made by a copolymer inculcating both – have such a long circulating time that they penetrated the brain more than any other modifications, including coating by polysorbate. This uptake was increased in pathological situations with, presumably, higher blood–brain barrier permeability.

Modeling of nanoparticle delivery and effects has raised the question of "diffusional instability". Vittorio Cristini et al. have shown that the delivery of cytotoxic agents to tumors, particularly anti-angiogenic drugs, might fractionate the lesion into multiple satellite neoplasms [127]. This is likely due to the rearrangement of the sources of oxygen and nutrient supply because of the anti-angiogenic therapy.

The other difficulty is the issue of control. Especially with synthetic methods derived from microencapsulation techniques, there may be considerable difficulty in encapsulating the drug and then controlling its release after encapsulation.

To summarize, a nanoparticle designed for targeted delivery of anticancer agents will have three essential components: (a) the capsule or matrix or core that provides the platform to bind or contain; (b) the drug or anticancer agent, which binds on the surface; and (c) the targeting molecule. This whole arrangement must be made immune from the attacks of the immune system by steric stabilization, usually by the attachment of PEG to the surface.

Mauro Ferrari has labeled these modified nanoparticles as "nanovectors" [1]. The name is appropriate as an indication of their function as carriers of large therapeutic payloads to target sites. He points out that while antibody conjugated therapies can only deliver single molecules of drug per recognition event, nanovectors can deposit a much larger amount. This makes nanoparticles particularly attractive when dealing with toxic drugs.

In the described tripartite arrangement, an optimum combination of matrix, drug and surface recognition element is required. As stated above and described elsewhere, many matrix polymers have been tested for their drug uptake and surface modification abilities (for a review see Ref. [128]). Targeting has ranged from "active" targeting, using antibodies, radiation-induced vascular neo-antigens, folic acid receptors and transferrin, to passive targeting that depends on surface modification with PEG and enhanced fenestrations of tumor vasculature to selectively deliver the drug.

The role of the surface-modified matrix and the targeting molecule is to effectively deliver the drug to the target site. For this, they must provide stability to peptides and others drugs with short half-lives in circulation, avoid uptake by the MPS or RES, circumvent the endothelial and blood–brain barrier, be non-toxic to normal body tissues and, finally, deliver the drug to the tumor cells, avoiding the enhanced osmotic pressure in tumor regions [129] and ensuring uptake and action.

However, a nanoparticle may have yet further uses. As demonstrated by Baker et al., dendrimer-based nanoparticles can be modified to not only avoid obstacles and target tumors but also to "report" on the presence and extent of the tumor by the means of an attached "reporter" molecule, in this case fluorescein [35]. In a similar vein, Loo et al. have demonstrated the dual functional ability of immunotargeted nanoshells for integrated cancer imaging and therapy [130]. They define nanoshells as "a novel class of optically tunable nanoparticles that consist of a dielectric core surrounded by a thin gold shell". The relative dimensions of the shell thickness and core radius allow nanoshells to either scatter and/or absorb light over a broad spectral range, including the near-infrared (NIR). The NIR is a wavelength region that provides maximal penetration of light through human tissue. Thus, with attachment of suitable antibodies, these nanoshells offer the opportunity to design vehicles that provide, in a single nanoparticle, both diagnostic and therapeutic capabilities.

Targeting can be enhanced or achieved by other factors, too. One of these is the use of drugs that act preferentially on tumor cells. While most conventional chemotherapeutic drugs now in use have greater or lesser degrees of tumor selectivity (usually by targeting the rapid proliferation of tumor cells), greater selectivity may be achieved by using siRNA that are specific for tumor antigens (see below). Another type of targeting, demonstrated by Potineni, et al. [131], describes a method to utilize pH differences to release drugs at tumor sites. They demonstrated the *in vitro* release of the anticancer drug paclitaxel by biodegradable poly(ethylene oxide)-modified poly(β-amino ester) nanoparticles. This can theoretically be reproduced at cancer sites, which have high metabolic rates and altered pH.

10.6
Conclusion

As far back as the late 19th century, Paul Ehrlich (1854–1915) – Nobel laureate for Physiology or Medicine in 1908 for his pioneering work in immunology and chemotherapy and for the discovery of the first effective therapy for syphilis – propounded and popularized the concept of a "magic bullet". This is a drug that targets and destroys diseased cells selectively, to the exclusion of all others, like a bullet that finds its mark every time. His invention, preparation 606, later called Salvarsan, was highly effective against syphilis and harmless despite a large arsenic content. This drug was also found to be successful in curing several other diseases. Indeed, it seemed that a "magic bullet" had indeed been found [132]. We, of course, realize that despite its effectiveness, arsenical-based therapy of syphilis can not accurately qualify for the label of magic bullet. Indeed, for years scientists have searched in vain for such a drug for cancer, but as our understanding of the complexities of the cancer cell and the diverse nature of the neoplastic process grows, more than one researcher has denounced this concept as an unachievable objective. The search for the magic bullet for cancer has become a sort of Holy Grail for cancer researchers worldwide – to be sought for, but not to be found. Each new advance and discovery in this field has been accompanied by hopes that finally the grail has been found, but in all cases these hopes have been shattered by the realities of failure rates. On the bright side, very large strides have been made in cancer treatment, and average life expectancy after detection has increased for most cancers, and several treatments have chalked up impressive cure rates.

The field of nanotechnology was famously introduced by another Nobel laureate, Richard P. Feynmann, in his lecture "There is plenty of room at the bottom" at the annual meeting of the American Physical Society at the California Institute of Technology in 1959. Many nanotechnology researchers consider this talk to be the inspiration for their work. Feynmann, in this lecture, proposed two challenges to future nanotechnologists, but neither, however, was related to the field of biology.

Lipid based formulations, especially liposome encapsulated doxorubicin, first raised visions of a route to the magic bullet. However, it was still just a drug modification that allowed the transfer of toxic but effective drugs in the serum. The introduction of monoclonal antibodies suddenly made active targeting an attractive possibility. However, the high cost of these antibodies, the large amounts needed to counter their rapid removal from circulation, and the single molecule of drug they carried, all presented technical difficulties for their widespread use. An ant can carry 10–20× its body weight of food; imagine how ineffective it would be if it could carry only a mouthful at a time! Nanoparticles – introduced in the 1970s – seemed to be the probable replacement for liposomes. However, they soon came up against the same problem that monoclonal antibodies faced: they were rapidly cleared from circulation by the mononuclear phagocytic system in the liver and spleen. Scientists innovatively tried to turn this drawback into an advantage, by using nanoparticles to target the liver specifically, e.g., to transfect genes to produce essential proteins, but in the end the goal of targeted delivery seemed to be as far away as before.

The discovery of the steric stabilization effect of poly(ethylene glycol) (PEG) solved that major problem. While "PEGylating" nanoparticles increased their circulation time dramatically, it also reaped an unexpected harvest. The local environment of the tumor supports extravasation of nanoparticles through large endothelial fenestrations and disordered neoplastic tissue architecture. Their retention is ensured by a low lymphatic drainage. This enhanced permeation and retention (EPR) effect had a dramatic influence on the distribution of long-circulating nanoparticles: they collected at tumor sites to the exclusion of other body tissues even without any "active" targeting molecules. Long time of circulation and the EPR effect ensured that nanoparticles circumvented the blood–brain barrier, with important implications for the treatment of resistant tumors with cell sanctuaries in the brain.

Innovations continue to increase almost daily. Not satisfied with costly monoclonal antibodies for tumor targeting, researchers have turned to such mundane molecules as folic acid, transferrin and biotin for enhanced selective uptake in tumors. Folic acid in particular has a relatively long history of tumor targeting abilities, and preferential uptake in a large number of solid tumors.

The drug carried to the tumor site is not restricted to the more traditional doxorubicins and paclitaxels. There had been considerable interest in a possible gene therapy of cancer. Indeed, at one time it had been hyped as the elusive "magic bullet". *In vitro* studies have conclusively demonstrated repetitively the effectiveness of this approach against a wide variety of cancer. The obstacle had always been an appropriate vehicle that can effectively deliver sufficient amounts of the gene to the cancer cells. Direct injection of the naked DNA into muscle cells (the so-called "gene gun") showed considerable uptake and expression. But for the ultimate delivery vehicle, man decided to trust the vector that Nature's evolutionary process has skillfully crafted over millions of years for this very purpose: viruses. Initial experiments with harmless adenoviridae *in vitro* promised success; however, clinical trials (to supplement the mutated p53 in ovarian cancer) failed to demonstrate any advantage over conventional therapy [133]. On September 17, 1999, eighteen-year-old Jesse Gelsinger died during a gene therapy clinical trial for ornithine transcarbamylase (OTC) deficiency, a rare metabolic disorder that is marked by dangerous levels of ammonia in the bloodstream. He died after a vector injected into his liver triggered an immune response that led to multiple organ failure. This incident seemed to ring the death knell to virus-delivered gene therapy.

Gene therapists turned to non-viral means of delivery – the obvious solution appeared to be nanoparticles. They can be modified to carry biological macromolecules, are themselves harmless, and can help protect the gene from the immune system till the cancerous tissue is reached. The complex of genetic material, especially siRNA, and targeted nanoparticle vector brought together a double selectivity process: the targeting ligand ensures selective uptake to cancer cells, and the siRNA ensures that even if some non-cancerous cells have taken up the particle, only the cancer cells are affected.

The excellent results with nanoparticulate gene vector complexes brought back the old nanoparticle workhorse, liposomes, to the forefront. Ease of DNA uptake by the cationic liposomes, and their uptake by fusion to membranes, ensures efficient transfection. With added targeting ligands, the complex promises to be the answer to many cancer therapy problems. Another nano-sized colloidal drug delivery system made its entry into the field more recently: block copolymer micelles or dendrimers.

With nearly all the components finally at hand, the assembly of the actively targeted nanoparticle against cancer was but a matter of time. Several such complexes have already shown promise. In a natural spin off, several start-up companies have already begun early research into human applications. Endocyte, Inc. has taken up folic acid as their solution, while Baker has established the Center for Biologic Nanotechnology at the University of Michigan to espouse the cause of dendrimers. Arguably the most exciting is the performance of Epeius Biotechnologies Corporation's liposomal based active targeting system that delivers genetic material to treat several cancers, including pancreatic head carcinoma, which has one of the worst prognoses among all neoplasms.

It may be that nanoparticles fail to provide us the "magic bullet" against all cancers. However, if current interest, effort and progress in the field of nanoparticulate targeted drug delivery systems are any indication, we may be heading towards an era of more effective chemotherapy with less morbidity and higher cure rates.

References

1 FERRARI, M., Cancer nanotechnology: Opportunities and challenges, *Nat. Rev. Cancer*, **2005**, 5, 161–171.
2 JAIN, K. K., Nanotechnology-based drug delivery for cancer, *Technol. Cancer Res. Treat.*, **2005**, 4, 407–416.
3 JAIN, K. K., Nanotechnology in clinical laboratory diagnostics, *Clin. Chim. Acta*, **2005**, 358, 37–54.
4 MCNEIL, S. E., Nanotechnology for the biologist, *J. Leukoc. Biol.*, **2005**.
5 FORTINA, P., KRICKA, L. J., SURREY, S., GRODZINSKI, P., Nanobiotechnology: The promise and reality of new

approaches to molecular recognition, *Trends Biotechnol.*, **2005**, 23, 168–173.

6   LEUSCHNER, C., KUMAR, C. S. S. R., in *Nanofabrication Towards Biomedical Applications*, eds. KUMAR C. S. S. R., HORMES J., LEUSCHNER C., Wiley-VCH, Verlag, Weinheim, **2005**, pp. 289–326.

7   COURVREUR, P. G. L., LENAERTS, V., BRASSEUR, F., GUIOT, P., BIORNACKI, A., *Biodegradable Polymeric Nanoparticles as Drug Carrier for Antitumor Agents*, ed. GUIOT, P. C. P., CRC Press, Boca Raton, **1986**.

8   FLORENCE, A. T., HUSSAIN, N., Transcytosis of nanoparticle and dendrimer delivery systems: Evolving vistas, *Adv. Drug Deliv. Rev.*, **2001**, 50(Suppl 1), S69–89.

9   KAWASHIMA, Y., Nanoparticulate systems for improved drug delivery, *Adv. Drug Deliv. Rev.*, **2001**, 47, 1–2.

10 MATSUMURA, Y., MAEDA, H., A new concept for macromolecular therapeutics in cancer chemotherapy: Mechanism of tumoritropic accumulation of proteins and the antitumor agent smancs, *Cancer Res.*, **1986**, 46, 6387–6392.

11 SLEDGE, G. W., JR., MILLER, K. D., Exploiting the hallmarks of cancer: The future conquest of breast cancer, *Eur. J. Cancer*, **2003**, 39, 1668–1675.

12 TEICHER, B. A., Molecular targets and cancer therapeutics: Discovery, development and clinical validation, *Drug Resist. Updat.*, **2000**, 3, 67–73.

13 SHENOY, D. B., AMIJI, M. M., Poly(ethylene oxide)-modified poly(epsilon-caprolactone) nanoparticles for targeted delivery of tamoxifen in breast cancer, *Int. J. Pharm.*, **2005**, 293, 261–270.

14 CHAWLA, J. S., AMIJI, M. M., Biodegradable poly(epsilon-caprolactone) nanoparticles for tumor-targeted delivery of tamoxifen, *Int. J. Pharm.*, **2002**, 249, 127–138.

15 BOULIKAS, T., STATHOPOULOS, G. P., VOLAKAKIS, N., VOUGIOUKA, M., Systemic Lipoplatin infusion results in preferential tumor uptake in human studies, *Anticancer Res.*, **2005**, 25, 3031–3039.

16 WEI, J. S. et al., Temperature- and pH-sensitive core-shell nanoparticles self-assembled from poly(n-isopropylacrylamide-co-acrylic acid-co-cholesteryl acrylate) for intracellular delivery of anticancer drugs, *Front Biosci.*, **2005**, 10, 3058–3067.

17 ALEXIOU, C. et al., Magnetic drug targeting – biodistribution of the magnetic carrier and the chemotherapeutic agent mitoxantrone after locoregional cancer treatment, *J. Drug Target*, **2003**, 11, 139–149.

18 GILCHRIST, R. K. et al., Selective inductive heating of lymph nodes, *Ann. Surg.*, **1957**, 146, 596–606.

19 LUBBE, A. S. et al., Clinical experiences with magnetic drug targeting: A phase I study with 4'-epidoxorubicin in 14 patients with advanced solid tumors, *Cancer Res.*, **1996**, 56, 4686–4693.

20 ALEXIOU, C. et al., Locoregional cancer treatment with magnetic drug targeting, *Cancer Res.*, **2000**, 60, 6641–6648.

21 MOORE, A., MARECOS, E., BOGDANOV, A., JR., WEISSLEDER, R., Tumoral distribution of long-circulating dextran-coated iron oxide nanoparticles in a rodent model, *Radiology*, **2000**, 214, 568–574.

22 LUBBE, A. S., ALEXIOU, C., BERGEMANN, C., Clinical applications of magnetic drug targeting, *J. Surg. Res.*, **2001**, 95, 200–206.

23 BROWNING, M., *The Cancer Cell and the Immune System*, ed. VILE, R. G., John Wiley & Sons, New York, **1995**.

24 HAKOMORI, S., *Tumor-associated Carbohydrate Markers: Chemical and Physical Basis and Cell Biological Implications*, ed. SELL, S., The Humana Press, Totowa, NJ, **1992**.

25 TAYLOR-PAPADIMITRIOU, J., BURCHELL, J., MILES, D. W., DALZIEL, M., MUC1 and cancer, *Biochim. Biophys. Acta*, **1999**, 1455, 301–313.

26 COLCHER, D., MILENIC, D. E., FERRONI, P., ROSELLI, M., SCHLOM, J., In vivo and in vitro clinical applications of monoclonal antibodies against TAG-72, *Int. J. Rad. Appl. Instrum. B*, **1991**, 18, 395–401.

27 HIRABAYASHI, Y. et al., Syngeneic

monoclonal antibody against melanoma antigen with interspecies cross-reactivity recognizes GM3, a prominent ganglioside of B16 melanoma, *J. Biol. Chem.*, **1985**, 260, 13 328–13 333.

28 HAKOMORI, S., Tumor malignancy defined by aberrant glycosylation and sphingo(glyco)lipid metabolism, *Cancer Res.*, **1996**, 56, 5309–5318.

29 CONEY, L. R. et al., Cloning of a tumor-associated antigen: MOv18 and MOv19 antibodies recognize a folate-binding protein, *Cancer Res.*, **1991**, 51, 6125–6132.

30 GARIN-CHESA, P. et al., Trophoblast and ovarian cancer antigen LK26. Sensitivity and specificity in immuno-pathology and molecular identification as a folate-binding protein, *Am. J. Pathol.*, **1993**, 142, 557–567.

31 HATTORI, Y., MAITANI, Y., Enhanced in vitro DNA transfection efficiency by novel folate-linked nanoparticles in human prostate cancer and oral cancer, *J. Controlled Release*, **2004**, 97, 173–183.

32 HOLM, J., HANSEN, S. I., HOIER-MADSEN, M., High-affinity folate binding in human liver membranes, *Biosci. Rep.*, **1991**, 11, 139–145.

33 MATTES, M. J. et al., Patterns of antigen distribution in human carcinomas, *Cancer Res.*, **1990**, 50, 880s–884s.

34 OYEWUMI, M. O., YOKEL, R. A., JAY, M., COAKLEY, T., MUMPER, R. J., Comparison of cell uptake, biodistribution and tumor retention of folate-coated and PEG-coated gadolinium nanoparticles in tumor-bearing mice, *J. Controlled Release*, **2004**, 95, 613–626.

35 QUINTANA, A. et al., Design and function of a dendrimer-based therapeutic nanodevice targeted to tumor cells through the folate receptor, *Pharm. Res.*, **2002**, 19, 1310–1316.

36 ROSS, J. F., CHAUDHURI, P. K., RATNAM, M., Differential regulation of folate receptor isoforms in normal and malignant tissues in vivo and in established cell lines. Physiologic and clinical implications, *Cancer*, **1994**, 73, 2432–2443.

37 TOFFOLI, G. et al., Overexpression of folate binding protein in ovarian cancers, *Int. J. Cancer*, **1997**, 74, 193–198.

38 WEITMAN, S. D., FRAZIER, K. M., KAMEN, B. A., The folate receptor in central nervous system malignancies of childhood, *J. Neurooncol.*, **1994**, 21, 107–112.

39 WEITMAN, S. D. et al., Distribution of the folate receptor GP38 in normal and malignant cell lines and tissues, *Cancer Res.*, **1992**, 52, 3396–3401.

40 WEITMAN, S. D. et al., Cellular localization of the folate receptor: Potential role in drug toxicity and folate homeostasis, *Cancer Res.*, **1992**, 52, 6708–6711.

41 COUSSENS, L. et al., Tyrosine kinase receptor with extensive homology to EGF receptor shares chromosomal location with neu oncogene, *Science*, **1985**, 230, 1132–1139.

42 BARGMANN, C. I., HUNG, M. C., WEINBERG, R. A., The neu oncogene encodes an epidermal growth factor receptor-related protein, *Nature*, **1986**, 319, 226–230.

43 YAMAMOTO, T. et al., Similarity of protein encoded by the human c-erb-B-2 gene to epidermal growth factor receptor, *Nature*, **1986**, 319, 230–234.

44 SLAMON, D. J. et al., Human breast cancer: Correlation of relapse and survival with amplification of the HER-2/neu oncogene, *Science*, **1987**, 235, 177–182.

45 BORG, A. et al., HER-2/neu amplification predicts poor survival in node-positive breast cancer, *Cancer Res.*, **1990**, 50, 4332–4337.

46 KEER, H. N. et al., Elevated transferrin receptor content in human prostate cancer cell lines assessed in vitro and in vivo, *J. Urol.*, **1990**, 143, 381–385.

47 ELLIOTT, R. L., ELLIOTT, M. C., WANG, F., HEAD, J. F., Breast carcinoma and the role of iron metabolism. A cytochemical, tissue culture, and ultrastructural study, *Ann. New York Acad. Sci.*, **1993**, 698, 159–166.

48 LeBIEN, T. W., McCORMACK, R. T., The common acute lymphoblastic leukemia antigen (CD10) –

emancipation from a functional enigma, *Blood*, **1989**, 73, 625–635.
49 CARREL, S., ZOGRAFOS, L., SCHREYER, M., RIMOLDI, D., Expression of CALLA/CD10 on human melanoma cells, *Melanoma Res.*, **1993**, 3, 319–323.
50 VERVOORDELDONK, S. F., MERLE, P. A., VAN LEEUWEN, E. F., VON DEM BORNE, A. E., SLAPER-CORTENBACH, I. C., Preclinical studies with radiolabeled monoclonal antibodies for treatment of patients with B-cell malignancies, *Cancer*, **1994**, 73, 1006–1011.
51 KOHLER, G., MILSTEIN, C., Continuous cultures of fused cells secreting antibody of predefined specificity, *J. Immunol.*, **2005**, 174, 2453–2455.
52 BUSKE, C., FEURING-BUSKE, M., UNTERHALT, M., HIDDEMANN, W., Monoclonal antibody therapy for B cell non-Hodgkin's lymphomas: Emerging concepts of a tumour-targeted strategy, *Eur. J. Cancer*, **1999**, 35, 549–557.
53 JONES, P. T., DEAR, P. H., FOOTE, J., NEUBERGER, M. S., WINTER, G., Replacing the complementarity-determining regions in a human antibody with those from a mouse, *Nature*, **1986**, 321, 522–525.
54 RIECHMANN, L., CLARK, M., WALDMANN, H., WINTER, G., Reshaping human antibodies for therapy, *Nature*, **1988**, 332, 323–327.
55 MARKS, J. D. et al., By-passing immunization. Human antibodies from V-gene libraries displayed on phage, *J. Mol. Biol.*, **1991**, 222, 581–597.
56 ABOU-JAWDE, R., CHOUEIRI, T., ALEMANY, C., MEKHAIL, T., An overview of targeted treatments in cancer, *Clin Ther.*, **2003**, 25, 2121–2137.
57 GLENNIE, M. J., VAN DE WINKEL, J. G., Renaissance of cancer therapeutic antibodies, *Drug Discov Today*, **2003**, 8, 503–510.
58 ZHOU, J. et al., Preparation of arsenic trioxide-loaded albuminutes immuno-nanospheres and its specific killing effect on bladder cancer cell in vitro, *Chin. Med. J. (Engl.)*, **2005**, 118, 50–55.
59 BECERRIL, B., POUL, M. A., MARKS, J. D., Toward selection of internalizing antibodies from phage libraries, *Biochem. Biophys. Res. Commun.*, **1999**, 255, 386–393.
60 NIELSEN, U. B. et al., Therapeutic efficacy of anti-ErbB2 immunoliposomes targeted by a phage antibody selected for cellular endocytosis, *Biochim. Biophys. Acta*, **2002**, 1591, 109–118.
61 LIU, B., CONRAD, F., COOPERBERG, M. R., KIRPOTIN, D. B., MARKS, J. D., Mapping tumor epitope space by direct selection of single-chain Fv antibody libraries on prostate cancer cells, *Cancer Res.*, **2004**, 64, 704–710.
62 LI, K. C., PANDIT, S. D., GUCCIONE, S., BEDNARSKI, M. D., Molecular imaging applications in nanomedicine, *Biomed. Microdevices*, **2004**, 6, 113–116.
63 FOLKMAN, J., Successful treatment of an angiogenic disease, *N. Engl. J. Med.*, **1989**, 320, 1211–1212.
64 CHEN, Q. R., ZHANG, L., GASPER, W., MIXSON, A. J., Targeting tumor angiogenesis with gene therapy, *Mol. Genet. Metab.*, **2001**, 74, 120–127.
65 HOOD, J. D. et al., Tumor regression by targeted gene delivery to the neovasculature, *Science*, **2002**, 296, 2404–2407.
66 HALLAHAN, D. et al., Integrin-mediated targeting of drug delivery to irradiated tumor blood vessels, *Cancer Cell*, **2003**, 3, 63–74.
67 SCHIFFELERS, R. M. et al., Transporting silence: Design of carriers for siRNA to angiogenic endothelium, *J. Controlled Release*, **2005**.
68 FAROKHZAD, O. C. et al., Nanoparticle-aptamer bioconjugates: A new approach for targeting prostate cancer cells, *Cancer Res.*, **2004**, 64, 7668–7672.
69 LI, L. et al., A novel antiangiogenesis therapy using an integrin antagonist or anti-Flk-1 antibody coated 90Y-labeled nanoparticles, *Int. J. Radiat. Oncol. Biol. Phys.*, **2004**, 58, 1215–1227.
70 SANTRA, S. et al., Folate conjugated fluorescent silica nanoparticles for labeling neoplastic cells, *J. Nanosci. Nanotechnol.*, **2005**, 5, 899–904.
71 HUEBERS, H. A., FINCH, C. A., The physiology of transferrin and

transferrin receptors, *Physiol. Rev.*, **1987**, 67, 520–582.
72 PONKA, P., Tissue-specific regulation of iron metabolism and heme synthesis: Distinct control mechanisms in erythroid cells, *Blood*, **1997**, 89, 1–25.
73 PUN, S. H. et al., Targeted delivery of RNA-cleaving DNA enzyme (DNAzyme) to tumor tissue by transferrin-modified, cyclodextrin-based particles, *Cancer Biol. Ther.*, **2004**, 3, 641–650.
74 BELLOCQ, N. C., PUN, S. H., JENSEN, G. S., DAVIS, M. E., Transferrin-containing, cyclodextrin polymer-based particles for tumor-targeted gene delivery, *Bioconj. Chem.*, **2003**, 14, 1122–1132.
75 NA, K. et al., Self-assembled nanoparticles of hydrophobically-modified polysaccharide bearing vitamin H as a targeted anti-cancer drug delivery system, *Eur. J. Pharm. Sci.*, **2003**, 18, 165–173.
76 MARUYAMA, K. et al., Characterization of in vivo immunoliposome targeting to pulmonary endothelium, *J. Pharm. Sci.*, **1990**, 79, 978–984.
77 HUTCHINSON, F. J., FRANCIS, S. E., LYLE, I. G., JONES, M. N., The characterisation of liposomes with covalently attached proteins, *Biochim. Biophys. Acta*, **1989**, 978, 17–24.
78 SARKAR, D. P., BLUMENTHAL, R., The role of the target membrane structure in fusion with Sendai virus, *Membr. Biochem.*, **1987**, 7, 231–247.
79 YU, D. et al., The specific delivery of proteins to human liver cells by engineered bio-nanocapsules, *FEBS J.*, **2005**, 272, 3651–3660.
80 GORDON, E. M., HALL, F. L., Nanotechnology blooms, at last (Review), *Oncol. Rep.*, **2005**, 13, 1003–1007.
81 GORDON, E. M. et al., First clinical experience using a "pathotropic" injectable retroviral vector (Rexin-G) as intervention for stage IV pancreatic cancer, *Int. J. Oncol.*, **2004**, 24, 177–185.
82 PARK, J. W., BENZ, C. C., MARTIN, F. J., Future directions of liposome- and immunoliposome-based cancer therapeutics, *Semin. Oncol.*, **2004**, 31, 196–205.
83 SENIOR, J. H., Fate and behavior of liposomes in vivo: A review of controlling factors, *Crit. Rev. Ther. Drug Carrier Syst.*, **1987**, 3, 123–193.
84 KLIBANOV, A. L., MARUYAMA, K., TORCHILIN, V. P., HUANG, L., Amphipathic polyethyleneglycols effectively prolong the circulation time of liposomes, *FEBS Lett.*, **1990**, 268, 235–237.
85 GABIZON, A. et al., Targeting folate receptor with folate linked to extremities of poly(ethylene glycol)-grafted liposomes: In vitro studies, *Bioconj. Chem.*, **1999**, 10, 289–298.
86 STORM, G., TEN KATE, M. T., WORKING, P. K., BAKKER-WOUDENBERG, I. A., Doxorubicin entrapped in sterically stabilized liposomes: Effects on bacterial blood clearance capacity of the mononuclear phagocyte system, *Clin. Cancer Res.*, **1998**, 4, 111–115.
87 MUGGIA, F. M. et al., Phase II study of liposomal doxorubicin in refractory ovarian cancer: Antitumor activity and toxicity modification by liposomal encapsulation, *J. Clin. Oncol.*, **1997**, 15, 987–993.
88 LASIC, D. D., PAPAHADJOPOULOS, D., Liposomes revisited, *Science*, **1995**, 267, 1275–1276.
89 ARAGNOL, D., LESERMAN, L. D., Immune clearance of liposomes inhibited by an anti-Fc receptor antibody in vivo, *Proc. Natl. Acad. Sci. U.S.A.*, **1986**, 83, 2699–2703.
90 DERKSEN, J. T., MORSELT, H. W., SCHERPHOF, G. L., Uptake and processing of immunoglobulin-coated liposomes by subpopulations of rat liver macrophages, *Biochim. Biophys. Acta*, **1988**, 971, 127–136.
91 MARUYAMA, K. et al., Targetability of novel immunoliposomes modified with amphipathic poly(ethylene glycol)s conjugated at their distal terminals to monoclonal antibodies, *Biochim. Biophys. Acta*, **1995**, 1234, 74–80.
92 MARUYAMA, K., ISHIDA, O., TAKIZAWA, T., MORIBE, K., Possibility of active

targeting to tumor tissues with liposomes, *Adv. Drug Deliv. Rev.*, **1999**, 40, 89–102.

93 HUWYLER, J., WU, D., PARDRIDGE, W. M., Brain drug delivery of small molecules using immunoliposomes, *Proc. Natl. Acad. Sci. U.S.A.*, **1996**, 93, 14 164–14 169.

94 BERINSTEIN, N., MATTHAY, K. K., PAPAHADJOPOULOS, D., LEVY, R., SIKIC, B. I., Antibody-directed targeting of liposomes to human cell lines: Role of binding and internalization on growth inhibition, *Cancer Res.*, **1987**, 47, 5954–5959.

95 GOREN, D. et al., Targeting of stealth liposomes to erbB-2 (Her/2) receptor: In vitro and in vivo studies, *Br. J. Cancer*, **1996**, 74, 1749–1756.

96 PARK, J. W. et al., Development of anti-p185HER2 immunoliposomes for cancer therapy, *Proc. Natl. Acad. Sci. U.S.A.*, **1995**, 92, 1327–1331.

97 PARK, J. W. et al., Anti-HER2 immunoliposomes for targeted therapy of human tumors, *Cancer Lett.*, **1997**, 118, 153–160.

98 POUL, M. A., BECERRIL, B., NIELSEN, U. B., MORISSON, P., MARKS, J. D., Selection of tumor-specific internalizing human antibodies from phage libraries, *J. Mol. Biol.*, **2000**, 301, 1149–1161.

99 WINTER, G., GRIFFITHS, A. D., HAWKINS, R. E., HOOGENBOOM, H. R., Making antibodies by phage display technology, *Annu. Rev. Immunol.*, **1994**, 12, 433–455.

100 SCHIER, R. et al., Isolation of picomolar affinity anti-c-erbB-2 single-chain Fv by molecular evolution of the complementarity determining regions in the center of the antibody binding site, *J. Mol. Biol.*, **1996**, 263, 551–567.

101 PARK, J. W., Liposome-based drug delivery in breast cancer treatment, *Breast Cancer Res.*, **2002**, 4, 95–99.

102 SPRINGER, C. J., NICULESCU-DUVAZ, I., Approaches to gene-directed enzyme prodrug therapy (GDEPT), *Adv. Exp. Med. Biol.*, **2000**, 465, 403–409.

103 VINGERHOEDS, M. H. et al., A new application for liposomes in cancer therapy. Immunoliposomes bearing enzymes (immuno-enzymosomes) for site-specific activation of prodrugs, *FEBS Lett.*, **1993**, 336, 485–490.

104 LEE, R. J., LOW, P. S., Delivery of liposomes into cultured KB cells via folate receptor-mediated endocytosis, *J. Biol. Chem.*, **1994**, 269, 3198–3204.

105 LEE, R. J., LOW, P. S., Folate-mediated tumor cell targeting of liposome-entrapped doxorubicin in vitro, *Biochim. Biophys. Acta*, **1995**, 1233, 134–144.

106 KIRPOTIN, D. et al., Sterically stabilized anti-HER2 immuno-liposomes: Design and targeting to human breast cancer cells in vitro, *Biochemistry*, **1997**, 36, 66–75.

107 MORRISON, S. L., Transfectomas provide novel chimeric antibodies, *Science*, **1985**, 229, 1202–1207.

108 FEERO, W. G. et al., Selection and use of ligands for receptor-mediated gene delivery to myogenic cells, *Gene Ther.*, **1997**, 4, 664–674.

109 XU, L., PIROLLO, K. F., TANG, W. H., RAIT, A., CHANG, E. H., Transferrin-liposome-mediated systemic p53 gene therapy in combination with radiation results in regression of human head and neck cancer xenografts, *Hum Gene Ther.*, **1999**, 10, 2941–2952.

110 LEE, R. J., HUANG, L., Folate-targeted, anionic liposome-entrapped polylysine-condensed DNA for tumor cell-specific gene transfer, *J. Biol. Chem.*, **1996**, 271, 8481–8487.

111 FRÉCHET, J. M. J., TOMALIA, D. A., *Dendrimers Dendritic Polym.*, **2001**.

112 ESFAND, R., TOMALIA, D. A., Poly(amidoamine) (PAMAM) dendrimers: From biomimicry to drug delivery and biomedical applications, *Dendrimers Dendritic Polym.*, **2001**, 6, 427–436.

113 MALIK, N. et al., Dendrimers: Relationship between structure and biocompatibility in vitro, and preliminary studies on the biodistribution of 125I-labelled polyamidoamine dendrimers in vivo, *J. Controlled Release*, **2000**, 65, 133–148.

114 BOSMAN, A. W., VESTBERG, R., HEUMANN, A., FRECHET, J. M., HAWKER, C. J., A modular approach

toward functionalized three-dimensional macromolecules: From synthetic concepts to practical applications, *J. Am. Chem. Soc.*, **2003**, 125, 715–728.

115 DEWEY, R. S. et al., The synthesis of peptides in aqueous medium. VII. The preparation and use of 2,5-thiazolidinediones in peptide synthesis, *J. Org. Chem.*, **1971**, 36, 49–59.

116 MALIK, N., EVAGOROU, E. G., DUNCAN, R., Dendrimer-platinate: A novel approach to cancer chemotherapy, *Anticancer Drugs*, **1999**, 10, 767–776.

117 GREENFIELD, R. S. et al., Evaluation in vitro of adriamycin immuno-conjugates synthesized using an acid-sensitive hydrazone linker, *Cancer Res.*, **1990**, 50, 6600–6607.

118 PADILLA DE JESUS, O. L., IHRE, H. R., GAGNE, L., FRECHET, J. M., SZOKA, F. C., JR., Polyester dendritic systems for drug delivery applications: In vitro and in vivo evaluation, *Bioconj. Chem.*, **2002**, 13, 453–461.

119 KONO, K., LIU, M., FRECHET, J. M., Design of dendritic macromolecules containing folate or methotrexate residues, *Bioconj. Chem.*, **1999**, 10, 1115–1121.

120 BIRRENBACH, G., SPEISER, P. P., Polymerized micelles and their use as adjuvants in immunology, *J. Pharm. Sci.*, **1976**, 65, 1763–1766.

121 GILBERT, L. et al., Molecular and structural characterization of fluorescent human parvovirus B19 virus-like particles, *Biochem. Biophys. Res. Commun.*, **2005**, 331, 527–535.

122 Mo, Y., LIM, L. Y., Preparation and in vitro anticancer activity of wheat germ agglutinin (WGA)-conjugated PLGA nanoparticles loaded with paclitaxel and isopropyl myristate, *J. Controlled Release*, **2005**.

123 BRIGGER, I., DUBERNET, C., COUVREUR, P., Nanoparticles in cancer therapy and diagnosis, *Adv. Drug Deliv. Rev.*, **2002**, 54, 631–651.

124 SCHWAB, G. et al., Antisense oligonucleotides adsorbed to polyalkylcyanoacrylate nanoparticles specifically inhibit mutated Ha-ras-mediated cell proliferation and tumorigenicity in nude mice, *Proc. Natl. Acad. Sci. U.S.A.*, **1994**, 91, 10 460–10 464.

125 LAMBERT, G. et al., EWS fli-1 antisense nanocapsules inhibits Ewing sarcoma-related tumor in mice, *Biochem. Biophys. Res. Commun.*, **2000**, 279, 401–406.

126 MORIN, C., BARRATT, G., FESSI, H., DEVISSAGUET, J. P., PUISIEUX, F., Improved intracellular delivery of a muramyl dipeptide analog by means of nanocapsules, *Int. J. Immuno-pharmacol.*, **1994**, 16, 451–456.

127 SINEK, J., FRIEBOES, H., ZHENG, X., CRISTINI, V., Two-dimensional chemotherapy simulations demonstrate fundamental transport and tumor response limitations involving nanoparticles, *Biomed. Microdevices*, **2004**, 6, 297–309.

128 DUNCAN, R., The dawning era of polymer therapeutics, *Nat. Rev. Drug Discov.*, **2003**, 2, 347–360.

129 NETTI, P. A., BAXTER, L. T., BOUCHER, Y., SKALAK, R., JAIN, R. K., Time-dependent behavior of interstitial fluid pressure in solid tumors: Implications for drug delivery, *Cancer Res.*, **1995**, 55, 5451–5458.

130 LOO, C., LOWERY, A., HALAS, N., WEST, J., DREZEK, R., Immunotargeted nanoshells for integrated cancer imaging and therapy, *Nano Lett.*, **2005**, 5, 709–711.

131 POTINENI, A., LYNN, D. M., LANGER, R., AMIJI, M. M., Poly(ethylene oxide)-modified poly(beta-amino ester) nanoparticles as a pH-sensitive biodegradable system for paclitaxel delivery, *J. Controlled Release*, **2003**, 86, 223–234.

132 EHRLICH, PAUL, Encyclopaedia Britannica, **2006**, Encyclopaedia Britannica Premium Service, 31 Mar 2006, http://www.britannica.com/eb/article-9032103.

133 ZEIMET, A. G., MARTH, C., Why did p53 gene therapy fail in ovarian cancer? *Lancet Oncol.*, **2003**, 4, 415–422.

134 ZHANG, Y., HUANG N., Intracellular uptake of CdSe-ZnS/polystyrene nanobeads, *J. Biomed. Mater. Res. B, Appl. Biomater.*, **2006**, 76(1), 161–168.

11
Colloidal Systems for the Delivery of Anticancer Agents in Breast Cancer and Multiple Myeloma

Sébastien Maillard, Elias Fattal, Véronique Marsaud, Brigitte Sola, and Jack-Michel Renoir

11.1
Introduction

The success of cancer treatments depends on an active drug concentration at the tumor sites. In fact, drug dispatch remains the objective of clinicians and drug delivery has been the subject of a tremendous amount of research during the past four decades [1, 2]. Indeed, most used anticancer drugs have great toxicity, including for normal tissues, and it is crucial to minimize massive damage to the rest of the body after their administration. Moreover, despite poor water solubility, owing to a high hydrophobicity, some chemotherapeutic drugs were initially given orally. However, drug administration, particularly for antitumor drugs, increasingly required intravenous perfusion, leading to a poor amount reaching the tumors and to important side effects because high doses are often administered. This poor specificity creates a toxicological problem that constitutes a serious obstacle to effective antitumor therapy. Thus, a crucial problem remains the addressing of the drug to the site of action, i.e., being more tumor selective.

Undoubtedly, intravenous administration of anticancer drugs could be of great help but it remains limited by the capture of the drug by macrophages of the mononuclear phagocytes system (MPS), leading to drug destruction in the liver, spleen, lung and bone marrow. If the propensity of MPS macrophages for endocytosis/phagocytosis provides an opportunity to deliver efficiently therapeutic agents to these organs, the targeting of non-MPS organs remains a priority. One strategy consists of associating antitumor drugs with colloidal systems engineered to resist the MPS and to increase the selectivity of drugs towards cancer cells while reducing their toxicity towards normal tissues. Various injectable delivery systems, nanoparticles, emulsions and liposomes (conventional drug delivery systems, CDDS), have been developed but only a few are clinically used. To avoid their recognition by opsonins (immunoglobulins, proteins from the complement) and their degradation by the MPS organs, hydrophilic polymers may be grafted [2, 3] at the surface of these colloidal systems [4, 5], rendering them capable of long circulation times in the blood stream, thus constituting "Stealth" drug delivery systems (SDDS) [6].

Another important issue is the capacity of formulations to cross the vascular endothelium (extravasation process). Vascularization of solid tumors is heterogeneous, with regions of necrosis and other areas that are densely vascularized in order to supply the tumor cells with nutrients for tumor growth. The formation of new capillaries from pre-existing vessels (angiogenesis) also contributes to tumor growth. Tumor blood vessels present various abnormalities (increased tortuosity, deficiency in pericytes, aberrant basement membrane formation [1, 7]) that result in an enhanced permeability of tumor vasculature. This enhancement is thought to be regulated by various mediators, such as vascular endothelium growth factor (VEGF), nitric oxide, prostaglandins and metalloproteinases [8, 9]. Macromolecules cross tumor vessels through open gaps (interendothelial junctions and transendothelial channels), vesicular endothelial organelles and fenestrations [10]. The pore size cut-off of various solid tumor models usually lies in the range 380–800 nm [2, 9–13], which is compatible with the extravasation of SDDS.

Thus, to deliver therapeutic agents to tumor cells following i.v. injection one must: (a) avoid the opsonization process and resulting destruction by the MPS; (b) realize a stealth formulation incorporating high amounts of drug; (c) dispatch a high and active drug concentration at the tumor level; (d) overcome drug resistance mechanisms. This chapter focuses on two types of cancer, which differ in the sort of cells they affect and in their morphology, i.e., breast cancers and multiple myeloma, and we describe how new antiestrogen drug delivery devices strongly inhibit tumor growth in xenograft models.

Besides their essential function in female reproduction, estrogens exert various responses in target cells, including promotion of tissue differentiation, morphogenesis, mitogenic activity and development of the mammary gland. They are also beneficial hormones in the regulation of bone mineral density, through effects on the osteoclast/osteoblast balance, in the prevention of atheromatous plaque, and in the nervous system. Importantly, they are also implicated in oncogenesis and maintenance of some tumor growth. In fact, estrogens are regulators of several proto-oncogenes coding for nuclear proteins [14]. Estrogens act on cells via interaction with two types of intracellular receptors, ERα [15, 16] and ERβ [17–19]. Like all steroid hormone receptors, ERs are trans-acting transcription enhancer factors that are activated by binding of their ligands. They become, therefore, capable of interacting with specific cis-acting enhancer elements that are usually located with the 5′-flanking regions of target genes. This classical mechanism of action was thought for many years to be unique. However, following recent findings reporting the membrane localization of ER, "non-genomic" mechanisms involving mitogen-activated protein kinase (MAPK) activation have emerged [20, 21].

Half of breast cancers are intrinsically estrogen sensitive and, therefore, respond to antiestrogens [22, 23]. To treat these estrogen-dependent cancers, pharmacologists have aimed to develop antiestrogens, i.e., compounds capable of blocking the effects of estradiol without displaying any estrogenic activity on their own. Numer-

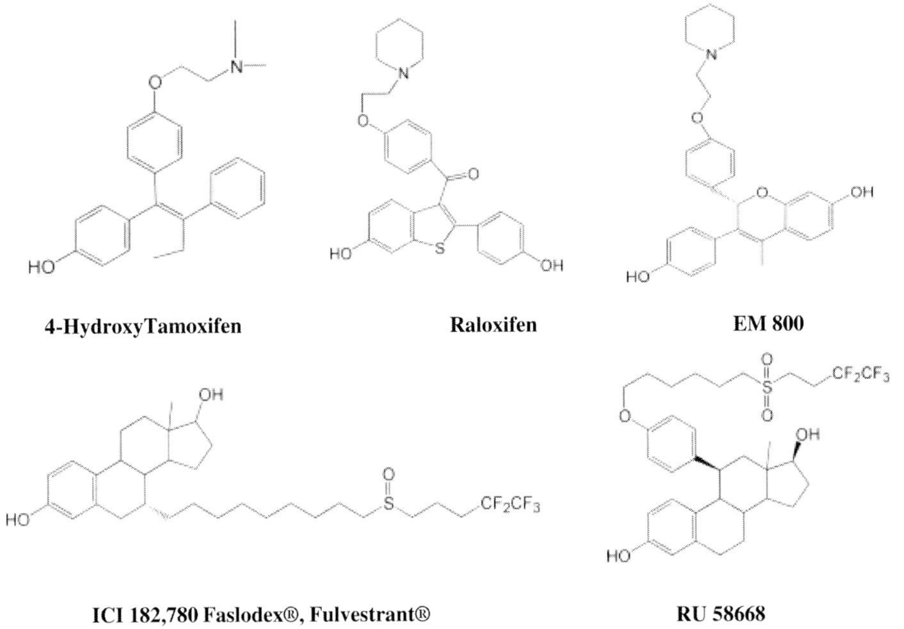

Fig. 11.1. Structure of the most commonly studied antiestrogens: The most commonly used SERMs (Tamoxifen and raloxifen) and three pure antiestrogens (ICI 182,780, RU 58 668 and EM 800) are shown. Among them only ICI 182,780 (Faslodex®) is used clinically, by deep i.m. oily injection in tamoxifen-resistant breast cancers. RU 58 668 development has been stopped and EM 800 is in phase II clinical trials [183, 184].

ous molecules (Fig. 11.1) behave both as agonists or antagonists of estradiol, depending on the tissue, the promoter and the type of the ER [22, 24]. These compounds have been classified as "selective estrogen receptor modulators" (SERMs). Recently, a few compounds able to bind ER and to abrogate all effects of estrogens in a competitive manner have been elaborated [24]. These so-called "pure antiestrogens" were expected to be very useful for estradiol-dependent cancer treatment. In addition to their ability to inhibit all the effects of estradiol, one other important characteristic is their capacity to target ERs to proteasome-mediated degradation in human breast cancer cells [25–27]. On the basis of this property, "pure antiestrogens" have been called selective ER down-regulators (SERDs). However, recent findings in various cancer cell lines prompted reconsideration of this notion, because it has been observed that any kind of antiestrogen, both SERMs and "pure antiestrogens", behave as inducers of transcription through the activating protein 1 (AP_1) pathway [28–30].

Multiple myeloma (MM) is a B-cell malignancy characterized by the clonal expansion of tumoral plasma cells in the bone marrow. This accumulation of malignant cells synthesizing immunoglobulins results in hyperproteinemia, renal dysfunctions, bone lesions and immunodeficiency [31]. This pathology, which accounts for 2% of all cancer deaths per year in occidental countries, remains largely incurable despite novel therapeutic approaches involving the targeting of both MM cells and the bone-marrow environment [32]. Previous studies have established that both ERα and β isotypes are expressed in some MM cells [33–36]. Surprisingly, and differing strongly from breast cancer cells, estradiol as well as SERMs and SERDs induce a decrease of MM cell proliferation, and at a micromolar concentration preclude a classical recruitment of ERs. However, SERMs and SERDs induce breast cancer and MM cell cycle arrest as well as apoptosis. Therefore, they all can be considered as potent anticancer drugs.

Other work relating to the use of anticancer drug delivery systems has shown the importance of this approach in developing new anticancer treatments. However, most work describes the use of drug delivery systems containing highly cytotoxic drugs devoid of specific cellular target, leading to ubiquitous effects and high toxicity [37, 38]. A physicochemical DDS containing an anticancer drug implies knowledge of the physicochemical properties of the drug, its exact mechanism of action in the pathology to be treated, and the physiology of the pathology itself. Here, we point out the importance of delivering a molecule endowed with a very specific intracellular target.

Thus, we focus first (Section 11.2.1) on the importance of estrogens in breast cancer development. The different roles of estrogens will be discussed by presenting the classical ER-ligand and ERE-dependant, ERE-independent, ER-ligand independent, and "non-genomic" pathways. Knowledge of these precise mechanisms has led to the use of various families of antiestrogens (AEs), the properties of which differ (Section 11.2.2). The benefits of AE encapsulation, to improve antitumoral activities *in vivo* and to limit side effects, are discussed in Section 11.2.3.

In Section 11.3, we focus on another pathology, multiple myeloma (MM), describing current treatments and new biological therapies as well as the promising use of AEs.

Whatever the concerned pathology, we will develop in Section 11.4 the requirement for a specific drug delivery system. Thus, the use of passive targeting long-circulating drug delivery systems in different types of xenografts are detailed, such as nanoparticles for breast cancer (Section 11.4.1) and liposomes for MM (Section 11.4.2). Finally, we discus the potential improvement of these systems on acquiring a specific tumoral recognition element (Section 11.4.3).

11.2
Hormone Therapy in Breast Cancers

Two important general ways have been imagined for eliminating the mitogenic activity of estradiol in estrogen-dependent breast cancers: blocking the site of an estrogenic specific target (the estradiol receptor) or blocking the synthesis of estradiol. For the first case, antiestrogenic compounds are used, and for the second case aromatase inhibitors have been developed [39].

11.2.1
Molecular Mechanisms of Estrogen Action in Breast Cancers

11.2.1.1 Classical ER-ligand and ERE-dependent Mechanism

The general mechanism by which estrogens transduce their mitogenic activity results from their binding to the two nuclear receptors from the steroid/thyroid superfamily of transcription factors, ERα and ERβ [40, 41]. In addition to binding estrogens with a high affinity, ERs bind to specific DNA sequences called estrogen responsive elements (EREs), possess defined activating functions (AF1 in the N-terminal domain and AF2 in the ligand binding domain), and interact with various protein partners [22, 24, 42–44]. In target cells unexposed to estradiol (E_2) or growth factors, ERα is part of a large complex in which molecular chaperones such as heat shock proteins (hsp) and co-chaperones are present (for a review see Ref. [44]). The major chaperones that favors the efficient folding of ERα are hsp90 and hsp70 [44–47]. Another important partner of this supramolecular complex is p23 [48, 49], a N-terminal interacting client protein of hsp90 [50], the overexpression of which decreases the transactivation capacity and ligand binding of ERα [51]. In fact, this p23 is a prostaglandin-E_2 synthase and its degradation is an apoptosis indicator [52, 53].

Following E_2 binding, a change in ERα conformation occurs, promoting the dissociation of hsp90 and other proteins from the molecular chaperone complex. Phosphorylation of ERα increases in response to estradiol binding prior to transcriptional enhancement [54]. This rapid process leads to dimerization of ERα, to its tight association with the nucleus [55] and to its interaction with specific nucleotide sequences of the specific EREs in gene promoters. This leads to enhancement of transcription through the recruitment of co-activators, some of which carry a histone acetyl transferase (HAT) activity (Fig. 11.2A), like SRC_1 and CBP/p300 [41, 56, 57]. Histone acetylation leads to chromatin opening and to repositioning of nucleosomes [58]. Other modifications of the ER state and of some co-regulators and the specific DNA sequences (like phosphorylations, methylation, ubiquitination, acetylation) also take place in the sequentially and processivity of transactivation [O'Malley, B. W., NURSA e-journal (2003) 1, ID# 3.11052003.1; ISSN 1550-7629]. This interaction of ER with DNA leads to activation of transcription [40, 42]. Recently, the concept of a "transcriptional clock" that directs and achieves the sequential and combinatorial assembly of 46 co-activators at a specific E_2-inducible (pS2) promoter in human breast cancer MCF-7 cells was defined [59, 60]. Following the pS2 promoter transactivation, the E_2-ER complex is polyubiquitinated, leading to its proteasome-mediated destruction [60, 61].

With binding of a SERM, cofactors different from those recruited upon E_2 binding associate with the ERα/DNA complex (Fig. 11.2A). Co-repressors, such as N-COR, SMRT and REA (specific for ERα), are present in these complexes [56, 57, 62]. The co-repressors are associated with a histone deacetylase activity (HDAC), which re-compacts chromatin followed by nucleosome re-positioning, leading to blockade of transcription of specific genes. It has also been proposed that, in the absence of ligand, ERα could be sequestered in complexes containing

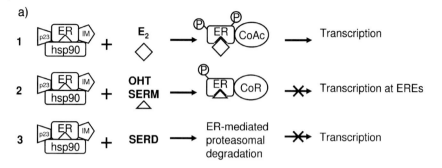
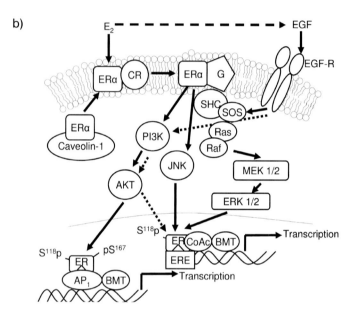

Fig. 11.2. Mechanisms of action of estradiol and antiestrogens: (A) In the classical genomic mechanism of ER, the inactivated receptor (1) is sequestered in a supramolecular chaperone complex (organized around hsp90, p23 and immunophilins). E_2 binds to inactivated ER, changes its conformation, promotes both its phosphorylation by different kinases and its release from the molecular chaperones, helps in the sequential recruitment of a set of co-activators, among which are histone acetyl transferases (HAT), which modify the chromatin structure en route to transcription. When a SERM like OH-tamoxifen (OHT) binds to ER (2), the induced conformational change and the phosphorylation state of ER differ from those induced by E_2 binding. Co-repressors instead of co-activators are recruited, among which are histone deacetylases (HDAC), which re-compact the chromatin and preclude interaction with the basal machinery of transcription (BMT) at an ERE, allowing activation of the polymerase and leading to inhibition of the transcription. With an AP1-containing promoter, the SERM-ERβ complex activates transcription while the E_2-ER complex inhibits AP1-mediated transcription. If a SERD like RU or Faslodex complexes with ERα (3) a delocalization of ER and its fast proteasome-mediated destruction occurs, precluding binding to DNA.

these repressors (or some of them), maintaining the receptor in an inactive state [63].

11.2.1.2 ERE-independent Pathway

In addition to the classical transactivation pathway that results in binding of the E_2/ERα-complex to EREs, tamoxifen and other SERMs also activate target genes (like collagenase gene) through binding at an activating protein-1 (AP-1) site; this is observed in uterine cells but not in breast tumor cells [28]. It occurs through direct binding of ER-E_2 complex to the heterodimer c-fos/c-jun at AP-1 binding sites and is mainly mediated via ERβ, the major ER form in the endometrium (Fig. 11.2B). Remarkably, the pharmacology of ER ligands bound to ERβ is reversed at an AP-1 element. Bound to ERβ, the antiestrogens, including the "pure antiestrogens" Faslodex® and RU 58668 (RU) (Gougelet, A., Marsaud, V., Renoir, J.-M., in preparation), act as transcriptional activators and E_2 behaves as a transcription inhibitor. These features may explain the incidence of uterine cancers following long-term tamoxifen treatment of E_2-dependent breast cancers, and could be involved in the acquisition of tamoxifen resistance in breast tumors [24]. Similarly, other genes containing GC-rich promoter sequences are activated via an ER-SP1 complex [64, 65].

11.2.1.3 ER-ligand-independent Pathway

There is accumulating evidence that ER-mediated transcription activation occurs also in the absence of ER-ligand binding in response to cell surface growth factor receptors and the intracellular signaling cascades that they activate [66–68]. Indeed, growth factors or cyclic adenosine monophosphate (cAMP) activate intracellular kinase pathways, leading to phosphorylation and activation of ER at ERE-containing promoters in a ligand-independent manner, possibly via p42/44 MAPK activation [69–71]. Similarly, phorbol-12-myristate-13-acetate (PMA) through the stimulation of adenylate cyclase also increases the ER-mediated transcription. Therefore, the genomic activity of ER can be induced by numerous different stimuli, ER agonists themselves but also activators of various kinases of the EGF receptor family as well as activators of PI3-K, PKA and PKC [54]. These variable processes target ER to different sequences present in E_2- and/or PMA-inducible promoters to affect gene expression. However, the crosstalk between the classical activation E_2-dependent pathways and other pathways has considerably helped our understanding of the complicated multifaceted mechanism of action of estrogens.

(B) ERα can also be rapidly phosphorylated on Ser118 by the MAPKinase pathway following activation of the EGF receptor which, through the Ras,Raf, MEK, ERK cascade, leads to an ERE-dependent, ER-ligand-independent transactivation activity, or an ERE-independent, ER-ligand-independent-PI3K/AKT-dependent activation process. The "non-genomic" activation process by which ERα can be activated proceeds through binding of E_2 to a cytosoluble ER bound to caveolin-1, to translocation of the caveolin-1/ERα complex to the membrane, leading to association with a caveolin-1 receptor (CR), allowing contact with small G proteins, which acts to activate the MAPK cascade and gene expression without any DNA binding of ERα.

11.2.1.4 "Non-genomic" Pathway

Besides this general and well-accepted genomic mechanism (representing direct or indirect binding of ER to DNA), estrogens may induce effects occurring within minutes ("non-genomic effects"), which is incompatible with a process requiring DNA and transcription (Fig. 11.2B). For instance, estrogen-induced vasodilatation is rapid and relies on the activation of nitric oxide synthase by a mechanism involving trimeric G proteins [72] and Akt via the PI3-K pathway [73]. Direct interaction between ER and PI3-K has been demonstrated [74]. Rapid activation by E_2 of protein kinase C in MCF-7 ER-positive (ER+) cells as well as in other human breast cancers lacking ER is membrane mediated and inhibited by tamoxifen [75]. Several groups have reported that E_2 induces a rapid activation of MAPK/ERK in breast cancer cells [76–78] as well as a rapid non-transcriptional activation of PI3-K [79]. However, this E_2-induced activation of both MAPK and PI3-K in human ER+ breast cancer cells such as MCF-7 and ZR75.1 has been severely criticized by several laboratories, who have questioned the non-genomic mitogenic activity of E_2 [80–83]. In fact, it is now accepted that a small amount of ERα exists as a functional non-traditional G-protein coupled receptor (GPCR) at the plasma membrane [78, 84–86]. Such a membrane localization of ERα is thought to be the consequence of the S522A mutation, which convert the receptor into a dominant-negative form [86]. This GPCR is now known as the endoplasmic reticulum GPR30 [84], which under direct E_2 binding results in intracellular calcium mobilization and synthesis of phosphatidylinositol 3,4,5-triphosphate in the nucleus [87].

11.2.2
Differential Activity of Antiestrogens

Evidently, from several reports emanating from various laboratories, an efficient way to inhibit the E_2-mediated mitogenic effects in breast tumors could be the suppression of the biological activity of ER. This is possible through different pathways that block ER-mediated transcription, as realized with SERMs and SERDs, but also by inhibiting ER protein expression. Introduced as a new approach for targeted therapy with few side effects compared with traditional cytotoxic chemotherapy, hormone therapy is based on the use of non-steroidal antiestrogens, the archetype of which is tamoxifen (Tam), which has been used clinically with success since the 1970s. Initially investigated in the late 1960s these antiestrogens are partial estrogen antagonists or SERMs (Fig. 11.1).

In terms of breast cancer growth, Tam as well as several other SERMs are antiestrogenic, whereas they exert estrogen-like actions in bone, the cardiovascular system, and in the regulation of circulating low density lipoproteins (LDL) [22, 23]. Unfortunately, the SERMs are agonist in the uterus and can enhance the growth of human endometrial tumors implanted into immune deficient mice [88]. SERMs such as tamoxifen and raloxifen induce an accumulation of ERα in breast cancer cells [25–27, 89, 90]. In contrast to tamoxifen, both ICI 182,780 (Faslodex®) and RU 58668 treatment of human breast cancer cells results in rapid delocalization and a fast destruction of endogenous ERα through the 26S proteasome pathway [25–27, 61, 86, 89, 90]. Such pure antiestrogen-induced down-regulation of ERα expression takes place in a nuclear compartment from which ERα is difficult to extract [27, 78, 91] and is recovered with the nuclear matrix [55]. The CSN5/Jab1, a subunit of the COP9 signalosome, is involved in ligand-dependent degradation of ERα by the proteasome [92]. Interestingly, any alteration of the activity of various kinases known to affect signaling involved in cell cycle progression targets ERα to the 26S proteasome-mediated degradation [27]. Similarly, several co-activators involved in the activation of ER-mediated transcription are also degraded through the proteasome pathway [61]. Contrarily to ERα, ERβ stably expressed in ER-negative MDA-MB-231 cells is more slowly degraded by the SERDs than the ERα isotype and still conserves a high expression level [93].

In addition to SERDs, hsp90 inhibitors are also able to trigger ERα to proteasomal degradation [94]. Both N-terminal hsp90 ligands such as geldanamycin and radicicol, as well as C-terminal hsp90 ligands such as novobiocin, target not only ERα but also ERβ [93] to degradation through the proteasome pathway. Thus, destabilization of steroid receptors by hsp90-binding drugs constitutes a ligand-independent approach to hormonal therapy of breast cancer [95, 96]. In general, any compound with high inhibitory ability of the activity of one (or more) of the proteins implicated in the mechanism of action of ERs could be a potential anticancer drug (SERDs for ERα, HDAC inhibitors, proteasome inhibitors, hsp90 inhibitors, immunosuppressors, p23 inhibitors).

11.2.3
The Need to Encapsulate Antiestrogens

We have already mentioned the large spectrum of activity of SERMs. Actually, their use in postmenopausal women treated for estrogen-dependent breast cancer is based on a daily oral administration of Tam (20 mg day^{-1}). Sometimes, detrimental side effects occur, leading often to hot flushes and more rarely to endometrium cancers (1–2%). Nevertheless, tamoxifen is still clinically used and, in addition, it is of benefit for osteoporosis and cardiovascular disease due to its agonistic activity in bones and vessels. Pure antiestrogens were first reported nearly 20 years ago [97]. Nevertheless, concern about the increased risk of osteoporosis and coronary heart disease, as well as problems with drug delivery, is similar to that caused by aromatase inhibitors developed as an alternative strategy for antiestrogen therapy [98]. Delivery of these molecules to their site of action at the desired rate is a challenge because their transport through compartmental barriers (endothelium or epithelium) in the body is inefficient and/or because they are rapidly metabolized. For controlled release or for site-specific delivery, new delivery systems will be required (as only oral tablets and an oil solution for i.m. injection of Faslodex® are currently available). Furthermore, alkyl estradiol derivatives are difficult to synthesize, and their oral bioavailability is very low. Therefore, a targeted drug delivery system of both SERMs and SERDs for breast cancer treatment could be of benefit.

11.3
Multiple Myeloma

Despite new insights into the biology of MM, the major prognostic for such pathology was not significantly improved. Thus, for several years, research into novel approaches, implying new therapeutic targets, has increased. We focus here on the new biological strategies and, particularly, on the promising outcomes of the estrogens and antiestrogens regimen.

11.3.1
Current Treatments

Addition of glucocorticoids (which trigger apoptosis) to the oral alkylating agent melphalan [melphalan-prednisone (MP) regimen] constituted the mainstay of therapy for many years, although remissions were rare (<5%) and median survival did not exceed three years. Addition of other alkylating agents such as anthracyclines and vinca alkaloids to MP did not improve patient survival. The further developed therapeutic management of MM for the last two decades has mainly involved regimens based on the use of glucocorticoids (dexamethasone) and cytotoxic chemotherapeutic drugs [99, 100]. The standard therapy for MM, at least for younger patients, is considered to be autologous peripheral blood stem cell (PBSC)-supported high-dose melphalan [101].

11.3.2
New Biological Therapies for MM Treatment

The need for efficient therapy has become obvious and several new agents seem very promising [102, 103]. Among them, SU 5416 [104], an inhibitor of VEGF-induced MM proliferation, and thalidomide, an inhibitor of tumor necrosis factor (TNFα) and VEGF production, which is in phase II trials, could be of great interest [105–107]. As many regulators of cell proliferation or apoptosis are degraded by the ubiquitin-proteasome pathway, proteasome inhibitors such as bortezomib (PS-341) [108–111] (Velcade; Millennium, Cambridge, MA) may be promising. Inhibiting the Ras/Raf/MAPK signaling pathway with farnesyltransferase inhibitors like R115777 [112–114] or promoting growth arrest and apoptosis by HDAC [115] and 2-methoxyestradiol [116, 117], which induce apoptosis in numerous cancers by producing reactive oxygen species as well as hsp90 ATPase inhibitors, are other areas of clinical development. All these drugs/inhibitors represent potential anticancer agents [93, 95, 96] alone and in combination with dexamethasone or other drugs. However, they are highly toxic and great care must be given to their administration so as to avoid healthy tissues. For example, the strong cardio-toxicity of anthracyclines must be suppressed as much as possible. This has been successfully realized with the incorporation of doxorubicin in numerous stealth liposomes. Among them caelix/doxil [118] is actually clinically used for different cancer therapies (breast, ovary, myeloma) alone or in combination with vincristine and dexamethasone [119]. Liposomal vincristine and liposomal cisplatin have also been clinically evaluated on various cancers [120]. Despite progress in delineating the activity of such regimens, at either conventional or high dose, MM has remained incurable. No substantial improvement of the median survival was obtained, even from new promising attempts.

11.3.3
Incidence of Estrogens and Antiestrogens on Multiple Myeloma

Both isotypes of ER have been identified in several MM cell lines as well as in cells isolated from patients [33–36], and both SERMs like OH-tamoxifen, the active metabolite of tamoxifen, and two pure antiestrogens (Faslodex® and RU 58668) have been shown to block the cell cycle progression of MM cells and to induce apoptosis [34, 36, 121]. Interestingly, 4-OH-tamoxifen-induced apoptosis in several ER-positive MM cells is a mitochondrial process [121]. Such an event, in conjunction with the high doses needed to obtain apoptosis in these MM cells (as compared with lower doses giving the same effect in ER-positive breast cancer cells), raised the question of the ER specificity of antiestrogens in MM. Even if the pathway(s) that conduct MM cells to apoptosis under antiestrogens is (are) not totally elucidated at present, this has prompted us to evaluate the efficacy of liposomal formulations loaded with RU in a xenograft model of MM.

11.4
Colloidal Systems for Antiestrogen Delivery

11.4.1
Nanoparticles Charged with AEs in Breast Cancer

The first strategy employed to obtain a controlled release of the 11β-alkylestradiol derivatives manufactured by Schering was the use of a polyacrylate-based matrix transdermal system [122, 123]. Pretreatment of the skin with a fluid permeation enhancer of propylene glycol–lauric acid should enhance absorption of transdermally administered antiestrogens. Drugs that permeate easily through the skin must be small molecules with moderate lipophilicity. Nanoparticles for intravenous administration of drugs have since been developed. Because the usefulness of conventional nanoparticles is limited by their rapid and massive capture by macrophages of the MPS after i.v. administration, different nanoparticle devices have been considered to target tumors that are not localized in the MPS organs. Several studies have been devoted to the development of so-called "stealth" particles, which are "invisible" to macrophages [6, 124]. These stealth nanoparticles are characterized by a prolonged half-life in the blood compartment, allowing them to extravasate in pathological sites like tumors or inflamed regions with a leaky vasculature [125, 126]. The size of the colloidal carriers and their surface characteristics are the keys for the biological fate of nanoparticles, since these parameters can prevent

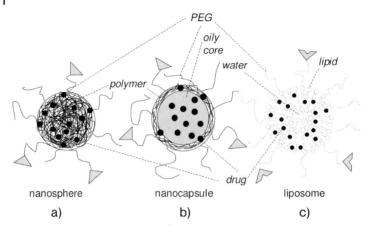

Fig. 11.3. Schematic structures of colloidal systems used for antiestrogen incorporation: Nanospheres (A), nanocapsules (B) and liposomes (C) are the drug delivery systems used to encapsulate antiestrogens. Nanospheres are composed of a polymeric matrix in which the ligand (•) can be encapsulated; nanocapsules consist of the same polymer surrounding an oily core, allowing solubilization of hydrophobic compounds. PEG-PLA or PEG-PLGA or PERG-ε-caprolactone copolymers were also used. Liposomes (C) made of various phospholipid compositions containing a water core were also employed. The antiestrogens were presumed to be located in the lipid bilayer. Tumor recognition molecules are shown as shaded triangles (A and B), and the ligand and antibody as shaded solid Vs (C).

their uptake by the MPS. A high curvature, small size (<100–200 nm) and a hydrophilic surface (as opposed to the hydrophobic surface of conventional nanoparticles) reduce opsonization and subsequent clearance by macrophages [2, 124].

A major breakthrough came with the use of hydrophilic polymers [poly(ethylene glycol) (PEG), poloxamines, poloxamers, polysaccharides] to coat the surface of nanoparticles [6, 127] (Fig. 11.3). Such coatings provide a dynamic "cloud" of hydrophilic and neutral chains at the surface of the particles, repelling plasma proteins and avoiding opsonization. Two ways have been used to introduce hydrophilic polymers at the surface of nanoparticles: either adsorption of surfactants or the use of branched copolymers [6, 127–130]. The second strategy, consisting of the covalent linkage of amphiphilic copolymers, is generally preferred for obtaining a protective hydrophilic cloud on nanoparticles, as it avoids the possibility of rapid coating desorption upon dilution or after contact with blood components. This approach has been employed with poly(lactic acid) (PLA), poly(caprolactone) and poly(cyanoacrylate) polymers, which were chemically coupled to PEG [125, 131–134]. This type of DDS has been used to incorporate antiestrogens in the studies described here.

The "pure antiestrogen" RU 58 668 [135, 136], referred as to RU below, is highly hydrophobic and could be trapped in the lipidic skin layer, avoiding its access to targets localized in other tissues. Colloidal particles may be used as drug carriers to obtain a site-specific drug delivery. The vascular endothelium in solid tumors is discontinuous [10], and nanoparticles are able to cross this discontinuous endothelium in a process called extravasation. Such a strategy has been already employed in the case of Tam, but the drug release was too fast and its loading too low to obtain significant effects *in vivo*. In fact, poly(ε-caprolactone) and poly(MePEGcyanoacrylate-*co*-hexadecyl cyanoacrylate) nanoparticles failed to encapsulate Tam correctly, with a maximum loading efficiency of 64% and most of the drug being adsorbed onto the particle surface [137–139]. In contrast, the highly promising pure antiestrogen RU has been successfully incorporated with a high loading efficiency in polyester-PEG nanospheres [140]. These small nanoparticulate systems (~150–200 nm) showed reduced drug release *in vitro*, and reduced protein adsorption in the presence of serum, two features compatible with increased persistence in blood [141]. These parenteral delivery systems inhibited estradiol-promoted tumor growth *in vivo* that correlated with an arrest of cell cycle proliferation in MCF-7 cells [142]. Accumulation of these nanoparticles in tumor sites following i.v. administration has been demonstrated by the use of radioactive polymer [142].

The recent antiestrogen-containing formulations we have synthesized represent, to our knowledge, the first drug delivery systems loaded with an antiestrogen that have been successfully evaluated biologically in animal models. They are made of copolymers that are organized as nanospheres or nanocapsules (Fig. 11.3); in the latter case, the antiestrogen is solubilized in an oily core. Two types of copolymers were employed, affording nanoparticles with a small size that is compatible with good extravasation (Table 11.1). Tamoxifen-loaded PEG-poly(alkyl cyanoacrylate) (PEG-PACA) nanospheres contained only a small amount of drug (1.8 µg mL^{-1}) [137], whereas RU-loaded PEG-poly(lactic acid) (PEG-PLA) nanocapsules (NC) and nanospheres (NS) contained at least 3× more drug (5 and 33 µg mL^{-1}, respectively) [140, 142]. In addition, PEG-PLA nanoparticles are devoid of any intrinsic toxicity, which is not the case for the PACA-based copolymer [143, 144]. RU-charged PEG-PLA nanoparticles in MCF-7 cells show greater toxicity than free RU, and flow cytometry experiments have revealed a strong increase of MCF-7 cells in apoptosis after treatment with RU-loaded NS and NC. Interestingly, following i.v. administration to mice bearing MCF-7 breast cancer xenografts, nanospheres loaded with RU were highly efficient at inhibiting E_2-induced tumor growth at a dosage 50 to 100× lower than that at which free RU is active [142, 145] (Fig. 11.4). Comparatively, RU-loaded nanocapsules possessed a strong potency to reduce the E_2-induced tumor growth in this model as well as in xenografts of MCF-7/ras tumors [142]. They also prolonged the anti-uterotrophic activity of RU [142] at a low dose.

Tamoxifen-loaded PEG-PLA nanospheres also possess a high potency for inhibiting E_2-induced growth in MCF-7 tumors (Fig. 11.4). Analysis of the cell cycle proteins and proteins involved in apoptosis in tumor extracts from mice injected with nanospheres or nanocapsules containing RU indicated that both cyclin dependent

Tab. 11.1. Physicochemical characteristics and biological parameters of antiestrogen-loaded colloidal systems. The size and zeta potential of nanospheres nanocapsules and liposomes loaded with either RU or OH-Tam are indicated, together with the encapsulation rates of the drugs. Interestingly, the change in zeta potential for nanoparticles loaded with OH-Tam indicates that some drug is located at their surface, unlike RU-nanoparticles and liposomes.

	RU	4HT	Mean diameter (nm)	Zeta potential (mV)	Encapsulation efficiency	Encapsulation percentage (%)	Estrogen-induced inhibitory activity half-life (min)
Nanoparticles					μg AE per mg polymer		
PLA NS	–	–	200 ± 50	–28 ± 3	nd	nd	
PLA NS	+	–	170 ± 30	–3.1 ± 0.7	32.8	≥98	
PLA NS	–	+	172 ± 33	4.1 ± 0.6	19	98	
PEG-PLA NS	–	–	133 ± 48	–26.1 ± 0.9	nd	nd	
PEG-PLA NS	+	–	160 ± 30	–24.7 ± 0.6	32.8	96	1900
PEG-PLA NS	–	+	137 ± 47	1.2 ± 0.7	18.1	97.6	1800
PLA NC	–	–	245 ± 90	–50 ± 1.1	nd	nd	
PLA NC	+	–	233 ± 75	–5.2 ± 4	5.3	≥99	
PEG-PLA NC	–	–	233 ± 67	–44.4 ± 1.3	nd	nd	
PEG-PLA NC	+	–	245 ± 87	–42.2 ± 2.3	5.3	98	2300
Liposomes					mM AE × 100 mM lipids		
EPC/DSPE-PEG^{2000} (94:6)	–	–	110 ± 33	–23.5 ± 5.8	nd	nd	
EPC/DSPE-PEG^{2000} (94:6)	+	–	111 ± 31	–24.0 ± 5.8	3.609	≥90	90
EPC/CHOL/DSPE-PEG^{2000} (94:30:6)	–	–	106 ± 29.7	–25.6 ± 5.9	nd	nd	
EPC/CHOL/DSPE-PEG^{2000} (64:30:6)	+	–	94.7 ± 32.2	–26.5 ± 5.8	3.205	≥90	300

Half-lives were determined following a 24 h-treatment of MELN cells with E_2 (0.1 nM) in the presence or not of 0.5 nM RU; results represent the times at which 50% luciferase expression was inhibited. Empty liposomes had no effects [154].
nd: not defined.

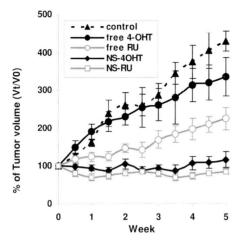

Fig. 11.4. Tumor evolution of MCF-7 breast cancer xenografts i.v. injected with RU- and OH-Tam-nanospheres. Bi-weekly i.v. injections of RU-PEG-PLA nanospheres (NS) and of 4-OH-Tam-PEG-PLA nanospheres (12 mg kg^{-1} week^{-1} each) were performed in mice ($n = 10$ per group) bearing MCF-7 tumors (ranging from 0.8 to 1 cm^3). Identical amounts of free antiestrogen in 5% glucose were i.v. injected as controls. All groups received 0.5 mg-E$_2$ kg^{-1} week^{-1} following skin deposition of an E$_2$ solution in ethanol. Tumors were measured each week.

kinase inhibitors p21$^{WAF-1/CIP1}$ and p27^{kip1} (CDKIs) were dramatically augmented (Ameller, T., Marsaud, V., Legrand, P., Renoir, J.-M., manuscript in preparation). This was not seen in tumor extracts from mice having received encapsulated 4-OHT. This finding supports the concept that delivery of RU incorporated in long-circulating formulations is able to arrest tumor growth by a mechanism in which both CDKIs are overexpressed [142]. This agrees with the need for an elevated level of p27^{kip1} for cells to remain quiescent, an effect produced by pure antiestrogens in human breast cancer cells [146] but not by SERMs such as 4-OHT. Additionally, immunological analysis of MCF-7 cells tumors exposed to both nanospheres and nanocapsules containing RU revealed a strong destruction of the ERα content, contrary to tumors exposed to nanospheres loaded with OH-Tamoxifen. In the former, TUNEL experiments indicated a strong apoptosis while a weaker one was detected in the latter. These data correlate with the proteasome-mediated degradation of ERα under RU, concomitantly with apoptosis induction.

Finally, s.c. administration of trapped RU in mice bearing MCF-7 tumors was as efficient as the i.v. route at the same dose of RU (Fig. 11.5). This finding led to the hypothesis that nanospheres could behave like a reservoir from which the drug could diffuse and release slowly in the interstitium before reaching the tumor. In that case it is intriguing that the efficiency was similar to that obtained by i.v. More research is necessary to elucidate this mechanism, in particular to find out how the administration of an antiestrogen through a stealth DDS affects its pharmacokinetics and biodistribution as well as its bioavailability.

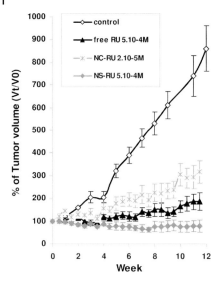

Fig. 11.5. Tumor evolution of MCF-7 breast cancer xenografts s.c. injected with RU-nanospheres and RU-nanocapsules. Bi-weekly subcutaneous injection of PEG-PLA-RU loaded nanocapsules and of PEG-PLA-RU-loaded nanospheres [containing, respectively, 2×10^{-5} M (0.3 mg kg^{-1} week^{-1}) and 5×10^{-4} M RU (4.3 mg kg^{-1} week^{-1})] was performed in nude mice bearing MCF-7 tumors ($n = 8$ mice per group), in addition to 0.5 mg kg^{-1} week^{-1} by skin deposition. Free RU at the highest concentration was s.c. injected at 4.3 mg kg^{-1} week^{-1}. Control group received only E_2.

11.4.2
Liposomes Charged with RU 58668 in MM

Several liposomal formulations have been elaborated for the delivery of highly cytotoxic anticancer drugs such as doxorubicin and platinum-based agents [4, 147–151]. These formulations utilized liposomes capable of circulating with very long half-lives (referred to as sterically stabilized liposomes) [152, 153] that avoid the opsonization process when injected in the blood stream. Similarly to nanoparticles, opsonization is reduced by coupling phospholipids to hydrophilic polymers such as PEG. PEG-liposomes can be injected either intravenously or subcutaneously.

From *in vitro* experiments, we have established that high antiestrogen amounts are able to trigger cell cycle arrest in MM cells as well as to induce apoptosis [121, 154]. We then incorporated RU in liposomes known to have a better incorporation capacity than nanoparticles. Table 11.1 summarizes the physicochemical characteristics of the different formulations obtained and their biological properties in terms of drug loading and *in vitro* release. At least 10× more RU was incorporated into liposomes of EPC/DSPE-PEG^{2000} than in nanospheres of PEG-PLA. Coupling of PEG did not modify any of the characteristics of liposomes but addition of cholesterol enhanced vesicle stability of the liposomes and strongly decreased the RU release measured *in vitro*. RU release from liposomes was comparatively faster than that of RU from nanoparticles (Table 11.1). Flow cytometry experiments carried out on various MM cell lines (LP1, NCI-H929, RPMI8226, OPM-2 and U266) exposed to increasing amounts of RU revealed that in all MM cells RU blocks the cell cycle progression in the G0/G1 phase and/or induced apoptosis (except in OPM-2). This occurred at up to 5 µM RU, a concentration far above (100×) that at which the antiestrogen induced apoptosis in breast cancer ER+ cancer cells. Interestingly, the extent of apoptosis in cells exposed to RU-loaded liposomes is strongly augmented in all MM cells (Fig. 11.6).

Intravenously injected in a RPMI8226 MM xenograft model, RU-charged liposomes inhibited strongly the tumor growth even when injected at 12 mg kg^{-1} week^{-1} in a single injection (Fig. 11.7). Conversely, free RU at this dose had no effect, as well as empty liposomes. This suggests that the release of RU is slow and that its potency in inhibiting the growth of the MM tumor is enhanced by incorporation in the liposome system. Another interesting observation made in RPMI8226 xenografts was the enhanced anti-angiogenesis activity following RU-charged liposomes (Maillard, S., Gauduchon, J., Gouilleux, F. Marsaud, V., Connault, E., Opolon, P., Fattal, E., Sola, B., Renoir, J.-M., submitted for publication), which is consistent with the inhibition of VEGF by RU in the MM cells (our unpublished work) as well as in human breast cancer cells [155, 156], an effect mediated by ERα contained in the endothelium membranes. In addition, in all of the MM and breast cancer xenograft models we used, no apparent toxicity (liver, lung, bone) was noticed. Interestingly, the size of the xenograft uteri injected with RU-PEG-PLA nanospheres was not decreased, contrary to that of healthy mature mice having received RU at 12 mg kg^{-1} week^{-1} for at least 5 weeks (RPMI8226 xenografts) and 4.3 mg kg^{-1} week^{-1} (MCF-7 xenografts) [142, 145, 154]. This strongly suggest that, although being passive, some tumor targeting is produced when stealth DDS are employed.

11.4.3
Tumor-targeted Drug-loaded Colloidal Systems

Targeting of drugs to specific tissues of the body has been the major focus of research in recent decades in an attempt to improve selectivity in cancer treatment. The PEG stabilizing effect results from local surface concentration of highly hydrated groups that sterically inhibit both hydrophobic and electrostatic interactions of various blood components at the carrier surface [152, 153, 157]. "Active targeting" of either nanoparticles or liposomes is generally attempted by conjugating ligands, to the carrier surface, that possess high affinity for the tumor cells [158]. Several types of ligands have been used, including antibody fragments [159–161], vitamins [162–164], glycoproteins [165], peptides (RGD-sequences) [166], and oligonucleotides aptamers [158]. The concept is actually over 25 years old, but only small improvements have been obtained. In fact many difficulties are encountered. The major one, besides the complexity of such conjugation chemistry, is the dilemma of preserving the stealth property of the formulation, by leaving enough

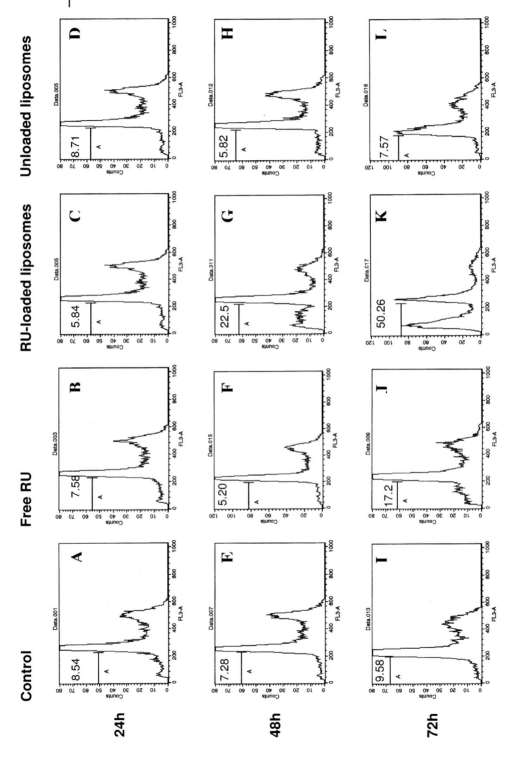

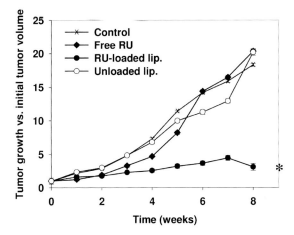

Fig. 11.7. Tumor evolution of multiple myeloma xenografts i.v. injected with RU-liposomes. RPMI8226 xenografted nude mice received, or not, a weekly injection of free RU or RU-loaded liposomes (12 mg per kg per week of RU) or unloaded liposomes at the same lipidic concentration as that containing RU. Tumor volume was measured and tumor evolution plotted as a function of the time of injection. *$p < 0.01$ with regard to control.

free PEG chains grafted at the surface, while also obtaining enough tumor grafted ligand molecules capable of binding to the tumor cell receptors (Fig. 11.3).

In some cases, "active targeting" may take advantage of the overexpression of a membrane receptor. This is, in particular, the case with breast cancer cells that are insensitive to antiestrogens because they are ER-negative. In that case, a tyrosine kinase membrane receptor of the EGFR family, HER2-Neu or Erb-B2, is overexpressed [24]. Overexpression of Erb-B2 remains a major risk factor in non-metastatic breast cancers treated with high-dose alkylating agents and autologous stem cell transplantation. The activity of Erb-B2 is blocked by a specific antibody, herceptin (trastuzumab), which enhances tumor necrosis factor-related apoptosis-inducing ligand-mediated apoptosis in breast and ovarian cancer cell lines [167]. In MM cells, Erb-B2 is also overexpressed, representing a major risk [168]. Then, coupling trastuzumab on PEG chains of PEGylated nanoparticles and liposomes loaded with an antiestrogen could be of great benefit for breast cancers and MM since both the "piloting" grafted recognition molecule and the encapsulating drug

Fig. 11.6. Flow cytometry: NCI-H929 MM cells (50% confluence) were grown and exposed or not to RU (1 µM) free or encapsulated, or to empty liposomes (at the same lipid concentration as that used in RU-loaded liposomes) for various periods. Cells (10^4) were then analyzed by FACS. Cell cycle profiles and the percentage of cells in the sub-G1 fraction obtained without treatment (control) are indicated in panels A, E, I for 24, 48 and 72 h, respectively. Panels B, F, J show the same parameters obtained after treatment with free RU. Similarly, panels C, G, K and D, H, L show the profiles obtained for cells treated with RU-loaded liposomes and unloaded liposomes, respectively.

can conduct tumor cells to apoptosis. Several sterically stabilized anti-Erb-B2 immunoliposomes have been synthesized to date [161, 169–173], all of which seem promising.

Another interesting and simple molecule to couple at the end of PEG chains is folic acid [174, 175]. Several malignant cells, including KB3-1 cells, endometrial Ishikawa cells and also some MM cells (but, unfortunately, not breast cancer cells), overexpress a folate receptor [176] that mediates internalization. Folate linked to PEGylated cyanoacrylate nanoparticles show a ten-fold higher apparent affinity for the folate membrane receptor than free folate does. The particles represent a multivalent form of free folic acid and the folate receptor(s) is (are) often disposed in clusters [175]. As a result, conjugated nanoparticles could display a multivalent and hence stronger interaction with the surface of the malignant cells [177]. Moreover, confocal microscopy has demonstrated that folate nanoparticles, compared with non-conjugated nanoparticles, were selectively taken up by the folate receptor-bearing KB3-1 cells, but not by MCF-7 cells. In the former case, the folate nanoparticles were located in the cell cytoplasm, as a consequence of receptor-mediated endocytosis [176]. Very recently, the improved therapeutic response of folate-nanoparticles charged with methotrexate demonstrated a marked decrease in toxicity and increase in antitumor capacity [178].

11.5
Conclusions and Perspectives

The potential use of pure antiestrogens in anticancer therapy leads again to the question of how they must be administered to avoid side effects. Altogether, the data obtained with AE-loaded colloidal systems demonstrate that drug delivery of this type of anticancer drug either i.v. or s.c. administered enhances the apoptotic activity of the drug, decreases the side effects and behaves as long-lasting delivery systems, although probably acting through different process. For example, while it is likely that PEGylated nanoparticles concentrate at tumor sites via a passive targeting, liposomes are believed to deliver the encapsulated RU through an endocytosis process (Fig. 11.8). Several novel drug delivery systems for steroid hormones have been developed, including intra-uterine delivery systems, implants and steroid-loaded vaginal rings [145]. However, what is needed for the delivery of anticancer drugs in general, and for antiestrogens in particular, is a specific targeting of tumor cells with maximum avoidance of normal cells. The challenge is not to reach the pathological cells with a high dose of drug but rather to incorporate at these sites the optimal drug concentration necessary to completely inhibit the activity of the target. If this condition is attained, little of the drug will be disseminated at other sites of the organism, thus minimizing undesirable side effects.

In fact, another way to inhibit ERα activity has emerged recently from Gustafsson's group, who have described ERβ as a potent dominant negative variant of ERα [179, 180]. This has been confirmed by other groups [181, 182]. Thus, delivery of the cDNA encoding ERβ or of silencing RNA directed against ERα concomi-

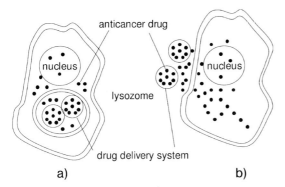

Fig. 11.8. Antiestrogen-release mechanisms from colloidal systems. The antiestrogen (●) is located inside the lipid bilayer of liposomes (A), which are internalized in lysosomes in cells capable of endocytosis. The drug is then released in the cell after degradation of the lysosomes. In nanospheres and in nanocapsules (B) the antiestrogen is trapped in the polymer matrix and/or in the oily core; they also can be submitted to endocytosis as in (A). The encapsulated drug can be released near the tumor cells, allowing its passive diffusion. In both cases, an increased intracellular antiestrogen concentration occurs.

tantly with RU could be a helpful approach since it can reduce, on the one hand, the activity of ERα and, on the other hand, destroys the ERα protein itself, preserving ERβ.

The results obtained are encouraging and we strongly believe that the drug delivery approach is promising not only for the administration of antiestrogens in estrogen-dependent breast cancers and MM but also for the delivery of much more toxic anticancer agents such as taxol, thalidomide, bortezomib, VEGF inhibitors, farnesyltransferase inhibitors, histone transferase inhibitors and hsp90 inhibitors.

Acknowledgments

We greatly appreciate the support of the Ligue Nationale contre le Cancer (Cher, Indre and Manche departments Comities), P. Van de Velde for the gift of RU 58 668 and the SCEA of IGR (Villejuif, France) for animal experiments.

References

1 JAIN, R. K., Delivery of molecular and cellular medicine to solid tumors, *Adv. Drug Deliv. Rev.*, **1997**, 26, 71–90.
2 MOGHIMI, S. M., SZEBENI, J., Stealth liposomes and long circulating nanoparticles, critical issues in pharmacokinetics, opsonization and protein-binding properties, *Prog. Lipid Res.*, **2003**, 42, 463–478.
3 DRUMMOND, D. C., MEYER, O., HONG, K., KIRPOTIN, D. B., PAPAHADJOPOULOS, D., Optimizing

liposomes for delivery of chemotherapeutic agents to solid tumors, *Pharmacol Rev.*, **1999**, 51, 691–743.

4 GABIZON, A., PAPAHADJOPOULOS, D., Liposome formulations with prolonged circulation time in blood and enhanced uptake by tumors, *Proc. Natl. Acad. Sci. U.S.A.*, **1988**, 85, 6949–6953.

5 ALLEN, T. M., HANSEN, C., MARTIN, F., REDEMANN, C., YAU-YOUNG, A., Liposomes containing synthetic lipid derivatives of poly(ethylene glycol) show prolonged circulation half-lives in vivo, *Biochim. Biophys. Acta*, **1991**, 1066, 29–36.

6 STORM, G., BELLIOT, S. O., DAEMEN, T., LASIC, D. D., Surface modification of nanoparticles to oppose uptake by the mononuclear phagocyte system, *Adv. Drug Deliv. Rev.*, **1995**, 17, 31–48.

7 SEYMOUR, L. W., Passive tumor targeting of soluble macromolecules and drug conjugates, *Crit. Rev. Ther. Drug Carrier Syst.*, **1992**, 9, 135–187.

8 BABAN, D. F., SEYMOUR, L. W., Control of tumour vascular permeability, *Adv. Drug Deliv. Rev.*, **1998**, 34, 109–119.

9 MAEDA, H., The enhanced permeability and retention (EPR) effect in tumor vasculature, the key role of tumor-selective macromolecular drug targeting, *Adv. Enzyme Regul.*, **2001**, 41, 189–207.

10 HOBBS, S. K., MONSKY, W. L., YUAN, F., ROBERTS, W. G., GRIFFITH, L., TORCHILIN, V. P., JAIN, R. K., Regulation of transport pathways in tumor vessels, role of tumor type and microenvironment, *Proc. Natl. Acad. Sci. U.S.A.*, **1998**, 95, 4607–4612.

11 YUAN, F., DELLIAN, M., FUKUMURA, D., LEUNIG, M., BERK, D. A., TORCHILIN, V. P., JAIN, R. K., Vascular permeability in a human tumor xenograft, molecular size dependence and cutoff size, *Cancer Res.*, **1995**, 55, 3752–3956.

12 YUAN, F., LEUNIG, M., HUANG, S. K., BERK, D. A., PAPAHADJOPOULOS, D., JAIN, R. K., Microvascular permeability and interstitial penetration of sterically stabilized (stealth) liposomes in a human tumor xenograft, *Cancer Res.*, **1994**, 54, 3352–3356.

13 MONSKY, W. L., FUKUMURA, D., GOHONGI, T., ANCUKIEWCZ, M., WEICH, H. A., TORCHILIN, V. P., YUAN, F., JAIN, R. K., Augmentation of transvascular transport of macromolecules and nanoparticles in tumors using vascular endothelial growth factor, *Cancer Res.*, **1999**, 59, 4129–4135.

14 WEISZ, A., BRESCIANI, F., Estrogen regulation of proto-oncogenes coding for nuclear proteins, *Crit. Rev. Oncog.*, **1993**, 4, 361–388.

15 JENSEN, E. V., On the mechanism of estrogen action, *Perspect. Biol. Med.*, **1962**, 6, 47–59.

16 TOFT, D., GORSKI, J., A receptor molecule for estrogens, isolation from the rat uterus and preliminary characterization, *Proc. Natl. Acad. Sci. U.S.A.*, **1966**, 55, 1574–1581.

17 KUIPER, G. G., ENMARK, E., PELTO-HUIKKO, M., NILSSON, S., GUSTAFSSON, J. A., Cloning of a novel receptor expressed in rat prostate and ovary, *Proc. Natl. Acad. Sci. U.S.A.*, **1996**, 93, 5925–5930.

18 MOSSELMAN, S., POLMAN, J., DIJKEMA, R., ER beta, identification and characterization of a novel human estrogen receptor, *FEBS Lett.*, **1996**, 392, 49–53.

19 KATZENELLENBOGEN, B. S., KORACH, K. S., A new actor in the estrogen receptor drama – enter ER-beta, *Endocrinology*, **1997**, 138, 861–862.

20 KOUSTENI, S., BELLIDO, T., PLOTKIN, L. I., O'BRIEN, C. A., BODENNER, D. L., HAN, L., HAN, K., DIGREGORIO, G. B., KATZENELLENBOGEN, J. A., KATZENELLENBOGEN, B. S., ROBERSON, P. K., WEINSTEIN, R. S., JILKA, R. L., MANOLAGAS, S. C., Nongenotropic, sex-nonspecific signaling through the estrogen or androgen receptors, dissociation from transcriptional activity, *Cell*, **2001**, 104, 719–730.

21 MARINO, M., ACCONCIA, F., BRESCIANI, F., WEISZ, A., TRENTALANCE, A., Distinct nongenomic signal transduction pathways controlled by 17beta-

estradiol regulate DNA synthesis and cyclin D(1) gene transcription in HepG2 cells, *Mol. Biol. Cell.*, **2002**, 13, 3720–3729.

22 MacGregor, J. I., Jordan, V. C., Basic guide to the mechanisms of antiestrogen action, *Pharmacol Rev.*, **1998**, 50, 151–196.

23 Clarke, R., Leonessa, F., Welch, J. N., Skaar, T. C., Cellular and molecular pharmacology of antiestrogen action and resistance, *Pharmacol Rev.*, **2001**, 53, 25–71.

24 Jordan, V. C., Antiestrogens and selective estrogen receptor modulators as multifunctional medicines. 2. Clinical considerations and new agents, *J. Med. Chem.*, **2003**, 46, 1081–1111.

25 Wijayaratne, A. L., McDonnell, D. P., The human estrogen receptor-alpha is a ubiquitinated protein whose stability is affected differentially by agonists, antagonists, and selective estrogen receptor modulators, *J. Biol. Chem.*, **2001**, 276, 35 684–35 692.

26 El Khissiin, A., Leclercq, G., Implication of proteasome in estrogen receptor degradation, *FEBS Lett.*, **1999**, 448, 160–166.

27 Marsaud, V., Gougelet, A., Maillard, S., Renoir, J. M., Various phosphorylation pathways, depending on agonist and antagonist binding to endogenous estrogen receptor alpha (ERalpha), differentially affect ERalpha extractability, proteasome-mediated stability, and transcriptional activity in human breast cancer cells, *Mol. Endocrinol.*, **2003**, 17, 2013–2027.

28 Webb, P., Lopez, G. N., Uht, R. M., Kushner, P. J., Tamoxifen activation of the estrogen receptor/AP-1 pathway, potential origin for the cell-specific estrogen-like effects of antiestrogens, *Mol. Endocrinol.*, **1995**, 9, 443–456.

29 Paech, K., Webb, P., Kuiper, G. G., Nilsson, S., Gustafsson, J., Kushner, P. J., Scanlan, T. S., Differential ligand activation of estrogen receptors ERalpha and ERbeta at AP1 sites, *Science*, **1997**, 277, 1508–1510.

30 Kushner, P. J., Agard, D. A., Greene, G. L., Scanlan, T. S., Shiau, A. K., Uht, R. M., Webb, P., Estrogen receptor pathways to AP-1, *J. Steroid Biochem. Mol. Biol.*, **2000**, 74, 311–317.

31 Hallek, M., Bergsagel, P. L., Anderson, K. C., Multiple myeloma, increasing evidence for a multistep transformation process, *Blood*, **1998**, 91, 3–21.

32 Hideshima, T., Anderson, K. C., Molecular mechanisms of novel therapeutic approaches for multiple myeloma, *Nat. Rev. Cancer*, **2002**, 2, 927–937.

33 Danel, L., Vincent, C., Rousset, F., Klein, B., Bataille, R., Flacher, M., Durie, B. G., Revillard, J. P., Estrogen and progesterone receptors in some human myeloma cell lines and murine hybridomas, *J. Steroid Biochem.*, **1988**, 30, 363–367.

34 Treon, S. P., Teoh, G., Urashima, M., Ogata, A., Chauhan, D., Webb, I. J., Anderson, K. C., Anti-estrogens induce apoptosis of multiple myeloma cells, *Blood*, **1998**, 92, 1749–1757.

35 Otsuki, T., Yamada, O., Kurebayashi, J., Moriya, T., Sakaguchi, H., Kunisue, H., Yata, K., Uno, M., Yawata, Y., Ueki, A., Estrogen receptors in human myeloma cells, *Cancer Res.*, **2001**, 60, 1434–1441.

36 Gauduchon, J., Gouilleux, F., Maillard, S., Marsaud, V., Renoir, M. J., Sola, B., The selective estrogen receptor modulator 4-hydroxy tamoxifen induces G1 arrest and apoptosis of multiple myeloma cell lines, *Ann. New York Acad. Sci.*, **2003**, 1010, 321–325.

37 Brigger, I., Dubernet, C., Couvreur, P., Nanoparticles in cancer therapy and diagnosis, *Adv. Drug Deliv. Rev.*, **2002**, 54, 631–651.

38 Brannon-Peppas, L., Blanchette, J. O., Nanoparticle and targeted systems for cancer therapy, *Adv. Drug Deliv. Rev.*, **2004**, 56, 1649–1659.

39 Smith, I. E., Dowsett, M., Aromatase inhibitors in breast cancer, *N. Engl. J. Med.*, **2003**, 348, 2431–2442.

40 Beato, M., Herrlich, P., Schutz, G., Steroid hormone receptors, many

actors in search of a plot, *Cell*, **1995**, 83, 851–857.

41 MCKENNA, N. J., O'MALLEY, B. W., Combinatorial control of gene expression by nuclear receptors and coregulators, *Cell*, **2002**, 108, 465–474.

42 TSAI, M. J., O'MALLEY, B. W., Molecular mechanisms of action of steroid/thyroid receptor superfamily members, *Annu. Rev. Biochem.*, **1994**, 63, 451–486.

43 NILSSON, S., MAKELA, S., TREUTER, E., TUJAGUE, M., THOMSEN, J., ANDERSSON, G., ENMARK, E., PETTERSSON, K., WARNER, M., GUSTAFSSON, J. A., Mechanisms of estrogen action, *Physiol. Rev.*, **2001**, 81, 1535–1565.

44 PRATT, W. B., TOFT, D. O., Steroid receptor interactions with heat shock protein and immunophilin chaperones, *Endocrinol. Rev.*, **1997**, 18, 306–360.

45 JOAB, I., RADANYI, C., RENOIR, M., BUCHOU, T., CATELLI, M. G., BINART, N., MESTER, J., BAULIEU, E. E., Common non-hormone binding component in non-transformed chick oviduct receptors of four steroid hormones, *Nature*, **1984**, 308, 850–853.

46 CATELLI, M. G., BINART, N., JUNG-TESTAS, I., RENOIR, J. M., BAULIEU, E. E., FERAMISCO, J. R., WELCH, W. J., The common 90-kd protein component of non-transformed '8S' steroid receptors is a heat-shock protein, *EMBO J.*, **1985**, 4, 3131–3135.

47 PRATT, W. B., The role of the hsp90-based chaperone system in signal transduction by nuclear receptors and receptors signaling via MAP kinase, *Annu. Rev. Pharmacol. Toxicol.*, **1997**, 37, 297–326.

48 JOHNSON, J. L., TOFT, D. O., A novel chaperone complex for steroid receptors involving heat shock proteins, immunophilins, and p23, *J. Biol. Chem.*, **1994**, 269, 24 989–24 993.

49 JOHNSON, J. L., BEITO, T. G., KRCO, C. J., TOFT, D. O., Characterization of a novel 23-kilodalton protein of unactive progesterone receptor complexes, *Mol. Cell Biol.*, **1994**, 14, 1956–1963.

50 GRENERT, J. P., SULLIVAN, W. P., FADDEN, P., HAYSTEAD, T. A., CLARK, J., MIMNAUGH, E., KRUTZSCH, H., OCHEL, H. J., SCHULTE, T. W., SAUSVILLE, E., NECKERS, L. M., TOFT, D. O., The amino-terminal domain of heat shock protein 90 (hsp90) that binds geldanamycin is an ATP/ADP switch domain that regulates hsp90 conformation, *J. Biol. Chem.*, **1997**, 272, 23 843–23 850.

51 FREEMAN, B. C., FELTS, S. J., TOFT, D. O., YAMAMOTO, K. R., The p23 molecular chaperones act at a late step in intracellular receptor action to differentially affect ligand efficacies, *Genes Dev.*, **2000**, 14, 422–434.

52 TANIOKA, T., NAKATANI, Y., SEMMYO, N., MURAKAMI, M., KUDO, I., Molecular identification of cytosolic prostaglandin E2 synthase that is functionally coupled with cyclooxygenase-1 in immediate prostaglandin E2 biosynthesis, *J. Biol. Chem.*, **2000**, 275, 32 775–32 782.

53 MOLLERUP, J., KROGH, T. N., NIELSEN, P. F., BERCHTOLD, M. W., Properties of the co-chaperone protein p23 erroneously attributed to ALG-2 (apoptosis-linked gene 2), *FEBS Lett.*, **2003**, 555, 478–482.

54 WEIGEL, N. L., Steroid hormone receptors and their regulation by phosphorylation, *Biochem. J.*, **1996**, 319(Pt 3), 657–667.

55 STENOIEN, D. L., PATEL, K., MANCINI, M. G., DUTERTRE, M., SMITH, C. L., O'MALLEY, B. W., MANCINI, M. A., FRAP reveals that mobility of oestrogen receptor-alpha is ligand- and proteasome-dependent, *Nat. Cell. Biol.*, **2001**, 3, 15–23.

56 KLINGE, C. M., Estrogen receptor interaction with co-activators and co-repressors, *Steroids*, **2000**, 65, 227–251.

57 ROBYR, D., WOLFFE, A. P., WAHLI, W., Nuclear hormone receptor coregulators in action, diversity for shared tasks, *Mol. Endocrinol.*, **2000**, 14, 329–347.

58 SPENCER, T. E., JENSTER, G., BURCIN, M. M., ALLIS, C. D., ZHOU, J.,

Mizzen, C. A., McKenna, N. J., Onate, S. A., Tsai, S. Y., Tsai, M. J., O'Malley, B. W., Steroid receptor coactivator-1 is a histone acetyltransferase, *Nature*, **1997**, 389, 194–198.

59 Metivier, R., Penot, G., Hubner, M. R., Reid, G., Brand, H., Kos, M., Gannon, F., Estrogen receptor-alpha directs ordered, cyclical, and combinatorial recruitment of cofactors on a natural target promoter, *Cell*, **2003**, 115, 751–763.

60 Reid, G., Hubner, M. R., Metivier, R., Brand, H., Denger, S., Manu, D., Beaudouin, J., Ellenberg, J., Gannon, F., Cyclic, proteasome-mediated turnover of unliganded and liganded ERalpha on responsive promoters is an integral feature of estrogen signaling, *Mol. Cell.*, **2003**, 11, 695–707.

61 Lonard, D. M., Nawaz, Z., Smith, C. L., O'Malley, B. W., The 26S proteasome is required for estrogen receptor-alpha and coactivator turnover and for efficient estrogen receptor-alpha transactivation, *Mol. Cell.*, **2000**, 5, 939–948.

62 Montano, M. M., Ekena, K., Delage-Mourroux, R., Chang, W., Martini, P., Katzenellenbogen, B. S., An estrogen receptor-selective coregulator that potentiates the effectiveness of antiestrogens and represses the activity of estrogens, *Proc. Natl. Acad. Sci. U.S.A.*, **1999**, 96, 6947–6952.

63 Chen, H., Lin, R. J., Schiltz, R. L., Chakravarti, D., Nash, A., Nagy, L., Privalsky, M. L., Nakatani, Y., Evans, R. M., Nuclear receptor coactivator ACTR is a novel histone acetyltransferase and forms a multimeric activation complex with P/CAF and CBP/p300, *Cell*, **1997**, 90, 569–580.

64 Porter, W., Saville, B., Hoivik, D., Safe, S., Functional synergy between the transcription factor Sp1 and the estrogen receptor, *Mol. Endocrinol.*, **1997**, 11, 1569–1580.

65 Saville, B., Wormke, M., Wang, F., Nguyen, T., Enmark, E., Kuiper, G., Gustafsson, J. A., Safe, S., Ligand-, cell-, and estrogen receptor subtype (alpha/beta)-dependent activation at GC-rich (Sp1) promoter elements, *J. Biol. Chem.*, **2000**, 275, 5379–5387.

66 Lee, H., Jiang, F., Wang, Q., Nicosia, S. V., Yang, J., Su, B., Bai, W., MEKK1 activation of human estrogen receptor alpha and stimulation of the agonistic activity of 4-hydroxytamoxifen in endometrial and ovarian cancer cells, *Mol. Endocrinol.*, **2000**, 14, 1882–1896.

67 Hall, J. M., Couse, J. F., Korach, K. S., The multifaceted mechanisms of estradiol and estrogen receptor signaling, *J. Biol. Chem.*, **2001**, 276, 36 869–36 872.

68 Campbell, R. A., Bhat-Nakshatri, P., Patel, N. M., Constantinidou, D., Ali, S., Nakshatri, H., Phosphatidylinositol 3-kinase/AKT-mediated activation of estrogen receptor alpha, a new model for antiestrogen resistance, *J. Biol. Chem.*, **2001**, 276, 9817–9824.

69 Aronica, S. M., Kraus, W. L., Katzenellenbogen, B. S., Estrogen action via the cAMP signaling pathway, stimulation of adenylate cyclase and cAMP-regulated gene transcription, *Proc. Natl. Acad. Sci. U.S.A.*, **1994**, 91, 8517–8521.

70 Bunone, G., Briand, P. A., Miksicek, R. J., Picard, D., Activation of the unliganded estrogen receptor by EGF involves the MAP kinase pathway and direct phosphorylation, *EMBO J.*, **1996**, 15, 2174–2183.

71 Joel, P. B., Traish, A. M., Lannigan, D. A., Estradiol-induced phosphorylation of serine 118 in the estrogen receptor is independent of p42/p44 mitogen-activated protein kinase, *J. Biol. Chem.*, **1998**, 273, 13 317–13 323.

72 Wyckoff, M. H., Chambliss, K. L., Mineo, C., Yuhanna, I. S., Mendelsohn, M. E., Mumby, S. M., Shaul, P. W., Plasma membrane estrogen receptors are coupled to endothelial nitric-oxide synthase through Galpha(i), *J. Biol. Chem.*, **2001**, 276, 27 071–27 076.

73 Hisamoto, K., Ohmichi, M., Kanda, Y., Adachi, K., Nishio, Y., Hayakawa, J., Mabuchi, S.,

74 SIMONCINI, T., HAFEZI-MOGHADAM, A., BRAZIL, D. P., LEY, K., CHIN, W. W., LIAO, J. K., Interaction of oestrogen receptor with the regulatory subunit of phosphatidylinositol-3-OH kinase, *Nature*, **2000**, 407, 538–541.

TAKAHASHI, K., TASAKA, K., MIYAMOTO, Y., TANIGUCHI, N., MURATA, Y., Induction of endothelial nitric-oxide synthase phosphorylation by the raloxifene analog LY117018 is differentially mediated by Akt and extracellular signal-regulated protein kinase in vascular endothelial cells, *J. Biol. Chem.*, **2001**, 276, 47 642–47 649.

75 BOYAN, B. D., SYLVIA, V. L., FRAMBACH, T., LOHMANN, C. H., DIETL, J., DEAN, D. D., SCHWARTZ, Z., Estrogen-dependent rapid activation of protein kinase C in estrogen receptor-positive MCF-7 breast cancer cells and estrogen receptor-negative HCC38 cells is membrane-mediated and inhibited by tamoxifen, *Endocrinology*, **2003**, 144, 1812–1824.

76 MIGLIACCIO, A., PICCOLO, D., CASTORIA, G., DI DOMENICO, M., BILANCIO, A., LOMBARDI, M., GONG, W., BEATO, M., AURICCHIO, F., Activation of the Src/p21ras/Erk pathway by progesterone receptor via cross-talk with estrogen receptor, *EMBO J.*, **1998**, 17, 2008–2018.

77 IMPROTA-BREARS, T., WHORTON, A. R., CODAZZI, F., YORK, J. D., MEYER, T., MCDONNELL, D. P., Estrogen-induced activation of mitogen-activated protein kinase requires mobilization of intracellular calcium, *Proc. Natl. Acad. Sci. U.S.A.*, **1999**, 96, 4686–4691.

78 RAZANDI, M., PEDRAM, A., GREENE, G. L., LEVIN, E. R., Cell membrane and nuclear estrogen receptors (ERs) originate from a single transcript, studies of ERalpha and ERbeta expressed in Chinese hamster ovary cells, *Mol. Endocrinol.*, **1999**, 13, 307–319.

79 CASTORIA, G., MIGLIACCIO, A., BILANCIO, A., DI DOMENICO, M., DE FALCO, A., LOMBARDI, M., FIORENTINO, R., VARRICCHIO, L., BARONE, M. V., AURICCHIO, F., PI3-kinase in concert with Src promotes the S-phase entry of oestradiol-stimulated MCF-7 cells, *EMBO J.*, **2001**, 20, 6050–6059.

80 CARISTI, S., GALERA, J. L., MATARESE, F., IMAI, M., CAPORALI, S., CANCEMI, M., ALTUCCI, L., CICATIELLO, L., TETI, D., BRESCIANI, F., WEISZ, A., Estrogens do not modify MAP kinase-dependent nuclear signaling during stimulation of early G(1) progression in human breast cancer cells, *Cancer Res.*, **2001**, 61, 6360–6366.

81 LOBENHOFER, E. K., MARKS, J. R., Estrogen-induced mitogenesis of MCF-7 cells does not require the induction of mitogen-activated protein kinase activity, *J. Steroid. Biochem. Mol. Biol.*, **2000**, 75, 11–20.

82 LOBENHOFER, E. K., HUPER, G., IGLEHART, J. D., MARKS, J. R., Inhibition of mitogen-activated protein kinase and phosphatidylinositol 3-kinase activity in MCF-7 cells prevents estrogen-induced mitogenesis, *Cell Growth Differ.*, **2000**, 11, 99–110.

83 GABEN, A. M., SAUCIER, C., BEDIN, M., REDEUILH, G., MESTER, J., Mitogenic activity of estrogens in human breast cancer cells does not rely on direct induction of mitogen-activated protein kinase/extracellularly regulated kinase or phosphatidylinositol 3-kinase, *Mol. Endocrinol.*, **2004**, 18, 2700–2713.

84 FILARDO, E. J., QUINN, J. A., BLAND, K. I., FRACKELTON, A. R., JR., Estrogen-induced activation of Erk-1 and Erk-2 requires the G protein-coupled receptor homolog, GPR30, and occurs via trans-activation of the epidermal growth factor receptor through release of HB-EGF, *Mol. Endocrinol.*, **2000**, 14, 1649–1660.

85 FILARDO, E. J., QUINN, J. A., FRACKELTON, A. R., JR., BLAND, K. I., Estrogen action via the G protein-coupled receptor, GPR30, stimulation of adenylyl cyclase and cAMP-mediated attenuation of the epidermal growth factor receptor-to-MAPK signaling axis, *Mol. Endocrinol.*, **2002**, 16, 70–84.

86 RAZANDI, M., ALTON, G., PEDRAM, A.,

GHONSHANI, S., WEBB, P., LEVIN, E. R., Identification of a structural determinant necessary for the localization and function of estrogen receptor alpha at the plasma membrane, *Mol. Cell Biol.*, 2003, 23, 1633–1646.

87 REVANKAR, C. M., CIMINO, D. F., SKLAR, L. A., ARTERBURN, J. B., PROSSNITZ, E. R., A transmembrane intracellular estrogen receptor mediates rapid cell signaling, *Science*, 2005, 307, 1625–1630.

88 GOTTARDIS, M. M., ROBINSON, S. P., SATYASWAROOP, P. G., JORDAN, V. C., Contrasting actions of tamoxifen on endometrial and breast tumor growth in the athymic mouse, *Cancer Res.*, 1988, 48, 812–815.

89 NAWAZ, Z., LONARD, D. M., DENNIS, A. P., SMITH, C. L., O'MALLEY, B. W., Proteasome-dependent degradation of the human estrogen receptor, *Proc. Natl. Acad. Sci. U.S.A.*, 1999, 96, 1858–1862.

90 ALARID, E. T., BAKOPOULOS, N., SOLODIN, N., Proteasome-mediated proteolysis of estrogen receptor, a novel component in autologous down-regulation, *Mol. Endocrinol.*, 1999, 13, 1522–1534.

91 GIAMARCHI, C., CHAILLEUX, C., CALLIGE, M., ROCHAIX, P., TROUCHE, D., RICHARD-FOY, H., Two antiestrogens affect differently chromatin remodeling of trefoil factor 1 (pS2) gene and the fate of estrogen receptor in MCF7 cells, *Biochim. Biophys. Acta*, 2002, 1578, 12–20.

92 CALLIGE, M., KIEFFER, I., RICHARD-FOY, H., CSN5/Jab1 is involved in ligand-dependent degradation of estrogen receptor {alpha} by the proteasome, *Mol. Cell Biol.*, 2005, 25, 4349–4358.

93 GOUGELET, A., BOUCLIER, C., MARSAUD, V., MAILLARD, S., MUELLER, S. O., KORACH, K. S., RENOIR, J. M., Estrogen receptor alpha and beta subtype expression and transactivation capacity are differentially affected by receptor-, hsp90- and immunophilin-ligands in human breast cancer cells, *J. Steroid Biochem. Mol. Biol.*, 2005, 94, 71–81.

94 SEGNITZ, B., GEHRING, U., The function of steroid hormone receptors is inhibited by the hsp90-specific compound geldanamycin, *J. Biol. Chem.*, 1997, 272, 18694–18701.

95 BAGATELL, R., KHAN, O., PAINE-MURRIETA, G., TAYLOR, C. W., AKINAGA, S., WHITESELL, L., Destabilization of steroid receptors by heat shock protein 90-binding drugs, a ligand-independent approach to hormonal therapy of breast cancer, *Clin. Cancer Res.*, 2001, 7, 2076–2084.

96 NECKERS, L., Hsp90 inhibitors as novel cancer chemotherapeutic agents, *Trends Mol. Med.*, 2002, 8, S55–61.

97 WAKELING, A. E., BOWLER, J., Steroidal pure antioestrogens, *J. Endocrinol.*, 1987, 112, R7–10.

98 BRODIE, A., Aromatase inhibitors in breast cancer, *Trends Endocrinol. Metab.*, 2002, 13, 61–65.

99 ATTAL, M., HAROUSSEAU, J. L., Randomized trial experience of the Intergroupe Francophone du Myeloma, *Semin. Hematol.*, 2001, 38, 226–230.

100 DESIKAN, R., BARLOGIE, B., SAWYER, J., AYERS, D., TRICOT, G., BADROS, A., ZANGARI, M., MUNSHI, N. C., ANAISSIE, E., SPOON, D., SIEGEL, D., JAGANNATH, S., VESOLE, D., EPSTEIN, J., SHAUGHNESSY, J., FASSAS, A., LIM, S., ROBERSON, P., CROWLEY, J., Results of high-dose therapy for 1000 patients with multiple myeloma, durable complete remissions and superior survival in the absence of chromosome 13 abnormalities, *Blood*, 2000, 95, 4008–4010.

101 BARLOGIE, B., ZANGARI, M., SPENCER, T., FASSAS, A., ANAISSIE, E., BADROS, A., CROMER, J., TRICOT, G., Thalidomide in the management of multiple myeloma, *Semin. Hematol.*, 2001, 38, 250–259.

102 BRUNO, B., ROTTA, M., GIACCONE, L., MASSAIA, M., BERTOLA, A., PALUMBO, A., BOCCADORO, M., New drugs for treatment of multiple myeloma, *Lancet Oncol.*, 2004, 5, 430–442.

103 RICHARDSON, P. G., MITSIADES, C. S., HIDESHIMA, T., ANDERSON, K. C., Novel biological therapies for the

104 Zangari, M., Anaissie, E., Stopeck, A., Morimoto, A., Tan, N., Lancet, J., Cooper, M., Hannah, A., Garcia-Manero, G., Faderl, S., Kantarjian, H., Cherrington, J., Albitar, M., Giles, F. J., Phase II study of SU5416, a small molecule vascular endothelial growth factor tyrosine kinase receptor inhibitor, in patients with refractory multiple myeloma, *Clin. Cancer Res.*, **2004**, 10, 88–95.

105 Hideshima, T., Chauhan, D., Shima, Y., Raje, N., Davies, F. E., Tai, Y. T., Treon, S. P., Lin, B., Schlossman, R. L., Richardson, P., Muller, G., Stirling, D. I., Anderson, K. C., Thalidomide and its analogs overcome drug resistance of human multiple myeloma cells to conventional therapy, *Blood*, **2000**, 96, 2943–2950.

106 Osman, K., Comenzo, R., Rajkumar, S. V., Deep venous thrombosis and thalidomide therapy for multiple myeloma, *N. Engl. J. Med.*, **2001**, 344, 1951–1952.

107 Zervas, K., Dimopoulos, M. A., Hatzicharissi, E., Anagnostopoulos, A., Papaioannou, M., Mitsouli, C., Panagiotidis, P., Korantzis, J., Tzilianos, M., Maniatis, A., Primary treatment of multiple myeloma with thalidomide, vincristine, liposomal doxorubicin and dexamethasone (T-VAD doxil), a phase II multicenter study, *Ann. Oncol.*, **2004**, 15, 134–138.

108 Adams, J., Palombella, V. J., Sausville, E. A., Johnson, J., Destree, A., Lazarus, D. D., Maas, J., Pien, C. S., Prakash, S., Elliott, P. J., Proteasome inhibitors, a novel class of potent and effective antitumor agents, *Cancer Res.*, **1999**, 59, 2615–2622.

109 Hideshima, T., Richardson, P., Chauhan, D., Palombella, V. J., Elliott, P. J., Adams, J., Anderson, K. C., The proteasome inhibitor PS-341 inhibits growth, induces apoptosis, and overcomes drug resistance in human multiple myeloma cells, *Cancer Res.*, **2001**, 61, 3071–3076.

110 Mitsiades, C. S., Mitsiades, N., Richardson, P. G., Treon, S. P., Anderson, K. C., Novel biologically based therapies for Waldenstrom's macroglobulinemia, *Semin Oncol.*, **2003**, 30, 309–312.

111 Orlowski, R. Z., Proteasome inhibitors in cancer therapy, *Methods Mol. Biol.*, **2005**, 301, 339–350.

112 Le Gouill, S., Pellat-Deceunynck, C., Harousseau, J. L., Rapp, M. J., Robillard, N., Bataille, R., Amiot, M., Farnesyl transferase inhibitor R115777 induces apoptosis of human myeloma cells, *Leukemia*, **2002**, 16, 1664–1667.

113 Ochiai, N., Uchida, R., Fuchida, S., Okano, A., Okamoto, M., Ashihara, E., Inaba, T., Fujita, N., Matsubara, H., Shimazaki, C., Effect of farnesyl transferase inhibitor R115777 on the growth of fresh and cloned myeloma cells in vitro, *Blood*, **2003**, 102, 3349–3353.

114 Doisneau-Sixou, S. F., Cestac, P., Faye, J. C., Favre, G., Sutherland, R. L., Additive effects of tamoxifen and the farnesyl transferase inhibitor FTI-277 on inhibition of MCF-7 breast cancer cell-cycle progression, *Int. J. Cancer*, **2003**, 106, 789–798.

115 Mitsiades, C. S., Mitsiades, N. S., McMullan, C. J., Poulaki, V., Shringarpure, R., Hideshima, T., Akiyama, M., Chauhan, D., Munshi, N., Gu, X., Bailey, C., Joseph, M., Libermann, T. A., Richon, V. M., Marks, P. A., Anderson, K. C., Transcriptional signature of histone deacetylase inhibition in multiple myeloma, biological and clinical implications, *Proc. Natl. Acad. Sci. U.S.A.*, **2004**, 101, 540–545.

116 Banerjeei, S. K., Zoubine, M. N., Sarkar, D. K., Weston, A. P., Shah, J. H., Campbell, D. R., 2-Methoxyestradiol blocks estrogen-induced rat pituitary tumor growth and tumor angiogenesis, possible role of vascular endothelial growth factor, *Anticancer Res.*, **2000**, 20, 2641–2645.

117 Chauhan, D., Catley, L., Hideshima,

T., Li, G., Leblanc, R., Gupta, D., Sattler, M., Richardson, P., Schlossman, R. L., Podar, K., Weller, E., Munshi, N., Anderson, K. C., 2-Methoxyestradiol overcomes drug resistance in multiple myeloma cells, *Blood*, **2002**, 100, 2187–2194.

118 Harrington, K. J., Lewanski, C., Northcote, A. D., Whittaker, J., Peters, A. M., Vile, R. G., Stewart, J. S., Phase II study of pegylated liposomal doxorubicin (Caelyx) as induction chemotherapy for patients with squamous cell cancer of the head and neck, *Eur. J. Cancer*, **2001**, 37, 2015–2022.

119 Harrington, K. J., Liposomal cancer chemotherapy, current clinical applications and future prospects, *Expert Opin. Investig. Drugs*, **2001**, 10, 1045–1061.

120 Harrington, K. J., Lewanski, C. R., Northcote, A. D., Whittaker, J., Wellbank, H., Vile, R. G., Peters, A. M., Stewart, J. S., Phase I-II study of pegylated liposomal cisplatin (SPI-077) in patients with inoperable head and neck cancer, *Ann. Oncol.*, **2001**, 12, 493–496.

121 Gauduchon, J., Gouilleux, F., Maillard, S., Marsaud, V., Renoir, J. M., Sola, B., The 4-hydroxytamoxifen inhibits proliferation of multiple myeloma cells in vitro and in vivo through down-regulation of c-Myc, up-regulation of p27Kip1 and modulation of Bcl-2 family members, *Clin. Cancer Res.*, **2005**, 11, 2345–2354.

122 Funke, A. P., Schiller, R., Motzkus, H. W., Gunther, C., Muller, R. H., Lipp, R., Transdermal delivery of highly lipophilic drugs, in vitro fluxes of antiestrogens, permeation enhancers, and solvents from liquid formulations, *Pharm. Res.*, **2002**, 19, 661–668.

123 Funke, A. P., Gunther, C., Muller, R. H., Lipp, R., In-vitro release and transdermal fluxes of a highly lipophilic drug and of enhancers from matrix TDS, *J. Controlled Release*, **2002**, 82, 63–70.

124 Storm, G., Crommelian, D. J. A., Liposomes, quo vadis, *Pharm. Sci. Technol. Today*, **1998**, 1, 19–31.

125 Gref, R., Minamitake, Y., Peracchia, M. T., Trubetskoy, V., Torchilin, V., Langer, R., Biodegradable long-circulating polymeric nanospheres, *Science*, **1994**, 263, 1600–1603.

126 Moghimi, S. M., Hunter, A. C., Murray, J. C., Long-circulating and target-specific nanoparticles, theory to practice, *Pharmacol. Rev.*, **2001**, 53, 283–318.

127 Torchilin, V. P., Trubetskoy, V. S., Whiteman, K. R., Caliceti, P., Ferruti, P., Veronese, F. M., New synthetic amphiphilic polymers for steric protection of liposomes in vivo, *J. Pharm. Sci.*, **1995**, 84, 1049–1053.

128 Stolnik, S., Dunn, S. E., Garnett, M. C., Davies, M. C., Coombes, A. G., Taylor, D. C., Irving, M. P., Purkiss, S. C., Tadros, T. F., Davis, S. S. et al., Surface modification of poly(lactide-co-glycolide) nanospheres by biodegradable poly(lactide)-poly(ethylene glycol) copolymers, *Pharm. Res.*, **1994**, 11, 1800–1808.

129 Gref, R., Minamitake, Y., Peracchia, M. T., Domb, A., Trubetskoy, V., Torchilin, V., Langer, R., Poly(ethylene glycol)-coated nanospheres, potential carriers for intravenous drug administration, *Pharm. Biotechnol.*, **1997**, 10, 167–198.

130 Vauthier, C., Dubernet, C., Fattal, E., Pinto-Alphandary, H., Couvreur, P., Poly(alkylcyanoacrylates) as biodegradable materials for biomedical applications, *Adv. Drug Deliv. Rev.*, **2003**, 55, 519–548.

131 Landry, F. B., Bazile, D. V., Spenlehauer, G., Veillard, M., Kreuter, J., Degradation of poly(D,L-lactic acid) nanoparticles coated with albumin in model digestive fluids (USP XXII), *Biomaterials*, **1996**, 17, 715–723.

132 Bazile, D., Prud'homme, C., Bassoullet, M. T., Marlard, M., Spenlehauer, G., Veillard, M., Stealth Me.PEG-PLA nanoparticles avoid uptake by the mononuclear phagocytes system, *J. Pharm. Sci.*, **1995**, 84, 493–498.

133 Peracchia, M. T., Vauthier, C., Puisieux, F., Couvreur, P., Development of sterically stabilized poly(isobutyl 2-cyanoacrylate) nanoparticles by chemical coupling of poly(ethylene glycol), *J. Biomed. Mater. Res.*, **1997**, 34, 317–326.

134 Peracchia, M. T., Vauthier, C., Desmaele, D., Gulik, A., Dedieu, J. C., Demoy, M., d'Angelo, J., Couvreur, P., Pegylated nanoparticles from a novel methoxypolyethylene glycol cyanoacrylate-hexadecyl cyanoacrylate amphiphilic copolymer, *Pharm. Res.*, **1998**, 15, 550–556.

135 Van de Velde, P., Nique, F., Bouchoux, F., Bremaud, J., Hameau, M. C., Lucas, D., Moratille, C., Viet, S., Philibert, D., Teutsch, G., RU 58,668, a new pure antiestrogen inducing a regression of human mammary carcinoma implanted in nude mice, *J. Steroid Biochem. Mol. Biol.*, **1994**, 48, 187–196.

136 Van de Velde, P., Nique, F., Planchon, P., Prevost, G., Bremaud, J., Hameau, M. C., Magnien, V., Philibert, D., Teutsch, G., RU 58668, further in vitro and in vivo pharmacological data related to its antitumoral activity, *J. Steroid Biochem. Mol. Biol.*, **1996**, 59, 449–457.

137 Brigger, I., Chaminade, P., Marsaud, V., Appel, M., Besnard, M., Gurny, R., Renoir, M., Couvreur, P., Tamoxifen encapsulation within polyethylene glycol-coated nanospheres. A new antiestrogen formulation, *Int. J. Pharm.*, **2001**, 214, 37–42.

138 Chawla, J. S., Amiji, M. M., Biodegradable poly(epsilon-caprolactone) nanoparticles for tumor-targeted delivery of tamoxifen, *Int. J. Pharm.*, **2002**, 249, 127–138.

139 Chawla, J. S., Amiji, M. M., Cellular uptake and concentrations of tamoxifen upon administration in poly(epsilon-caprolactone) nanoparticles, *AAPS PharmSci.*, **2003**, 5, E3.

140 Ameller, T., Marsaud, V., Legrand, P., Gref, R., Barratt, G., Renoir, J. M., Polyester-poly(ethylene glycol) nanoparticles loaded with the pure antiestrogen RU 58668, physico-chemical and opsonization properties, *Pharm. Res.*, **2003**, 20, 1063–1070.

141 Ameller, T., Marsaud, V., Legrand, P., Gref, R., Renoir, J. M., Pure antiestrogen RU 58668-loaded nanospheres, morphology, cell activity and toxicity studies, *Eur. J. Pharm. Sci.*, **2004**, 21, 361–370.

142 Ameller, T., Marsaud, V., Legrand, P., Gref, R., Renoir, J. M., In vitro and in vivo biologic evaluation of long-circulating biodegradable drug carriers loaded with the pure antiestrogen RU 58668, *Int. J. Cancer*, **2003**, 106, 446–454.

143 Cruz, T., Gaspar, R., Donato, A., Lopes, C., Interaction between polyalkylcyanoacrylate nanoparticles and peritoneal macrophages, MTT metabolism, NBT reduction, and NO production, *Pharm. Res.*, **1997**, 14, 73–79.

144 Evans, C. E., Lees, G. C., Trail, I. A., Cytotoxicity of cyanoacrylate adhesives to cultured tendon cells, *J. Hand Surg. [Br]*, **1999**, 24, 658–661.

145 Ameller, T., Legrand, P., Marsaud, V., Renoir, J. M., Drug delivery systems for oestrogenic hormones and antagonists, the need for selective targeting in estradiol-dependent cancers, *J. Steroid Biochem. Mol. Biol.*, **2004**, 92, 1–18.

146 Carroll, J. S., Lynch, D. K., Swarbrick, A., Renoir, J. M., Sarcevic, B., Daly, R. J., Musgrove, E. A., Sutherland, R. L., p27(Kip1) induces quiescence and growth factor insensitivity in tamoxifen-treated breast cancer cells, *Cancer Res.*, **2003**, 63, 4322–4326.

147 Uziely, B., Jeffers, S., Isacson, R., Kutsch, K., Wei-Tsao, D., Yehoshua, Z., Libson, E., Muggia, F. M., Gabizon, A., Liposomal doxorubicin, antitumor activity and unique toxicities during two complementary phase I studies, *J. Clin. Oncol.*, **1995**, 13, 1777–1785.

148 Gabizon, A., Martin, F., Polyethylene glycol-coated (pegylated) liposomal doxorubicin. Rationale for use in solid

tumours, *Drugs*, **1997**, 54(Suppl 4), 15–21.

149 SYMON, Z., PEYSER, A., TZEMACH, D., LYASS, O., SUCHER, E., SHEZEN, E., GABIZON, A., Selective delivery of doxorubicin to patients with breast carcinoma metastases by stealth liposomes, *Cancer*, **1999**, 86, 72–78.

150 LOPES DE MENEZES, D. E., PILARSKI, L. M., BELCH, A. R., ALLEN, T. M., Selective targeting of immunoliposomal doxorubicin against human multiple myeloma in vitro and ex vivo, *Biochim. Biophys. Acta*, **2000**, 1466, 205–220.

151 BURGER, K. N., STAFFHORST, R. W., DE VIJLDER, H. C., VELINOVA, M. J., BOMANS, P. H., FREDERIK, P. M., DE KRUIJFF, B., Nanocapsules, lipid-coated aggregates of cisplatin with high cytotoxicity, *Nat. Med.*, **2002**, 8, 81–84.

152 PAPAHADJOPOULOS, D., ALLEN, T. M., GABIZON, A., MAYHEW, E., MATTHAY, K., HUANG, S. K., LEE, K. D., WOODLE, M. C., LASIC, D. D., REDEMANN, C. et al., Sterically stabilized liposomes, improvements in pharmacokinetics and antitumor therapeutic efficacy, *Proc. Natl. Acad. Sci. U.S.A.*, **1991**, 88, 11 460–11 464.

153 LASIC, D. D., MARTIN, F. J., GABIZON, A., HUANG, S. K., PAPAHADJOPOULOS, D., Sterically stabilized liposomes, a hypothesis on the molecular origin of the extended circulation times, *Biochim. Biophys. Acta*, **1991**, 1070, 187–192.

154 MAILLARD, S., AMELLER, T., GAUDUCHON, J., GOUGELET, A., GOUILLEUX, F., LEGRAND, P., MARSAUD, V., FATTAL, E., SOLA, B., RENOIR, J. M., Innovative drug delivery nanosystems improve the anti-tumor activity in vitro and in vivo of anti-estrogens in human breast cancer and multiple myeloma, *J. Steroid Biochem. Mol. Biol.*, **2005**, 94, 111–121.

155 BUTEAU-LOZANO, H., ANCELIN, M., LARDEUX, B., MILANINI, J., PERROT-APPLANAT, M., Transcriptional regulation of vascular endothelial growth factor by estradiol and tamoxifen in breast cancer cells, a complex interplay between estrogen receptors alpha and beta, *Cancer Res.*, **2002**, 62, 4977–4984.

156 SENGUPTA, K., BANERJEE, S., SAXENA, N., BANERJEE, S. K., Estradiol-induced vascular endothelial growth factor-A expression in breast tumor cells is biphasic and regulated by estrogen receptor-alpha dependent pathway, *Int. J. Oncol.*, **2003**, 22, 609–614.

157 WOODLE, M. C., LASIC, D. D., Sterically stabilized liposomes, *Biochim. Biophys. Acta*, **1992**, 1113, 171–199.

158 WILLIS, M., FORSSEN, E., Ligand-targeted liposomes, *Adv. Drug Deliv. Rev.*, **1998**, 29, 249–271.

159 HEATH, T. D., BRAGMAN, K. S., MATTHAY, K. K., LOPEZ-STRAUBINGER, N. G., PAPAHADJOPOULOS, D., Antibody-directed liposomes, the development of a cell-specific cytotoxic agent, *Biochem. Soc. Trans.*, **1984**, 12, 340–342.

160 ALLEN, T. M., BRANDEIS, E., HANSEN, C. B., KAO, G. Y., ZALIPSKY, S., A new strategy for attachment of antibodies to sterically stabilized liposomes resulting in efficient targeting to cancer cells, *Biochim. Biophys. Acta*, **1995**, 1237, 99–108.

161 HANSEN, C. B., KAO, G. Y., MOASE, E. H., ZALIPSKY, S., ALLEN, T. M., Attachment of antibodies to sterically stabilized liposomes, evaluation, comparison and optimization of coupling procedures, *Biochim. Biophys. Acta*, **1995**, 1239, 133–144.

162 LEE, R. J., LOW, P. S., Folate-mediated tumor cell targeting of liposome-entrapped doxorubicin in vitro, *Biochim. Biophys. Acta*, **1995**, 1233, 134–144.

163 MASTROBATTISTA, E., KONING, G. A., STORM, G., Immunoliposomes for the targeted delivery of antitumor drugs, *Adv. Drug Deliv. Rev.*, **1999**, 40, 103–127.

164 STEPHENSON, S. M., LOW, P. S., LEE, R. J., Folate receptor-mediated targeting of liposomal drugs to cancer cells, *Methods Enzymol.*, **2004**, 387, 33–50.

165 KIKUCHI, A., SUGAYA, S., UEDA, H.,

TANAKA, K., ARAMAKI, Y., HARA, T., ARIMA, H., TSUCHIYA, S., FUWA, T., Efficient gene transfer to EGF receptor overexpressing cancer cells by means of EGF-labeled cationic liposomes, *Biochem. Biophys. Res. Commun.*, **1996**, 227, 666–671.

166 GYONGYOSSY-ISSA, M. I., MULLER, W., DEVINE, D. V., The covalent coupling of Arg-Gly-Asp-containing peptides to liposomes, purification and biochemical function of the lipopeptide, *Arch. Biochem. Biophys.*, **1998**, 353, 101–108.

167 CUELLO, M., ETTENBERG, S. A., CLARK, A. S., KEANE, M. M., POSNER, R. H., NAU, M. M., DENNIS, P. A., LIPKOWITZ, S., Down-regulation of the erbB-2 receptor by trastuzumab (herceptin) enhances tumor necrosis factor-related apoptosis-inducing ligand-mediated apoptosis in breast and ovarian cancer cell lines that overexpress erbB-2, *Cancer Res.*, **2001**, 61, 4892–4900.

168 BRAUD, A. C., MATHOULIN PORTIER, M. P., BARDOU, V. J., BERTUCCI, F., GRAVIS, G., CAMERLO, J., BEGUE, M., HOUVENAEGHEL, G., MARANINCHI, D., JACQUEMIER, J., VIENS, P., Overexpression of erb B2 remains a major risk factor in non-metastatic breast cancers treated with high-dose alkylating agents and autologous stem cell transplantation, *Bone Marrow Transplant.*, **2002**, 29, 753–757.

169 PARK, J. W., HONG, K., CARTER, P., ASGARI, H., GUO, L. Y., KELLER, G. A., WIRTH, C., SHALABY, R., KOTTS, C., WOOD, W. I. et al., Development of anti-p185HER2 immunoliposomes for cancer therapy, *Proc. Natl. Acad. Sci. U.S.A.*, **1995**, 92, 1327–1331.

170 KIRPOTIN, D., PARK, J. W., HONG, K., ZALIPSKY, S., LI, W. L., CARTER, P., BENZ, C. C., PAPAHADJOPOULOS, D., Sterically stabilized anti-HER2 immunoliposomes, design and targeting to human breast cancer cells in vitro, *Biochemistry*, **1997**, 36, 66–75.

171 PARK, J. W., KIRPOTIN, D. B., HONG, K., SHALABY, R., SHAO, Y., NIELSEN, U. B., MARKS, J. D., PAPAHADJOPOULOS, D., BENZ, C. C., Tumor targeting using anti-her2 immunoliposomes, *J. Controlled Release*, **2001**, 74, 95–113.

172 PARK, J. W., HONG, K., KIRPOTIN, D. B., COLBERN, G., SHALABY, R., BASELGA, J., SHAO, Y., NIELSEN, U. B., MARKS, J. D., MOORE, D., PAPAHADJOPOULOS, D., BENZ, C. C., Anti-HER2 immunoliposomes, enhanced efficacy attributable to targeted delivery, *Clin. Cancer Res.*, **2002**, 8, 1172–1181.

173 NIELSEN, U. B., KIRPOTIN, D. B., PICKERING, E. M., HONG, K., PARK, J. W., REFAAT SHALABY, M., SHAO, Y., BENZ, C. C., MARKS, J. D., Therapeutic efficacy of anti-ErbB2 immunoliposomes targeted by a phage antibody selected for cellular endocytosis, *Biochim. Biophys. Acta*, **2002**, 1591, 109–118.

174 GABIZON, A., HOROWITZ, A. T., GOREN, D., TZEMACH, D., MANDELBAUM-SHAVIT, F., QAZEN, M. M., ZALIPSKY, S., Targeting folate receptor with folate linked to extremities of poly(ethylene glycol)-grafted liposomes, in vitro studies, *Bioconj. Chem.*, **1999**, 10, 289–298.

175 LU, Y., LOW, P. S., Folate-mediated delivery of macromolecular anticancer therapeutic agents, *Adv. Drug Deliv. Rev.*, **2002**, 54, 675–693.

176 STELLA, B., MARSAUD, V., COUVREUR, P., ARPICCO, S., PERACCHIA, M. T., GERAUD, G., IMMORDINO, M. L., CATTEL, L., RENOIR, J. M., Biological characterization of the folic acid-nanoparticles in cellular models, The 28th International Symposium on Controlled Release of Bioactive Materials, San Diego, June 2001, 23–27, Abstract N° 5200.

177 STELLA, B., ARPICCO, S., PERACCHIA, M. T., DESMAELE, D., HOEBEKE, J., RENOIR, M., D'ANGELO, J., CATTEL, L., COUVREUR, P., Design of folic acid-conjugated nanoparticles for drug targeting, *J. Pharm. Sci.*, **2000**, 89, 1452–1464.

178 KUKOWSKA-LATALLO, J. F., CANDIDO, K. A., CAO, Z., NIGAVEKAR, S. S., MAJOROS, I. J., THOMAS, T. P.,

Balogh, L. P., Khan, M. K., Baker, J. R., Jr., Nanoparticle targeting of anticancer drug improves therapeutic response in animal model of human epithelial cancer, *Cancer Res.*, **2005**, 65, 5317–5324.

179 Pettersson, K., Delaunay, F., Gustafsson, J. A., Estrogen receptor beta acts as a dominant regulator of estrogen signaling, *Oncogene*, **2000**, 19, 4970–4978.

180 Delaunay, F., Pettersson, K., Tujague, M., Gustafsson, J. A., Functional differences between the amino-terminal domains of estrogen receptors alpha and beta, *Mol. Pharmacol.*, **2000**, 58, 584–590.

181 Hall, J. M., McDonnell, D. P., The estrogen receptor beta-isoform (ERbeta) of the human estrogen receptor modulates ERalpha transcriptional activity and is a key regulator of the cellular response to estrogens and antiestrogens, *Endocrinology*, **1999**, 140, 5566–5578.

182 Lazennec, G., Bresson, D., Lucas, A., Chauveau, C., Vignon, F., ER beta inhibits proliferation and invasion of breast cancer cells, *Endocrinology*, **2001**, 142, 4120–4130.

183 Tremblay, A., Tremblay, G. B., Labrie, C., Labrie, F., Giguere, V., EM-800, a novel antiestrogen, acts as a pure antagonist of the transcriptional functions of estrogen receptors alpha and beta, *Endocrinology*, **1998**, 139, 111–118.

184 Labrie, F., Labrie, C., Belanger, A., Simard, J., Giguere, V., Tremblay, A., Tremblay, G., EM-652 (SCH57068), a pure SERM having complete antiestrogenic activity in the mammary gland and endometrium, *J. Steroid Biochem. Mol. Biol.*, **2001**, 79, 213–225.

Index

a

accumulation, ff-MWCNTs 322
aclacinomycin, cancer treatment 9
Acridine 138
– Immunoliposomes 137–138
Actinomycin D 190
active targeting 223–225
– circumstantial 221
– ligands 346–354
– molecular 221
– nanoparticles 8–10
administration, drug nanoparticles routes 10–12
advanced generation nanoparticles 172
adverse effects, cancer treatment 74–75
AE see antiestrogens
albumin-microspheres, local chemotherapy 160
alkylating agents, anticancer agents release 187–188
alpha particles, boron neutron capture therapy (BNCT) 122
ammonium-functionalized SWCNTs and MWCNTs 319
amphiphilic copolymers 8
angiogenic factors, cancer cell targeting 349–350
anisotropy, magnetic 283–284
anthracyclines 12–19
– MM treatment 380
– nucleic acids synthesis inhibition 211
anti-metabolites, cancer treatment 20–21
antibodies
– coupled to PEG 225, 356
– monoclonal 126–127
– targeting 225
anticancer agents
– colloidal systems 371–392
– controlled release 168, 186–191
– kinetics 181
– targeted delivery 338–364
anticancer antibiotics 190–191
anticancer drug delivery 3–5
– inorganic nanoparticles 180
– nanoparticles 12, 209–211
anticancer therapies, critical analysis 200–202
anticancer treatment strategies, ferromagnetic filled carbon nanotubes 259–324
antiestrogen delivery
– characteristics 384
– colloidal systems 381–390
antiestrogens (AE) 19–20, 373–374
– differential activity 378–379
– encapsulation 379
– multiple myeloma 381
– nanoparticles charged with 381–386
– pure 382–387
antimetabolic agent 188
arsenic trioxide, cancer treatment 26
Au@PNAL, fabrication strategy 178

b

^{10}B, neutron capture therapy 88
BBB see blood brain barrier
biocompatible drug-loaded nanoparticles 218
biodegradable nanoparticles
– polymeric 57
– drug-loaded 218
biodegradable polymers 5–12
– preparation 58–61
biodistribution
– conventional nanoparticles 5
– gadolinium 98–101
– photosensitizer-nanoparticle couples 70–71
biological drug delivery 208
biological therapies, multiple myeloma 380–381

biomedical applications, anticancer agents 170–174
biotin 354
bladder cancer, paclitaxel therapy 24
Bleomycin, anticancer antibiotics 190
block-copolymers, cancer chemotherapy 203
blood brain barrier
– anthracyclines 16
– boron neutron capture therapy (BNCT) 142
– cancer therapy 341
– drug delivery nanoparticulate systems 210
– magnetic drug nanovectors 219
– Paclitaxel 25
– targeting methods 343
BNCT see boron neutron capture therapy
boranes, liposomal encapsulation 134–136
boron clusters, monoclonal antibodies 126
boron compounds, delivery 141
boron-containing macromolecules, tumor delivery 142–143
boron delivery
– dextrans 139–141
– general requirements 124–125
– liposomes 130
– targeted liposomes 137–139
boron neutron capture therapy (BNCT) 88–93, 122–145
boronated dendrimers 130
– attachment 126–127
– boron neutron capture therapy (BNCT) 126–127
– receptor ligands 127
boronophenylalanine, liposomal encapsulation 133–136
brain metastases, paclitaxel therapy 25
brain tumors, boron neutron capture therapy (BNCT) 142–143
breast cancer
– cisplatin therapy 24
– colloidal systems 371
– hormone therapy 374–379
butyric acid, cancer treatment 26

c
calcium hydroxyapatite, drug delivery vectors design 180
camptothecins 20–21, 214
cancer-associated cells, targeting 223
cancer cells, multidrug resistance 17–19
cancer chemotherapy 339–341
cancer markers, targeting 224
cancer therapy
– critical analysis 199–232

– local 156–167
– neutron capture 87–113
– new experimental 214–215
– resistance mechanisms 227–229
cancer treatment, drug nanoparticles 1–40
cancer vasculature, targeting 223
carbohydrates, targeting 225
carbohydrates-ceramic nanoparticles 204
carbon nanotubes (CNTs)
– compared to other nanoparticles 315
– description 264–277
– ferromagnetic filled 259–324
– functionality 315
– heat generation 290–309
– magnetism 277–290
– morphology 315
carboranes, liposomal encapsulation 134–136
carmustine, anticancer agents release 188
carrier characteristics, cancer cells 63
CDDS see conventional drug delivery systems
CED see convection-enhanced delivery
cell models, polymeric nanoparticles 61–70
cell separation, magnetic 158
cellular "sanctuaries", chemotherapy 360
ceramic nanoparticles, photodynamic therapy 55–56
cetyltrimethylammonium bromide (CTAB) 28
chamber model, administration and characterization of ff-MWCNTs 313–314
chemical drug delivery 208
chemical vapor deposition (CVD) 312–313
chemotherapeutic drug nanoparticles 1–40
chemotherapy
– cancer 339–341
– local 156–158
chitosan 105–109, 180
– clinical drug delivery 212
– neutron capture therapy 68
– polymeric nanoparticles 180, 359
– preparation methods 104
chitosan nanoparticles, gadolinium delivery 104–113
chitosan-type polymers, gene therapy 215–216
chlorambucil, anticancer agents release 187–188
1-[p-chlorobenzoyl]-2-methyl-5-methoxy-3-indoleacetic acid 183
circumstantial targeting, active 221
cisplatin 213–214
– cancer treatment 21

Index

clinical considerations, boron neutron capture therapy (BNCT) 144–145
clinical use, nanoparticulate drug delivery 211–214
clusters, CNTs 319
CNTs see carbon nanotubes
Co, magnetic characteristics 280
co-surfactants
– gadolinium biodistribution 99
– Gd-nanoLE 98
coercivities, ferromagnetic filled multiwalled CNTs 289
colloidal Au, drug delivery vectors design 181
colloidal systems
– antiestrogen-loaded 384
– delivery of anticancer agents 371–392
– drug carriers 11
– drug delivery 209
– drug-loaded 387–390
concentration gradient, anticancer agents kinetics 182
controlled release, anticancer agents 168, 181
convection-enhanced delivery (CED), boron neutron capture therapy 143
conventional drug delivery systems (CDDS) 371
conventional nanoparticles, cancer treatment 5–12
core–shell nanoparticles, intelligent 219
coronary plaque removal, anticancer agents 173
cremophor EL 21–23
critical analysis
– anticancer therapies 200–202
– cancer therapy 199–232
– drug delivery systems 207–220
– nanoparticles 202–207
– nanovectors 207–220
– resistance to therapy 227–229
– targeting 220–227
– toxicity issues 229–230
CTAB see cetyltrimethylammonium bromide
CuX zeolite, drug delivery vectors design 180
CVD see chemical vapor deposition
cyclophosphamide, anticancer agents release 187–188
cystatins, cancer treatment 26
cytarabine, antimetabolic agent 188–189

d

daunorubicin, anticancer antibiotics 191
DDS see drug delivery system
delivery agents
– boron neutron capture therapy 122–145
– dendrimer-related 125–130
dendrimer-related delivery agents, boron neutron capture therapy 125–130
dendrimers 204
– boron neutron capture therapy 125–126
– boronated 126–127
– cancer cell targeting 357–359
– structure 132
– targeted delivery of anticancer agents 338, 342, 357–361
2-devinyl-2-(1-hexyloxyethyl) pyropheophorbide (HPPH) 55
dextran-coated magnetite-particles 244
dextrans, boron delivery 139–141
diagnostics, in vitro 338
diethylenetriaminepentaacetic acid, cancer treatment 26
diffusion model, anticancer agents kinetics 182–183
direct intracerebral delivery, brain tumors 142–143
dissolution model 183–186
DNA
– interaction with CNTs 318–319
– nanoparticle complexes 351–352
domains, magnetic 280–283
DOTAP 29
doxorubicin 12–19
drug biodistribution profile 5–6
drug concentration, photocytotoxicity 64
drug-conjugated delivery systems 209
drug delivery
– anticancer 209–211
– intracellular 210, 225–226
– nanoparticulate 211–214, 341–343
drug delivery systems (DDS) 90
– cancer treatment 4
– critical analysis 207–220
– future directions 191
– polymer materials 174–175
drug delivery target 2–4
drug delivery vectors, design 175–181
drug-loaded nanoparticles
– biocompatible and biodegradable 218
– colloidal systems 387–390
– different types 58
– toxicity 217
drug–nanocarrier linkers 227–228
drug nanoparticles
– chemotherapeutic 1–40
– routes 10–12
drug resistance mechanisms 211
drug targeting, magnetic 156–167

drug transport vectors, brain tumors 142–143
drugs
– new experimental 214–215
– tumor selectivity 362

e

eddy current losses, ff-MWCNTs 293
EGF *see* epidermal growth factors
EGFR targeted liposomes, Acridine 138
embolization hyperthermia, ferromagnetic 248–249
emulsion/solid lipid nanoparticles, drug delivery 209
endocytosis
– Fe-MWCNTs 309–311
– ff-MWCNT uptake 309–311
endothelial growth factor, vascular 129–130
enhanced permeation and retention (EPR) effect 3, 343–344, 363
epidermal growth factors (EGF) 128–129
EPR *see* enhanced permeation and retention (EPR) effect
ER *see* estrogen receptor
ER-ligands, molecular mechanisms 375–377
ERE *see* estrogen responsive elements
estrogen action in breast cancers, molecular mechanisms 375–378
estrogen receptor (ER) 344
– colloidal systems 372–396
estrogen responsive elements (EREs) 375–377
estrogens
– breast cancer 372–374
– multiple myeloma 381

f

f-MWCNTs *see* filled multiwalled CNTs
FACS *see* FSC-SSC scatter flow cytometric analysis
Fe, magnetic characteristics 280
feasibility studies, thermotherapy 249–254
ferrimagnetic particles, hyperthermia 297
ferrofluids, anticancer agents
ferromagnetic embolization hyperthermia 248–249
ferromagnetic filled carbon nanotubes 259–324
ferromagnetic filled multiwalled CNTs (ff-MWCNTs) 259
– accumulation 322
– evaluation for anticancer applications 313–324
– *in vitro* and *in vivo* applications 309–324
– lipid-mediated uptake 309–311
– magnetic properties 285–290
– production 312–313
ferromagnetic particles, hyperthermia 297
ferromagnetism 278
ff-MWCNTs *see* ferromagnetic filled multiwalled CNTs
fff-MWCNTs *see* functionalized and ferromagnetic filled multiwalled CNTs
Fick's second law, anticancer agents kinetics 182
filled carbon nanotubes, ferromagnetic 259–324
filled multiwalled CNTs (f-MWCNTs)
– filling and closing 275–277
– growth 271–275
– structure 269–270
– synthesis 266–269
first generation nanoparticles, anticancer agents 171
flow cytometric analysis *see* FSC-SSC scatter flow cytometric analysis
flower-state magnetization 281–282
fluorescence 41
fluorouracil (FU) 213
– antimetabolic agent 189
folate, targeting 224
folate-overexpressing HeLa cancer cells 316–317
folate receptor (FR)
– boron neutron capture therapy 129
– targeted liposomes 138
folic acid, cancer cell targeting 350–352
FR *see* folate receptor
FR expressing tumor cells, laser radiation 317
FSC-SSC scatter flow cytometric analysis (FACS) 309–310
FU *see* fluorouracil
functional nanoparticles, anticancer agents 172
functionality, CNTs 315
functionalization, CNTs 316
functionalized and ferromagnetic filled multiwalled CNTs (fff-MWCNTs) 259
functionalized MWCNTs, improved 313–324
functionalized SWNTs, selective targeting 317

g

G-protein coupled receptor (GPCR) 378
gadobutrol, GdNCT 94
gadolinium biodistribution 98–101
– intravenous administration 101–104

gadolinium delivery
- chitosan nanoparticles 104–113
- lipid emulsion 95–104
gadolinium neutron capture therapy (GdNCT) 87
GBM see glioblastoma multiforme
^{157}Gd, neutron capture therapy 88
Gd accumulation, tumor tissue 102
Gd concentration, *in vivo* growth suppression 110
Gd-DTPA-SA 96
Gd-nanoCP 104–113
- bioadhesion 109–113
- Gd-DTPA release property 107–108
- preparation 105–107
- uptake 109–113
Gd-nanoLE see gadolinium delivery
- intraperitoneal administration 98–101
- intravenous administration 101–104
- preparation 95–98
GdNCT see gadolinium neutron capture therapy
- therapeutic potential 93–112
- typical research 94–95
gel drug carriers 219
gene therapy 27, 215–216
- cancer 363–364
Gliadel®, anticancer agents release 188
glioblastoma multiforme (GBM)
- boron neutron capture therapy 123
- thermotherapy 247, 250–251
gliomas
- boron neutron capture therapy 123
- cyclophosphamide 187
- paclitaxel therapy 25
- thermotherapy 244, 250
GPCR see G-protein coupled receptor
Greene's melanoma, gadolinium biodistribution 99
growth suppression, melanoma solid tumor 108–109

h

Haematoporphyrin derivative (HpD) 45
HCO-60, gadolinium biodistribution 99–100
heat generation, carbon nanotubes 290–309
HeLa cancer cells, folate-overexpressing 316–317
high molecular weight (HMW) delivery agents 122–145
- boron 125–130
HMW see high molecular weight (HMW) delivery agents
hormone therapy, breast cancer 374–379

HpD see Haematoporphyrin derivative
HPPH see 2-devinyl-2-(1-hexyloxyethyl) pyropheophorbide
HT see hyperthermia therapy
hydrophile–lipophile balance value, photosensitizers 44, 66
hyperthermia
- AC-heating 322–323
- development of materials 296–300
- embolization 248–249
- magnetic particle 263
hyperthermia therapy (HT), synergetical effects 54
hysteresis loop
- Fe-MWCNTs 302
- ferromagnetic materials 279, 282, 287–289
- hyperthermia 299–300
- well- and ill-conditioned 300
hysteresis losses, ff-MWCNTs 293

i

ICP-AES see inductively coupled plasma atomic emission spectroscopy
ill-conditioned hysteresis loop 300
IMC see indomethacin (IMC) release
immunoliposomes 357
- boron neutron capture therapy (BNCT) 137–139
immunoreaction, gadolinium biodistribution 102
in situ polymerization, biodegradable polymeric nanoparticles 58–59
in vitro efficacy assay, polymeric nanoparticles 62
in vitro molecular diagnostics 338
incubation medium, cell viability 63
incubation time, *in vivo* growth suppression 110
indomethacin (IMC) release, kinetics 183
inductively coupled plasma atomic emission spectroscopy (ICP-AES) 99
inorganic components, nano-sized 170
inorganic nanoparticles, drug delivery vectors 180–180
intelligent core–shell nanoparticles 219
internalization, ferromagnetic filled multiwalled CNTs (ff-MWCNTs) 309–311
intracellular drug delivery 210, 225–226
intracellular localization, photosensitizers 68
intracerebral delivery, brain tumors 142–143
intravenous administration, gadolinium biodistribution 101–104
intravenously injected nanoparticles 205–206
iron oxide crystals, superparamagnetic 204

iron oxide nanoparticles, superparamagnetic 169–173

k
Kaplan–Meier survival curves, doxorubicin 15

l
laser radiation, folate receptor expressing tumor cells 317
ligands, active targeting 346–354
linkers, drug–nanocarrier 227–228
lipid addition, *in vitro* 309–311
lipid emulsion, gadolinium delivery 95–104
liposomal encapsulation, boron neutron capture therapy 133–136
liposomes
– antiestrogen incorporation 382
– boron neutron capture therapy 130
– cancer cell targeting 355–357
– cancer chemotherapy 203
– charged with RU 58668 386–388
– drug delivery 209
– non-targeted 133–134
– receptor-targeted 138
– structure 132
– targeted 137–139, 338–364
liquid source chemical vapor deposition (LSCVD) 312–313
LMW *see* low molecular weight
local cancer therapy 156–167
local chemotherapy 156–158
locoregional adjuvant therapy 219
low molecular weight (LMW) boron delivery agents 122–125
LSCVD *see* liquid source chemical vapor deposition

m
mAb *see* monoclonal antibodies
macromolecular transport pathways, cancer treatment 2–4
macromolecules, brain tumors 142–143
macrophages, targeting 223
MagForce nanotherapy 249–254
magic bullet 362–364
magnetic anisotropy 283–284
magnetic cell separation, local chemotherapy 158
magnetic domains 280–283
magnetic drug delivery, local chemotherapy 158–163
magnetic drug nanovectors 219–220
magnetic drug targeting (MDT) 156–167

magnetic fluid hyperthermia thermotherapy (MFH) 323
magnetic microspheres, local chemotherapy 159
magnetic nanoparticle therapy, clinical experiences 249–254
magnetic nanoparticles 156–167, 244–247
magnetic particle hyperthermia 263
magnetic reversal 284–285
magnetic targeting 345–346
magnetism, nano-sized materials 277–290
magnetite-particles, dextran-coated 244
magnetization distribution, nanoparticles 279–283
Magnetofection™, local chemotherapy 158
magnetoliposomes, local chemotherapy 159
Magnevist®
– comparative studies 112
– *in vivo* growth suppression 108
malignant tissue suppression 43
MDT *see* magnetic drug targeting
melanoma cells, GdNCT 94
melanoma solid tumor, *in vivo* growth suppression 108–109
metallic nanoparticles
– drug delivery vectors design 181
– photodynamic therapy (PDT) 54–55
metastatic lymph nodes, local chemotherapy 160
methotrexate 214
– antimetabolic agent 189–190
methoxy poly(ethylene glycol)-poly(dl-lactic acid), antitumor characteristics 20
MFH *see* magnetic fluid hyperthermia thermotherapy
micelles
– cancer chemotherapy 203
– polymeric 209
microspheres, magnetic 159
miscellaneous agents, cancer treatment 26
mitoxantrone 27
MM *see* multiple myeloma
model chamber, administration and characterization of ff-MWCNTs 313–314
molecular targeting, active 221
monoclonal antibodies
– boron neutron capture therapy 126–127
– boronated dendrimers 126–127
– cancer cell targeting 346–350
– conjugation scheme 128
mononuclear phagocytic system (MPS) 360, 371
morphology, CNTs 315
MPS *see* mononuclear phagocytic system

m-tetra(hydroxyphenyl)chlorin (mTHPC) 55
– cell uptake 67
mTHPC *see m*-tetra(hydroxyphenyl)chlorin
multi-domain particles, magnetic 280–283
multidrug resistance 6
– cancer cells 17–19
multifunctional nanocontainers 260
multiple myeloma (MM)
– colloidal systems 371
– new biological therapies 380–381
multiwalled CNTs (MWCNTs) 259
– ammonium-functionalized 319
– functionalized 313–324
– *see also* ferromagnetic filled multiwalled CNTs
MWCNT model 263
MWCNTs *see* multiwalled CNTs

n

nano-sized materials, magnetism 277–290
nanocapsules 59
– advanced generation nanoparticles 172
– anticancer therapies 202
– antiestrogen incorporation 382
– colloidal systems 381–386
– polymerization 58
nanocarriers, thermosensitive 177
nanoencapsulation, polymeric nanoparticles 61
nanomagnets 278, 283–285
nanomedicine 231
nanoparticle targeting, passive or active 206–207
nanoparticle–DNA complexes 351
nanoparticles
– actively targetable 8–10
– antiestrogen incorporation 381–386
– biocompatible and biodegradable 218
– biodegradable polymeric 57
– cancer treatment 1–40
– carbohydrates-ceramic 204
– compared to CNTs 315
– critical analysis 202–220
– drug delivery in cancer 341–343
– emulsion/solid lipid 209
– generations 171–172
– intelligent core–shell 219
– intravenously injected 205–206
– magnetic 244–247
– magnetic drug targeting 156–167
– magnetic targeting 345–346
– non-biodegradable 54–57
– oxidation-sensitive 219
– polymeric 338–364

– sterically stabilized 6–8
– superparamagnetic iron oxide 219–220
– targeting methods 343–364
– thermoresponsive 219
– thermotherapy 242–254
nanoparticulate drug delivery systems
– clinical use 211–214
– targeted drugs 210
– types 342
nanoshells 361
nanospheres
– antiestrogen incorporation 382
– cancer chemotherapy 204
– polymeric 209
nanotechnology 338
nanotube–protein conjugates 318
nanotubes, carbon *see* carbon nanotubes
nanovectors 203–205, 361
– anticancer agents 173
– critical analysis 207–220
nanovehicles, boron neutron capture therapy (BNCT) 122–145
NCT *see* neutron capture therapy
neutron capture cross section 88
neutron capture therapy (NCT) 87–113
– principle 88–89
NI, magnetic characteristics 280
non-biodegradable nanoparticles 54–57
non-biodegradable polymers 56–57
non-chitosan-type polymers, gene therapy 216
nucleic acids, interaction with CNTs 318–319

o

oils, Gd-nanoLE 97
oxidation-sensitive nanoparticles 219

p

P-glycoprotein 3
– multidrug resistance 17–19
PAA *see* polyacrylamide
PACA *see* poly(alkyl cyanoacrylates)
paclitaxel 212–213
– cancer treatment 21
PAMAM *see* polyamidoamine dendrimers
– structure 126
particle size, polymeric nanoparticles 65
passive targeting 221–223
PCa *see* prostate cancer
PDT *see* photodynamic therapy
PEG *see* poly(ethylene glycol)
PEGylation 133, 206, 355–356, 390
penetration depth, magnetic field 293–294
PEO *see* poly(ethylene oxide)

pharmacokinetics, photosensitizer-
    nanoparticle couples 70–71
photodetection 41–55
photodiagnosis 41
photodynamic therapy (PDT) 215
– biodegradable polymeric nanoparticles 57
– cancer therapy 40–89
– induced damage 72
– limitations 51–54
– non-biodegradable nanoparticles 54–57
– polymeric nanoparticles 61–70
– polymeric nanoparticles *in vivo*
    relevance 70–74
Photofrin® 45, 56
photosensitizers (PS) 45
– adverse effects 52
– biodistribution 70–71
– cancer therapy 40–41
– clinical trials 48–49
– conventional 45–47
– coupled to nanoparticles 70–71
– formulations 53–54
– intracellular localization 68
– pharmacokinetics 70–71
– photodynamic therapy (PDT) 215
– tumor uptake 43
– uptake 66–70
phototoxic effects 41
physicochemical drug delivery 208
PLA *see* poly(lactic acid)
PLGA *see* poly(lactic-co-glycolic acid)
PNAL *see* poly[(NIPAAm-r-AAm)-co-lactic
    acid]
poly(alkyl cyanoacrylates) (PACA) 58
poly(ethylene glycol) (PEG)
– cancer targeting 355
– cancer treatment 8–11, 19–21
– chains 7, 389
– CNT stabilization 343–364
– coating 205
– drug delivery 217–218
– drug delivery systems 213
– folate receptor targeting agents 129
– nanoparticle coating 382
– non-biodegradable polymers 56
– passive targeting 344
– photosensitizers uptake 69
– RU coupled 386
– steric stabilization effect 363
– targetable nanoparticles 8
poly(ethylene oxide) (PEO), nanoparticle
    coating 344
poly(lactic acid) (PLA)
– biodegradable polymeric nanoparticles 59

– nanoparticle coating 382
poly(lactic-co-glycolic acid) (PLGA),
    biodegradable polymeric nanoparticles 60
poly[(NIPAAm-r-AAm)-co-lactic acid]
    (PNAL) 177
poly[β-(1 → 4)-2-amino-2-deoxy-D-
    glucopyranose] *see* chitosan
polyacrylamide (PAA) 56
polyamidoamine (PAMAM) dendrimers 358
polymer gels 219
polymer materials, anticancer agents
    174–175
polymer modifications, polymeric
    nanoparticles 63
polymer therapeutics 202
polymeric micelles 209
polymeric nanoparticles
– biodegradable 57
– cancer cell targeting 359–362
– drug delivery vectors design 175–180
– *in vitro* relevance 61–70
– *in vivo* relevance 70–74
– targeted delivery of anticancer agents 338,
    342, 361–362
polymeric nanospheres, drug delivery 209
polymerization, *in situ* 58–59
polymers
– biological characteristics 204
– dispersion 59–61
– non-biodegradable 56–57
porous materials, drug delivery vectors
    design 180
porphyrin derivatives, polymeric nanoparticles
    in vivo 72
porphyrins, photosensitizer structure 46
preformed polymer, dispersion 59–61
prostate cancer, description 260–264
prostate carcinoma, recurrent 253–254
protein–nanotube conjugates 318
PS *see* photosensitizers
PSA, serum marker 261–262
pure antiestrogen, RU 58 668 382–387

q
QD *see* quantum dots
quantum dots (QDs) 47
– cancer cell targeting 351–352

r
radiation-producing element, ^{10}B 88
radiotherapy, neutron capture therapy 87
reactive oxygen species (ROS) 42
– cell uptake 68
receptor ligands, boronated dendrimers 127

receptor-mediated tumor cell targeting   137
recurrent glioblastoma multiforme,
   thermotherapy   250–251
recurrent prostate carcinoma   253–254
release, controlled   168
residual tumors, thermotherapy   251–253
resistance to therapy, critical analysis
   227–229
reticuloendothelial system (RES)   129
– nanocarrier uptake   69
reversal, magnetic   284–285
ROS   see reactive oxygen species
routes, drug-loaded nanoparticles   10–12
RU 58668   382–387

s

SAR   see specific absorption rate
SDDS   see stealth drug delivery systems
second generation nanoparticles, anticancer
   agents   171–172
secondary tumors, chemotherapy   157
selective estrogen receptor down-regulators
   (SERDs)   373–374
selective estrogen receptor modulators
   (SERMS)   373–374
selective tumor-targeting   313, 317
self-regulating thermoseeds,
   hyperthermia   297
semiconductors, photosensitizers   47
SERD   see selective ER down-regulators
SERM   see selective estrogen receptor
   modulators
serum marker, PSA   261–262
"shell-in-shell" structures   176
silicon substrate, ferromagnetic filled
   multiwalled CNTs   287–288
single-domain particles, magnetic   280–283
single-walled carbon nanotubes
   (SWCNTs)   264–265
– ammonium-functionalized   319
– functionalized   315–319
– uptake   316–318
skin damage, *in vivo* growth suppression
   108
skin photosensitivity scoring system   74
SLN   see solid lipid nanoparticles
sodium borocaptate, liposomal
   encapsulation   133–136
solid lipid nanoparticles (SLN)   , 20
solid source CVD (SSCVD)   312
specific absorption rate (SAR)
   300–309
SPIONs   see superparamagnetic iron oxide
   nanoparticles

SPM   see superparamagnetic material
SSCVD   see solid source CVD
stabilized nanoparticles, cancer treatment
   6–8
stealth drug delivery systems (SDDS)   372
Stealth particles, biodegradable polymeric
   nanoparticles   60–61
steric stabilization effect, PEG   363
sterically stabilized nanoparticles   6–8
superparamagnetic iron oxide crystals   204
superparamagnetic iron oxide nanoparticles
   (SPIONs)   169–173, 219–220
superparamagnetic material (SPM)
   298–299
superparamagnetic nanoparticles   160
– hyperthermia   297
superparamagnetism   282
surface modifications, polymeric
   nanoparticles   63
survival rate, cancer cells   63
SWCNTs   see single-walled carbon nanotubes

t

TAA   see tumorassociated antigens
tamoxifen   19, 213
targeted delivery, anticancer agents   199
targeted drug delivery nanoparticulate
   systems   210
targeted liposomes
– boron delivery   137–139
– EGFR   138
targeted nanoparticles, biodegradable
   polymeric   60–61
targeting
– critical analysis   220–227
– nanoparticles in cancer therapy   206–207,
   343–364
taxol   see paclitaxel
thermoresponsive nanoparticles   219
thermoseeds, self-regulating   297
thermosensitive nanocarriers   177
thermotherapy
– clinical experiences   249–254
– feasibility studies   249–254
– ferromagnetic embolization
   hyperthermia   248–249
– MagForce nanotherapy   249–254
– magnetic nanoparticles   244–247
– nanoparticles   242–254
– recurrent glioblastoma multiforme
   250–251
– recurrent prostate carcinoma   253–254
toxicity
– doxorubicin   13

- drug-loaded nanoparticles 217
- MWCNTs 319–323
toxicity issues, critical analysis 229–230
transferrin
- cancer cell targeting 352–354
- cancer treatment 9
transferrin receptors, targeting 225
transmission electron microscopy (TEM), *in vivo* growth suppression 110
TSA *see* tumor-specific antigens
tumor-associated antigens (TAA) 346
tumor selectivity, drugs 362
tumor-specific antigens (TSA) 346–350
tumor suppression effects, *in vivo* efficacy 73–74
tumor-targeting, selective 313, 317
tumor tissues
- characterization 44
- ligth penetration 53
- oxygenation 53
tumor uptake, selective 43

u

uptake, ferromagnetic filled multiwalled CNTs 309–311

v

vascular effects, polymeric nanoparticles *in vivo* 71
vascular endothelium growth factor (VEGF) 129–130
- angiogenic process 323
- colloidal systems 371
- preparation 133
- targeting cancer vasculature 223
VEGF *see* vascular endothelium growth factor
vitamin H *see* biotin
vortex structure, magnetic domains 281

w

well-conditioned hysteresis loop 300